Random Field Models in Earth Sciences

Random Field Models in Earth Sciences

GEORGE CHRISTAKOS

DOVER PUBLILCATIONS, INC.
Mineola, New York

Bibliographical Note

This Dover edition, first published in 2005, is a corrected republication of the work originally published by Academic Press, Inc., San Diego, in 1992.

International Standard Book Number: 0-486-43872-4

Manufactured in the United States of America
Dover Publications, Inc., 31 East 2nd Street, Mineola, N.Y. 11501

For Lana and Maria

Contents

"Nothing would be done at all if a man waited till he could do it so well that no one could find fault with it."

J. H. Newman

Chapter 3
The Intrinsic Spatial Random Field Model

Chapter 4
The Factorable Random Field Model

Chapter 5
The Spatiotemporal Random Field Model

Chapter 6
Space Transformations of Random Fields

Chapter 7
Random Field Modeling of Natural Processes

Chapter 8
Simulation of Natural Processes

Chapter 9
Estimation in Space and Time

Chapter 10
Sampling Design

Foreword

The emphasis on modern research in the earth sciences has shifted from the explanatory to the predictive. In the past, geologists, hydrologists, oceanographers, and other students of the earth were content to observe and describe natural phenomena and to expostulate on how things came to be the way they are. Now these scientists not only theorize about the operation of natural processes, they must devise ways to test their theories and predict the possible consequences implicit in the theories. The human race is no longer restricted to a minor role as a gatherer of the wealth created by nature. Instead, modern society exerts an increasingly important influence on the physical and chemical processes that operate on and near the surface of the earth and within the atmosphere and oceans of our planet. To understand these processes requires that we devise models that describe the dynamics of their behavior and manipulate these models in order to estimate the consequences of our activities.

Geological models have evolved from purely deterministic to probabilistic to geostatistical models incorporating the hybrid concept of regionalized variables. Geostatistical methods are among the most powerful techniques presently available for constructing models of the spatial variation in natural properties. However, these geostatistical models are only static representations of the state of nature at an instant in time; they must be combined with other models (in the current state of the art, usually deterministic) that simulate dynamic behavior. Spatiotemporal models, as described in this seminal volume by George Christakos, represent the next evolutionary step in geological modeling, because such models combine dynamic processes with spatial variability and incorporate the inevitable uncertainties that result from incomplete knowledge of both spatial patterns and dynamic behavior.

This book will place serious demands on the reader because it is both heavily mathematical and very philosophical. George perhaps could have written a more approachable treatment of spatiotemporal models by concentrating on methodological nuts and bolts, emphasizing anecdotes, examples, and case studies. However, such an approach would not have served the dual purpose of this volume: (a) to lay a foundation for spatiotemporal modeling as a fundamental methodology in the earth sciences and (b) to fit spatiotemporal modeling into the general framework of the scientific method. George's objective has necessitated the careful exposition of assumptions and the painstaking development of the philosophical justifications behind them, as well as the detailed derivations of consequences that spring from the fundamental assumptions. Of course, it may also be true that his concern for philosophical underpinnings reflects in part the Christakos heritage, rooted as it is in the land of Greece, the birthplace of western philosophy.

After completing undergraduate work at the University of Athens, George Christakos continued engineering studies at the University of Birmingham (U.K.), receiving a master's degree in soil mechanics, and at M.I.T., where he obtained a master's degree in civil engineering. This was followed by a one-year stay at the Centre de Géostatistique in Fontainebleau, France, where he acquired an extensive background in geostatistics. He then returned to Athens, where he incorporated his knowledge of both deterministic and geostatistical modeling into his doctoral research in mining engineering. Noting the criticisms that have been directed at early geostatisticians for their failure to demonstrate the many connections between geostatistics and other forms of mathematical modeling, George, in this book, has taken great pains to demonstrate how the random field model relates to the various aspects of geostatistics, to the classical models of time series analysis and stochastic processes, and to the other variant models of physical phenomena.

Following his graduate studies, George came to the United States as a visiting research scientist at the Kansas Geological Survey. For two years, the Kansas Survey provided an environment where he could organize his thoughts and consolidate them into the manuscript that eventually resulted in this book. With the cooperation of the Survey, preparation of the text continued over the following two years while George pursued a second Ph.D. and conducted research in the Division of Applied Sciences at Harvard University. The Mathematical Geology Section of the Kansas Geological Survey takes special pride in George's contribution and is pleased to have been instrumental in the preparation and publication of this volume. This book is not only a valuable contribution to science but also a testimony to the benefits that come from our support and encouragement of the exchange of scientists between nations. George Christakos is one in a succession of international scholars that have worked and studied at the Kansas Geological Survey; all have brought new ideas and viewpoints and have made lasting contributions to both our organization and to science in general. It is our hope that George's book will direct researchers into new lines of investigation and increase the interaction between geoscientists and environmental scientists in all parts of the world. After all, the problems that must be addressed in energy, natural resources, and environment are global in nature. Fortunately, George Christakos has assembled a powerful collection of tools with which to address them.

John C. Davis
Kansas Geological Survey
Lawrence, Kansas

Preface

"If you do not fix your foot outside the earth, you will never make it to stay on her."

O. Elytis

This book is about modeling as a principal component of scientific investigations. In general terms, modeling is the fundamental process of combining intellectual creativity with physical knowledge and mathematical techniques in order to learn the properties of the mechanisms underlying a physical phenomenon and make predictions. The book focuses on a specific class of models, namely, random field models and certain of their physical applications in the context of a stochastic data analysis and processing research program. The term *application* is considered here in the sense wherein the mathematical random field model is shaping, but is also being shaped by, its objects.

Since the times of Bacon, Mill, Whewell, Pierce, and other great methodologists of science, it has been recognized that in scientific reasoning it is as important to operate with the right concepts and models as it is to perform the right experiments. Conceptual innovation and model building have always been central to any major advance in the physical sciences. Scientific reasoning employs to a large extent probabilistic concepts and stochastic notions. Indeed, scientific induction (in the Baconian tradition) is concerned with hypotheses about physical situations as well as with the gradation of the inductive support that experimental results give to these hypotheses. Such a gradation is needed, for these hypotheses are expected to generate predictions that extrapolate beyond the existing experimental data. The gradation is also necessary because all models possess some evidential support and counterexamples. To choose between them, the degree of support must be addressed. In scientific hypothetico-deduction (in the sense of Pierce and Popper), on the other hand, one first formulates a hypothesis and then exposes it and its logical consequences to criticism, usually by experimentation. In both cases (induction and hypothetico-deduction), the fact that one makes hypotheses implies that one is not dealing with certain knowledge but rather with probable knowledge where the gradation of the support that the experimental results give to hypotheses is achieved by means of probabilistic (stochastic) terms. In Poincaré's words: "Predictable facts can only be probable." Lastly, modeling is an important component of sophisticated instrumentation, which forms the conditions for and is the mediator of much of modern scientific knowledge (as is emphasized by instrumental realists, like Ihde, Hacking, and others).

Unfortunately, it seems that these fundamental truths are not always well appreciated nowadays. In particular, unlike physics where conceptualization–modeling and observation–experimentation are closely linked to each other following parallel paths, in some geological and environmental fields measurement is heavily overemphasized, while little attention is given to important modeling issues; even less attention is given to the problem of the rationality of model testing. Undoubtedly, such an approach, besides being very unpleasant to one's sense of symmetry, violates the most central concepts of scientific reasoning, the latter being considered hypothetico-deductive, neo-inductive, or instrumental-realistic. As a consequence, it may be a particularly inefficient and costly approach, which provides poor representations of the actual physical situations and leads to serious misinterpretations of experimental findings.

Furthermore, it is sometimes argued that conceptual innovation and advanced modeling are not likely to be practical and, hence, one should restrict oneself to classical methods and techniques that have been in use for long periods of time. Besides being distinctively opposed to the very essence of scientific progress, this view grossly misinterprets the real meaning of both terms, "practical" and "classical." Regarding the former term, it suffices to state Whittle's own words: "The word 'practical' is one which is grossly and habitually misused and the common antithesis between it and the word 'theoretical' is largely false. A practical solution is surely one which, for all it may be approximate, is approximate in an enlightened sense, shows insight and gets to the bottom of things. However the term is much more often used for a solution which is quick and provisional—quick and dirty might be nearer the mark. The world being what it is, we may need quick, provisional solutions, but to call these 'practical' is surely degradation of an honourable term." In fact, before any meaningful practical solution to a physical problem is obtained, the fundamental conceptual and physical aspects of the problem must be first completely understood and a powerful theory must be developed. A good example is the problem of fluid flow turbulence. Despite its great practical importance and intensive applied research over several decades, a completely satisfactory practical solution to the problem is still not available. And this is largely due to the fact that the fundamental theoretical aspects of turbulence are still unresolved.

As regards the term "classical," the most influencial writers in the area of modern scientific methodology (such as Kuhn, Lakatos, and Nagel) have repeatedly emphasized the fact that scientific achievement is but an endless series of historical data where problems that could not be handled by the then classical approaches were solved in terms of novel concepts and mathematically more advanced methods; which then became classical themselves, only to be replaced in turn by more powerful new models and techniques. This has always been the way that science progresses. In fact, according to Lakatos, all great scientific achievements had one characteristic in common: They were all based on new concepts and models, and they all predicted novel facts, facts that had been either undreamt of or had indeed been contradicted by previous classical theories. Scientific progress is a revolutionary process in the sense of Kuhn. Moreover, as Rescher stated: "Progress in basic natural science is a matter of constantly rebuilding from the very foundations. Significant progress is generally a matter, not of adding further facts, but of changing the framework itself. Science in the main does not develop by sequentially filling-in of certain basically fixed positions in greater and greater detail." Modern quantum mechanics, for example, has predicted and explained an enormous number of effects in physics and chemistry that could not be predicted or explained in terms of classical mechanics. Quantum mechanics, however, is not a refined or extended version of classical mechanics; it is rather a revolutionary step toward changing the classical framework itself (e.g., the renowned von Neumann's world, which is entirely quantum, contains no classical physics at all).

In earth sciences and environmental engineering, important problems nowadays include the assessment of the space–time variability of hydrogeologic magnitudes for use in analytical and numerical models; the elucidation of the spatiotemporal evolution characteristics of the earth's surface temperature and the prediction of extreme conditions; the estimation of atmospheric pollutants at unmeasured points in space and time; the study of transport models that are the backbone of equations governing atmospheric and groundwater flow as well as pollutant fate in all media; the quantitative modeling and simulation of rainfall for satellite remote-sensing studies; the design of optimal sampling networks for meteorological observations; and the simulation of oil reservoir characteristics as a function of the spatial position and the production time.

These are all problems where the development and implementation of the appropriate model is of great significance. The importance of the modeling aspect becomes even more profound in physical situations at large space–time scales where controlled experimentation is very difficult or even impossible. Furthermore, all the above problems are characterized by the significant amount of uncertainties in the behavior of the natural processes involved. Such uncertainties constitute an essential part of many controversial scientific investigations and policy responses. A good, timely example is the global warming problem. Global warming from the increase in greenhouse gases has become a major scientific and political issue during the past decade. This is due mainly to the huge uncertainty involved in all global warming studies. For example, forecasts of the space–time variability of natural processes, such as soil moisture or precipitation, have large uncertainties. Also, uncertainties in the future changes in greenhouse gas concentrations and feedback processes that are not properly accounted for in the models could produce greater or smaller increases in the surface temperature. Policy responses are delayed because scientists are not able to properly quantify these uncertainties and use them in the context of climate models. Therefore, modeling tools and approaches leading to satisfactory solutions to these difficult problems are extremely important. And this is why the problem of uncertainty, probability, and probable knowledge is not a problem of armchair philosophers. It has grave scientific, ethical, and political implications and is of vital social and economic relevance.

Of course, the fact that research in earth sciences and environmental engineering faces a series of difficult problems nowadays should by no means dishearten us. On the contrary, it should encourage us to reconsider the usefulness of many of our traditional approaches and techniques and develop novel, more sophisticated models. At the same time, we should increase our confidence in pursuing difficult problems. This last issue is very important, for it provides the surest guarantee for the continuing vitality and rapid growth of any scientific discipline. The great significance of the confidence issue in scientific research underlies Einstein's aphorism: "I have little patience with scientists who take a board of wood, look for the thinnest part and drill a great number of holes where drilling is easy."

In the light of the above considerations, this book is concerned with the study of problems of earth and environmental sciences by means of theoretical models that have as an essential basis a purely random (stochastic) element. In particular, the term *stochastic data analysis and processing* refers here to the study of spatial and spatiotemporal natural processes in terms of the random field model. As we saw above, spatial and spatiotemporal natural processes occur in nearly all the areas of earth sciences and environmental engineering, such as hydrogeology, environmental engineering, climate predictions and meteorology, and oil reservoir engineering. In such a framework, geostatistics, stochastic hydrology, and environmetrics are all considered as subdomains of the general stochastic research program.

From a mathematical viewpoint, random fields (spatial or spatiotemporal) constitute an area that studies random (nondeterministic) functions. This is an area of mathematics that is usually called stochastic functional analysis and deals with any topic covered by the ordinary (deterministic) theory of functions. In addition, the existence of the random component makes stochastic functional analysis a much larger, considerably more complex, and also more challenging subject than the ordinary theory of functions. The mathematical theory of random fields works in all these physical situations, where traditional (deterministic) models do not, because (a) it has the clearest theoretical justification and captures important characteristics of the underlying natural processes that traditional methods do not; and (b) it has a superb analytical apparatus and is able to solve complex physical problems on which the traditional methods fail.

In fact, as one probes more deeply into the origin of this highly mathematical discipline, it becomes quite clear that the pioneers of the random field theory are in fact hardheaded realists, driven to develop the new approach only because of the failure of the esteemed traditional approaches to provide an accurate description of nature in problems such as those described above. By now, the stochastic approach has been applied successfully to several engineering applications. As Medawar could have stated it, engineering is complex, richly various, and challenging—just like real life. Perhaps it travels slower nowadays than quantum physics or nuclear chemistry, but it travels nearer to the ground. Hence, it should give us a specially direct and immediate insight into science in the making.

Certainly the choice between the various possible models depends partly on one's guess about the outcome of future experiments in earth sciences and environmental engineering and partly on one's philosophical view about the world. For the mathematical methods, though, to become operational and to obtain an objective meaning in the sense of positive sciences, it is necessary to be associated with the empirical theses and computational notions of the stochastic data analysis and processing research program. Certainly the utilization of the stochastic research program in practice raises technical questions that can be answered by way of a framework weighting all sorts of data and knowledge available, describing the specific objectives, and choosing the appropriate technique. The computer as a research instrument provides the powerful means of implementing stochastic data analysis and processing in complex physical situations. The technology that emerges from the use of computers has exciting implications as regards the traditional relationship between theory–modeling and observation–experimentation.

More specifically, the book is organized as follows: We start with discussions of the science of the probable, the various theories of probability, and the physical significance of the random field model. Random field representations of unique natural processes are rich in physical content and can account for phenomena possessing complex macroevolution and microevolution features. In this context, various problems in earth sciences, where the use of the random field approach is completely justified, are reviewed throughout the book. The subject of random fields is vast. Inevitably we have to restrict our interests to reasonably specific areas. The choice of these areas is directly dependent on their importance in the context of applied environmental sciences. The entropy-related sysketogram function is introduced and its advantages over the traditional correlation functions are discussed. Following a critical and concise summary of the fundamental concepts and results of the general random field theory, the intrinsic spatial random field model, which describes generally nonhomogeneous distributions in space, is established in terms of generalized functions. The latter involve more complex mathematical concepts and tools. However, it pays here to use more sophisticated mathematics, since this provides us with a more complete description of random fields, which strengthens the theoretical support of the intrinsic model and leads to novel results. The power of the underlying mathematical structure lies in its capacity to capture essential features of complex physical processes, to replace assumptions regarding physical processes by more powerful and realistic ones, and to pave the way for establishing important connections between the intrinsic spatial random field model and stochastic differential equations. The study of natural processes in space–time is achieved by introducing the spatiotemporal random field concept. More precisely, a theory of ordinary as well as generalized random fields is built on the appropriate space–time structure. The results obtained act then as the theoretical support to practical space–time variability models and optimal space–time estimation methods. The concept of factorable random fields provides the means for studying an important set of problems of nonlinear systems analysis and estimation in several dimensions. Space transformation is an operation that can solve multidimensional problems by transferring them to a suitable unidimensional setting. The underlying concept has both substance and depth, possessing elegant and comprehensive representations in both the physical and frequency domains. It can be used as a valuable tool in testing the permissibility of spatial and spatiotemporal correlation functions, in the study of differential equation models governing subsurface processes, as well as in the simulation of environmental properties. The spatial and spatiotemporal estimation problems are solved in all generality. A heuristic adopted to the stochastic research program yields a Bayesian–maximum-entropy approach to the spatial estimation problem, which incorporates into analysis prior information and knowledge that are highly relevant to the spatial continuity of the natural process under estimation. The Bayesian–maximum-entropy concept may have significant applications in multiobjective decision analysis and in artificial intelligence studies. Interesting solutions can be obtained concerning certain important time-series-related problems, such as system nonlinearity. These time series are involved in a variety of water resources and environmental problems, including streamflow forecasting, flood estimation, and environmental pollution monitoring and control. Multidimensional simulation is a valuable tool in applied sciences. In the book various random field simulation techniques are reviewed and their relative advantages are discussed. Lastly the sampling design problem is discussed. An estimation variance factorization scheme with attractive properties is studied, which leads to an efficient and quick multiobjective sampling design method. Several other sampling methods of considerable importance in earth sciences and environmental engineering are reviewed too.

This work has been influenced by discussions with many friends, colleagues, and even certain theoretical opponents. My sincere thanks are due to Drs. M. B Fiering, J. J. Harrington, and P. P. Rogers of Harvard University; A. G. Journel of Stanford University; J. C. Davis, R. A. Olea, and M. Sophocleous of Kansas

Geological Survey; C. Panagopoulos, I. Ikonomopoulos, K. Mastoris, N. Apostolidis, and P. Paraskevopoulos of National Technical University of Athens; P. Whittle and J. Skilling of University of Cambridge, England; G. B. Baecher and D. Veneziano of Massachusetts Institute of Technology; G. Baloglou of State University of New York at Oswego; M. David of Montreal University, Canada; R. Dimitrakopoulos of McGill University, Canada; and V. Papanicolaou of Duke University. The author is grateful to all these friends, as well as to Mrs. C. Cowan, who typed the original manuscript, and Mrs. R. Hensiek, who drafted several of the illustrations.

George Christakos

1 Prolegomena

"Common sense is the layer of prejudice laid down in the mind prior to the age of eighteen."

A. Einstein

1. The Science of the Probable and the Random Field Model

There are numerous phenomena in the physical world a direct (deterministic) study of which is not possible. In fact, physics, geology, meteorology, hydrology, and environmental engineering have introduced us to a realm of phenomena that cannot give rise to certainty in our knowledge. However, a scientific knowledge of these phenomena is possible by replacing the study of individual natural processes by the study of statistical aggregates to which these processes may give rise. A statistical aggregate is a configuration of possibilities relative to a certain natural process. The properties of such aggregates are expressed in terms of the concept of probability, more specifically, under the form of a probability law. It is important to recognize that the probability law is a perfectly determined concept. The difference between a probabilistic and a deterministic law is that, while in the deterministic law the states of the system under consideration directly characterize an individual natural process, in a probabilistic law these states characterize a set of possibilities regarding the process.

The application of the mathematical theory of probability to the study of real phenomena is made through statistical concepts. Therefore, it is essentially in the form of statistical knowledge that a science of the probable is constituted. This implies that the science of the probable replaces a direct study of natural processes by the study of the set of possibilities to which these processes may give rise.

Since knowledge regarding these processes is achieved only indirectly by way of statistical concepts, it will be characterized as *probable* or *stochastic knowledge.* Modern science provides convincing evidence that such probable knowledge is no less exact than certain knowledge. Naturally, the constitution of a science of the probable raises two fundamental philosophical problems:

(a) The elucidation of the content of the concept of probability.

(b) The foundation of probable or stochastic knowledge as a concept directly related to the application of the concept of probability to the study of real phenomena.

Clearly, these two problems are closely related to each other. For example, the concept of probability must be such that it can be used in the empirical world. Each one of these problems, however, possesses certain distinct aspects that deserve to be studied separately.

Problem (a) can be considered under the light of either a *subjectivist* explanation of the notion of probability, taken in itself, or an *objectivist* explanation. More specifically, according to the subjectivist explanation, probability is a measure attached to the particular state of knowledge of the subject. It may correspond to a degree of certitude or to a degree of belief (say, about where the actual state of nature lies); or to the attitude with which a rational person will approach a given situation that is open to chance (say, the attitude with which one is willing to place a bet on an event whose outcome is not definitely known in advance). According to explanations of the objectivist type, probability is a measure attached to certain objective aspects of reality. Such a measure may be regarded as the ratio of the number of particular outcomes, in a specific type of experiment, over the total number of possible outcomes; or as the limit of the relative frequency of a certain event in an infinite sequence of repeated trials; or as a characteristic property (particularly, the propensity) of a certain experimental arrangement. The main point of the last view is that it takes as fundamental the probability of the outcome of a single experiment with respect to its conditions, rather than the frequency of outcomes in a sequence of experiments. (See, e.g., Keynes, 1921; von Mises, 1928; Popper, 1934, 1972; Jeffreys, 1939; Savage, 1954; Byrne, 1968.)

In conclusion, there exist more than one meaning of probability. Figure 1.1 merely represents a skeleton outline of various complex and diversified analyses of the notion of probability considered over the years by several eminent mathematicians, scientists, and philosophers (see Poincaré, 1912, 1929; Borel, 1925, 1950; Kolmogorov, 1933; Reichenbach, 1935; Nagel, 1939; Boll, 1941; Gendre, 1947; Servien, 1949; Carnap, 1950; Polya, 1954; Polanyi, 1958; Fisher, 1959; Russel, 1962; Jaynes, 1968; de Finetti, 1974).

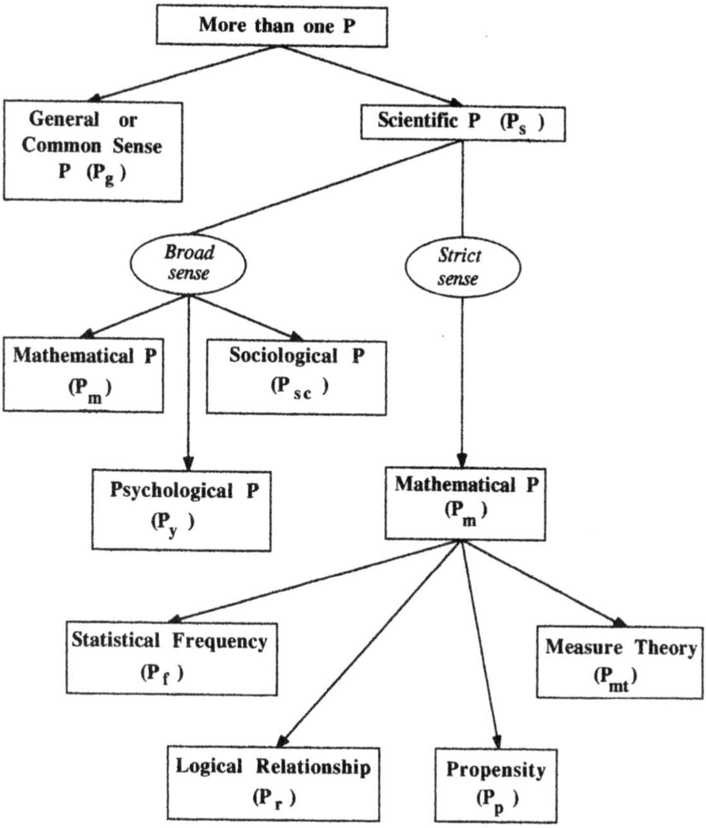

Figure 1.1 The various meanings of the concept of probability *P*

A subjective (sociological, P_{sc}, or psychological, P_y) concept of probability is of importance in social and psychological sciences, but it cannot serve as a basis for inductive logic or a calculus of probability applicable as a general tool of science. An objective (logical relationship, P_r, or propensity P_p) notion of probability is based on the assumption that objectivity is equivalent to formalization. It is, however, open to question whether formal logic can achieve the goals of the P_r concept, as has been demonstrated by Graig's theorem regarding empirical logic and by Godel's theorem on the limitation of formalization. Moreover, modern approaches to logic argue that the world obeys a nonhuman kind of reasoning and, hence, to cope with natural processes we must scrap our mode of human reasoning in favor of a new so-called quantum logic. The P_p concept, on the other hand, has been seriously criticized on the basis of the argument that what people understand by probability is broader than mathematical formalization. Last, to restrict probability to a mathematical meaning (P_m) is for many philosophers an ineffective approach, because the notion of probability transcends the bounds of mathematics (e.g., Byrne, 1968; Benenson, 1984).

Naturally, this variability of theoretical viewpoints reflects to an analogous variability in the practical implementation of the theory of probability. It seems that in the various fields of science and engineering, people do not stick to a unique meaning of probability. Occasionally, they prefer to choose what they consider to be the most appropriate meaning for the specific problem at hand.

The problem of probable knowledge [problem (b) above] is closely related to important modern scientific areas such as, for example, artificial intelligence and expert systems. With regard to this problem, two types of answers have been given: One is related to a *subjectivist* interpretation of probable knowledge, and the other is related to an *objectivist* interpretation. According to the former, the phenomena we are studying with the aid of probability theory are in themselves entirely determined and, therefore, they could be, ideally, the object of certain knowledge. And, if we are obliged to restrict ourselves to a probable knowledge of these phenomena, it is merely because we have at our disposal only incomplete information. The limitation of our information can be conceived either as purely contingent (due to insufficient sources of knowledge, inadequate measuring instruments and computers, etc.), or as a limitation in principle (because our capacities are inherently limited). The former point of view is used by classical statistical mechanics, while the latter is used in the context of the so-called orthodox theory of quantum mechanics. On the other hand, the objective interpretation of probable knowledge assumes that the incompleteness of our information is due to the object itself. Probable knowledge is then the expression of an objective contingency of the real phenomena. This contingency reflects either a principle of chance that exists in the very elementary components of the physical phenomena, or the lack of access to the various processes that determine these phenomena.

Evidently, there is a mutual relationship between the two aforementioned sets of problems: the subjective (objective) explanation of probability is well suited to the subjective (objective) explanation of the foundation of probable knowledge. But this is not always the case. For example, a subjective explanation of probability can well be used in the context of an objective interpretation of probable knowledge. Therefore, it is necessary that these two sets of problems be distinguished one from the other.

In any case, satisfactory answers to problems (a) and (b) above clearly belong to the field of epistemology and, thus, they require access to multidimensional philosophical considerations. In particular, any argument concerning the objectiveness or the subjectiveness of the concept of probability demands a deeper understanding of human nature and knowledge. To adopt the subjectivist or the objectivist interpretation of probable knowledge is a decision closely related to understanding of the nature of the world. In this regard, all the aforementioned attempts to answer the fundamental philosophical problems (a) and (b) are without doubt inadequate. It is far from being evident that we have at our disposal today the philosophical tools necessary to obtain a true understanding of the concepts of probability and probable knowledge. There is, perhaps, in the probabilistic concept the emergence of a type of knowledge very different from that considered by traditional schools of philosophy. In fact, people begin to realize that full consciousness of what is involved in knowledge of this sort is going to oblige us to modify fundamental concepts such as truth, knowledge, and experience.

In view of the above considerations, in this book we will not define probability as a concept in itself and will not indulge the epistemological problematics of probable knowledge. Had we decided to do so we would then have had the extremely difficult task of providing sound justification for a number of issues: If we had adopted the subjectivist explanation, we should explain why and how we have the right to suppose that the natural phenomena are entirely determined in themselves, and also why and how our knowledge, supposedly inadequate (be it in principle or merely in fact), turns out nevertheless to be quite adequate at the level of the statistical aggregates. If we had chosen the objectivist interpretation, we should justify why and how contingency appears in the physical phenomena, and why and how phenomena supposedly undetermined in themselves can give rise to statistical aggregates that are, for their part, entirely determined.

This book will focus attention on the language of probability, which is not at all constituted from some given epistemology, but from certain concrete problems that the traditional methods were not able to solve. Our concern will be on issues of application of the science of the probable in the context of the so-called *random field* (*RF*) model. In particular, the RF model will be considered a statistical aggregate about which we will make two *a priori* assumptions:

(i) Randomness is not a property of reality itself but, instead, a property of the RF model used to describe reality.

(ii) Probable knowledge cannot be considered as an image of reality. Through it we aim at reality and we learn something about it, but the relationship between our knowledge and its object becomes indirect and remote.

Under the light of assumptions (i) and (ii), the concept of probability in all its richness of content is of far greater importance for real world applications than the words and terms used to express it. By using probable knowledge the real is approached only through an abstract construction that involves the possible and is rather like a detecting device through which one grasps certain aspects of reality. Thanks to the detecting device, one can register certain reactions of reality and thus know it, not through an image of it, but through the answers it gives to one's questions.

The RF formalism does not restrict our concept of probable science to a physical theory of natural phenomena, governed by "randomness" or "chance"; or to a logical theory of plausible reasoning in the presence of incomplete knowledge. These theories, as well as several others, are viewed as potential modeling tools and detecting devices, rather than as unique realities. Within the RF context, we are looking at real problems that are in principle very subtle and complex. Therefore, like many other human enterprises, the practice of the science of the probable requires a constantly shifting balance between a variety of theories and methods, such as stochastic calculus, probability and statistics, logic and information theory.

As a matter of fact, RF methods have proven themselves very useful, although the notion of probability itself has not been philosophically well defined. This is true for almost all scientific theories. For example, despite the fact that terms such as mass, energy, and atom are philosophically undefined or ill-defined, theories based on these terms have led to extremely valuable applications in science and engineering.

In this book we study the use of RF models in the context of stochastic data analysis and processing. More precisely, the term *stochastic data analysis and processing* refers to the study of a natural process on the basis of a series of observations measured over a sample region of space (spatial series), over a sample period of time (time series), or over a spatial region for a sample time period (space–time series). In general, the aim of such a study is to evaluate and reconstruct the properties of the underlying unique physical process from fragmentary sampling data, by accommodating the probabilistic notion of RF. Hence, before proceeding with the description of the stochastic data analysis and processing research program, it is appropriate to discuss the physical content implicit in the RF representation of a natural process.

2. The Physical Significance of the Random Field Model

In this section our efforts will be focused on an exposition of theses and arguments that justify the use of RF models to represent physical processes that vary in space and/or time. Let χ_i, $i = 1, 2, \ldots, m$ be a spatial series of values of a physical variate χ. For illustration, a porosity profile (%; Christakos, 1987b) is depicted in Fig. 1.2; also, a lead concentration surface (Pb in ppb; Journel, 1984) around a Dallas smelter site is shown in Fig.

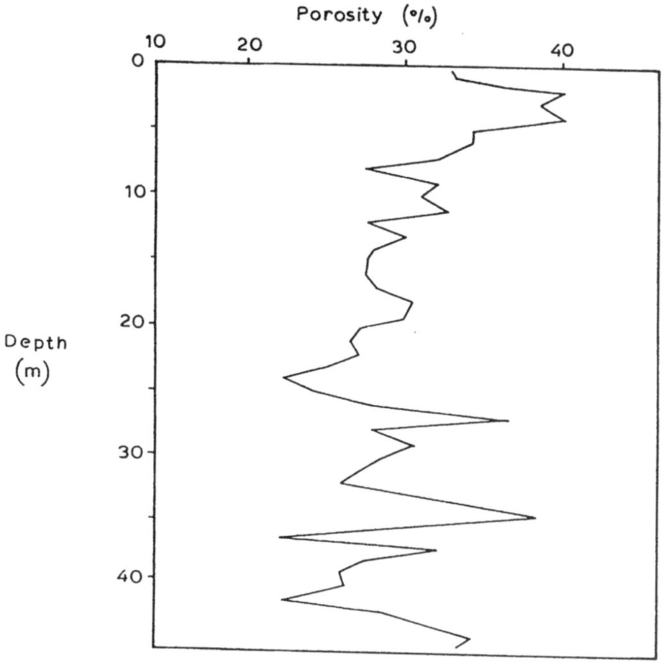

Figure 1.2 A typical soil porosity profile

1.3. The pattern of change of these series in space constitutes an evolution process. Particularly, by careful examination of these figures one notices two important descriptive features of the evolution process:

(i) A well-defined spatial *structure* at the macroscopic level (i.e., well-defined trends in the spatial variability of the porosity; also, a high dome centered at the smelter site, an NE trend of high lead values corresponding to the direction of prevailing winds, areas where changes in lead are rapid, areas with less rapid change, etc.).

(ii) A very *irregular* character at the microscopic level (that is, complex variations of the porosity within short distances; erratic local fluctuations in the lead surface, etc.).

These macroevolution and microevolution features are equally relevant to the understanding of the evolution process of a natural variable. (The reader may detect some similarities between these features and the wavelike and particlelike properties of matter and electromagnetic radiation.) The coexistence of macroscopic and microscopic properties in a natural variable is called macro–micro duality.

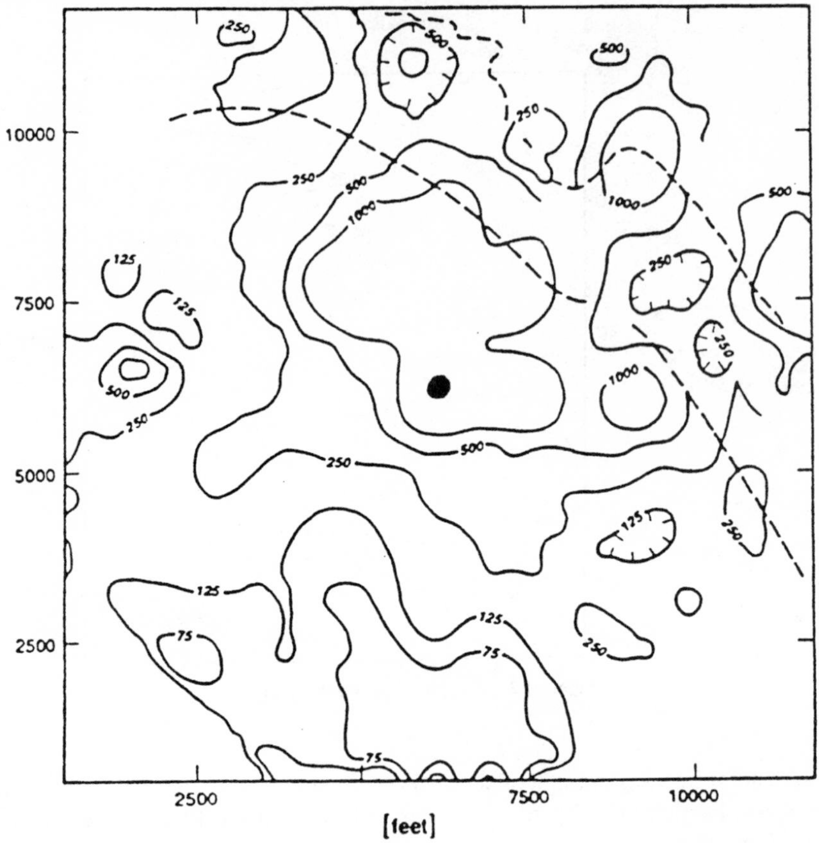

Figure 1.3 A lead concentration surface (in ppb; ● denotes the smelter site)

The origins of these properties depend on the nature of the phenomenon under study. In subsurface hydrology, for example, the macroscopic and the microscopic properties are closely related to the natural hierarchy of distinct spatial and temporal observation scales (e.g., atomic, hydrodynamic, pore, representative elementary volume, lab and field). The very concept of such a hierarchy has old as well as deep philosophical roots in the Baconian tradition: there is a hierarchy of explanatory laws to be discovered in physical sciences, and the scientific investigator should expect to make a gradual ascent to more and more comprehensive laws, each law leading to new experiments strictly within the domain and scale of its validity. It is clear that any adequate model of the above spatial series should take into account both features (i) and (ii). The model, in turn, should give rise to a number of fundamental considerations. For example, one might ask if the model should be continuous in its spatial coordinates; or how sound is the assumption that the evolution process underlying this model is differentiable, and in what sense; or if the model accounts for rough and erratic changes in the morphology of the spatial series. Additional considerations include the directionality of the model in space, and the complexity of the trends (local or global) involved. From the application point of view, a large number of important issues about spatial series are intimately tied up with the above modeling considerations.

A purely deterministic model of the spatial series is not appropriate for the case, for such a model will involve the estimation of an extremely large number of parameters, which is not practically possible due to the limited number of samples available. In the classical theory of probability, a *random variable* $x = x(u)$ is a function of elementary events $u \in \Omega$ defined on Ω (Ω is the sample space of a random experiment) and assuming real values (rigorous definitions are given in Chapter 2). Hence, one may consider the possibility of modeling the above spatial series by applying the classical theory of probability, where each observation χ_i is considered the outcome of a random variable $x_i = x_i(u)$, $i = 1, 2, \ldots, m$. However, by definition the random variables x_i do not change in space, and the relative distances and the geometric configuration of the observation points in space do not enter the analysis of the correlation structure of the spatial series. Consequently, the random variable model does not determine any law of change of the underlying natural variable in space, and therefore it does not constitute an adequate description of the evolution process. One may also try to model spatial series by a finite set of random variables (also called *vector* random variables). In reality, however, an infinite set of random variables is needed to describe the evolution process at every location in space and, as is well known, classical probability theory cannot study such sets.

In fact, it is precisely the concept of spatial evolution that guides us to represent a spatial series by means of a theoretical process, which we call the *spatial random field* (*SRF*). The specific nature of SRF, which makes it an appropriate model for the spatial evolution process, manifests itself when regarding it as a function whose properties are coordinated with the algebraic structure of the space, viz., $X(\mathbf{s}) = X(u, \mathbf{s})$, where $u \in \Omega$ and $\mathbf{s} \in R^n$ (R^n is the n-dimensional Euclidean space). More specifically, in the SRF formalism $\mathbf{s}$ accounts for the spatial structure of the evolution process at the macroscopic level and u accounts for the random character at the microscopic level. It is important to realize that the concept of randomness is used here as an intrinsic part of the spatial evolution of the physical variate, and not only as a statistical description of possible states. As a result of its functional structure, the SRF model is fully equipped with the necessary mathematics to account for all modeling considerations mentioned earlier. In relation to these considerations, the following methodological hypothesis is made: "The series of values χ_i, $i = 1, 2, \ldots, m$ is assumed to constitute a *single realization* from the infinite number of realizations that constitute a particular SRF $X(\mathbf{s})$." In other words, while in classical statistics population and sample are the two vital concepts, in the above setting the equivalent concepts are the (theoretical) SRF and the realization or observed spatial series. Also, the methodological hypothesis of many SRF realizations is quite similar to the so-called many worlds interpretation of quantum physics, according to which reality consists of a steadily increasing number of parallel universes (e.g., DeWitt and Graham, 1973).

The SRF paradigm can be extended in the space–time domain. A given space–time series $\chi_{i,t}$, $i = 1, 2, \ldots, m$ and $t \in T \subseteq R^1_{+,\{0\}}$ is assumed to be a single sample from a particular *spatiotemporal random field* (S/TRF) $X(\mathbf{s}_i, t) = X(u; \mathbf{s}_i, t)$, where $u \in \Omega$ and $(\mathbf{s}_i, t) \in R^n \times T$. (Notice that, by definition, the S/TRF is an RF whose arguments vary over some subset of the product set "Euclidean n-dimensional space × time axis.") The procedure by which $\chi_{i,t}$ is generated from $X(\mathbf{s}_i, t)$ indicates the manner in which the spatiochronological series is formed at each location/instant $(\mathbf{s}_i, t)$ but, due to its stochastic nature, it does not determine the actual value of the space–time series at any location/instant. In this sense, S/TRF are fundamentally different from vector time series: The latter do not constitute an adequate model for the combined space–time evolution process, for reasons similar to those according to which the aforementioned vector random variable did not form an appropriate representation of the spatial evolution process.

On the basis of the discussion above, the laws of nature for which the RF concept constitutes an appropriate model include not only (i) causal laws, but also (ii) noncausal laws, as well as (iii) laws dealing with the relationships between (i) and (ii). This classification is, indeed, in accordance with the fact that a causal law is not applicable for all possible events in nature; to the contrary, its applicability is limited by means of Heisenberg's uncertainty relation. In the RF context, the notion of causal connections is a modeling approximation that offers a partial treatment of certain aspects of the macrostructure of the natural process. The notion of noncausal contingencies is another approximation that deals with certain aspects of the microstructure of the process. These two notions must be completed by a consideration of their interconnections.

From the prediction point of view, when working with RF one considers essentially models of limited predictability to which one assigns some measure of accuracy. This is the method to which modern science is moving. It uses no principle other than that of predicting with as much assurance as possible, but with no more than is possible. That is, it idealizes the future from the outset, not as completely determined, but as determined within a defined area of uncertainty.

3. The Mathematics of Random Fields

From a mathematical viewpoint, random fields (SRF or S/TRF) constitute an area that studies random (nondeterministic) functions. This is an area of mathematics usually called *stochastic functional analysis* and deals with any topic covered by the ordinary (deterministic) theory of functions. In addition, the existence of the random component makes stochastic functional analysis a much larger, considerably more complex, and, also, more challenging subject than the ordinary theory of function (see, e.g., Einstein, 1905; Langevin, 1908; Wiener, 1930; Heisenberg, 1930; Kolmogorov, 1941; Chandrasekhar, 1943; Levy, 1948; von Neumann, 1955; Yaglom, 1962;

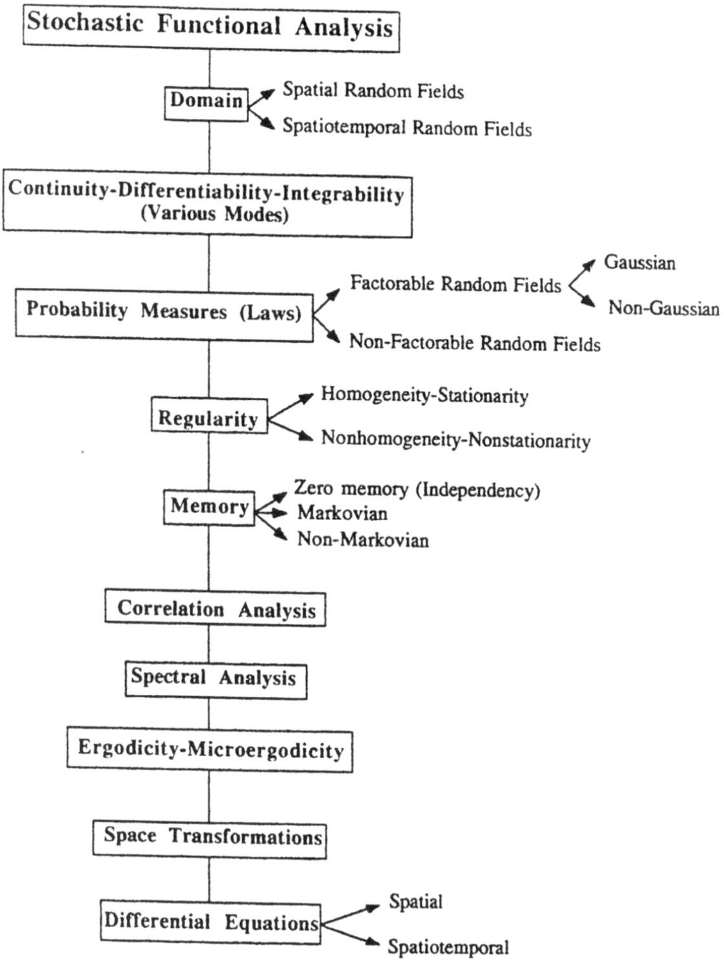

Figure 1.4 The mathematics of the RF model

Bohr, 1963; Kallianpur, 1980). Furthermore, as happens with ordinary functional analysis, the RF theory in several spatial or spatiotemporal dimensions deals with significantly more complicated problems than the one-dimensional theory of random processes, where the single argument is usually time (for an interesting discussion of the significant differences between time series analysis and RF theory see Ripley, 1988). Figure 1.4 summarizes certain of the most important mathematical topics of stochastic functional analysis. For the definitions of the various terms in Fig. 1.4 see later chapters of the book.

The SRF theory, in particular, has led to startling advances on literally every scientific front (see, e.g., Matern, 1960; Gandin, 1963; Matheron, 1965; Beran, 1968; Panchev, 1971; Journel and Huijbregts, 1978; Yaglom, 1986; Dagan, 1989). However, it is widely admitted that SRF is tough to work with, mainly for two reasons: the difficulty of the mathematics involved, and the nonexistence of systematic books on the subject. The published literature is notably lacking in completeness and coherence. It is, therefore, useful to construct an SRF model that incorporates all stochastic concepts and mathematical tools necessary for the effective implementation of stochastic data analysis and processing notions and techniques. In addition, the emergence of more complex physical problems requires that the SRF model be significantly extended and that novel notions and tools be developed. The situation is even more uncomfortable as regards S/TRF. The mathematical literature devoted to S/TRF is definitely very sparse and incomplete. In this book an attempt is made to elaborate to a certain extent toward a mathematical theory of specific classes of S/TRF, all of which are of practical importance.

In conformity with the above considerations, an adequate approach to the stochastic data analysis and processing problem should combine efforts

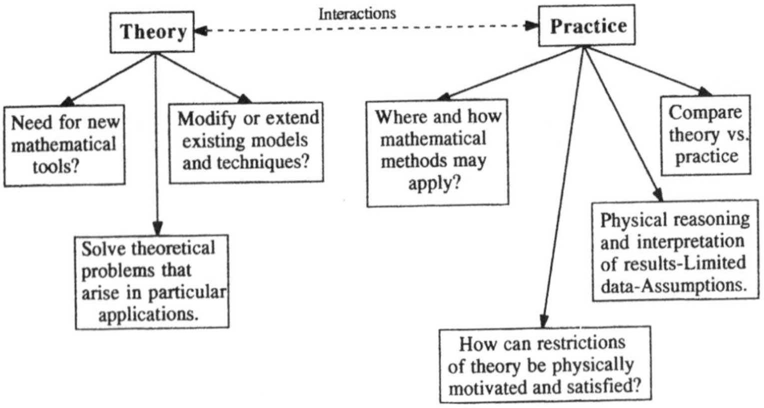

Figure 1.5 Theoretical and practical aspects of the stochastic research approach

on both grounds of theoretical and experimental investigations; this is illustrated in Fig. 1.5. To avoid misuse, the theory should not be used merely as a tool or as a procedure without a deeper understanding of its underlying mathematical structure. Moreover, these mathematical results obtain empirical meaning and substance only if they are associated with the philosophical theses of the stochastic research program.

4. The Philosophical Theses of the Stochastic Research Program

The stochastic data analysis and processing research program can be described in terms of certain philosophical theses on the methodology of scientific research programs (e.g., Lakatos, 1970). In particular, the stochastic research program consists of (see also Fig. 1.6)

(a) A *hard core* of fundamental concepts and constitutive hypotheses in terms of RF theory, which are not subject to direct experimental test (e.g., the hypothesis "the spatial series available constitutes a realization of an SRF").

(b) A set of *auxiliary hypotheses* and *model parameters* linking the RF theory with the observed phenomenon. Certain of these auxiliary hypotheses and model parameters are *testable*, in the sense that they possess *real counterparts* that can be observed, measured, and tested after the event (e.g., the hypothesis of local homogeneity, or certain correlation parameters expressed in terms of spatial integrals). In such a circumstance, and in accordance with Popper's criterion of falsification (Popper, 1934, 1972), these hypotheses and parameters have an objective meaning (see, also, Matheron, 1978). In addition to being compatible with the data over the sampled area, the testable hypotheses and parameters must establish the desirable *duality relations* between part (a) and the real phenomenon, so that physical inferences about the latter can be made by means of the former. The auxiliary hypotheses may also contain testable information themselves. The latter is not included in the data and comes from knowledge and experience with the physics of the particular problem. On the other hand, *nontestable* are these auxiliary hypotheses and parameters that do not possess real counterparts that can be observed, measured, and tested after the event. They are, however, valuable in the process of choosing a model that is compatible with the nature of the real process it represents (e.g., homogeneous, multi-Gaussian, or factorable RF). Or, they are related to certain sources of qualitative information. Nontestable auxiliary hypotheses can only be judged by the successes they lead to.

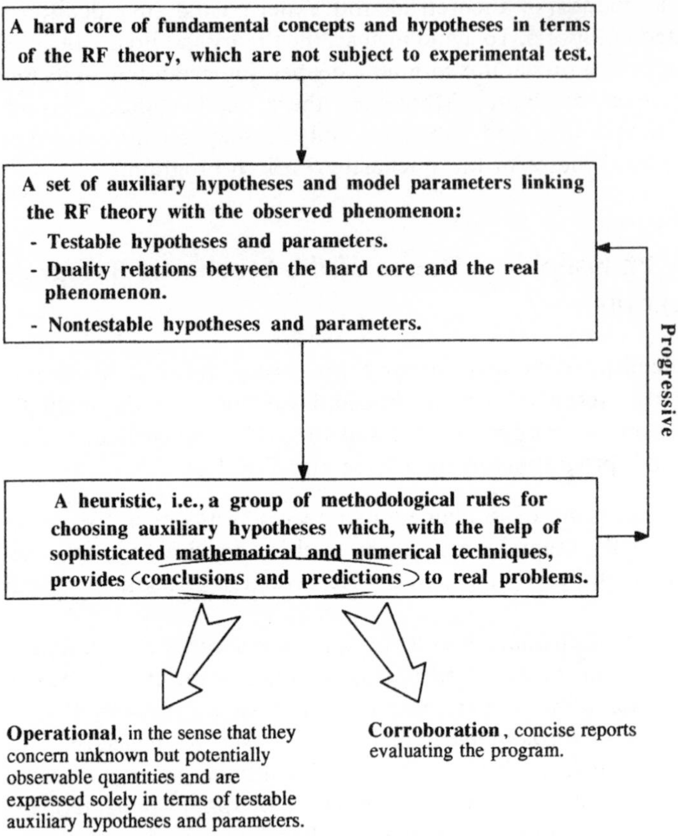

Figure 1.6 The methodology of the stochastic research program

(c) A *heuristic*, that is, group of methodological rules for choosing auxiliary hypotheses, that, with the help of sophisticated mathematical and numerical techniques, provide conclusions and predictions with regard to real problems. The choice of auxiliary hypotheses can only be made intelligently if there is significant physical understanding.

In the stochastic data analysis and processing context the final conclusions and predictions are *operational*, in the sense that they concern unknown but potentially observable quantities and are expressed solely in terms of testable auxiliary hypotheses and parameters. For example, the minimum mean square error estimator of a natural process at an unsampled location in a soil deposit is a function of the measurements available at known locations and the experimentally calculated spatial correlation functions.

At this point the stochastic research program is consistent with the Bohrian position expressed as, "If a parameter cannot be measured, it cannot be used for prediction" (Bohr, 1963). If the conclusions in (c) are refuted one must go back to part (b) and choose another set of auxiliary hypotheses. This fact is, in turn, consistent with the Lakatosian concept, according to which prediction-testing is important to revise a research program, while sticking to its hard core. Part (b) must be revised—by means of the heuristic part (c)—in a *progressive* way. This means that the replacement successfully anticipates previously refuted predictions and leads to valuable novel conclusions and advances, which were not possible by previous research programs.

It is instructive to mention some progressive cases carried out by the stochastic research program:

(i) The use of stochastic functional analysis throughout the book gives a sound interpretation to important features of physical quantities, such as spatial variability and continuity, which are not detected by, say, classical statistics.

(ii) The concept of generalized spatiotemporal random fields (Chapter 5) provides the means for successfully representing natural processes that have irregular, spatially nonhomogeneous, and time-nonstationary properties. Such a representation is not possible in terms of traditional time series methods. Moreover, the spatiotemporal estimation scheme takes into account time-related information and, therefore, provides improved results compared to those obtained by purely spatial estimation techniques.

(iii) The operations of space transformations (Chapter 6) materialize the intuitively attractive concept of simplifying the study of a natural process in several dimensions by "conveying" analysis to a suitable one-dimensional space. Space transformations provide solutions to practical problems, such as the establishment of comprehensive permissibility criteria for multidimensional correlation functions, the simulation of anisotropic natural processes in the space domain, and the solution of multidimensional stochastic differential equations.

(iv) The Bayesian, maximum-entropy-based estimation formalism (Chapter 9) enables one to use a variety of sources of prior information and, at the same time, to significantly restrict the range of arbitrariness; this is not possible by means of traditional spatial estimation approaches. The formalism relies heavily on the assumption that a significant part of physical sciences is related closely to the concept of information.

(v) Hypotheses associated with the notion of factorability (Chapter 4) can lead to significant extensions and offer solutions to certain model-related problems of nonlinear geostatistics.

These five cases emphasize the progressive nature of the stochastic research program in general, and the crucial role of part (c) in particular. If the heuristic is well-established and operational in the context of a specific physical problem, the difficulties associated with the application of the stochastic data analysis and processing program become merely mathematical.

The stochastic research program is also equipped with the concept of *corroboration.* By the latter we mean concise reports evaluating the program with respect to the way it solves its problems, its degree of testability, the severity of tests it has undergone, and the way it has withstood these tests. In the long run these tests are part of what is usually called the sanction of practice. And scientists know that the latter acts somehow like natural selection: It mercilessly eliminates inadequate models and insufficient approaches. In the short run corroboration is essentially comparative. One can claim only that the stochastic program has a higher degree of corroboration than a competing program, in the light of critical considerations that include testing up to some time *t*. However, the degree of corroboration at time *t* says nothing about the degree of corroboration of the program at a time later than *t*. In this sense corroboration does not verify any program; it shows only its heuristic power.

5. The Practice of the Stochastic Research Program and the Spectrum of Its Applications

The methodology of the stochastic research program applies to a variety of real-life phenomena that occur in space and/or in time. For illustration let us mention a few of them:

(i) Transport processes in porous media
(ii) Space–time concentrations of atmospheric pollutants
(iii) Spatial distribution of hydrologic data
(iv) Fluid flow and rock mechanics studies
(v) Agricultural crop yield
(vi) Turbulent fluctuations of meteorologic elements
(vii) Safety assessment of earth dams
(viii) Image processing and remote sensing
(ix) Random sea surfaces
(x) Oil reservoir characterization parameters
(xi) Ore reserve evaluation and grade control processes

In such a framework, geostatistics, stochastic hydrology, environmetrics, etc., are all considered subdomains of the general stochastic data analysis and processing area. It is a widely recognized fact that in earth sciences traditional (deterministic, etc.) mathematical models of complex physical phenomena and natural processes are giving way to stochastic mathematical models, because only the latter provide the theoretical concepts and practical tools needed to describe quantitatively the combinations of complexities, heterogeneities, and uncertainties involved in these phenomena. In hydrogeology, for example, the incorporation of the spatial and temporal variability of soil and hydraulic properties in the study of fluid flow and solute transport can be achieved only by means of stochastic models. In petroleum engineering, probability functions of the fracture geometry underly the study of geologic media. In meteorology, due to complex inherent fluctuations and lack of information, atmospheric processes resolved by deterministic models rely on unrealistic conceptual representations of the underlying physical mechanisms. The modeling of the spatiotemporal evolution of mesoscale storm systems is possible only in terms of random field concepts (for an excellent discussion see NRC, 1991).

Practical stochastic data analysis and processing generally involves data collections, analysis, and interpretation. In particular, the practical use of stochastic data analysis and processing is first to discover possible laws in the spatial and/or chronological evolution of the natural process, then to express these laws by means of the appropriate models, and finally to allow predictions. Examples of such predictions are given in Table 1.1

The completion of these predictions can be approached in an efficient, cost-effective manner or in an inefficient, cost-ineffective manner. As a matter of fact, a fundamental problem of practical stochastic data analysis and processing is the widespread lack of appreciation concerning what the different RF models and techniques can actually accomplish. Accordingly, the choice of RF models and techniques discussed in this book has been governed primarily by their usefulness in practical stochastic data analysis and processing.

Model construction is a primary constituent of stochastic data analysis and processing. It involves intellectual creativity, physical knowledge, and

Table 1.1 Examples of Stochastic Analysis and Processing Problems

(i) Quantitative assessment of spatial and temporal variability in the hydrologic properties of crustal rocks and fluid pressures; such assessments constitute a crucial prerequisite for a deeper understanding of the role of pore fluids in tectonic processes.

(ii) Reconstruction of the whole field of an environmental process in terms of the fragmentary space-time data available.

(iii) Simulation of oil reservoir characteristics, such as permeability and porosity, as a function of the spatial coordinates and the production phase.

(iv) Study of the relations between the mechanisms of earth dam failures and random hydrologic processes such as streamflow, hydraulic conductivity, seepage velocity, and precipitation.

(v) Modeling and simulation of rock fracture networks.

(vi) Optimal dynamic sampling design of meteorological observations.

mathematical techniques by means of a two-fold operation: sorting out the relevant parameters, which may or may not be directly observable as a part of the available data, and trying to discover possible causal and noncausal laws as well as laws dealing with the relationships between them. In practice, model building may be seen as an operation that establishes a quantitative description of the relationship between a set of observable natural processes and a set of free parameters to be fitted to the data. The aim of stochastic data analysis and processing is to learn more or to learn more efficiently, and to produce information that can be applied in decision making or problem solving. In other words, the stochastic data analysis and processing task may not be the actual or ultimate problem in need of a solution. Completing the stochastic data analysis and processing task produces information for solving one or more other problems. Once the data processing has been performed, it is the job of the decision maker to use the information obtained. For instance, the simulation of oil reservoir characteristics produces information for decision making; it does not produce decisions. Similarly, the completion of a land-use inventory with remotely sensed data produces land-use information for decision making. In view of the foregoing remarks, the outcomes of stochastic data analysis and processing may be considered as input information to the decision-making part. Such considerations include the possibility of incorporating highly corroborable models and informative statements, which are available under a format that can be used in the stochastic context.

Before applying any of the stochastic analysis and processing techniques to a particular problem, it is necessary to develop a *framework* for weighting all sorts of data and knowledge available and choosing the appropriate technique:

1. *Preliminary stage*: In this stage it must be decided if the stochastic analysis and processing can actually provide information that will help the solution of the problem. For example, geotechnical procedures for predicting soil performance consist of three parts, namely (a) constitutive models; (b) stochastic data analysis and processing methods for the estimation of soil parameters used in these models; and (c) numerical approaches to apply the models in practice. The accuracy in estimating the soil parameters in (b) obviously affects the reliability of any prediction made, and is, therefore, of significant practical consequence.

Certainly, decisions regarding the appropriateness of stochastic analysis are based on experience and, therefore, it is difficult to generalize. Familiarity with the problem's specific discipline is at least as important as experience with stochastic procedures. For example, confronted with a groundwater flow problem, a stochastic data analysis and processing expert with a hydrologic background will respond better than a structural engineer.

2. *Determination of the objectives of stochastic data analysis and processing*: This part raises certain questions concerning the proper specification of stochastic data analysis and processing objectives. For example, the geologists may want to find out if there exist rough and erratic changes in the morphology of the bedrock. The petroleum engineer may be interested in the spatial continuity of the oil reservoir depths. Additional considerations may include the directionality of the natural process in space, and the complexity of the trends (local or global) involved. In many situations the objective may be the optimal design of a rainfall observation network. Another objective may be the exploration level of the unknown object, say an oil deposit. Is it sufficient? If it is not, by which means and to what extent should the exploration be continued? Are the results of stochastic analysis and processing going to be used in the context of groundwater modeling? (For example, estimated log transmissivities and hydraulic heads can be combined by means of inverse modeling to obtain improved estimates of the former.) Answers to such questions help in selecting one data processing technique over another.

3. *Data and resources available*: It is necessary that enough data are collected to ensure that the stochastic technique will function properly and that the immediate and foreseeable objectives of stochastic analysis and processing will be completed. The quality of the data is also important; some stochastic data analysis and processing techniques cannot produce results of some specified reliability and accuracy unless all input information is at least of that reliability and accuracy. Some of the data may have to be disregarded for technical reasons or due to unacceptably low quality.

Under certain circumstances, it may be possible to take into account prior information that is highly relevant to the spatial variability of the natural processes involved. Such information may be, for example, knowledge of the physics of the underlying phenomena, geological interpretations, intuition, and experience with similar site conditions. Of course, the data and the resources required depend on the objectives of the stochastic data analysis and processing. For example, to assure a high level of exploration accuracy it may be necessary to take an increased number of observations. Finally, easy access to computer facilities will favor the use of stochastic methods.

4. *Choice and implementation of the appropriate technique*: In general, the choice of the "best" stochastic approach will depend on the following factors: (i) the objectives (part 2 above), (ii) the data and resources available (part 3 above), and (iii) the choices to be made, as well as parameters to be evaluated *on-line* during the specific project.

For example, one will choose, among the various types of theoretical RF, the one that best represents the natural process of interest, on the basis of scientific understanding as well as data analysis. Before making a decision concerning the use of a specific estimation technique, the earth scientist should be asked to define precisely the physical quantity of interest: Is it the observed process, or is it another quantity related to the observed process by means of some physical model? In the latter case, estimation may provide numerically more accurate and physically more meaningful results by incorporating the model in the analysis. Also, the choice of an estimation technique depends on the previous identification of the spatial variability.

The selected stochastic approach should be robust, and it should provide consistency throughout the various steps of implementation. The latter must be adopted to the ever-changing needs of the project. Nonfeedback methods are inappropriate.

Effective implementation depends on the clear understanding of the RF concept, as well as experience with the particular scientific discipline. In addition to these, an important issue is the development of libraries of interactive computer programs with multiple optional forms and automatic procedures. Computers constitute a very important part of stochastic data analysis and processing. For example, computerized procedures are necessary to simulate several equiprobable alternative maps of saturations, porosities, etc., to be used as inputs to reservoir flow models. These maps have in common whatever information is available and by studying their differences—using manipulation methods such as visualization, transformation, reduction, etc.—valuable insight is gained regarding spatial uncertainties. Automatic procedures may be used to assign values to random field-related parameters that have no real counterparts, to solve huge systems of equations, etc.

In conclusion, the development of the stochastic research program represents an intellectual experience that cannot be considered simply a prolongation of well-established approaches; it is a procedure in which gradually novel concepts and insightful conclusions will be emerging. In the history of thought, actual experience always precedes human understanding of it. And this understanding emerges only very slowly. In general, a new experience is first interpreted in the framework of preexisting theories, and it is only step by step that its true nature appears and that there is discovered the true novelty it involves. And so it is for the stochastic research program.

2 The Spatial Random Field Model

"Models are to be used, but not to be believed."
H. Theil

1. Introduction

Due to their importance in almost any scientific discipline, spatial random fields (SRF) constitute an active area of current research. A lot of work has been done in the theory of SRF, but many important topics still remain to be studied. On the other hand, it is widely admitted that SRF are tough to work with, mainly for two reasons: the difficulty of the mathematics involved, and the absence of systematic books on the subject.

This chapter, therefore, is organized as follows: In the first few sections we develop a critical and concise summary of the fundamental concepts and results of the theory of SRF that have important applications in the stochastic analysis and processing research program. Although most of these results will be repeatedly used in subsequent sections of the treatise, certain proofs and other details will not, so they will not be discussed. Instead, instructive examples, illustrating the most important application-related aspects of these proofs, will be discussed. We consider both scalar and vector SRF. Scalar SRF represent physical processes characterized by a single quantity, such as soil porosity or hydraulic conductivity. Vector SRF represent processes that require more than one quantity; for example, the soil strength at a point within a soil layer may be characterized by the undrained active, passive, and direct shear strength.

Much attention is drawn to characterizing an SRF by means of its spatial as well as its spectral moments. By studying these moments, important insight is gained regarding the mathematical structure underlying the SRF concept. Additional information is obtained by studying the sample function stochastic properties of the SRF. In this regard, the usefulness of the geometrical features of the SRF in providing valuable visual clues is strongly advocated. In fact, this is precisely the material needed for the numerous potential applications in earth sciences, meteorology, environmental engineering, physics, image processing, and many other fields.

We will classify SRF in a number of different ways, each one of which reflects certain of their most important properties. These classifications are very useful from the stochastic analysis viewpoint, as well as for the efficient implementation of the theoretical results in the practical situations to be considered in this book. In connection with the latter, certain auxiliary hypotheses are explored in detail in the remainder of this chapter. These hypotheses are related to the classes of homogeneous SRF, isotropic SRF, and to a specific class of nonhomogeneous SRF with regard to which some new developments are discussed. These developments prove to be important in practical applications, and they also form the gist of a significant part of the theory to be presented in subsequent chapters.

Finally, the notions of ergodicity, quasi-ergodicity, and microergodicity are considered in light of the preceding theory, and their significance with regard to stochastic inferences is studied.

In a mathematical text the question of notation is always a crucial one. Here, random variables are denoted by lowercase letters, x, y, etc., random fields by uppercase letters X, Y, etc., but for their values Greek letters, χ, ψ, etc., are used.

2. Basic Notions

2.1 The Spatial Random Field Concept

The formal description of the basic notions in stochastic analysis is based on set-theoretic notions. Though the set-theoretic approach provides rather general concepts, it is by no means "art pour l'art." On the contrary, the notions it deals with are of fundamental importance in developing stochastic functional analysis. (For a more detailed treatment of the basics on stochastic analysis see, e.g., Yaglom, 1962; Gihman and Skorokhod, 1974a, b, and c.) In fact what we do is to translate set-theoretic concepts into probabilistic ones.

Let (Ω, F, P) be a *probability space*, where Ω is the *sample space*, F is a σ-field (or Borel field) of subsets of Ω, and P is a *probability measure* on the measurable space (Ω, F) satisfying *Kolmogorov's axioms*:

(a) $P(\Omega) = 1$;

(b) $0 \leq P(A_i) \leq 1$ for all sets $A_i \in F$; and

(c) if $A_i \cap A_j = \varnothing \quad (i \neq j)$, then $P\left(\bigcup_{i=1}^{\infty} A_i\right) = \sum_{i=1}^{\infty} P(A_i)$.

The sets A_i of the field F are called *events*. The probability space (Ω, F, P) serves as the basic model on which all stochastic calculations are performed.

Remark 1: A parenthetical remark may be appropriate at this point. In the context of stochastic functional analysis, axiomatic probability is considered a formal method of manipulating probabilities using the Kolmogorov axioms. To apply the theory, the probability space Ω must be defined and the probability measure assigned. These are *a priori* probabilities that, at this point, suffice to be considered purely mathematical notions lacking any objective or subjective meaning (see discussion in Chapter 1).

Example 1: Let χ_i be the air pollution concentration at a specific location i and let c_i be the permissible pollution level, which is determined on the basis of environmental and ecological (etc.) considerations. Consider the sample space $\Omega = \{(\chi_i, c_i): \chi_i, c_i \in R^1\}$, where R^1 is the set of real numbers (real line). The event A_i, "the permissible level c_i has been exceeded," is defined as the subset $A_i = \{(\chi_i, c_i): \chi_i > c_i\}$ of Ω, to which one can assign a probability measure $P(A_i)$ satisfying Kolmogorov's axioms (a), (b), and (c) above. This rather simple setup constitutes a very powerful construction for our future investigations.

An important concept in stochastic functional analysis is that of a random variable.

Definition 1: Let $(R^1, \mathfrak{F}^1)$ be a measurable space, where $\mathfrak{F}^1$ is a σ-field of Borel sets on the real line R^1. A (real-valued) *random variable* $x(u)$, where $u \in \Omega$ are elementary events, is a measurable mapping x from (Ω, F) into $(R^1, \mathfrak{F}^1)$, so that

$$\forall B \in \mathfrak{F}^1, \qquad x^{-1}(B) = \{u \in \Omega: x(u) \in B\} \in F \tag{1}$$

The terms probability measure and random variable never occur isolated from each other. Indeed, on the strength of Definition 1, $x(u)$ (or simply x) is said to be an F-measurable (real-valued) random variable, where measurability induces a probability measure μ_x on $(R^1, \mathfrak{F}^1)$ such as

$$\forall B \in \mathfrak{F}^1, \qquad \mu_x(B) = P[x^{-1}(B)] = P[x = x(u) \in B] \tag{2}$$

Naturally, the study of a random variable $x(u)$ can be accomplished by studying the probability measure μ_x on $(R^1, \mathfrak{F}^1)$. For real-valued random variables it is convenient to introduce the distribution function of the measure μ_x.

Definition 2: Let us define the set

$$I_\alpha = \{\chi : \chi \le \alpha, \alpha \in R^1\} \tag{3}$$

Then the function

$$F_x(\chi) = \mu_x(I_\chi) = P[x \le \chi] \tag{4}$$

is called the *distribution function* of the random variable x (or the distribution function of the measure μ_x).

In most cases of practical interest the distribution function (4) can be replaced by the probability density function.

Definition 3: If the probability measure of Eq. (4) is absolutely continuous and m is the Lebesgue measure on R^1, the function f_x defined on R^1 so that

$$\int_B f_x(\chi)\, dm(\chi) = \mu_x(B) \tag{5}$$

for each Borel set B, is called the ***probability density*** of the random variable $x(u)$.

Below we introduce a space of random variables that exhibits certain useful properties.

Definition 4: Let (Ω, F, P) be a probability space. An $L_p(\Omega, F, P)$ space (or, simply, an L_p-space), $p \ge 1$, is a linear normed space of random variables x on (Ω, F, P) that satisfy the condition

$$E|x|^p = \int |x(u)|^p P(du) < \infty \tag{6}$$

The corresponding norm is defined by the usual formula

$$\|x\| = \{E|x|^p\}^{1/p} \tag{7}$$

and L_p is a complete space.

In this book we consider random variables that satisfy Definition 4 for $p = 2$. These random variables are called *second-order* random variables. Notice that an L_2 space equipped with the scalar product

$$(x_1, x_2) = E[x_1 x_2] = \int x_1(u) x_2(u) P(du) \tag{8}$$

where x_1 and x_2 are random variables, is a *Hilbert* space.

Let $\mathbf{s} = (s_1, s_2, \ldots, s_n) \in R^n$, $n \ge 1$, be spatial coordinates such that

$$\mathbf{s}^\alpha = s_1^{\alpha_1}\; s_2^{\alpha_2} \;\ldots\; s_n^{\alpha_n}$$

and

$$|\mathbf{s}| = \sqrt{\sum_{i=1}^{n} s_i^2}$$

where $\boldsymbol{\alpha} = (\alpha_1, \alpha_2, \ldots, \alpha_n)$ is a multi-index of nonnegative integers such that $|\boldsymbol{\alpha}| = \sum_{i=1}^{n} \alpha_i$ and $\boldsymbol{\alpha}! = \alpha_1! \, \alpha_2! \ldots \alpha_n!$. We recall that a study of a real-valued random variable can be made by studying probability measures (2) or, equivalently, probability density functions (5).

However, when dealing with a natural process that is observable at several points within the Euclidean space R^n, such a study loses track of the spatial model underlying the process. If we consider several random variables $x_1, x_2, \ldots, x_m$ at points $\mathbf{s}_1, \mathbf{s}_2, \ldots, \mathbf{s}_m$ in R^n, the corresponding probability measures $\mu_{x_1}, \mu_{x_2}, \ldots, \mu_{x_m}$ are not by themselves sufficient to express all the important features of the random variables $x_1, x_2, \ldots, x_m$ and of their correlations. It seems, therefore, natural to consider $X = (x_1, x_2, \ldots, x_m)$ a measurable mapping from (Ω, F) into $(R^m, \mathfrak{F}^m)$, where $\mathfrak{F}^m$ is a suitably chosen σ-field of subsets of R^m, and define a suitable new measure by extending definition (2). Then, a complete study of the random variables $x_1, x_2, \ldots, x_m$ and of their spatial relations can be achieved on the basis of the measure of the random quantity X, which is the collection of the random variables under consideration. In addition, in most applications (such as statistical continuum problems) the set of points $\{\mathbf{s}_1, \mathbf{s}_2, \ldots\}$ is infinite. For example, the complete characterization of a turbulent velocity field requires the joint probability distribution over the infinite family of random variables $\{x_1, x_2, \ldots\}$. These ideas lead to the following definition of a spatial random field.

Definition 5: Let (Ω, F, P) be a probability space and let $(R^1, \mathfrak{F}^1)$ be a measurable space, both in the sense defined above. A *spatial random field* (*SRF*) $X(\mathbf{s})$, $\mathbf{s} \in R^n$ is a family of random variables $\{x_1, x_2, \ldots\}$ at points $\mathbf{s}_1, \mathbf{s}_2, \ldots$, where each random variable is defined on (Ω, F, P) and takes values in $(R^1, \mathfrak{F}^1)$.

The space of all SRF will be denoted by $\mathbf{Y}$. An SRF is termed continuous-parameter or discrete-parameter according to whether the argument $\mathbf{s}$ takes discrete or continuous values. In the light of the foregoing considerations and since one can define the random variable as a function of elementary events $u \in \Omega$ [i.e., $x = x(u)$], it follows that the SRF can be considered as a function of both the elementary events $u \in \Omega$ and the spatial positions $\mathbf{s} \in R^n$ [i.e., $X(\mathbf{s}) = X(\mathbf{s}, u)$]. To a family of random variables $\{x_1, x_2, \ldots, x_m\}$ we associate a family of probability measures of the form

$$\mu_x(B) = \mu_{\mathbf{s}_1, \ldots, \mathbf{s}_m}(B)$$
$$= P[X^{-1}(B)] = P[(x_1, \ldots, x_m) \in B] \tag{9}$$

for every $B \in \mathfrak{F}^m$. (Note that to emphasize the dependency of the probability measure on the spatial positions the notation $\mu_{\mathbf{s}_1, \ldots, \mathbf{s}_m}(B)$ is used.) According to Kolmogorov (1933), a necessary and sufficient condition for the existence of the SRF $X(\mathbf{s})$, $\mathbf{s} \in R^n$, is that the probability measures (9) satisfy the following conditions:

(a) *Symmetry condition*: Let

$$\mu_{\mathbf{s}_{i_1},\ldots,\mathbf{s}_{i_m}}(B) = P[(x_{i_1}, \ldots, x_{i_m}) \in B]$$

where $i_1, \ldots, i_m$ is a permutation of the indices $1, \ldots, m$. Symmetry requires that

$$\mu_{\mathbf{s}_1,\ldots,\mathbf{s}_m}(B) = \mu_{\mathbf{s}_{i_1},\ldots,\mathbf{s}_{i_m}}(B) \tag{10}$$

for any permutation.

(b) *Consistency condition*: It holds that

$$\mu_{\mathbf{s}_1,\ldots,\mathbf{s}_{m+k}}(B \times R^k) = \mu_{\mathbf{s}_1,\ldots,\mathbf{s}_m}(B) \tag{11}$$

for any m, $k \geq 1$ and $B \in \mathfrak{I}^m$.

Again, it is convenient to work in terms of the *distribution function* of the measure (9) defined as

$$F_{\mathbf{s}_1,\ldots,\mathbf{s}_m}(\chi_1, \ldots, \chi_m) = P[x_1 \leq \chi_1, \ldots, x_m \leq \chi_m] \tag{12}$$

for any m. Then, conditions (10) and (11) can be written, respectively,

$$F_{\mathbf{s}_{i_1},\ldots,\mathbf{s}_{i_m}}(\chi_{i_1}, \ldots, \chi_{i_m}) = F_{\mathbf{s}_1,\ldots,\mathbf{s}_m}(\chi_1, \ldots, \chi_m) \tag{13}$$

$$F_{\mathbf{s}_1,\ldots,\mathbf{s}_m,\mathbf{s}_{m+1},\ldots,\mathbf{s}_{m+k}}(\chi_1, \ldots, \chi_m, \infty, \ldots, \infty) = F_{\mathbf{s}_1,\ldots,\mathbf{s}_m}(\chi_1, \ldots, \chi_m) \tag{14}$$

Furthermore, we usually assume the existence of the *probability density functions* corresponding to (12) and denoted by

$$\begin{aligned} f_x(\chi_1, \ldots, \chi_m) &= f_{\mathbf{s}_1,\ldots,\mathbf{s}_m}(\chi_1, \ldots, \chi_m) \\ &= \frac{\partial^m}{\partial\chi_1 \cdots \partial\chi_m} F_{\mathbf{s}_1,\ldots,\mathbf{s}_m}(\chi_1, \ldots, \chi_m) \end{aligned} \tag{15}$$

In the following, both symbols $f_x(F_x)$ and $f_{\mathbf{s}}(F_{\mathbf{s}})$ will be used to denote probability densities (distributions) of SRF.

One may also define the *conditional probability densities* of an SRF as

$$\begin{aligned} f_x(\chi_{k+1}, \ldots, \chi_m \mid \chi_1, \ldots, \chi_k) &= f_{\mathbf{s}_1,\ldots,\mathbf{s}_m}(\chi_{k+1}, \ldots, \chi_m \mid \chi_1, \ldots, \chi_k) \\ &= \frac{f_{\mathbf{s}_1,\ldots,\mathbf{s}_m}(\chi_1, \ldots, \chi_m)}{f_{\mathbf{s}_1,\ldots,\mathbf{s}_k}(\chi_1, \ldots, \chi_k)} \end{aligned} \tag{16}$$

An interesting consequence of Eq. (16) is the expression

$$f_x(\chi_1, \ldots, \chi_m) = f_x(\chi_1) f_x(\chi_2 \mid \chi_1) \cdots f_x(\chi_m \mid \chi_1, \ldots, \chi_{m-1}) \tag{17}$$

Equation (17) has proven to be a very useful tool in a variety of random field applications such as, for example, spatial simulation of natural processes (see Chapter 8).

Last, the m-dimensional Fourier transform of the probability density (15) yields the so-called *characteristic function*

$$\begin{aligned}\phi_x(w_1, \ldots, w_m) &= E\left[\exp\left[i \sum_{k=1}^{m} w_k x_k\right]\right] \\ &= \int_{R^m} \exp\left[i \sum_{k=1}^{m} w_k \chi_k\right] f_x(\chi_1, \ldots, \chi_m)\, d\chi_1 \cdots d\chi_m \\ &= FT[f_x(\chi_1, \ldots, \chi_m)] \end{aligned} \tag{18}$$

where $i = \sqrt{-1}$. Equation (18) uniquely determines the probability density function $f_x(\chi_1, \ldots, \chi_m)$. The fact that the probability density function and the characteristic function constitute a Fourier transform pair implies that the narrower the former, the wider the latter. Other interesting properties of the characteristic function include the following:

$$|\phi_x(w_1, \ldots, w_m)| \leq 1 \tag{19}$$

$$\phi_x(0, \ldots, 0) = 1 \tag{20}$$

$$\phi_x(w_1, \ldots, w_{m-k}) = \phi_x(w_1, \ldots, w_{m-k}, w_{m-k+1} = 0, \ldots, w_m = 0) \tag{21}$$

The SRF $X(\mathbf{s})$, $\mathbf{s} \in R^n$ is specified completely by means of all finite dimensional probability measures (9), probability distributions (12), probability densities (15), or characteristic functions (18) of orders $m = 1, 2, \ldots$. Other approaches for defining an SRF include one that is analogous to the definition of a random variable and one in terms of the linear normed spaces discussed earlier.

Definition 6: Let $\mathbf{A}$ be the set of all real-valued functions in R^n and let $\mathbf{G}$ be a suitable σ-field of subsets of $\mathbf{A}$. An SRF $X(\mathbf{s})$, $\mathbf{s} \in R^n$ is a measurable mapping from (Ω, F) into $(\mathbf{A}, \mathbf{G})$.

Definition 7: Let $L_2(\Omega, F, P)$ be the Hilbert space of the random variables x at $\mathbf{s} \in R^n$. An SRF $X(\mathbf{s})$ is defined as a mapping on R^n with values in the Hilbert space $L_2(\Omega, F, P)$, viz.,

$$X : R^n \to L_2(\Omega, F, P) \tag{22}$$

Example 2: The case of a two-dimensional SRF $X(\mathbf{s})$, $\mathbf{s} \in R^2$, representing, say, a lead concentration surface, is illustrated in Fig. 2.1. Clearly, there are two alternative viewpoints from which one can look at an SRF:

(i) Vertically, as a collection of random variables $x(u)$ ($\mathbf{s} = (s_1, s_2)$ is fixed in R^2 and the generic element u is varying in Ω).

(ii) Horizontally, as a family of realizations $\chi(\mathbf{s})$ (u is given in Ω and $\mathbf{s}$ varies in R^2).

Viewpoint (i) is recommended for theoretical studies while (ii) is more appropriate in a modeling context.

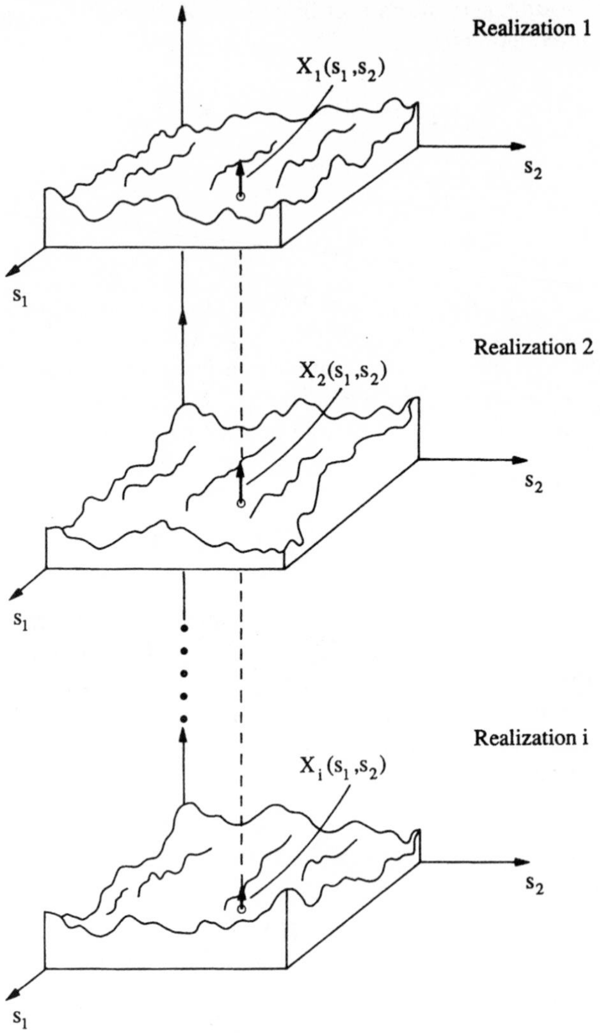

Figure 2.1 An illustration of an SRF $X(\mathbf{s})$, $\mathbf{s} \in R^2$

Remark 2: If instead of $\mathbf{s} \in R^n$, the argument is $s \in R^1$ (or time $t \in T$, T is the time axis $\subseteq R^1_{+,\{0\}}$), the $X(s)$ [or $X(t)$] is called a *random process* (RP).

Last, it is worth mentioning that a complete stochastic characterization of an SRF $X(\mathbf{s})$ can also be achieved through the concept of the characteristic functional. Let $q(\mathbf{s})$ be a nonrandom function such that the integral (continuous linear functional)

$$X(q) = \langle q(\mathbf{s}), X(\mathbf{s}) \rangle = \int_U q(\mathbf{s}) X(\mathbf{s})\, d\mathbf{s} \tag{23}$$

where $U \subseteq R^n$, exists for almost all realizations of $X(\mathbf{s})$. Then, the *characteristic functional* of the SRF $X(\mathbf{s})$ is defined as follows:

$$\Phi(q) = E\{\exp[iX(q)]\} \tag{24}$$

Clearly, the $\Phi(q)$ must be known for any $q(\mathbf{s})$. When the $\Phi(q)$ is available, one may derive the characteristic function (18), the corresponding probability density, and the moments of $X(\mathbf{s})$. Other interesting properties of the characteristic functionals are as follows:

(i) If $\Phi_1(q)$ and $\Phi_2[q]$ are characteristic functionals, and $a_1, a_2 \geq 0$ such that $a_1 + a_2 = 1$, then the $a_1\Phi_1[q] + a_2\Phi_2(q)$ and $\Phi_1[q]\Phi_2[q]$ are characteristic functionals, as well.

(ii) Let $\Phi_1[q]$ and $\Phi_2[q]$ be the characteristic functionals of the SRF $X_1(\mathbf{s})$ and $X_2(\mathbf{s})$, respectively. Suppose that

$$X_2(\mathbf{s}') = L[X_1(\mathbf{s}')] = \int_U f(\mathbf{s}, \mathbf{s}')X_1(\mathbf{s})\, d\mathbf{s}$$

where $f(\mathbf{s}, \mathbf{s}')$ is a suitable deterministic function. Then

$$\Phi_2[q] = \Phi_1\{L^*[q]\} = \Phi_1\left[\int_U f(\mathbf{s}', \mathbf{s})q(\mathbf{s})\, d\mathbf{s}\right]$$

where L^* is a conjugate to the operator L.

(iii) Let $\Phi_u[q]$ be a family of characteristic functionals that depend on the random variable u. Then, the

$$\Phi[q] = \int_U \Phi_u[q]\, dF_u(\upsilon) \tag{25}$$

where $F_u(\upsilon)$ is the probability distribution of u, is a characteristic functional too.

The functional characterization of SRF is closely related to the theory of intrinsic SRF (Chapter 3); as we will see, in this case the functions $q(\mathbf{s})$ are assumed to belong to the Schwartz spaces of functions. The functional treatment of SRF is particularly useful in the study of stochastic differential equations modeling physical systems (e.g., Beran, 1968).

2.2 Vector Spatial Random Fields

The foregoing theory deals with *scalar* SRF. However, it can be extended to include the case of a set of SRF stochastically correlated to one another.

Definition 8: A *vector SRF* $\mathbf{X}(\mathbf{s})$, $\mathbf{s} \in R^n$ is a set of SRF $X_1(\mathbf{s}), X_2(\mathbf{s}), \ldots, X_k(\mathbf{s})$ denoted by the vector notation

$$\mathbf{X}(\mathbf{s}) = [X_1(\mathbf{s}), X_2(\mathbf{s}), \ldots, X_k(\mathbf{s})]^{\mathrm{T}} \tag{26}$$

where the SRF $X_1(\mathbf{s})$, $X_2(\mathbf{s}), \ldots, X_k(\mathbf{s})$ are called *component-SRF* of the vector SRF $\mathbf{X}(\mathbf{s})$.

The specification of a vector SRF $\mathbf{X}(\mathbf{s})$ is made by analogy with the case of a single SRF $X(\mathbf{s})$. That is, all the multivariate probability densities of the component SRF $X_1(\mathbf{s})$, $X_2(\mathbf{s}), \ldots, X_k(\mathbf{s})$ should be determined.

2.3 Classifications of the Spatial Random Field Model

The SRF model may be classified in five distinct ways. Brief descriptions of these classifications are as follows (detailed discussions are given in later sections):

Classification A: The first way of classifying SRF depends on whether the space argument, $\mathbf{s} \in R^n$, and the realizations of the SRF $X(\mathbf{s})$ are *discrete* or *continuous*. In particular:

(i) discrete $\mathbf{s}$—discrete $X(\mathbf{s})$
(ii) discrete $\mathbf{s}$—continuous $X(\mathbf{s})$
(iii) continuous $\mathbf{s}$—discrete $X(\mathbf{s})$
(iv) continuous $\mathbf{s}$—continuous $X(\mathbf{s})$

Cases (i) and (ii) can be derived from cases (iii) and (iv), respectively, by discretization.

Classification B: The second way of classifying SRF depends on whether the space argument, $\mathbf{s} \in R^n$, and the SRF $X(\mathbf{s})$ are *scalar* or *vector*. More specifically four subclasses of SRF can be distinguished:

(i) Both the space argument and the SRF are scalar, viz., an RP $X(s)$ or $X(t)$.
(ii) The space argument is scalar but there are several SRF, viz., a *vector* RP $\mathbf{X}(s) = [X_1(s), X_2(s), \ldots, X_k(s)]^T$.
(iii) The space argument is vector but the SRF is scalar, viz., $X(\mathbf{s})$.
(iv) Both the space argument and the SRF are vectors, viz., a *vector* SRF $\mathbf{X}(\mathbf{s}) = [X_1(\mathbf{s}), X_2(\mathbf{s}), \ldots, X_k(\mathbf{s})]^T$.

Classification C: Another way of classifying SRF is by means of the *form of the corresponding probability laws.* Broadly speaking, we may consider Gaussian and non-Gaussian SRF. An SRF is called a *Gaussian* SRF if all its finite dimensional probability density functions are multivariate Gaussian. This sort of field is most frequently encountered in practical applications and it has several important properties: For example, a Gaussian SRF is completely characterized by its mean and covariance. Also, if an SRF is Gaussian, then any linear transformation of the SRF is a Gaussian SRF too. Much more will be said about the Gaussian SRF later. Important SRF are the so-called *factorable* SRF (Chapter 4).

Classification D: SRF can be classified according to the *spatial variability of* $X(\mathbf{s})$ over $\mathbf{s} \in R^n$. More specifically an SRF can be characterized as *spatially homogeneous* or *spatially nonhomogeneous*. The latter refers to SRF whose statistical properties depend on the space origin, while the former refers to SRF that are independent of the space origin. Classification D plays a crucial role in the stochastic research program [particularly parts (b) and (c); see Section 4 of Chapter 1].

Classification E: This sort of classification is based on the *memory* of the SRF. An SRF with *zero memory* (also called an *independent* SRF) is an SRF that is completely specified by means of the univariate density $f_{\mathbf{s}}(\chi)$, since in this case

$$f_{\mathbf{s}_1,\ldots,\mathbf{s}_m}(\chi_1,\ldots,\chi_m) = \prod_{i=1}^{m} f_{\mathbf{s}_i}(\chi_i) \tag{27}$$

for all m. Equation (27) represents *complete chaos.* An RP with a very useful structure is the *Markov* RP. According to the Markov assumption, knowledge of only the present determines the future. In terms of the conditional probabilities and assuming that $t_1 < t_2 < \cdots < t_m$, Eq. (17) gives

$$f_{t_1,t_2,\ldots,t_m}(\chi_1,\ldots,\chi_m) = f_{t_1}(\chi_1) f_{t_2,t_1}(\chi_2|\chi_1) \cdots f_{t_m,t_{m-1}}(\chi_m|\chi_{m-1}) \tag{28}$$

It must be noted that each of the above classifications is independent of the others. That is, a Gaussian SRF can be either homogeneous or nonhomogeneous, Markov or non-Markov; a homogeneous SRF can be Gaussian or non-Gaussian, etc.

3. Characterization of Spatial Random Fields by Means of Their Second-Order Statistical Moments—Correlation Theory

The SRF considered are *second-order SRF* (scalar or vector), i.e., they consist of second-order random variables. In applied sciences usually it is not possible to completely characterize an SRF on the basis of its distribution functions; this is primarily due to the small number of available realizations. (In most cases we have only one sequence of measurements.) Thus, we are limited to a characterization of the RF in terms of its statistical moments of order up to two. The part of the general SRF theory that studies only the properties of SRF determined by their statistical moments of order up to two is called *correlation theory.* In this case there is no need to mention the relevant probability distributions. Throughout this work, all the above statistical quantities are assumed to be real-valued continuous functions in R^n. Also we will always assume that all the SRF under consideration have finite means and variances.

3.1 Scalar Statistical Moments up to Second Order

In the case of scalar SRF the corresponding statistical moments of order up to two are defined as follows.

Definition 1: The *mean value* $m_x(\mathbf{s})$ and the *covariance* $c_x(\mathbf{s}, \mathbf{s}')$ of an SRF $X(\mathbf{s})$ are defined by

$$m_x(\mathbf{s}) = E[X(\mathbf{s})] = \int \chi f_{\mathbf{s}}(\chi)\, d\chi \tag{1}$$

and

$$\begin{aligned} c_x(\mathbf{s}, \mathbf{s}') &= E[[X(\mathbf{s}) - m_x(\mathbf{s})][X(\mathbf{s}') - m_x(\mathbf{s}')]] \\ &= \iint [\chi - m_x(\mathbf{s})][\chi' - m_x(\mathbf{s}')] f_{\mathbf{s},\mathbf{s}'}(\chi, \chi')\, d\chi\, d\chi' \end{aligned} \tag{2}$$

respectively; $\mathbf{s}$ and $\mathbf{s}'$ are constants under the integration. The class of all covariances in R^n is denoted by C_n.

Remark 1: Equation (1) is also called the *first moment.* Equation (2) is the *central second moment*; in this sense the $E[X(\mathbf{s})X(\mathbf{s}')]$ is called the *noncentered second moment* (or *covariance*) and it follows that

$$c_x(\mathbf{s}, \mathbf{s}') = E[X(\mathbf{s})X(\mathbf{s}')] - m_x(\mathbf{s})m_x(\mathbf{s}') \tag{3}$$

The mean $m_x(\mathbf{s})$ may be constant or may be any function of $\mathbf{s}$. In the latter case it describes systematic variations of the SRF $X(\mathbf{s})$ and is also known as *trend.* Once the covariance is known we can define two more useful statistical moments:

(i) the *variance*

$$\sigma_x^2(\mathbf{s}) = c_x(\mathbf{s}, \mathbf{s}' = \mathbf{s}) \tag{4}$$

which describes local random variations; and

(ii) the *spatial correlation function*

$$\rho_x(\mathbf{s}, \mathbf{s}') = \frac{c_x(\mathbf{s}, \mathbf{s}')}{\sqrt{\sigma_x^2(\mathbf{s})}\sqrt{\sigma_x^2(\mathbf{s}')}} \tag{5}$$

which expresses spatial relations. The class of all correlation functions in R^n will be denoted L_n.

An SRF $X(\mathbf{s})$ is called an *uncorrelated* SRF if

$$c_x(\mathbf{s}, \mathbf{s}') = \begin{cases} \sigma_x^2(\mathbf{s}) & \text{if} \quad \mathbf{s} = \mathbf{s}' \\ 0 & \text{otherwise} \end{cases} \tag{6}$$

As we shall see later, such an SRF is also called a *white-noise* SRF. In general, it is not true that an uncorrelated SRF is independent, too; that is, Eq. (6) does not necessarily imply that $f_x(\mathbf{s}, \mathbf{s}') = f_x(\mathbf{s}) f_x(\mathbf{s}')$ (recall that as stated in Section 2.1 both symbols $f_x(\mathbf{s})$ and $f_{\mathbf{s}}(\chi)$ are used to denote probability densities). Hence, while Eq. (6) offers us a measure about the extent of the relationship between $X(\mathbf{s})$ and $X(\mathbf{s}')$, it does not represent the entire relationship given by $f_x(\mathbf{s}, \mathbf{s}')$. It is clear, however, that independency always implies lack of correlation.

To a specific covariance correspond more than one SRF, often quite dissimilar to each other. For example, let $X_2(\mathbf{s}) = vX_1(\mathbf{s})$, where v is a zero-mean random variable independent of $X_1(\mathbf{s})$ for any $\mathbf{s} \in R^n$ and $E[v^2] = 1$; it is easily seen that both SRF, $X_1(\mathbf{s})$, and $X_2(\mathbf{s})$, have the same covariance although their realizations may be quite different. However, the statistical mean and covariance functions have proven extremely useful in the correlation theory of SRF for at least the following reasons:

(a) A number of characteristics of an SRF can be specified completely in terms of the mean and the covariance.

(b) Under certain circumstances, which usually apply in practice, the mean and covariance can be calculated accurately and with much less effort than other probabilistic characteristics.

(c) The very important class of Gaussian SRF (Section 3.3 below) is completely characterized by the corresponding means and covariances. Therefore, it is expected that these two statistical moments will play a fundamental role in this presentation.

Within the framework of correlation theory, certain important properties of the covariance function are discussed next:

Property 1: A function $c_x(\mathbf{s}, \mathbf{s}')$ is a covariance [i.e., $c_x(\mathbf{s}, \mathbf{s}') \in C_n$] if and only if it is of the *nonnegative-definite* type; that is,

$$\sum_{i=1}^{m} \sum_{j=1}^{m} q_i q_j c_x(\mathbf{s}_i, \mathbf{s}_j) \geq 0 \tag{7}$$

for all nonnegative integers m, all points $\mathbf{s}_i, \mathbf{s}_j \in R^n$, and all real (or complex) numbers $q_1, q_2, \ldots, q_m$. Inequality (7) is a direct consequence of

$$E\left[\sum_{i=1}^{m} q_i [X(\mathbf{s}_i) - m_x(\mathbf{s}_i)]\right]^2$$

$$= \sum_{i=1}^{m} \sum_{j=1}^{m} q_i q_j \; E[X(\mathbf{s}_i) - m_x(\mathbf{s}_i)][X(\mathbf{s}_j) - m_x(\mathbf{s}_j)]\} \geq 0$$

Property 2: Each covariance $c_x(\mathbf{s}, \mathbf{s}') \in C_n$ is a symmetric function; that is,

$$c_x(\mathbf{s}, \mathbf{s}') = c_x(\mathbf{s}', \mathbf{s}) \in C_n \tag{8}$$

Property 3: It is valid that

$$\lim_{|\mathbf{s}-\mathbf{s}'|\to\infty} c_x(\mathbf{s}, \mathbf{s}') = 0 \tag{9}$$

Property 4: The class C_n of covariances is closed under addition, multiplication, and passages to the limit; in other words, given the sequence of covariances $\{c_{x,k}(\mathbf{s}, \mathbf{s}')\} \in C_n$, $k = 1, 2, \ldots$, the

$$\sum_{\rho\leq k} c_{x,\rho}(\mathbf{s}, \mathbf{s}') \in C_n \tag{10}$$

and

$$\lim_{k\to\infty} c_{x,k}(\mathbf{s}, \mathbf{s}') = c_x(\mathbf{s}, \mathbf{s}') \in C_n \tag{11}$$

assuming that the above limit exists for all pairs $(\mathbf{s}, \mathbf{s}')$.

Property 5: If $\mu(u)$ is a measure on a space U and $c_u(\mathbf{s}, \mathbf{s}' \cdot u)$ is a function integrable on a subspace V of U for each pair $(\mathbf{s}, \mathbf{s}')$, then

$$\int_V c_u(\mathbf{s}, \mathbf{s}' \cdot u)\, \mu(du) = c_x(\mathbf{s}, \mathbf{s}') \in C_n \tag{12}$$

Property 6: Every covariance $c_x(\mathbf{s}, \mathbf{s}') \in C_n$ is also the covariance of a Gaussian SRF.

According to Property 1, some nonnegative-definite function can be the covariance function of some SRF. Property 3 means that when the distance between $\mathbf{s}$ and $\mathbf{s}'$ tends to infinity the correlation between $X(\mathbf{s})$ and $X(\mathbf{s}')$ tends to zero. Properties 4 and 5 provide interesting means of constructing covariances; let us discuss a few examples.

Example 1: If the $c_{x,k}(\mathbf{s}, \mathbf{s}')$, $k = 1, 2, \ldots$ are covariances, the quantities

$$c_x(\mathbf{s}, \mathbf{s}') = \sum_{\rho\leq k} a_\rho [c_{x,\rho}(\mathbf{s}, \mathbf{s}')]^\rho \tag{13}$$

and

$$c_x(\mathbf{s}, \mathbf{s}') = \lim_{k\to\infty} \sum_{\rho\leq k} a_\rho [c_{x,\rho}(\mathbf{s}, \mathbf{s}')]^\rho \tag{14}$$

where $a_\rho \geq 0$ and $\rho \leq k$ are covariances too.

Example 2: In Eq. (12) let $c_x(\mathbf{s}, \mathbf{s}') = c_x(r)$, where

$$r = |\mathbf{s} - \mathbf{s}'|$$

$$c_u(\mathbf{s}, \mathbf{s}' \cdot u) = c_u(r, u) = \sigma^2 \exp\left[-\frac{r^2}{u^2}\right]$$

and

$$\mu(du) = \phi(u)\, du$$

Then

$$c_x(r) = \sigma^2 \int_0^\infty \exp\left[-\frac{r^2}{u^2}\right] \phi(u)\, du \tag{15}$$

where the function $\phi(u)$ must be such that $\int_0^\infty \phi(u)\, du = 1$. If we choose

$$\phi(u) = \frac{1}{\sqrt{\pi}\, a} \exp\left[-\left(\frac{u}{2a}\right)^2\right]$$

Eq. (15) yields $c_x(r) = \sigma^2 \exp[-u/a]$. For $\phi(u) = \delta(u-a)$, Eq. (15) gives $c_x(r) = \sigma^2 \exp[-u^2/a^2]$.

Property 6 is of particular interest: First, it is easily seen that a similar property holds for every mean value $m_x(\mathbf{s})$, as well. Then, an important implication is that the study of any SRF in terms of its mean and covariance is equivalent to the study of a Gaussian SRF that has the same mean and covariance. The consequences of this fact are significant in various problems of the SRF theory, such as estimation and simulation of SRF on the basis of their means and covariances (Chapters 8 and 9).

Remark 2: The construction of multidimensional non-Gaussian probability distributions on the basis of the mean value and the covariance function can lead to inconsistent results. However, as already mentioned, in the context of the correlation theory this fact does not create any problem, for there is no need to deal with the form of the multidimensional probability distribution.

Higher-order moments may also be defined. For example, a third-order moment that is useful in applications (e.g., analysis of historical hydrologic series) is the *skewness function* defined as

$$\lambda(\mathbf{s}) = E[X(\mathbf{s}) - m_x(\mathbf{s})]^3 \tag{16}$$

The skewness expresses the lack of symmetry of the probability distribution. Usually, the skewness function is divided by $\sigma_x^3(\mathbf{s})$, leading to the *coefficients of skewness*

$$\gamma(\mathbf{s}) = \frac{E[X(\mathbf{s}) - m_x(\mathbf{s})]^3}{\sigma_x^3(\mathbf{s})} \tag{17}$$

Moreover, moments of any order can be derived by differentiating the multivariate characteristic function $\phi_x(w_1, \ldots, w_m)$ and setting $\mathbf{w} = (w_1, \ldots, w_m) = (0, \ldots, 0) = \mathbf{0}$; that is,

$$E[X^{\rho_1}(\mathbf{s}_1) X^{\rho_2}(\mathbf{s}_2) \cdots X^{\rho_m}(\mathbf{s}_m)] = i^{-\sum_{k=1}^m \rho_k} \left[\frac{\partial^{\sum_{k=1}^m \rho_k}}{\partial w_1^{\rho_1}\, \partial w_2^{\rho_2} \cdots \partial w_m^{\rho_m}} \phi_x(\mathbf{w})\right]_{\mathbf{w}=\mathbf{0}} \tag{18}$$

3.2 Vector Statistical Moments up to Second Order

Under certain circumstances it is possible to study two or more SRF that are related to each other. We then define an additional statistical moment, namely the spatial cross-covariance.

Definition 2: Let $X_1(\mathbf{s})$ and $X_2(\mathbf{s}')$, $\mathbf{s}$, $\mathbf{s}' \in R^n$ be two SRF. The *spatial cross-covariance function* is defined as

$$c_{x_1x_2}(\mathbf{s}, \mathbf{s}') = E[[X_1(\mathbf{s}) - m_{x_1}(\mathbf{s})][X_2(\mathbf{s}') - m_{x_2}(\mathbf{s}')]]$$
$$= \iint [\chi_1 - m_{x_1}(\mathbf{s})][\chi_2 - m_{x_2}(\mathbf{s}')] f_{\mathbf{s},\mathbf{s}'}(\chi_1, \chi_2)\, d\chi_1\, d\chi_2 \qquad (19)$$

where $f_{\mathbf{s},\mathbf{s}'}(\chi_1, \chi_2)$ is now the joint probability density of $X_1(\mathbf{s})$ and $X_2(\mathbf{s}')$.

By analogy to Eq. (5), the *coefficient of spatial cross-correlation* is defined as

$$\rho_{x_1x_2}(\mathbf{s}, \mathbf{s}') = \frac{c_{x_1x_2}(\mathbf{s}, \mathbf{s}')}{\sqrt{\sigma^2_{x_1}(\mathbf{s})}\sqrt{\sigma^2_{x_2}(\mathbf{s}')}} \qquad (20)$$

where $\sigma^2_{x_1}(\mathbf{s})$ and $\sigma^2_{x_2}(\mathbf{s}')$ are the variances of the SRF $X_1(\mathbf{s})$ and $X_2(\mathbf{s}')$, respectively.

The definition above can be easily extended to more than two SRF. Let $\mathbf{X}(\mathbf{s}) = [X_1(\mathbf{s}), X_2(\mathbf{s}), \ldots, X_k(\mathbf{s})]^T$ be a vector SRF. The corresponding matrix of cross-covariances between the component SRF is

$$\mathbf{C_X} = [c_{x_px_q}(\mathbf{s}_i, \mathbf{s}_j)] \qquad (21)$$

where $p, q = 1, 2, \ldots, k$. Some interesting properties of the cross-covariance matrix (21) are summarized below:

Property 1: A matrix $\mathbf{C_X}$ is *nonnegative-definite* type, that is,

$$\mathbf{q}^T\mathbf{C_X}\mathbf{q} \geq 0 \qquad (22)$$

for all vectors $\mathbf{q}^T = [q_1, q_2, \ldots, q_k]$. This is because it must hold

$$\mathrm{Var}[\mathbf{q}^T\mathbf{X}(\mathbf{s})] \geq 0.$$

Property 2: The $\mathbf{C_X}$ is a symmetric matrix, since by definition

$$c_{x_px_q}(\mathbf{s}_i, \mathbf{s}_j) = c_{x_qx_p}(\mathbf{s}_j, \mathbf{s}_i)$$

However there is no symmetry, in general, with respect to $\mathbf{s}_i$ and $\mathbf{s}_j$.

Property 3: A straightforward application of the Schwartz's inequality yields

$$|c_{x_px_q}(\mathbf{s}_i, \mathbf{s}_j)| \leq \sqrt{\sigma^2_{x_p}(\mathbf{s}_i)}\,\sqrt{\sigma^2_{x_q}(\mathbf{s}_j)} \qquad (23)$$

for all p, $q = 1, 2, \ldots, k$ and $i, j = 1, 2, \ldots, m$.

To clarify some aspects of the matrix algebra involved consider the following example.

Example 3: Let $k=2$ so that

$$\mathbf{C_X}=\begin{bmatrix} c_{x_1}(\mathbf{s}_i,\mathbf{s}_j) & c_{x_1x_2}(\mathbf{s}_i,\mathbf{s}_j) \\ c_{x_2x_1}(\mathbf{s}_j,\mathbf{s}_i) & c_{x_2}(\mathbf{s}_i,\mathbf{s}_j) \end{bmatrix}$$

If we define the SRF

$$Y(\mathbf{s}_i)=\sum_{k=1}^{2} q_k X_k(\mathbf{s}_i)$$

where $q_k \in R^1$ ($k=1$ and 2), the corresponding covariance is given by

$$c_Y(\mathbf{s}_i,\mathbf{s}_j)=\sum_{k=1}^{2}\sum_{k'=1}^{2} q_k q_{k'} c_{x_k x_{k'}}(\mathbf{s}_i,\mathbf{s}_j)$$

which must be a nonnegative-definite function.

Let $\mathbf{X}=[x_1,\ldots,x_m]^{\mathrm{T}}$ and $\mathbf{Y}=[y_1,\ldots,y_m]^{\mathrm{T}}$ be vectors of random variables from the SRF $X(\mathbf{s})$ and $Y(\mathbf{s})$, respectively. The conditional mean of $\mathbf{X}$ given $\mathbf{Y}=\mathbf{\Psi}=[\psi_1,\ldots,\psi_m]^{\mathrm{T}}$ is the deterministic vector

$$m_{\mathbf{X}|\mathbf{\Psi}}=E[\mathbf{X}|\mathbf{\Psi}]=\int_{R^m}\chi f_{\mathbf{X}|\mathbf{Y}}(\chi|\mathbf{\Psi})\,d\chi \tag{24}$$

The conditional covariance of $\mathbf{X}$ given $\mathbf{\Psi}$ is the deterministic matrix

$$\begin{aligned}\mathbf{C}_{\mathbf{X}|\mathbf{\Psi}}&=E\{[\mathbf{X}-m_{\mathbf{X}|\mathbf{\Psi}}][\mathbf{X}-m_{\mathbf{X}|\mathbf{\Psi}}]^{\mathrm{T}}|\mathbf{\Psi}\}\\ &=\int_{R^m}[\chi-m_{\mathbf{X}|\mathbf{\Psi}}][\chi-m_{\mathbf{X}|\mathbf{\Psi}}]^{\mathrm{T}}f_{\mathbf{X}|\mathbf{Y}}(\chi|\mathbf{\Psi})\,d\chi\end{aligned} \tag{25}$$

If $\mathbf{Y}$ is regarded as random, the conditional mean vector and the covariance matrix become random quantities as well, namely

$$m_{\mathbf{X}|\mathbf{Y}}=\int_{R^m}\chi f_{\mathbf{X}|\mathbf{Y}}(\chi|\mathbf{Y})\,d\chi \tag{26}$$

and

$$\begin{aligned}\mathbf{C}_{\mathbf{X}|\mathbf{Y}}&=E\{[\mathbf{X}-m_{\mathbf{X}|\mathbf{Y}}][\mathbf{X}-m_{\mathbf{X}|\mathbf{Y}}]^{\mathrm{T}}|\mathbf{Y}\}\\ &=\int_{R^m}[\chi-m_{\mathbf{X}|\mathbf{Y}}][\chi-m_{\mathbf{X}|\mathbf{Y}}]^{\mathrm{T}}f_{\mathbf{X}|\mathbf{Y}}(\chi|\mathbf{Y})\,d\chi\end{aligned} \tag{27}$$

Remark 3: A useful relationship is provided by the *chain rule* for conditional expectations of SRF

$$E_Y\{E_X[X(\mathbf{s})|\,Y(\mathbf{s})]\}=E_X[X(\mathbf{s})] \tag{28}$$

where the symbols E_X and E_Y denote expectation with respect to the SRF $X(\mathbf{s})$ and $Y(\mathbf{s})$, respectively.

3.3 Gaussian and Related Spatial Random Fields

The family of the so-called *Gaussian SRF* is very important for a variety of reasons, such as:

(i) Members of this family have convenient mathematical properties that greatly simplify calculations involving linear transformations; in fact, many results can be worked out only for Gaussian SRF.

(ii) It is the limit approached by the superposition of a large number of other non-Gaussian SRF; this is a famous result of the well-known central limit theorem (e.g., Feller, 1966).

(iii) It has been found that the distribution of many natural processes can be approximated very satisfactorily by means of Gaussian SRF.

(iv) For any SRF with finite first and second-order moments it is always possible to construct a Gaussian SRF with the same mean and covariance. This is not true for non-Gaussian SRF. In fact, the construction of multivariate non-Gaussian probability densities on the basis of first and second-order moments can lead to inconsistent results.

(v) The Gaussian SRF maximizes the probabilistic entropy function subject to the constraints associated with the first and second-order moments (for more details, see Section 13 below; also, Chapters 7 and 9).

Definition 3: A *multivariate Gaussian* SRF is an SRF $X(\mathbf{s})$ for which the multivariate probability density function of the vector random variable $\mathbf{X}=[x_1,\ldots,x_m]^T$ for all m is given by

$$f_{\mathbf{X}}(\boldsymbol{\chi})=\frac{1}{(2\pi)^{m/2}|\mathbf{C_X}|^{1/2}}\exp\left[-\frac{[\boldsymbol{\chi}-\boldsymbol{m}_{\mathbf{X}}]^T\mathbf{C}_{\mathbf{X}}^{-1}[\boldsymbol{\chi}-\boldsymbol{m}_{\mathbf{X}}]}{2}\right] \tag{29}$$

where $\boldsymbol{\chi}=[\chi_1,\ldots,\chi_m]^T$ and $\mathbf{C_X}=[c_x(\mathbf{s}_i,\mathbf{s}_j),\ i,\ j=1,\ 2,\ldots,m]$. If $m=2$, $X(\mathbf{s})$ is called *bivariate* (or *two-dimensional*) Gaussian SRF; if $m=3$, it is called a *trivariate* (or *three-dimensional*) Gaussian SRF.

Moreover, a multivariate Gaussian SRF has the following important properties:

Property 1: A multivariate SRF $X(\mathbf{s})$ is completely characterized by its first and second-order moments

Property 2: The conditional probability density for any two random vectors $\mathbf{X}$ and $\mathbf{Y}$ from a multivariate Gaussian SRF is also multivariate Gaussian and is given by [notation as in Eqs. (24) through (28) above]

$$f_{\mathbf{X}|\mathbf{Y}}(\boldsymbol{\chi}|\boldsymbol{\Psi})=\frac{1}{(2\pi)^{m/2}|\mathbf{C}_{\mathbf{X}|\boldsymbol{\Psi}}|^{1/2}}\exp\left[-\frac{[\boldsymbol{\chi}-m_{\mathbf{X}|\boldsymbol{\Psi}}]^T\mathbf{C}_{\mathbf{X}|\boldsymbol{\Psi}}^{-1}[\boldsymbol{\chi}-m_{\mathbf{X}|\boldsymbol{\Psi}}]}{2}\right] \tag{30}$$

where

$$m_{\mathbf{X}|\boldsymbol{\Psi}}=\mathbf{C_{XY}}\mathbf{C}_{\mathbf{Y}}^{-1}[\boldsymbol{\Psi}-m_{\mathbf{Y}}]+m_{\mathbf{X}}$$

and

$$\mathbf{C}_{\mathbf{X}|\Psi} = \mathbf{C}_{\mathbf{X}} - \mathbf{C}_{\mathbf{XY}}\mathbf{C}_{\mathbf{Y}}^{-1}\mathbf{C}_{\mathbf{XY}}^{T}$$

Property 3: A multivariate Gaussian SRF $X(\mathbf{s})$ is an independent SRF if and only if it is an uncorrelated SRF; that is, if $\mathbf{C}_{\mathbf{XY}} = [0]$.

Property 4: It is valid that

$$\begin{aligned} &E\{[X(\mathbf{s}_1) - m_x(\mathbf{s}_1)][X(\mathbf{s}_2) - m_x(\mathbf{s}_2)] \cdots [X(\mathbf{s}_m) - m_x(\mathbf{s}_m)]\} \\ &\quad = 0 \quad \text{if} \quad m = \text{odd} \\ &\quad = \sum_{i_1, i_2, \ldots, i_m} c_x(\mathbf{s}_{i_1}, \mathbf{s}_{i_2}) c_x(\mathbf{s}_{i_3}, \mathbf{s}_{i_4}) \cdots c_x(\mathbf{s}_{i_{m-1}}, \mathbf{s}_{i_m}) \\ &\quad \text{if} \quad m = \text{even} \end{aligned} \tag{31}$$

for all distinct pairs of subscripts $(i_1, i_2, \ldots, i_m)$ that are permutations of $(1, 2, \ldots, m)$.

Property 5: If $X(\mathbf{s})$ is a Gaussian SRF, then so are its derivatives in the mean square sense. (These derivatives are defined as limits of linear combinations of Gaussian variables; see next section.) In addition, the joint distribution of $X(\mathbf{s})$ and its derivatives are multivariate Gaussian as well.

Of considerable importance in applications is the *lognormal* SRF defined as follows. Consider a nonlinear function of the form

$$Y(\mathbf{s}) = \exp[X(\mathbf{s})] \tag{32}$$

Its inverse is

$$X(\mathbf{s}) = \log Y(\mathbf{s}) \tag{33}$$

where $Y(\mathbf{s}) > 0$.

Assume that $X(\mathbf{s})$ is a Gaussian SRF with univariate density

$$f_x(\chi) = \frac{1}{\sqrt{2\pi}\sigma_x} \exp\left[-\frac{(\chi - m_x)^2}{2\sigma_x^2}\right]$$

Then, the $Y(\mathbf{s})$ is a lognormal SRF with a univariate density of the form

$$f_Y(\psi) = \frac{1}{\sqrt{2\pi}\sigma_x \psi} \exp\left[-\frac{(\log \psi - m_x)^2}{2\sigma_x^2}\right] \tag{34}$$

Moreover, the following classical relationships hold true,

$$E[Y^k(\mathbf{s})] = E\{\exp[kX(s)]\} = \exp[km_x + \tfrac{1}{2}k^2\sigma_x^2] \tag{35}$$

$$c_Y(\mathbf{h}) = m_Y^2\{\exp[c_x(\mathbf{h})] - 1\} \tag{36}$$

and

$$\gamma_Y(\mathbf{h}) = [1 + c_Y(\mathbf{0})]\{1 - \exp[-\gamma_x(\mathbf{h})]\} \tag{37}$$

where $\gamma(\cdot)$ is the so-called *semivariogram function* to be defined in Section 11 below. Lognormal SRF occur in hydrology, turbulence, biology, mining, etc.

4. Certain Geometrical Properties of Spatial Random Fields

The study of the behavior of an SRF with respect to mathematical notions such as continuity, differentiability, and integrability can lead to conclusions regarding the geometry of the SRF and, consequently, the regularity, homogeneity, etc of the natural process modeled by the SRF (see, also, Chapter 7). Just as for deterministic function $f(\mathbf{s})$, where questions about the geometrical properties of $f(\mathbf{s})$ are transferred to questions about the convergence of the series $\{f(\mathbf{s}_n)\}$, as $\mathbf{s}_n \xrightarrow[n\to\infty]{} \mathbf{s}$, geometrical questions concerning the SRF $X(\mathbf{s})$ reduce to the concept of *stochastic convergence* of the corresponding sequence of random variables,

$$\{x_n = X(\mathbf{s}_n)\}, \qquad n = 1, 2, \ldots$$

4.1 Stochastic Convergence

There are several types of convergence in a stochastic framework.

Definition 1: Let $\{x_n\}$ be a sequence of random variables of $L_2(\Omega, F, P)$. The $\{x_n\}$ is said to converge to the random variable x:

(a) In the *mean square sense* (*m.s.s.*), $x_n \xrightarrow{\text{m.s.}} x$, or $\text{l.i.m.}_{n\to\infty}\, x_n = x$, if

$$\lim_{n\to\infty} E|x_n - x|^2 = 0 \tag{1}$$

(b) *Almost surely* (*a.s.*) or *with probability one* (*w.p.*1), $x_n \xrightarrow{\text{a.s.}} x$, if

$$P[\lim_{n\to\infty} x_n = x] = 1 \tag{2}$$

(or, $\exists S \subseteq \Omega$ satisfying $P[S] = 0$, and

$$\lim_{n\to\infty} |x_n(u) - x(u)| = 0$$

for all $u \notin S$).

(c) In *probability* (P), $x_n \xrightarrow{P} x$, or $\text{l.i.p.}_{n\to\infty}\, x_n = x$, if for all $\varepsilon > 0$

$$\lim_{n\to\infty} P[|x_n - x| > \varepsilon] = 0 \tag{3}$$

(d) *Weakly or in distribution* (F), $x_n \xrightarrow{F} x$, if

$$\lim_{n\to\infty} F_{x_n}(\chi) = F_x(\chi) \tag{4}$$

on the continuity set of F_x.

The types of convergence above are related to each other. Particularly

$$\left.\begin{aligned} x_n &\xrightarrow{\text{m.s.}} x \\ \\ x_n &\xrightarrow{\text{a.s.}} x \end{aligned}\right\} \Rightarrow x_n \xrightarrow{P} x \Rightarrow x_n \xrightarrow{F} x \tag{5}$$

Remark 1: Moreover, the following assertion holds (e.g., Sobczyk, 1991). If for some $k>0$,

$$\sum_{n=1}^{\infty} E|x_n - x|^k < \infty$$

then $x_n \xrightarrow{\text{a.s.}} x$.

Since the (second-order) random variable $x_n \in L_2(\Omega, F, P)$, the L_2-convergence is convergence in the m.s.s. Let $x_n \xrightarrow{\text{m.s.}} x$ and $y_m \xrightarrow{\text{m.s.}} y$. Then,

$$E[(x_n y_m - xy)] = E[(x_n - x)(y_m - y)] + E[(x_n - x)y] + E[x(y_m - y)]$$

From Schwartz's inequality,

$$|E[(x_n - x)(y_m - y)]| \leq \sqrt{E|x_n - x|^2}\sqrt{E|y_m - y|^2} \to 0 \qquad \text{as} \quad n, m \to \infty$$

$$|E[(x_n - x)y]| \to 0$$

and

$$|E[x(y_m - y)]| \to 0$$

which lead to the following.

Proposition 1: If $x_n \xrightarrow{\text{m.s.}} x$ and $y_m \xrightarrow{\text{m.s.}} y$, then $E[x_n] \to E[x]$ and $E[x_n y_m] \to E[xy]$. Thus, the operators "l.i.m." and "$E[.]$" commute.

In accordance with the analysis above, the SRF $X(\mathbf{s})$ is said to converge to $X(\mathbf{s}_0)$ when $\mathbf{s} \to \mathbf{s}_0$, in the sense of one of the aforementioned types, if the corresponding sequence of random variables $\{x_n = X(\mathbf{s})\}$ at $\mathbf{s} = \mathbf{s}_1, \mathbf{s}_2, \ldots, \mathbf{s}_n, \ldots$ tends to $x_0 = X(\mathbf{s}_0)$ as $n \to \infty$. For the SRF theory, convergence in the m.s.s. and a.s.s. are the two most important types (e.g., Loeve, 1953).

Proposition 2 (*M.s. convergence criterion*): Let $X(\mathbf{s})$ be an SRF and let $\mathbf{s}_0$ be a fixed point in R^n. Then $X(\mathbf{s}) \xrightarrow{\text{m.s.}} X(\mathbf{s}_0)$, if and only if $E[X(\mathbf{s})X(\mathbf{s}')] \to E[X(\mathbf{s}_0)]^2$ when $\mathbf{s}, \mathbf{s}' \to \mathbf{s}_0$. (Note that the convergence of $X(\mathbf{s})$ is in the m.s.s., while that of $E[X(\mathbf{s})X(\mathbf{s}')]$ is in the ordinary sense.)

4.2. Stochastic Continuity

Definition 2: A second-order SRF $X(\mathbf{s})$ is *continuous in the m.s.s.* at $\mathbf{s} \in R^n$ if $X(\mathbf{s}') \xrightarrow{\text{m.s.}} X(\mathbf{s})$ as $\mathbf{s}' \to \mathbf{s}$ or, which is the same, if

$$X(\mathbf{s}+\mathbf{h}) \xrightarrow{\text{m.s.}} X(\mathbf{s}) \qquad \text{as} \quad \mathbf{h} \to \mathbf{0} \tag{6}$$

When Eq. (6) holds for all $\mathbf{s}$ in the domain of $X(\mathbf{s})$, the latter is m.s. continuous everywhere.

On the strength of Proposition 2 we obtain the proposition below.

Proposition 3: An SRF $X(\mathbf{s})$ is continuous in the m.s.s. at $\mathbf{s} \in R^n$, if and only if its covariance $c_x(\mathbf{s}, \mathbf{s}')$ is continuous at $(\mathbf{s}, \mathbf{s}' = \mathbf{s}) \in R^n \times R^n$. $X(\mathbf{s})$ is everywhere continuous if $c_x(\mathbf{s}, \mathbf{s}')$ is continuous at every diagonal point $(\mathbf{s}, \mathbf{s}' = \mathbf{s})$.

Proposition 3 shows that the m.s. continuity of a second-order SRF is determined by the ordinary continuity of the corresponding covariances. A stronger form of continuity, which is also of particular importance within the context of this book, is continuity of the realizations of the SRF, namely almost sure or sample function continuity.

Definition 3: An SRF $X(\mathbf{s})$ is *almost surely (a.s.) continuous* at $\mathbf{s} \in R^n$ if

$$X(\mathbf{s}+\mathbf{h}) \xrightarrow{\text{a.s.}} X(\mathbf{s}) \qquad \text{as} \quad \mathbf{h} \to \mathbf{0} \tag{7}$$

When Eq. (7) holds for all $\mathbf{s}$ in the domain of $X(\mathbf{s})$, the latter is a.s. continuous everywhere. This type of continuity is also called *sample function continuity.*

It can be shown (Belyaev, 1972; Adler, 1980) that if

$$E|X(\mathbf{s}+\mathbf{h}) - X(\mathbf{s})|^{\lambda} \leq \frac{\alpha|\mathbf{h}|^{2n}}{|\log|\mathbf{h}||^{1+\beta}} \tag{8}$$

where α is a positive constant, $\lambda > 0$, and $\beta > \lambda$, then the SRF $X(\mathbf{s})$, $\mathbf{s} \in R^n$ is a.s. continuous over any compact set $C \subset R^n$. In view of this result and by setting $\lambda = 2$, it is not difficult to prove the following proposition, which provides sufficient conditions for sample function continuity.

Proposition 4: Let $c_x(\mathbf{s}, \mathbf{s}')$ be the covariance function of the SRF $X(\mathbf{s})$, $\mathbf{s} \in R^n$. If for all $\mathbf{s}, \mathbf{h} \in C$, it is true that

$$c_x(\mathbf{s}+\mathbf{h}, \mathbf{s}+\mathbf{h}) - c_x(\mathbf{s}+\mathbf{h}, \mathbf{s}) - c_x(\mathbf{s}, \mathbf{s}+\mathbf{h}) + c_x(\mathbf{s}, \mathbf{s}) \leq \frac{\alpha|\mathbf{h}|^{2n}}{|\log|\mathbf{h}||^{1+\beta}} \tag{9}$$

where $\beta > 2$, the $X(\mathbf{s})$ is a.s. continuous.

M.s. continuity is the one usually applied in second-order SRF studies, and it does not imply sample function continuity. A classical example is the *Poisson* process. While it is m.s. continuous, almost all of its realizations have discontinuities. Nevertheless, a stochastic research program uses both forms of stochastic continuity to study the geometrical properties of the underlying SRF.

4.3 Stochastic Differentiation

Definition 4: A second-order SRF $X(\mathbf{s})$ is *differentiable in the m.s.s.* with respect to the component s_i of the point $\mathbf{s}=(s_1, \ldots, s_i, \ldots, s_n) \in R^n$, if there exists an SRF $X_{(i)}(\mathbf{s})$ such that

$$\frac{X(\mathbf{s}+h\boldsymbol{\varepsilon}_i)-X(\mathbf{s})}{h} \xrightarrow{\text{m.s.}} X_{(i)}(\mathbf{s}) \qquad \text{as } h \to 0 \tag{10}$$

where $\boldsymbol{\varepsilon}_i$ is the unit vector along direction i; i.e., a vector whose ith element is 1 with all the others being zero $(1 \leq i \leq n)$. SRF $X_{(i)}(\mathbf{s})=\partial X(\mathbf{s})/\partial s_i$ is called the ith *partial derivative* of $X(\mathbf{s})$ at $\mathbf{s}$ and is also denoted by

$$\frac{\partial X(\mathbf{s})}{\partial s_i}=\underset{h\to 0}{\text{l.i.m.}}\, \frac{X(\mathbf{s}+h\boldsymbol{\varepsilon}_i)-X(\mathbf{s})}{h} \tag{11}$$

which means that

$$\lim_{h\to 0} E\left[\frac{X(\mathbf{s}+h\boldsymbol{\varepsilon}_i)-X(\mathbf{s})}{h}-\frac{\partial X(\mathbf{s})}{\partial s_i}\right]^2=0 \tag{12}$$

When Eq. (10) holds for all $\mathbf{s}$ in the domain of $X(\mathbf{s})$, the latter is m.s. differentiable everywhere.

We can define higher-order derivatives $\partial^{\nu} X(\mathbf{s})/(\partial s_{i_1} \cdots \partial s_{i_\nu})$ in a similar manner. For example, the *second-order i, j*th *partial derivative* is given by

$$\frac{\partial^2 X(\mathbf{s})}{\partial s_i\, \partial s_j}=\underset{h,r\to 0}{\text{l.i.m.}}\, \frac{1}{hr}[X(\mathbf{s}+h\boldsymbol{\varepsilon}_i+r\boldsymbol{\varepsilon}_j)$$

$$-X(\mathbf{s}+h\boldsymbol{\varepsilon}_i)-X(\mathbf{s}+r\boldsymbol{\varepsilon}_j)+X(\mathbf{s})] \tag{13}$$

where $1 \leq i, j \leq n$.

Definition 4 is not always useful, since it includes the unknown SRF $X_{(i)}(\mathbf{s})$. An alternative definition is as follows: A second-order SRF $X(\mathbf{s})$ is differentiable in the m.s.s. with respect to the component s_i of the point $\mathbf{s}=(s_1, \ldots, s_i, \ldots, s_n) \in R^n$ if

$$\lim_{h_1,h_2\to 0} E\left[\frac{X(\mathbf{s}+h_1\boldsymbol{\varepsilon}_i)-X(\mathbf{s})}{h_1}-\frac{X(\mathbf{s}+h_2\boldsymbol{\varepsilon}_i)-X(\mathbf{s})}{h_2}\right]^2=0 \tag{14}$$

One of the important consequences of the alternative definition (14) is that it leads to the following proposition (e.g., Loeve, 1953).

Proposition 5: An SRF $X(\mathbf{s})$ is m.s. differentiable, if and only if, (a) the mean value $E[X(\mathbf{s})]$ is differentiable and (b) the covariance

$$\operatorname{cov}\left(\frac{\partial X(\mathbf{s})}{\partial s_i}, \frac{\partial X(\mathbf{s}')}{\partial s'_i}\right)$$

$$= \frac{\partial^2 c_x(\mathbf{s}, \mathbf{s}')}{\partial s_i \, \partial s'_i}$$

$$= \lim_{h,r \to c} \frac{1}{hr} [c_x(\mathbf{s}+h\boldsymbol{\varepsilon}_i, \mathbf{s}'+r\boldsymbol{\varepsilon}_i) - c_x(\mathbf{s}, \mathbf{s}'+r\boldsymbol{\varepsilon}_i)$$

$$- c_x(\mathbf{s}+h\boldsymbol{\varepsilon}_i, \mathbf{s}') + c_x(\mathbf{s}, \mathbf{s}')] \tag{15}$$

exists and is finite at all diagonal points $\mathbf{s} = \mathbf{s}'$. Since we are assuming zero mean SRF, condition (a) may be relaxed.

Remark 2: M.s. differentiability of an SRF $X(\mathbf{s})$ at $\mathbf{s} \in R^n$ implies m.s. continuity of $X(\mathbf{s})$ at $\mathbf{s} \in R^n$, for

$$\lim_{h \to 0} E|X(\mathbf{s}+h\boldsymbol{\varepsilon}_i) - X(\mathbf{s})|^2$$

$$= \lim_{h \to 0} h^2 \lim_{h \to 0} \frac{1}{h^2} E|X(\mathbf{s}+h\boldsymbol{\varepsilon}_i) - X(\mathbf{s})|^2$$

$$= 0 \times \lim_{h \to 0} \frac{1}{h^2} [c_x(\mathbf{s}+h\boldsymbol{\varepsilon}_i, \mathbf{s}+h\boldsymbol{\varepsilon}_i)$$

$$- 2c_x(\mathbf{s}+h\boldsymbol{\varepsilon}_i, \mathbf{s}) + c_x(\mathbf{s}, \mathbf{s})] = 0 \times \frac{\partial^2 c_x(\mathbf{s}, \mathbf{s})}{\partial s_i^2} = 0$$

where the second derivative of the covariance is finite by hypothesis. Generalizations of the above results are straightforward.

Proposition 6: The $\partial^\nu X(\mathbf{s})/(\partial s_{i_1} \cdots \partial s_{i_\nu})$ exists in the m.s.s., if and only if the

$$\operatorname{cov}\left(\frac{\partial^\nu X(\mathbf{s})}{\partial s_{i_1} \cdots \partial s_{i_\nu}}, \frac{\partial^\nu X(\mathbf{s}')}{\partial s'_{i_1} \cdots \partial s'_{i_\nu}}\right) = \frac{\partial^{2\nu} c_x(\mathbf{s}, \mathbf{s}')}{\partial s_{i_1} \cdots \partial s_{i_\nu} \, \partial s'_{i_1} \cdots \partial s'_{i_\nu}} \tag{16}$$

exists and is finite at all diagonal points $\mathbf{s} = \mathbf{s}'$.

Example 1: Consider the partial derivative $\partial^\nu X(\mathbf{s})/\partial s_i^\nu$: It exists in the m.s.s. if and only if

$$\operatorname{cov}\left(\frac{\partial^\nu X(\mathbf{s})}{\partial s_i^\nu}, \frac{\partial^\nu X(\mathbf{s}')}{\partial s_i'^\nu}\right) = \frac{\partial^{2\nu} c_x(\mathbf{s}, \mathbf{s}')}{\partial s_i^\nu \, \partial s_i'^\nu} \tag{17}$$

exists and is finite at all $\mathbf{s} = \mathbf{s}'$. Furthermore, if the m.s. derivatives $\partial^\nu X(\mathbf{s})/(\partial s_{i_1} \cdots \partial s_{i_\nu})$ and $\partial^\mu X(\mathbf{s}')/(\partial s'_{i_1} \cdots \partial s'_{i_\mu})$ exist, then

$$\operatorname{cov}\left(\frac{\partial^\nu X(\mathbf{s})}{\partial s_{i_1} \cdots \partial s_{i_\nu}}, \frac{\partial^\mu X(\mathbf{s}')}{\partial s'_{i_1} \cdots \partial s'_{i_\mu}}\right) = \frac{\partial^{\nu+\mu} c_x(\mathbf{s}, \mathbf{s}')}{\partial s_{i_1} \cdots \partial s_{i_\nu} \, \partial s'_{i_1} \cdots \partial s'_{i_\mu}} \tag{18}$$

Similarly, if $X_1(\mathbf{s})$ and $X_2(\mathbf{s})$ are two SRF and the derivatives $\partial^\nu X_1(\mathbf{s})/(\partial s_{i_1} \cdots \partial s_{i_\nu})$ and $\partial^\mu X_2(\mathbf{s}')/(\partial s'_{i_1} \cdots \partial s'_{i_\mu})$ exist, the cross-covariance is

$$\operatorname{cov}\left(\frac{\partial^\nu X_1(\mathbf{s})}{\partial s_{i_1} \cdots \partial s_{i_\nu}}, \frac{\partial^\mu X_2(\mathbf{s}')}{\partial s'_{i_1} \cdots \partial s'_{i_\mu}}\right) = \frac{\partial^{\nu+\mu} c_{x_1 x_2}(\mathbf{s}, \mathbf{s}')}{\partial s_{i_1} \cdots \partial s_{i_\nu} \partial s'_{i_1} \cdots \partial s'_{i_\mu}} \tag{19}$$

Example 2: Let $X(s)$, $s \in R^1$ be a stationary RP. If

$$Y(s) = \frac{dX(s)}{ds} = \underset{h \to 0}{\text{l.i.m.}} \frac{X(s+h) - X(s)}{h}$$

exists it is a stationary RP too. By the definition of the m.s. derivative,

$$\lim_{h \to 0} E\left[\frac{X(s+h) - X(s)}{h} - Y(s)\right]^2 = 0$$

which implies

$$\lim_{h \to 0} \left\{\frac{2}{h^2}[c_x(0) - c_x(h)] + c_Y(0) - 2E\left[\frac{X(s+h) - X(s)}{h} Y(s)\right]\right\} = 0$$

or

$$\lim_{h \to 0} \left\{\frac{2}{h^2}[c_x(0) - c_x(h)] + c_Y(0)\right\} - 2c_Y(0) = 0$$

or

$$c_x(h) = c_x(0) - \frac{c_Y(0)}{2} h^2$$

as $h \to 0$. Clearly, if $X(s)$ is m.s. differentiable the last equation implies

$$\left.\frac{dc_x(h)}{dh}\right|_{h=0} = 0$$

Moreover, the following related results can be derived without any difficulty.

$$E[Y(s)] = E\left[\underset{h \to 0}{\text{l.i.m.}} \frac{X(s+h) - X(s)}{h}\right]$$

$$= \lim_{h \to 0} \frac{1}{h}\{E[X(s+h)] - E[X(s)]\}$$

$$= \frac{d}{ds} E[X(s)]$$

$$E[Y(s)X(s')] = E\left[\underset{h\to 0}{\text{l.i.m.}}\frac{1}{h}[X(s+h)-X(s)]X(s')\right]$$

$$= \lim_{h\to 0}\frac{1}{h}[c_x(s+h, s') - c_x(s, s')]$$

$$= \frac{\partial}{\partial s}c_x(s, s')$$

and

$$E[Y(s)Y(s')] = E\left[\underset{h,r\to 0}{\text{l.i.m.}}\frac{1}{hr}[X(s+h)-X(s)][X(s'+r)-X(s')]\right]$$

$$= \lim_{h\to 0}\frac{1}{h}\lim_{r\to 0}\frac{1}{r}[c_x(s+h, s'+r) - c_x(s+h, s')$$

$$- c_x(s, s'+r) + c_x(s, s')]$$

$$= \lim_{h\to 0}\frac{1}{h}\left[\frac{\partial c_x(s+h, s')}{\partial s'} - \frac{\partial c_x(s, s')}{\partial s'}\right]$$

$$= \frac{\partial^2}{\partial s\,\partial s'}c_x(s, s')$$

Just as for the concept of stochastic continuity, another important form of SRF stochastic differentiability is the differentiability of the realizations of the SRF, also termed almost surely differentiability.

Definition 5: A second-order SRF $X(\mathbf{s})$ is *almost surely (a.s.) differentiable* at $\mathbf{s}\in R^n$, if there exists an SRF $X_{(i)}(\mathbf{s})$ such that

$$\frac{X(\mathbf{s}+h\boldsymbol{\varepsilon}_i)-X(\mathbf{s})}{h}\xrightarrow{\text{a.s.}} X_{(i)}(\mathbf{s}) \qquad \text{as} \quad h\to 0 \tag{20}$$

which is also called SRF *sample function differentiability.*

Assume that the m.s. derivative of the SRF $X(\mathbf{s})$ is

$$X_{(i)}(\mathbf{s}) = \frac{\partial X(\mathbf{s})}{\partial s_i}$$

and its covariance is

$$c_{x_{(i)}}(\mathbf{s}, \mathbf{s}') = \frac{\partial^2 c_x(\mathbf{s}, \mathbf{s}')}{\partial s_i\,\partial s_i'}$$

Then, using the result of Proposition 4 we conclude that the $X_{(i)}(\mathbf{s})$ is a.s. continuous (i.e., it has continuous realizations) if

$$c_{x_{(i)}}(\mathbf{s}+\mathbf{h}, \mathbf{s}+\mathbf{h}) - c_{x_{(i)}}(\mathbf{s}+\mathbf{h}, \mathbf{s}) - c_{x_{(i)}}(\mathbf{s}, \mathbf{s}+\mathbf{h}) + c_{x_{(i)}}(\mathbf{s}, \mathbf{s}) \le \frac{\alpha|\mathbf{h}|^{2n}}{|\log|\mathbf{h}||^{1+\beta}} \tag{21}$$

where $\beta > 2$, for all $\mathbf{s}, \mathbf{h} \in C$. Sufficient conditions with regard the sample function continuity of higher-order derivatives can be obtained in a similar way. Just as with sample function continuity, the m.s. differentiability does not imply a.s. differentiability.

4.4 The Central Limit Theorem

The central limit theorem is, perhaps, the most renowned theorem in statistics and probability theory. Generally speaking, it states that under certain conditions a probability distribution (typically the distribution of the sum of a large number of independent random variables) will tend to approach the Gaussian (or normal) probability distribution.

More specifically, let us consider the limit theorem associated with the *convergence of probability distributions.* Assume that $\{x_k\}$, $k=1,2,\ldots$ is a sequence of independent random variables with mean values m_k, variances $\sigma_k^2 \neq 0$, and $E|x_k - m_k|^3 < \infty$. Let

$$\left\{ y_n = \frac{\sum_{k=1}^{n} (x_k - m_k)}{\sqrt{\sum_{k=1}^{n} \sigma_k^2}} \right\} \tag{22}$$

be a sequence of random variables with probability distributions $\{F_n(\psi)\}$. If

$$\lim_{n\to\infty} \frac{\sqrt[3]{\sum_{k=1}^{n} E|x_k - m_k|^3}}{\sqrt{\sum_{k=1}^{n} \sigma_k^2}} = 0 \tag{23}$$

then the $\{F_n(\psi)\}$ tends to the standard Gaussian distribution $N(0,1)$ as $n \to \infty$.

This is one type of limit theorem. For a detailed treatment of the subject see Cramer (1946), Loeve (1953), and Rosenblatt (1956). An illustrative example follows.

Example 3: Let $x_1, x_2, \ldots, x_k$ be identically distributed independent random variables with mean m and variance σ^2. If $S_k = \sum_{i=1}^{k} x_i$, then

$$\lim_{k\to\infty} P\left[\frac{S_k - km}{\sigma\sqrt{k}} \leq \chi\right] = \frac{1}{\sqrt{2\pi}} \int_{-\infty}^{\chi} \exp\left[-\frac{u^2}{2}\right] du$$

for all $\chi \in R^1$; that is,

$$\frac{S_k - km}{\sigma\sqrt{k}} \xrightarrow[k\to\infty]{F} N(0,1)$$

4.5 Stochastic Integration

Let us first consider m.s. Riemann integration.

Definition 6: Let $X(\mathbf{s})$ be an SRF. The integral

$$Z(\mathbf{w}) = \int_V \alpha(\mathbf{w}; \mathbf{s}) X(\mathbf{s})\, d\mathbf{s} \tag{24}$$

where $V \subset R^n$ and $\alpha(\mathbf{w}; \mathbf{s})$ is a deterministic bounded and piecewise continuous function, exists in the m.s.s. if the limit of the sum

$$Z_m(\mathbf{w}) = \sum_{i=1}^{m} \alpha(\mathbf{w}; \mathbf{s}_i) X(\mathbf{s}_i)\, \boldsymbol{\Delta}(\mathbf{s}_i) \tag{25}$$

exists for $m \to \infty$ (also in the m.s.s.). This limit is called the *m.s. Riemann integral* of $X(\mathbf{s})$. Here $\Delta(\mathbf{s}_i)$ is an infinitesimal volume centered at $\mathbf{s}_i$, where the measure of the largest $\Delta(\mathbf{s}_i)$ tends to zero, and the sum of all $\Delta(\mathbf{s}_i)$ equals V. (Clearly, both $Z(\mathbf{w})$ and $Z_m(\mathbf{w})$ are random variables.) Then it is valid that

$$Z(\mathbf{w}) = \underset{m \to \infty}{\text{l.i.m.}}\, Z_m(\mathbf{w}) \tag{26}$$

and the SRF $X(\mathbf{s})$ is said to be *integrable in the m.s.s.*

Working along the lines of Definition 6 the proposition below can be proven (e.g., Gihman and Skorokhod, 1974a).

Proposition 7: An SRF $X(\mathbf{s})$ is integrable in the m.s. Riemann sense if and only if

$$E[Z^2(\mathbf{w})] = \int_V \int_V \alpha(\mathbf{w}; \mathbf{s}) \overline{\alpha(\mathbf{w}; \mathbf{s}')} c_x(\mathbf{s}, \mathbf{s}')\, d\mathbf{s}\, d\mathbf{s}' < \infty \tag{27}$$

where $\overline{\alpha(\mathbf{w}; \mathbf{s}')}$ denotes the complex conjugate of $\alpha(\mathbf{w}; \mathbf{s}')$, and $V \subset R^n$ as before.

Under certain circumstances one may need to define the so-called Riemann–Stieltjes integral.

Definition 7: Consider the integral

$$Z(\mathbf{w}) = \int_V \alpha(\mathbf{w}; \mathbf{s})\, d\aleph_x(\mathbf{s}) \tag{28}$$

where $\aleph_x$ is a random additive set function associated with the SRF $X(\mathbf{s})$, $\mathbf{s} \in V \subset R^n$; the $\aleph_x$ is defined on some class $\mathscr{A}$ of sets $S_i \subset R^n$. Let $V = \bigcup_{i=1}^{n_m} S_i^{(m)}$ $(m = 1, 2, \ldots)$ be such a sequence of partitions $\{P_m\}$ of V that

$$\Delta_m = \max_i \sup_{\mathbf{s}, \mathbf{s}' \in S_i^{(m)}} |\mathbf{s} - \mathbf{s}'| \xrightarrow[m \to \infty]{} 0$$

In this case, the counterpart of the sum in Eq. (25) is

$$Z_m(\mathbf{w}) = \sum_{i=1}^{n_m} \alpha(\mathbf{w}; \mathbf{s}_i^{(m)}) \aleph_x(S_i^{(m)}) \tag{29}$$

where $\mathbf{s}_i^{(m)}$ is any point of the region $S_i^{(m)}$. Then, the *m.s. Riemann-Stieltjes integral* (28) is defined as

$$Z(\mathbf{w}) = \underset{m\to\infty}{\text{l.i.m.}}\, Z_m(\mathbf{w}) \tag{30}$$

Proposition 7 above also holds true in the m.s. Riemann-Stieltjes sense where, though, the integral in Eq. (27) must be replaced by one in the ordinary Riemann-Stieltjes sense (see, e.g., Loeve, 1953; Pugachev and Sinitsyn, 1987).

Of significant importance is the construction of the so-called *integral canonical representation* of an SRF $X(\mathbf{s})$, namely

$$X(\mathbf{s}) = m_x(\mathbf{s}) + \int_V \alpha(\mathbf{w}; \mathbf{s})\, d\aleph_x(\mathbf{w}) \tag{31}$$

where the $\aleph_x$ has now zero mean. To Eq. (31) one can associate an integral canonical representation of its covariance.

In applications, it is sometimes useful to express stochastic integrals as integrals containing white noise SRF, viz.

$$\int_V \alpha(\mathbf{s}'; \mathbf{s})\, d\aleph_x(\mathbf{s}') = \int_V \alpha(\mathbf{s}'; \mathbf{s})\, e(\mathbf{s}')\, d\mathbf{s}' \tag{32}$$

in which $e(\mathbf{s}')$ is a zero-mean white-noise SRF (for more details on white-noise SRF see Chapter 3). On the basis of Eq. (32) the following integral canonical representation of the SRF $X(\mathbf{s})$ may be constructed.

$$X(\mathbf{s}) = m_x(\mathbf{s}) + \int_V \alpha(\mathbf{s}'; \mathbf{s})\, e(\mathbf{s}')\, d\mathbf{s}' \tag{33}$$

Representation (33), when possible, has important consequences in applications related to the stochastic differential equation modeling of earth systems.

M.s. integrals have the formal properties of ordinary integrals. Some other results of stochastic integration will be discussed below in the context of specific classes of SRF.

5. Spectral Characteristics of Spatial Random Fields

5.1 Scalar Spatial Random Fields

The stochastic integration analysis of the previous section are used here by setting

$$\alpha(\mathbf{w}; \mathbf{s}) = \frac{\exp[-i\mathbf{w}\cdot\mathbf{s}]}{(2\pi)^{n/2}}$$

where $i=\sqrt{-1}$. More specifically one may define an SRF in the frequency domain as follows (without loss of generality we assume zero mean SRF):

Definition 1: An SRF $X(\mathbf{s})$ is said to be a *harmonizable* SRF if there exists an SRF $\tilde{X}(\mathbf{w})$, $\mathbf{w}\in R^n$ such that

$$X(\mathbf{s})=\int_{R^n} \exp[i\mathbf{w}\cdot\mathbf{s}]\,\tilde{X}(\mathbf{w})\,d\mathbf{w} \tag{1}$$

Equation (1) is a Riemann integral representation of the SRF $X(\mathbf{s})$.

By applying the theory of the previous section we find that

$$\tilde{X}(\mathbf{w})=\frac{1}{(2\pi)^{n/2}}\int_{R^n} \exp[-i\mathbf{w}\cdot\mathbf{s}]\,X(\mathbf{s})\,d\mathbf{s}$$

exists in the m.s.s. if and only if

$$|E[\tilde{X}(\mathbf{w}_i)\overline{\tilde{X}(\mathbf{w}_j)}]|=\left|\frac{1}{(2\pi)^n}\int_{R^n}\int_{R^n} \exp[-i(\mathbf{w}_i\cdot\mathbf{s}_i-\mathbf{w}_j\cdot\mathbf{s}_j)]\, c_x(\mathbf{s}_i,\mathbf{s}_j)\,d\mathbf{s}_i\,d\mathbf{s}_j\right|<\infty \tag{2}$$

for all $\mathbf{w}_i$, $\mathbf{w}_j\in R^n$, where $\overline{\tilde{X}(\mathbf{w}_j)}$ denotes the complex conjugate of the SRF $\tilde{X}(\mathbf{w}_j)$. On the basis of the foregoing, besides the characterization of SRF on the basis of statistical moments of order up to two, an equivalent characterization can be made in terms of spectral moments of order up to two. These are the n-fold Fourier transforms of the statistical moments of order up to two. As happened with the statistical moments, the spectral moments are assumed to be continuous functions in R^n.

Definition 2: The *spectral density function* of an SRF $X(\mathbf{s})$ is defined by

$$\begin{aligned}C_x(\mathbf{w}_i,\mathbf{w}_j)&=E[\tilde{X}(\mathbf{w}_i)\overline{\tilde{X}(\mathbf{w}_j)}]\\ &=\frac{1}{(2\pi)^n}\int_{R^n}\int_{R^n} \exp[-i(\mathbf{w}_i\cdot\mathbf{s}_i-\mathbf{w}_j\cdot\mathbf{s}_j)]c_x(\mathbf{s}_i,\mathbf{s}_j)\,d\mathbf{s}_i\,d\mathbf{s}_j\end{aligned} \tag{3}$$

assuming that the integral exists.

Remark 1: In certain applications (e.g., statistical continuum media), one may need to define Fourier transforms (FT) of statistical moments of higher orders, such as

$$\mathrm{FT}\{E[X(\mathbf{s}_1)X(\mathbf{s}_2)\cdots X(\mathbf{s}_k)]\}=E[\tilde{X}(\mathbf{w}_1)\tilde{X}(\mathbf{w}_2)\cdots\tilde{X}(\mathbf{w}_k)]$$

The $C_x(\mathbf{w}_i,\mathbf{w}_j)$ in Eq. (3) may be viewed as the covariance function of the SRF $\tilde{X}(\mathbf{w})$ and is such that

$$\int_V\int_V C_x(\mathbf{w}_i,\mathbf{w}_j)\,d\mathbf{w}_i\,d\mathbf{w}_j\geq 0 \tag{4}$$

for all $V \subset R^n$. In view of the above definition, $c_x(\mathbf{s}_i, \mathbf{s}_j)$ can also be defined as the *inverse Fourier transform* of the spectral density $C_x(\mathbf{w}_i, \mathbf{w}_j)$; that is,

$$c_x(\mathbf{s}_i, \mathbf{s}_j) = \int_{R^n} \int_{R^n} \exp[i(\mathbf{w}_i \cdot \mathbf{s}_i - \mathbf{w}_j \cdot \mathbf{s}_j)] C_x(\mathbf{w}_i, \mathbf{w}_j) \, d\mathbf{w}_i \, d\mathbf{w}_j \tag{5}$$

Certain important special classes of SRF, like the *homogeneous* SRF (Section 7 below), do not possess a Fourier transform and, therefore, they do not admit the Riemann integral representation (1). However, it can be shown that all classes of SRF, including the homogeneous ones, admit the Fourier–Stieltjes representation

$$X(\mathbf{s}) = \int_{R^n} \exp[i\mathbf{w} \cdot \mathbf{s}] \, d\aleph_x(\mathbf{w}) \tag{6}$$

where $\aleph_x(\mathbf{w})$ is a random field that is not necessarily differentiable (also, Section 7 later). If $\aleph_x(\mathbf{w})$ is differentiable, the Fourier–Stieltjes integral reduces to the Riemann integral. But the functions $\aleph_x(\mathbf{w})$ associated to homogeneous SRF do not satisfy this property.

The covariance function $c_x(\mathbf{s}_i, \mathbf{s}_j)$ of the SRF (6) can be written

$$c_x(\mathbf{s}_i, \mathbf{s}_j) = E\left[\int_{R^n} \exp[i\mathbf{w}_i \cdot \mathbf{s}_i] \, d\aleph_x(\mathbf{w}_i) \int_{R^n} \exp[-i\mathbf{w}_j \cdot \mathbf{s}_j] \, \overline{d\aleph_x(\mathbf{w}_j)}\right]$$

$$= \int_{R^n} \int_{R^n} \exp[i(\mathbf{w}_i \cdot \mathbf{s}_i - \mathbf{w}_j \cdot \mathbf{s}_j)] E[d\aleph_x(\mathbf{w}_i) \, \overline{d\aleph_x(\mathbf{w}_j)}] \tag{7}$$

$$= \int_{R^n} \int_{R^n} \exp[i(\mathbf{w}_i \cdot \mathbf{s}_i - \mathbf{w}_j \cdot \mathbf{s}_j)] \, dQ_x(\mathbf{w}_i, \mathbf{w}_j) \tag{8}$$

where $Q_x(\mathbf{w}_i, \mathbf{w}_j)$ is the so-called *spectral distribution function* of $X(\mathbf{s})$ and is not necessarily differentiable; note that $Q_x(\mathbf{w}_i, \mathbf{w}_j)$ can be also viewed as the covariance of the random field $\aleph_x(\mathbf{w})$. By comparing Eqs. (5) and (7) one gets

$$C_x(\mathbf{w}_i, \mathbf{w}_j) \, d\mathbf{w}_i \, d\mathbf{w}_j = E[d\aleph_x(\mathbf{w}_i) \, \overline{d\aleph_x(\mathbf{w}_j)}] \tag{9}$$

Furthermore, if $Q_x(\mathbf{w}_i, \mathbf{w}_j)$ is differentiable,

$$C_x(\mathbf{w}_i, \mathbf{w}_j) = \frac{\partial^{2n} Q_x(\mathbf{w}_i, \mathbf{w}_j)}{\partial w_{i_1} \cdots \partial w_{i_n} \, \partial w_{j_1} \cdots \partial w_{j_n}} \tag{10}$$

and then Eq. (8) coincides with Eq. (5). When a covariance can take the form (5), it is said to be *harmonizable.* The proposition below establishes an important link between an SRF and its covariance (Loeve, 1953).

Proposition 1: An SRF is harmonizable if and only if its covariance is harmonizable.

According to Proposition 1 the stochastic integral representations (1) of an SRF exist if and only if the deterministic integral representation of the corresponding covariance, see Eq. (5), exists.

Remark 2: From mathematical analysis we find that there are certain conditions with respect to $c_x(\mathbf{s}_i, \mathbf{s}_j)$ and $C_x(\mathbf{w}_i, \mathbf{w}_j)$ for *convergence* of the deterministic Fourier integral (3) and the inverse Fourier integral (5), respectively. These conditions must be checked carefully when the above formulas are used (see, e.g., Sections 7 and 9, later).

Example 1: As an illustration of how a covariance can be represented in the form (8) consider in R^1 the SRF

$$X(s) = v \exp[is\alpha] \tag{11}$$

where v is a zero-mean random variable with $E[v^2] = c^2$ and α is a random variable independent of v having probability distribution

$$F_\alpha(w) = P[\alpha \leq w] \tag{12}$$

Then

$$E[X(s)] = E[v]E[\exp(is\alpha)] = 0$$

and

$$\begin{aligned} c_x(h) &= E[X(s)\overline{X(s+h)}] = E[v^2]E[\exp(ih\alpha)] \\ &= c^2 \int_{R^1} \exp[ihw]\, dF_\alpha(w) = \int_{R^1} \exp[ihw]\, dQ_x(w) \end{aligned} \tag{13}$$

where $Q_x(w) = c^2 F_\alpha(w)$ is an arbitrary nondecreasing bounded function such that $Q_x(w) \to 0$ when $w \to -\infty$.

Remark 3: In the case of real-valued covariance and spectral density functions, the exponentials in the above equations are replaced by cosine functions. Nevertheless, from a mathematical point of view, it is usually convenient to use exponential forms even in the real case.

Definition 3: The *second-order spectral moment* of a SRF $X(\mathbf{s})$ is defined by

$$\beta_x^{ij} = \int_{R^n} \int_{R^n} w_i w_j C_x(\mathbf{w}, \mathbf{w}')\, d\mathbf{w}\, d\mathbf{w}' \tag{14}$$

Higher-order spectral moments may be defined in a similar fashion.

Remark 4: It can be easily shown that

$$\beta_x^{ij} = \left. \frac{\partial^2 c_x(\mathbf{s}, \mathbf{s}')}{\partial s_i\, \partial s_j'} \right|_{\mathbf{s}=\mathbf{s}'=\mathbf{0}} \tag{15}$$

5.2 Linear Transformation of Spatial Random Fields

Let us consider the following definitions.

Definition 4: A *linear transformation* $\Im[\cdot]$ is a transformation which, when applied on an SRF $X(\mathbf{s})$, yields the new SRF

$$Y(\mathbf{s}) = \Im[X(\mathbf{s})] \tag{16}$$

such that if $X_i(\mathbf{s})$, $i = 1, 2, \ldots, m$ are SRF, then

$$\Im\left[\sum_{i=1}^{m} X_i(\mathbf{s})\right] = \sum_{i=1}^{m} \Im[X_i(\mathbf{s})] \tag{17}$$

for all nonnegative integers m.

Remark 5: The transformation $\Im[\cdot]$ is translation-invariant if

$$\Im[S_{\mathbf{s}'}X(\mathbf{s})] = S_{\mathbf{s}'}\Im[X(\mathbf{s})] \tag{18}$$

for all $\mathbf{s}' \in R^n$, where $S_{\mathbf{s}'}X(\mathbf{s}) = X(\mathbf{s}+\mathbf{s}')$ is the shift operator.

Example 2: Common transformations in the sense of Definition 4 are (i) the differentiation of mth order

$$\Im[\cdot] = \frac{\partial^m}{\partial s_1^{m_1}\, \partial s_2^{m_2} \cdots \partial s_i^{m_i}}[\cdot] \tag{19}$$

where $\sum_{j=1}^{i} m_j = m$; and (ii) the integration

$$\Im[\cdot] = \int_{R^n} f(\mathbf{s})[\cdot]\, d\mathbf{s} \tag{20}$$

where $f(\mathbf{s})$ is a suitable function in R^n.

Definition 5: The *transfer function* of the transformation $\Im[\cdot]$ is defined by

$$H(\mathbf{w}) = \frac{\Im[\exp[i\mathbf{w}\cdot\mathbf{s}]]}{\exp[i\mathbf{w}\cdot\mathbf{s}]} \tag{21}$$

Assume now that the SRF $X(\mathbf{s})$ is harmonizable in the sense of Eq. (6) above. Then we can write

$$Y(\mathbf{s}) = \Im[X(\mathbf{s})] = \Im\left[\int_{R^n} \exp[i\mathbf{w}\cdot\mathbf{s}]\, d\aleph_X(\mathbf{w})\right]$$

$$= \int_{R^n} \Im[\exp[i\mathbf{w}\cdot\mathbf{s}]]\, d\aleph_X(\mathbf{w}) \tag{22}$$

By comparing Eqs. (21) and (22) we obtain

$$Y(\mathbf{s}) = \int_{R^n} \exp[i\mathbf{w}\cdot\mathbf{s}]\, H(\mathbf{w})\, d\aleph_X(\mathbf{w}) \tag{23}$$

$$= \int_{R^n} \exp[i\mathbf{w}\cdot\mathbf{s}]\, d\aleph_Y(\mathbf{w}) \tag{24}$$

where $d\aleph_Y(\mathbf{w}) = H(\mathbf{w})\, d\aleph_x(\mathbf{w})$. It is easy to show that the spectral density functions of the SRF $X(\mathbf{s})$ and $Y(\mathbf{s})$ (if they exist) are related by

$$C_Y(\mathbf{w}_i, \mathbf{w}_j) = H(\mathbf{w}_i)\overline{H(\mathbf{w}_j)}C_x(\mathbf{w}_i, \mathbf{w}_j) \tag{25}$$

5.3 Evolutionary Mean Power Spectral Density Function

Suppose that $X(\mathbf{s})$ is a nonhomogeneous SRF represented by

$$X(\mathbf{s}) = \int_{R^n} H(\mathbf{s}, \mathbf{w})\, d\aleph_x(\mathbf{w}) \tag{26}$$

where $\aleph_x(\mathbf{w})$ is a random field with uncorrelated increments and variance $Q_x(\mathbf{w})$ and $H(\mathbf{s}, \mathbf{w})$ has the form

$$H(\mathbf{s}, \mathbf{w}) = \Theta(\mathbf{s}, \mathbf{w}) \exp[i\mathbf{w} \cdot \mathbf{s}] \tag{27}$$

where $\Theta(\mathbf{s}, \mathbf{w})$ is an amplitude-modulating function that varies slowly with $\mathbf{s}$.

On the basis of these spectral representations one can define another useful concept in the spectral analysis of nonhomogeneous SRF, namely the evolutionary mean power spectral density function (Veneziano, 1980; Priestley, 1981).

Definition 5: Let $X(\mathbf{s})$ be a nonhomogeneous SRF represented as in Eq. (26) above where the variance $Q_x(\mathbf{w})$ of the random field $\aleph_x(\mathbf{w})$ is differentiable; that is,

$$C_x(\mathbf{w}) = \frac{\partial^n Q_x(\mathbf{w})}{\partial w_1 \cdots \partial w_n}$$

Then, the *evolutionary mean power spectral density function* is defined by

$$C_x(\mathbf{s}, \mathbf{w}) = |\Theta(\mathbf{s}, \mathbf{w})|^2 C_x(\mathbf{w}) \tag{28}$$

Note that in this case the evolutionary mean power spectral density function $C_x(\mathbf{s}, \mathbf{w})$ is not the Fourier transform of the covariance function $c_x(\mathbf{s}_i, \mathbf{s}_j)$.

5.4 Vector Spatial Random Fields

We saw above that when one deals with two stochastically correlated SRF, the cross-covariance function should be involved into the analysis. Then the corresponding spectral cross-density function is defined as follows.

Definition 6: Let $X_p(\mathbf{s})$ and $X_q(\mathbf{s}')$, $\mathbf{s}, \mathbf{s}' \in R^n$ be two SRF. The *cross-spectral density function* writes

$$C_{x_p x_q}(\mathbf{w}_i, \mathbf{w}_j) = \frac{1}{(2\pi)^n} \int_{R^n} \int_{R^n} \exp[-i(\mathbf{w}_i \cdot \mathbf{s}_i - \mathbf{w}_j \cdot \mathbf{s}_j)] \times c_{x_p x_q}(\mathbf{s}_i, \mathbf{s}_j)\, d\mathbf{s}_i\, d\mathbf{s}_j \tag{29}$$

Let $\mathbf{C}_x$ be the symmetric matrix of cross-covariances of the vector SRF $\mathbf{X}(\mathbf{s}) = [X_1(\mathbf{s}), X_2(\mathbf{s}), \ldots, X_k(\mathbf{s})]^T$. The definition of the symmetric matrix of cross-spectral density functions is straightforward.

Definition 7: The *cross-spectral density matrix* of a vector SRF $\mathbf{X}(\mathbf{s})$ is defined as

$$\tilde{\mathbf{C}}_x = [C_{x_p x_q}(\mathbf{w}_i, \mathbf{w}_j)], \qquad p, q = 1, 2, \ldots, k \tag{30}$$

where the component cross-spectral density functions $C_{x_p x_q}(\mathbf{w}_i, \mathbf{w}_j)$ are given by Eq. (29), assuming that these integrals exist.

6. Auxiliary Hypotheses

The study of an SRF by means of its statistical moments up to second order (Section 3) requires certain *auxiliary hypotheses* so that the purely mathematical model of SRF is compatible with the nature of the phenomenon it describes and, at the same time, this model is applicable under practical circumstances. The first hypothesis that follows is a constitutive hypothesis necessary for inference.

Hypothesis 1: A natural process is modeled as a *homogeneous—in the wide sense—SRF.* This implies that

$$m_x(\mathbf{s}) = m \tag{1}$$

and

$$c_x(\mathbf{s}, \mathbf{s}') = c_x(\mathbf{h} = \mathbf{s} - \mathbf{s}') \tag{2}$$

that is, its mean value is a constant and its covariance depends only on the vector distance between two points in space.

The homogeneity property of the SRF $X(\mathbf{s})$, the latter considered as a function with values in the Hilbert space $L_2(\Omega, F, P)$, amounts to the fact that there exist in the closed linear subspace H spanned by the random variable x in $L_2(\Omega, F, P)$ a group of unitary operators $U_\mathbf{h}$ such that

$$U_\mathbf{h} X(\mathbf{s}) = X(\mathbf{s} + \mathbf{h}) \tag{3}$$

where $\mathbf{s}, \mathbf{h} \in R^n$. The physical meaning of homogeneity is that the large-scale characteristics (macrostructure) of the underlying physical variate do not change over space.

Remark 1: In the *one-dimensional* case, where the SRF becomes an RP $X(s)$, $s \in R^1$, homogeneity is equivalent to the assumption of *stationarity* (of time series analysis, etc.).

Remark 2: In general an SRF $X(\mathbf{s})$ is called *homogeneous in the strict sense* if the probability distributions of the sequences $x(\mathbf{s}_1), x(\mathbf{s}_2), \ldots, x(\mathbf{s}_k)$ for any integer k and all points $\mathbf{s}_1, \ldots, \mathbf{s}_k$ remain the same when all the points $\mathbf{s}_1, \mathbf{s}_2, \ldots, \mathbf{s}_k$ are translated by an arbitrary vector $\mathbf{h} \in R^n$. Clearly if the mean and covariance of such an SRF exist they will satisfy Eqs. (1) and (2). Wide-sense homogeneity, however, does not necessarily imply homogeneity in the strict sense. An important special case of SRF where wide and strict homogeneity imply each other is the Gaussian SRF. Nevertheless, strict homogeneity is very rarely applicable in practical situations and, thus, further on wide-sense homogeneous SRF will simply be called homogeneous SRF. The following hypothesis is an extension of the previous one in the sense that all directions in space are taken as equivalent.

Hypothesis 2: A natural process is modeled as an *isotropic—in the wide sense—SRF.* In this case the SRF is assumed to have constant mean and

$$c_x(\mathbf{s}, \mathbf{s}') = c_x(r = |\mathbf{s} - \mathbf{s}'|) \tag{4}$$

that is, its covariance depends only on the length of the vector distance between any two points in space.

Remark 3: One may also define an isotropic SRF in the *strict sense* similarly to the strictly homogeneous SRF discussed in Remark 2 above.

Our third hypothesis, following, is essential in the case that hypotheses 1 and 2 do not apply, that is, when the SRF under study is nonhomogeneous.

Hypothesis 3: A natural process is modeled as an *ordinary SRF with homogeneous increments of some order* ν (*OSRF-*ν) or an *intrinsic SRF of order* ν (*ISRF-*ν). This means that, although the SRF $X(\mathbf{s})$ itself is nonhomogeneous, there exists a linear transformation T such that the

$$Y(\mathbf{s}) = T[X(\mathbf{s})] \tag{5}$$

is a homogeneous SRF (Section 9 below and Chapter 3). The term OSRF-ν has its origin in the theory of random distributions (Ito, 1954; Gel'fand, 1955), while the term ISRF-ν is used in the geostatistical literature (Matheron, 1973). It can be shown that, in this case, the nonhomogeneous covariance function $c_x(\mathbf{s}_i, \mathbf{s}_j)$ of $X(\mathbf{s})$ consists of a homogeneous part $k_x(\mathbf{h})$, $\mathbf{h} = \mathbf{s}_i - \mathbf{s}_j$, called a *generalized spatial covariance of order* ν, (Matheron, 1973; the term generalized is used here to distinguish $k_x(\mathbf{h})$ from the ordinary covariances) and a polynomial part of order ν in $\mathbf{s}_i$ and $\mathbf{s}_j$, $p_\nu(\mathbf{s}_i, \mathbf{s}_j)$.

Remark 4: A homogeneous SRF is also an ISRF-ν for any value of ν, but the converse is not generally true. We will see below that in certain cases of nonhomogeneity one may make use of alternative tools of stochastic inference such as the *structure* or the *semivariogram function* (Section 11), and the aforementioned generalized spatial covariance function (Chapter 3).

Hypothesis 4: A natural process is assumed to be represented by an *ergodic* SRF $X(\mathbf{s})$. Ergodicity, which is a term borrowed from statistical mechanics (Khinchin, 1949), implies that the mean and covariance of $X(\mathbf{s})$ coincide with those calculated by means of the single available realization (more detailed definitions in Section 12 below).

The above hypothesis arises from the fact that in most practical problems we have only one available sequence of measurements. (As we shall see in Chapter 7 ergodicity is a *working* or *methodological hypothesis*, that is, a hypothesis which, while not necessarily always verified in practice, can be tested on the basis of the successes to which it leads.)

Remark 5: In many practical applications the validity of the above hypotheses is restricted to limited domains over space. Then, hypotheses 1, 3, and 4 are called *quasi-homogeneity, quasi-intrinsity*, and *quasi-ergodicity* (or *microergodicity*), respectively.

7. Homogeneous Spatial Random Fields

7.1 Scalar Homogeneous Spatial Random Fields

To study homogeneous SRF it is appropriate to start by defining a very important type of complex-valued SRF, namely the so-called SRF with *orthogonal increments*, say $\aleph_x(\mathbf{w})$, $\mathbf{w} \in R^n$. This SRF is such that for any pair of disjoint sets S_1, $S_2 \subset R^n$, it is valid that

$$E[\aleph_x(S_1)\overline{\aleph_x(S_2)}] = 0 \tag{1}$$

An important property of $\aleph_x(\mathbf{w})$ is that it determines a measure $Q_x(\mathbf{w})$ that satisfies $E[|\aleph_x(\mathbf{w})|^2] = Q_x(\mathbf{w})$. The foregoing considerations imply that a homogeneous SRF admits the Fourier–Stieltjes representations

$$X(\mathbf{s}) = \int_{R^n} \exp[i\mathbf{w} \cdot \mathbf{s}]\, d\aleph_x(\mathbf{w}) \tag{2}$$

From the definitions of Section 3 we obtain the following expression of the homogeneous covariance function.

$$\begin{aligned} c_x(\mathbf{h}) &= E[[X(\mathbf{s}) - m_x(\mathbf{s})][X(\mathbf{s}+\mathbf{h}) - m_x(\mathbf{s}+\mathbf{h})]] \\ &= \iint [\chi_1 - m_x(\mathbf{s})][\chi_2 - m_x(\mathbf{s}+\mathbf{h})] f_{\mathbf{s},\mathbf{s}+\mathbf{h}}(\chi_1, \chi_2)\, d\chi_1\, d\chi_2 \end{aligned} \tag{3}$$

Moreover, by setting $\mathbf{s}_i = \mathbf{s}_j + \mathbf{h}$, Eqs. (7) and (8) of Section 5 yield

$$c_x(\mathbf{h}) = \int_{R^n}\int_{R^n} \exp[i(\mathbf{w}_i \cdot (\mathbf{s}_j + \mathbf{h}) - \mathbf{w}_j \cdot \mathbf{s}_j)] E[d\aleph_x(\mathbf{w}_i)\, \overline{d\aleph_x(\mathbf{w}_j)}] \tag{4}$$

$$= \int_{R^n}\int_{R^n} \exp[i(\mathbf{w}_i \cdot (\mathbf{s}_j + \mathbf{h}) - \mathbf{w}_j \cdot \mathbf{s}_j)]\, dQ_x(\mathbf{w}_i, \mathbf{w}_j) \tag{5}$$

Assuming that the $Q_x(\mathbf{w}_i, \mathbf{w}_j)$ is differentiable and since

$$C_x(\mathbf{w}_j)\, \delta(\mathbf{w}_i - \mathbf{w}_j)\, d\mathbf{w}_i\, d\mathbf{w}_j = E[d\aleph_x(\mathbf{w}_i)\, \overline{d\aleph_x(\mathbf{w}_j)}] \tag{6}$$

where

$$C_x(\mathbf{w}) = \frac{\partial^n Q_x(\mathbf{w})}{\partial w_1 \cdots \partial w_n}$$

is the so-called *spectral density function*, Eq. (5) becomes

$$c_x(\mathbf{h}) = \int_{R^n} \exp[i\mathbf{w} \cdot \mathbf{h}]\, dQ_x(\mathbf{w}) \tag{7}$$

$$= \int_{R^n} \exp[i\mathbf{w} \cdot \mathbf{h}] C_x(\mathbf{w})\, d\mathbf{w} \tag{8}$$

By taking the inverse Fourier transform of Eq. (8) one gets

$$C_x(\mathbf{w}) = \frac{1}{(2\pi)^n} \int_{R^n} \exp[-i\mathbf{w} \cdot \mathbf{h}] c_x(\mathbf{h})\, d\mathbf{h} \tag{9}$$

These results lead to the following proposition.

Proposition 1: Any homogeneous SRF $X(\mathbf{s})$ admits a spectral representation of the form of Eq. (2), where $\aleph_x(\mathbf{w})$ is a complex-valued SRF with orthogonal increments. To $X(\mathbf{s})$ one can attach a spectral density function $C_x(\mathbf{w})$, provided that Eq. (6) is satisfied.

Remark 1: As in Remark 2, Section 5, there are certain conditions for convergence of the integral representation (8). In the case of homogeneous SRF these conditions lead to the conclusion that if the covariance $c_x(\mathbf{h})$ tends to zero fast enough with $|\mathbf{h}| \to \infty$, then

$$\int_{R^n} |c_x(\mathbf{h})|\, d\mathbf{h} < \infty \tag{10}$$

and the $c_x(\mathbf{h})$ can be represented by Eq. (8), where $C_x(\mathbf{w})$ is given by Eq. (9) above. In other words, if $c_x(\mathbf{h})$ is an absolutely integrable function in R^n, then the spectral density function $C_x(\mathbf{w})$ exists. In this case

$$\int_{R^n} C_x(\mathbf{w})\, d\mathbf{w} = c_x(\mathbf{0}) < \infty \tag{11}$$

that is, the spectral density is always integrable.

Remark 2: Usually we are dealing with real fields where both $c_x(\mathbf{h})$ and $C_x(\mathbf{w})$ are even functions. In this case, they are related by the Fourier cosine transform

$$c_x(\mathbf{h}) = \int_{R^n} \cos(\mathbf{w} \cdot \mathbf{h}) C_x(\mathbf{w})\, d\mathbf{w} \tag{12}$$

and

$$C_x(\mathbf{w}) = \frac{1}{(2\pi)^n} \int_{R^n} \cos(\mathbf{w} \cdot \mathbf{h}) c_x(\mathbf{h})\, d\mathbf{h} \tag{13}$$

It is always assumed that these integrals exist. For example, in R^1 the existence of the integral (12) requires that $\lim_{|w| \to 0} [wC_x(w)] = 0$ and $\lim_{|w| \to \infty} [wC_x(w)] = 0$.

Let $X(\mathbf{s})$ be a homogeneous SRF, and let

$$\rho_x(\mathbf{h}) = \frac{c_x(\mathbf{h})}{c_x(\mathbf{0})} \tag{14}$$

be the homogeneous *correlation function* corresponding to $c_x(\mathbf{h})$. Then the following properties are valid.

Property 1: The covariance and the correlation function are symmetric functions, that is,

$$c_x(\mathbf{h}) = c_x(-\mathbf{h}) \qquad \text{and} \quad \rho_x(\mathbf{h}) = \rho_x(-\mathbf{h}) \tag{15}$$

They also are bounded functions; that is,

$$c_x(\mathbf{h}) \leq c_x(\mathbf{0}) \tag{16}$$

$$|\rho_x(\mathbf{h})| \leq 1 \tag{17}$$

Property 2: At infinity we have

$$\lim_{|\mathbf{h}| \to \infty} \frac{c_x(\mathbf{h})}{|\mathbf{h}|^{(1-n)/2}} = 0 \tag{18}$$

Property 3: If $\rho_x(\mathbf{h}) \in \mathcal{H}_{n,0}$ ($\mathcal{H}_{n,0}$ is the class of the homogeneous correlation functions that are continuous everywhere except, perhaps, at the space origin), then

$$\rho_x(\mathbf{h}) = \alpha\, \delta(\mathbf{h}) + \beta \rho_b(\mathbf{h}) \tag{19}$$

where $\delta(\mathbf{h})$ denotes the Kronecker delta (the same symbol is also used to denote delta function, see Remark 3 below), $\rho_b(\mathbf{h}) \in \mathcal{H}_{n,C}$ ($\mathcal{H}_{n,C}$ is the class of homogeneous correlation functions in R^n that are everywhere continuous) and α, β are nonnegative coefficients. This is the case of the so-called *nugget-effect phenomenon* of geostatistics; for a detailed discussion about this phenomenon, as well as the SRF models used to describe it mathematically, see Chapter 7. Equation (19) implies that a homogeneous SRF $X(\mathbf{s})$ with $\rho_x(\mathbf{h}) \in \mathcal{H}_{n,0}$ can be written as

$$X(\mathbf{s}) = X_\alpha(\mathbf{s}) + X_b(\mathbf{s})$$

where $X_\alpha(\mathbf{s})$ is the "chaotic" component and $X_b(\mathbf{s})$ is the m.s. continuous component of $X(\mathbf{s})$.

Property 4: If $\rho_x(\mathbf{h}) \in \mathscr{H}_{n,C}$, then

$$\rho_x(\mathbf{h}) = E\{\exp[i\mathbf{w} \cdot \mathbf{h}]\} \tag{20}$$

where $\mathbf{w}$ is an n-dimensional random vector. That is, the $\rho_x(\mathbf{h})$ can be considered as the characteristic function of $\mathbf{w}$.

The proof of Property 1 is available in any reference on SRF (e.g., Gihman and Skorokhod, 1974a). For Property 2 see Christakos (1984b), and for Properties 3 and 4 see Matern (1960).

Remark 3: Associated with the nugget-effect phenomenon are the mathematical notions of (a) the delta (or Dirac) function, which is defined as $\int_{R^n} \delta(\mathbf{h})\, d\mathbf{h} = 1$; $\delta(\mathbf{h} \neq \mathbf{0}) = 0$; and (b) Kronecker delta, which is defined by

$$\delta(\mathbf{h}) = \begin{cases} 1 & \text{if} \quad \mathbf{h} = \mathbf{0} \\ 0 & \text{otherwise} \end{cases}$$

or by

$$\delta_{ij} = \begin{cases} 1 & \text{if} \quad i = j \\ 0 & \text{otherwise} \end{cases}$$

In this book, the symbol $\delta(\mathbf{h})$ will be used in both cases (a) and (b), and the appropriate interpretation will be obvious from the text. The delta function models nugget effects in the continuous-parameter case; the Kronecker delta is used to model nugget effects of discrete-parameter SRF (see also Chapter 7).

In the case of homogeneous SRF, the following formulas are straightforward consequences of the stochastic differentiation results obtained in Section 4.3:

$$\operatorname{cov}\left(\frac{\partial^\nu X(\mathbf{s})}{\partial s_{i_1}\, \partial s_{i_2} \cdots \partial s_{i_\nu}}, \frac{\partial^\nu X(\mathbf{s}')}{\partial s'_{i_1}\, \partial s'_{i_2} \cdots \partial s'_{i_\nu}}\right) = (-1)^\nu \frac{\partial^{2\nu} c_x(\mathbf{h})}{\partial h_{i_1}^2\, \partial h_{i_2}^2 \cdots \partial h_{i_\nu}^2} \tag{21}$$

and

$$\operatorname{cov}\left(\frac{\partial^\nu X(\mathbf{s})}{\partial s_i^\nu}, \frac{\partial^\mu X(\mathbf{s}')}{\partial s_j'^\mu}\right) = (-1)^\nu \frac{\partial^{\nu+\mu} c_x(\mathbf{h})}{\partial h_i^\nu\, \partial h_j'^\mu} \tag{22}$$

Example 1: Let $X(s)$ be a zero mean stationary RP and let $Y(s) = dX(s)/ds$, which is a stationary RP too. By differentiating the integrable representation (2) we find

$$Y(s) = \int_{R^1} \exp[iws]iw\, d\aleph_x(w) \tag{23}$$

The covariance is given by

$$c_Y(h) = \int_{R^1} \exp[iwh] w^2 \, dQ_x(w) \tag{24}$$

The same result is obtained by combining Eq. (21) above with the spectral representation

$$c_x(h) = \int_{R^1} \exp[iwh] \, dQ_x(w) \tag{25}$$

By comparing Eq. (24) with equation

$$c_Y(h) = \int_{R^1} \exp[iwh] \, dQ_Y(w) \tag{26}$$

we find $dQ_Y(w) = w^2 \, dQ_x(w)$. The integral (24) exists if the following condition holds true.

$$\int_{R^1} w^2 \, dQ_x(w) < \infty \tag{27}$$

A homogeneous but in general not isotropic SRF is characterized by

$$c_x(\mathbf{h}) = c_x(\sqrt{\mathbf{h}^T \mathbf{G} \mathbf{h}}) \tag{28}$$

where $\mathbf{G}$ is a nonnegative matrix. This situation is sometimes called *geometrical anisotropy.* Note that if $\mathbf{G}$ is the identity matrix, then $c_x(\mathbf{h}) = c_x(|\mathbf{h}|)$ and the SRF is isotropic. Interesting covariance models of homogeneous but anisotropic SRF are the

$$c_x(\mathbf{h}) = c_x\left(\frac{|h_1|}{a_1}, \frac{|h_2|}{a_2}, \frac{|h_3|}{a_3}\right) \tag{29}$$

where $a_1, a_2, a_3 > 0$, and the

$$c_x(\mathbf{h}) = c_x\left(\frac{c_1 h_1 + c_2 h_2 + c_3 h_3}{a}\right) \tag{30}$$

where $a > 0$.

Example 2: A special case of Eq. (29) is the anisotropic Gaussian covariance

$$c_x(\mathbf{h}) = c_x(\mathbf{0}) \exp\left[-\frac{1}{2}\left(\frac{h_1^2}{a_1^2} + \frac{h_2^2}{a_2^2} + \frac{h_3^2}{a_3^2}\right)\right] \tag{31}$$

A straightforward generalization of Eq. (30) leads to the covariance

$$c_x(\mathbf{h}) = c_x(\mathbf{0}) \exp\left[-\frac{1}{2} \sum_{i,j=1}^{3} b_{ij} h_i h_j\right] \tag{32}$$

with $\det \mathbf{B} \neq 0$, where $\mathbf{B}$ is the matrix with elements b_{ij}, $i, j = 1, 2$, and 3.

For future reference, we summarize here the most important classes of homogeneous correlation functions: (a) the class $\mathscr{H}_n$ of homogeneous, in general, correlation functions in R^n; (b) the class $\mathscr{H}_{n,O}$ of homogeneous correlation functions in R^n, which are everywhere continuous except, perhaps, at the origin; (c) the class $\mathscr{H}_{n,c}$ of homogeneous correlation functions in R^n, which are continuous everywhere. Clearly, $\mathscr{H}_{n,c} \subset \mathscr{H}_{n,O} \subset \mathscr{H}_n$.

7.2 The Geometry of Homogeneous Spatial Random Fields

Just as for arbitrary SRF, the geometry of an homogeneous SRF is related to the behavior of the covariance at the origin ($\mathbf{h} = \mathbf{O}$). More specifically, the proposition below is a straightforward consequence of the results of Section 4 above.

Proposition 2: Let $X(\mathbf{s})$ be a homogeneous SRF. (a) If $c_x(\mathbf{h})$ is continuous at $\mathbf{h} = \mathbf{O}$, it is continuous everywhere in R^n; then and only then the $X(\mathbf{s})$ is m.s. continuous. (b) If the

$$\operatorname{cov}\left(\frac{\partial^\nu X(\mathbf{s})}{\partial s_{i_1}\,\partial s_{i_2}\cdots\partial s_{i_\nu}}, \frac{\partial^\nu X(\mathbf{s}')}{\partial s'_{i_1}\,\partial s'_{i_2}\cdots\partial s'_{i_\nu}}\right)$$

$$= (-1)^\nu \frac{\partial^{2\nu} c_x(\mathbf{h})}{\partial h_{i_1}^2\,\partial h_{i_2}^2\cdots\partial h_{i_\nu}^2}$$

exists and is finite at $\mathbf{h} = \mathbf{O}$, then and only then the partial derivative

$$\frac{\partial^\nu X(\mathbf{s})}{\partial s_{i_1}\,\partial s_{i_2}\cdots\partial s_{i_\nu}}$$

exists in the m.s.s. (c) If $c_x(\mathbf{h})$ is continuous in R^n, the stochastic integral of Eq. (24), Section 4 exists in the m.s.s. When $c_x(\mathbf{h})$ is not continuous at $\mathbf{h} = \mathbf{O}$ (nugget effect), the homogeneous SRF $X(\mathbf{s})$ will not be m.s. continuous at any point $\mathbf{s} \in R^n$.

As a particular case of (b), the $\partial^\nu X(\mathbf{s})/\partial s_i^\nu$ exists in the m.s.s. if and only if the $(-1)^\nu(\partial^{2\nu} c_x(\mathbf{h})/\partial h_i^{2\nu})$ exists and is finite at $\mathbf{h} = \mathbf{O}$. A very interesting result is introduced by the following corollary.

Corollary 1: The homogeneous SRF $X(\mathbf{s})$ is m.s. differentiable (that is, the $\partial^\nu X(\mathbf{s})/\partial s_i^\nu$ exists in the m.s.s.). Then

$$\left.\frac{\partial^{2\nu-1} c_x(\mathbf{h})}{\partial h_i^{2\nu-1}}\right|_{\mathbf{h}=\mathbf{O}} = 0 \tag{33}$$

Example 3: Consider the spatial covariance

$$c_x(\mathbf{h}) = \exp\left[-\frac{|\mathbf{h}|^2}{a^2}\right]$$

where $a > 0$. It is continuous at $\mathbf{h} = \mathbf{O}$, where all its derivatives are defined and are finite. Therefore, the associated SRF $X(\mathbf{s})$ is m.s. continuous and differentiable of any order.

Example 4: The unidimensional covariance

$$c_x(h) = \exp\left[-\frac{h}{a}\right]\cos(bh), \qquad a, b > 0$$

is continuous at $h = 0$. However,

$$\left.\frac{dc_x(h)}{dh}\right|_{h=0} \neq 0$$

and therefore the associated RP $X(s)$, while continuous in the m.s.s., is not m.s. differentiable.

Example 5: Condition (27) is equivalent to the existence of

$$-\left.\frac{d^2c_x(h)}{dh^2}\right|_{h=0} < \infty \tag{34}$$

and, therefore, according to Proposition 2 above the RP $X(s)$ is m.s. differentiable. [It is worth noticing that condition (27) for the existence of the spectral representation (24) is equivalent to the expression (34), which involves the covariance function itself.]

Regarding Proposition 2, one may distinguish three major shapes of $c_x(\mathbf{h})$ near the origin ($\mathbf{h} = \mathbf{O}$):

(i) If $c_x(\mathbf{h})$ is discontinuous at $\mathbf{h} = \mathbf{O}$ (nugget effect), the SRF is not continuous in the m.s.s. and, thus, is very irregular; the SRF becomes more irregular if the covariance is discontinuous at the origin and then drops immediately to zero for $\mathbf{h} > \mathbf{O}$ (pure nugget effect).

(ii) If the covariance behaves linearly near the origin $\mathbf{h} = \mathbf{O}$, that is, the $c_x(\mathbf{h})$ is continuous and once differentiable, the SRF is continuous in the m.s.s. but not differentiable.

(iii) If the shape of $c_x(\mathbf{h})$ near the origin is parabolic, the $c_x(\mathbf{h})$ is continuous and twice differentiable; the corresponding SRF is also continuous and once differentiable in the m.s.s.

Remark 4: The following point must be stressed: The existence of the νth-order spatial derivatives $\nabla^\nu X(\mathbf{s})$ imposes certain requirements on the spectral density $C_x(\mathbf{w})$ of the SRF $X(\mathbf{s})$, viz., it must hold true that $E[\nabla^\nu X(\mathbf{s})]^2 < \infty$, or

$$\int_{R^n} |\mathbf{w}|^{2\nu} C_x(\mathbf{w})\, d\mathbf{w} < \infty$$

In R^3 the last inequality requires that the $C_x(\mathbf{w})$ decreases as $|\mathbf{w}| \to \infty$ faster than $|\mathbf{w}|^{-3-2\nu}$.

For example, the spectral density of an SRF with an exponential covariance [see Eq. (11) of Section 8 below] does not satisfy the above condition for $\nu = 1$ (the spectral density falls off with $|\mathbf{w}|$ as $|\mathbf{w}|^{-4}$ and hence the SRF is not differentiable).

As regards the stochastic integration of homogeneous SRF, the condition expressed by Eq. (27) of Section 4 becomes

$$E[Z^2(\mathbf{w})] = \int_V \alpha(\mathbf{w}; \mathbf{s})\, d\mathbf{s} \int_V \overline{\alpha(\mathbf{w}; \mathbf{s}')} c_x(\mathbf{s} - \mathbf{s}')\, d\mathbf{s}' < \infty \qquad (35)$$

Following Matheron (1965) and Dagan (1989) and assuming $\alpha(\mathbf{w}; \mathbf{s}) = 1$, the stochastic integral (24) of Section 4 and Eq. (35) can be written as

$$Z = \int_{R^n} I(\mathbf{s}) X(\mathbf{s})\, d\mathbf{s} \qquad (36)$$

and

$$E[Z^2] = \int_V \Lambda(\mathbf{h}) c_x(\mathbf{h})\, d\mathbf{h} < \infty \qquad (37)$$

respectively, where

$$I(\mathbf{s}) = \begin{cases} 1 & \text{if } \mathbf{s} \in V \\ 0 & \text{otherwise} \end{cases}$$

is the indicator function and

$$\Lambda(\mathbf{h}) = \int_{R^n} I(\mathbf{s}) I(\mathbf{s} + \mathbf{h})\, d\mathbf{s}$$

7.3 A Criterion of Permissibility for Spatial Covariances

Just as in the general case of SRF, the necessary and sufficient condition that a continuous function must satisfy to be a covariance $c_x(\mathbf{h})$ of a homogeneous SRF is the *nonnegative-definite* condition, namely,

$$\sum_{i=1}^{m} \sum_{j=1}^{m} q_i q_j c_x(\mathbf{h}) \geq 0 \qquad (38)$$

where $\mathbf{h} = \mathbf{s}_i - \mathbf{s}_j$, for all integers m, distance vectors $\mathbf{h}$, and coefficients $q_1, q_2, \ldots, q_m$.

Alternatively, since the application of (38) is practically impossible, one applies *Bochner's* theorem (Bochner, 1959): In general, a continuous function $c_x(\mathbf{h})$ is a nonnegative-definite function if and only if it can be expressed as in Eq. (7) above, where $dQ_x(\mathbf{w})$ is a nonnegative finite measure on the Borel sets of R^n. Furthermore, it is usually assumed that the necessary conditions are satisfied for a representation of the form of Eq. (8) or (12) to be valid.

Under these circumstances, the following *first criterion of permissibility (COP-1)* is much more comprehensive to apply: A continuous and symmetric function $c_x(\mathbf{h})$ in R^n will be called a *permissible* covariance of a homogeneous SRF; that is, it satisfies the nonnegative-definiteness condition (38), if and only if it is the n-fold Fourier transform of a nonnegative bounded function $C_x(\mathbf{w})$. In other words, we must have

$$C_x(\mathbf{w}) \geq 0 \tag{39}$$

for all $\mathbf{w} \in R^n$.

Remark 4: COP-1 imposes certain restrictions on $c_x(\mathbf{h})$, namely, it must be such that (i) the integral in Eq. (9) or (13) converges, and (ii) the integral is nonnegative for all $\mathbf{w}$. For more details on the practical implementation of COP-1, as well as other COPs to be discussed in the following sections, see Chapter 7.

Some illustrative examples are discussed below.

Example 6: Consider the RP introduced by Eq. (11) of Section 5.1 above. This is obviously a stationary process in R^1 and, as Khinchin (1934) has proved, the covariance of any stationary RP can be represented in the form of the integral (13), Section 5.1. Conversely, every function of the form of Eq. (13), Section 5.1 is a permissible covariance of a stationary RP.

Example 7: Let us define the function

$$c_x(\mathbf{h}) = a\,\delta(\mathbf{h}) \tag{40}$$

where $a > 0$, and $\delta(\mathbf{h})$ is the delta function. The function of Eq. (40) is such that

$$\int_{R^n} |c_x(\mathbf{h})|\, d\mathbf{h} = \int_{R^n} a\,\delta(\mathbf{h})\, d\boldsymbol{h} = a < \infty$$

and, hence, it can be represented by the integral (8) above, where

$$C_x(\mathbf{w}) = \frac{a}{(2\pi)^n} \geq 0$$

This implies that Eq. (40) is a permissible covariance function (the corresponding SRF is a white noise).

Example 8: Consider in R^1 the function

$$c_x(h) = \pi \exp[-h^2] \tag{41}$$

The spectral density is

$$C_x(w) = \frac{\sqrt{\pi}}{2} \exp\left[-\frac{w^2}{4}\right] \geq 0$$

for all w and, therefore, Eq. (41) is a permissible covariance in R^1.

7.4 Linear Transformation of Homogeneous Spatial Random Fields

The analysis of Section 5.2 above remains valid, where now both $\aleph_x(\mathbf{w})$ and $\aleph_Y(\mathbf{w})$ are SRF with uncorrelated increments. In addition

$$E[|\aleph_x(\mathbf{w})|^2] = Q_x(\mathbf{w})$$

and

$$E[|\aleph_Y(\mathbf{w})|^2] = Q_Y(\mathbf{w}) \tag{42}$$

The spectral density function of $Y(\mathbf{s})$ is

$$c_Y(\mathbf{h}) = \int_{R^n} \exp[i\mathbf{w}\cdot\mathbf{h}] E[d\aleph_Y(\mathbf{w})\, \overline{d\aleph_Y(\mathbf{w})}] \tag{43}$$

$$= \int_{R^n} \exp[i\mathbf{w}\cdot\mathbf{h}] |H(\mathbf{w})|^2 E[d\aleph_x(\mathbf{w})\, \overline{d\aleph_x(\mathbf{w})}] \tag{44}$$

By comparing Eqs. (43) and (44) we find that

$$E[d\aleph_Y(\mathbf{w})\, \overline{d\aleph_Y(\mathbf{w})}] = |H(\mathbf{w})|^2 E[d\aleph_x(\mathbf{w})\, \overline{d\aleph_x(\mathbf{w})}] \tag{45}$$

or

$$dQ_Y(\mathbf{w}) = |H(\mathbf{w})|^2\, dQ_x(\mathbf{w}) \tag{46}$$

Naturally, if the functions $Q_Y(\mathbf{w})$ and $Q_x(\mathbf{w})$ are differentiable, Eq. (46) can be expressed by means of the spectral densities, viz.,

$$C_Y(\mathbf{w}) = |H(\mathbf{w})|^2 C_x(\mathbf{w}) \tag{47}$$

7.5 Vector Homogeneous Spatial Random Fields

In the case of several correlated homogeneous SRF, the most important of the results of Section 3.2 above apply by restricting analysis in terms of distance vectors $\mathbf{h} = \mathbf{s} - \mathbf{s}'$. One can now define one more statistical moment of second order, the homogeneous cross-covariance.

Definition 2: Let $X_p(\mathbf{s})$ and $X_{p'}(\mathbf{s}')$, $\mathbf{s}$, $\mathbf{s}' \in R^n$ be two homogeneous SRF. The homogeneous cross-covariance is defined as

$$c_{x_p x_{p'}}(\mathbf{h}) = E\{[X_p(\mathbf{s}) - m_{x_p}(\mathbf{s})][X_{p'}(\mathbf{s}+\mathbf{h}) - m_{x_{p'}}(\mathbf{s}+\mathbf{h})]\}$$

$$= \int\int [\chi_p - m_{x_p}(\mathbf{s})][\chi_{p'} - m_{x_{p'}}(\mathbf{s}+\mathbf{h})] f_{\mathbf{s},\mathbf{s}+\mathbf{h}}(\chi_p, \chi_{p'})\, d\chi_p\, d\chi_{p'} \tag{48}$$

where $\mathbf{h}=\mathbf{s}-\mathbf{s}'$ and $f_{\mathbf{s},\mathbf{s}+\mathbf{h}}(\chi_p, \chi_{p'})$ is now the joint probability density of $X_p(\mathbf{s})$ and $X_{p'}(\mathbf{s}')$.

The coefficient of cross-correlation can be written

$$\rho_{x_p x_{p'}}(\mathbf{h})=\frac{c_{x_p x_{p'}}(\mathbf{h})}{\sqrt{\sigma^2_{x_p}(\mathbf{s})}\sqrt{\sigma^2_{x_{p'}}(\mathbf{s}+\mathbf{h})}} \tag{49}$$

where $\sigma^2_{x_p}(\mathbf{s})$ and $\sigma^2_{x_{p'}}(\mathbf{s}+\mathbf{h})$ are the variances of the SRF $X_p(\mathbf{s})$ and $X_{p'}(\mathbf{s}')$, respectively.

The definition above can be easily extended to more than two SRF. Let

$$\mathbf{X}(\mathbf{s})=[X_1(\mathbf{s}), X_2(\mathbf{s}), \ldots, X_k(\mathbf{s})]^{\mathrm{T}}$$

be a vector homogeneous SRF. The corresponding matrix of homogeneous spatial cross-covariances between the component SRF is

$$\mathbf{C_X}=[c_{x_p x_{p'}}(\mathbf{h}_{ij})] \tag{50}$$

where $\mathbf{h}_{ij}=\mathbf{s}_i-\mathbf{s}_j$ for all $p, p'=1, 2, \ldots, k$.

The properties below are straightforward consequences of the preceding analysis.

Property 1: The matrix $\mathbf{C}_{\mathbf{x}}$ is nonnegative-definite, namely,

$$\mathbf{q}^{\mathrm{T}}\mathbf{C}_{\mathbf{x}}\mathbf{q}\geq 0 \tag{51}$$

for all deterministic vectors $\mathbf{q}=[q_1, \ldots, q_k]^{\mathrm{T}}$. Just as for scalar SRF, this is an immediate consequence of the fact that if

$$Y(\mathbf{s}_i)=\sum_{p=1}^{k} q_p X_p(\mathbf{s}_i)$$

it must hold true that

$$\mathrm{Var}[Y(\mathbf{s}_i)]=\sum_{p=1}^{k}\sum_{p'=1}^{k} q_p q_{p'} c_{x_p x_{p'}}(\mathbf{h}_{ij})\geq 0$$

Property 2: The $\mathbf{C}_{\mathbf{x}}$ is a symmetric matrix, since

$$c_{x_p x_{p'}}(\mathbf{h}_{ij})=c_{x_{p'} x_p}(\mathbf{h}_{ji}) \tag{52}$$

where $\mathbf{h}_{ji}=-\mathbf{h}_{ij}$. However, in general,

$$c_{x_p x_{p'}}(\mathbf{h}_{ij})\neq c_{x_p x_{p'}}(\mathbf{h}_{ji}) \tag{53}$$

Property 3: A straightforward application of the Schwartz's inequality yields

$$|c_{x_p x_{p'}}(\mathbf{h}_{ij})|\leq\sqrt{\sigma^2_{x_p}(\mathbf{s}_i)}\sqrt{\sigma^2_{x_{p'}}(\mathbf{s}_j)} \tag{54}$$

for all p, $p'=1, 2, \ldots, k$ and $i, j=1, 2, \ldots, m$.

Given the matrix of homogeneous cross-covariances (50), the definition of the corresponding matrix of cross-spectral density functions is rather straightforward, namely, the symmetric matrix

$$\tilde{\mathbf{C}}_{\mathbf{X}} = [C_{x_p x_{p'}}(\mathbf{w}_{ij})] \tag{55}$$

where $\mathbf{w}_{ij} = \mathbf{w}_i - \mathbf{w}_j$, and $p, p' = 1, \ldots, k$; the component cross-spectral density functions $C_{x_p x_{p'}}(\mathbf{w}_{ij})$ are

$$C_{x_p x_{p'}}(\mathbf{w}_{ij}) = \frac{1}{(2\pi)^n} \int_{R^n} \exp[-i\mathbf{w}_{ij} \cdot \mathbf{h}_{ij}] c_{x_p x_{p'}}(\mathbf{h}_{ij})\, d\mathbf{h}_{ij} \tag{56}$$

assuming that these integrals exist.

The covariance corresponding to

$$Y(\mathbf{s}_i) = \sum_{p=1}^{k} q_p X_p(\mathbf{s}_i)$$

is

$$c_Y(\mathbf{h}_{ij}) = \sum_{p=1}^{k} \sum_{p'=1}^{k} q_p q_{p'} c_{x_p x_{p'}}(\mathbf{h}_{ij})$$

The $c_Y(\mathbf{h}_{ij})$ must be nonnegative-definite, which, according to COP-1, implies that its spectral density function must be $C_Y(\mathbf{w}_{ij}) \geq 0$ for all $\mathbf{w}_{ij} \in R^n$. This, in turn, implies that

$$C_Y(\mathbf{w}_{ij}) = \sum_{p=1}^{k} \sum_{p'=1}^{k} q_p q_{p'} C_{x_p x_{p'}}(\mathbf{w}_{ij}) \geq 0$$

for all q_p and $q_{p'}$, p, $p' = 1, \ldots, k$. Hence, the symmetric matrix (55) is nonnegative-definite. Conversely, for any vector of deterministic coefficients $\boldsymbol{\psi}^T = [\psi_1, \ldots, \psi_m]$ it is valid that

$$\sum_{j=1}^{m} \sum_{j'=1}^{m} \boldsymbol{\psi}_j^T \mathbf{C}_{\mathbf{X}} \boldsymbol{\psi}_{j'} = \int_{R^n} \bar{\boldsymbol{\alpha}}^{\mathrm{T}} \tilde{\mathbf{C}}_{\mathbf{X}} \boldsymbol{\alpha} \geq 0$$

where $\boldsymbol{\alpha} = \sum_{j=1}^{m} \exp[-i(\mathbf{w} \cdot \mathbf{s}_j)] \boldsymbol{\psi}_j$.

On the basis of these considerations, one obtains the following *second criterion of permissibility (COP-2)*: In order that the matrix (50) be the matrix of permissible cross-covariance functions of a homogeneous vector SRF, it is necessary and sufficient that the matrix (55) of the corresponding spectral cross-density functions be nonnegative-definite for all $\mathbf{w}_{ij} \in R^n$. By virtue of the matrix theory the latter means that the principal minor determinants of the matrix (55) must be nonnegative.

Let us elaborate a bit on this criterion by means of a simple example.

Example 8: Let $k=2$; then

$$\mathbf{C_X} = \begin{bmatrix} c_{x_1}(\mathbf{h}_{ij}) & c_{x_1x_2}(\mathbf{h}_{ij}) \\ c_{x_2x_1}(\mathbf{h}_{ji}) & c_{x_2}(\mathbf{h}_{ij}) \end{bmatrix}$$

If we define the SRF

$$Y(\mathbf{s}_i) = \sum_{p=1}^{2} q_p X_p(\mathbf{s}_i)$$

with covariance

$$c_Y(\mathbf{h}_{ij}) = \sum_{p=1}^{2} \sum_{p'=1}^{2} q_p q_{p'} c_{x_p x_{p'}}(\mathbf{h}_{ij})$$

the latter must be, naturally, a nonnegative-definite function. This happens if and only if $C_Y(\mathbf{w}_{ij}) \geq 0$ for all $\mathbf{w}_{ij} \in R^n$, or,

$$C_Y(\mathbf{w}_{ij}) = \sum_{p=1}^{2} \sum_{p'=1}^{2} q_p q_{p'} C_{x_p x_{p'}}(\mathbf{w}_{ij}) \geq 0$$

for any numbers q_1 and q_2. This implies that the symmetric matrix

$$\tilde{\mathbf{C}}_{\mathbf{X}} = \begin{bmatrix} C_{x_1}(\mathbf{w}_{ij}) & C_{x_1x_2}(\mathbf{w}_{ij}) \\ C_{x_2x_1}(\mathbf{w}_{ji}) & C_{x_2}(\mathbf{w}_{ij}) \end{bmatrix}$$

is nonnegative-definite. In terms of the principal minor determinants of the above matrix the latter requirement yields

$$C_{x_1}(\mathbf{w}_{ij}),\ C_{x_2}(\mathbf{w}_{ij}) \geq 0 \tag{57}$$

and

$$|C_{x_1x_2}(\mathbf{w}_{ij})|^2 = |C_{x_2x_1}(\mathbf{w}_{ji})|^2 \leq C_{x_1}(\mathbf{w}_{ij}) C_{x_2}(\mathbf{w}_{ij}) \tag{58}$$

Remark 5: Note that by using the theory of matrices (e.g., Horn and Johnson, 1985), one can derive several other useful results regarding the permissibility of $\mathbf{C}_x$.

8. Isotropic Spatial Random Fields

8.1 Basic Formulas

An SRF is called an *isotropic SRF* if its covariance function is a function of the distance $r = |\mathbf{h}|$ only, viz.,

$$c_x(\mathbf{h}) = c_x(r) \tag{1}$$

Similarly, in the frequency domain the spectral density function can be written

$$C_x(\mathbf{w}) = C_x(\omega) \tag{2}$$

where $\omega = |\mathbf{w}|$.

Obviously, if $X(\mathbf{s})$ is an isotropic SRF then it is an homogeneous SRF too. (However, the converse is not generally true.) Therefore, the results of Section 7 can be transformed into their isotropic equivalents by simply applying the n-dimensional spherical coordinates

$$\begin{aligned}
\mathbf{w} &= (w_1, \ldots, w_i, \ldots, w_n) \\
w_1 &= \omega \cos \theta_1 \\
w_i &= \omega \cos \theta_i \prod_{j=1}^{i-1} \sin \theta_j \qquad \text{for} \quad i = 2, 3, \ldots, n-1 \\
w_n &= \omega \prod_{j=1}^{n-1} \sin \theta_j
\end{aligned} \tag{3}$$

and by setting $\mathbf{w} \cdot \mathbf{h} = \omega r \cos \theta_1$.

The fundamental Eqs. (8) and (9) of Section 7, which relate the covariance and the spectral density functions, become

$$c_x(r) = (2\pi)^{n/2} \int_0^\infty \frac{J_{(n-2)/2}(\omega r)}{(\omega r)^{(n-2)/2}} C_x(\omega) \omega^{n-1} \, d\omega \tag{4}$$

and

$$C_x(\omega) = \frac{1}{(2\pi)^{n/2}} \int_0^\infty \frac{J_{(n-2)/2}(\omega r)}{(\omega r)^{(n-2)/2}} c_x(r) r^{n-1} \, dr \tag{5}$$

respectively, where $J_{(n-2)/2}$ is the Bessel function of $(n-2)/2$th order (Gradshteyn and Ryzhik, 1965). In order that $c_x(r)$ be a covariance function of an isotropic SRF, it is necessary and sufficient that this function admits a representation of the form (4), where $C_x(\omega)$ is a nonnegative bounded function.

Example 1: In R^2, Eqs. (4) and (5) yield, respectively,

$$c_x(r) = 2\pi \int_0^\infty J_0(\omega r) C_x(\omega) \omega \, d\omega \tag{6}$$

and

$$C_x(\omega) = \frac{1}{2\pi} \int_0^\infty J_0(\omega r) c_x(r) r \, dr \tag{7}$$

In R^3,

$$c_x(r) = 4\pi \int_0^\infty \frac{\sin(\omega r)}{r} C_x(\omega)\omega\, d\omega \tag{8}$$

and

$$C_x(\omega) = \frac{1}{2\pi^2} \int_0^\infty \frac{\sin(\omega r)}{\omega} c_x(r) r\, dr \tag{9}$$

Example 2: Useful covariances of isotropic SRF are (i) the *Gaussian* covariance

$$c_x(r) = c_x(0) \exp\left[-\frac{r^2}{a^2}\right] \tag{10}$$

where $a > 0$; (ii) the *exponential* covariance

$$c_x(r) = c_x(0) \exp\left[-\frac{r}{a}\right] \tag{11}$$

where $a > 0$; and (iii) the *spherical* covariance of geostatistics

$$c_x(r) = \begin{cases} c_x(0)\left[1 - \dfrac{3r}{2a} + \dfrac{r^3}{2a^3}\right] & \text{if} \quad r \in [0, a] \\ 0 \quad \text{if} \quad r \geq a \end{cases} \tag{12}$$

where $a > 0$.

8.2 The Geometry of Isotropic Spatial Random Fields

The geometrical properties of isotropic SRF are immediate consequences of the corresponding properties of homogeneous SRF. More specifically:

(a) The isotropic SRF $X(\mathbf{s})$ is m.s. continuous, if and only if the $c_x(r)$ is continuous at $r = 0$.

(b) The partial derivative

$$\frac{\partial^\nu X(\mathbf{s})}{\partial s_{i_1}\, \partial s_{i_2} \cdots \partial s_{i_\nu}}$$

exists in the m.s.s., if and only if the

$$\operatorname{cov}\left(\frac{\partial^\nu X(\mathbf{s})}{\partial s_{i_1}\, \partial s_{i_2} \cdots \partial s_{i_\nu}}, \frac{\partial^\nu X(\mathbf{s}')}{\partial s'_{i_1}\, \partial s'_{i_2} \cdots \partial s'_{i_\nu}}\right) = (-1)^\nu \frac{\partial^{2\nu} c_x(r)}{\partial r_{i_1}^2\, \partial r_{i_2}^2 \cdots \partial r_{i_\nu}^2}$$

exists and is finite at $r = 0$.

(c) If $c_x(r)$ is continuous in R^n, the stochastic integral of Eq. (24), Section 4 exists in the m.s.s. When $c_x(r)$ is not continuous at $r=0$, the isotropic SRF $X(\mathbf{s})$ will not be m.s. continuous at any point $\mathbf{s} \in R^n$. (This is the case of an isotropic nugget effect.)

The $\partial^\nu X(\mathbf{s})/\partial s^\nu$, in particular, exists in the m.s.s. if and only if the

$$(-1)^\nu \frac{\partial^{2\nu} c_x(r)}{\partial r^{2\nu}}$$

exists and is finite at $r=0$. Moreover, on the basis of Corollary 1 of Section 7 we find that the $\partial^\nu X(\mathbf{s})/\partial s^\nu$ exists in the m.s.s. if and only if

$$\left. \frac{\partial^{2\nu-1} c_x(r)}{\partial r^{2\nu-1}} \right|_{r=0} = 0 \tag{13}$$

Note that if $X(\mathbf{s})$ is an isotropic SRF, its partial derivatives are homogeneous but not necessarily isotropic SRF.

Example 3: The Gaussian covariance (10) is continuous at $r=0$, and all its derivatives are defined and are finite at $r=0$. Therefore, the associated SRF $X(\mathbf{s})$ is m.s. continuous and differentiable of any order.

Example 4: The covariance

$$c_x(r) = \exp\left[-\left(\frac{r}{a}\right)^\beta\right] \tag{14}$$

where $\beta \in (0, 1]$ is also continuous covariance at $r=0$, but

$$\left. \frac{dc_x(r)}{dr} \right|_{r=0} \neq 0$$

Hence, the corresponding SRF $X(\mathbf{s})$ is m.s. continuous but not m.s. differentiable. Similar geometrical properties possess the exponential covariance (11) and the covariance

$$c_x(r) = \exp\left[-\frac{r}{a}\right] \cos(br) \tag{15}$$

where a and b are suitable coefficients.

8.3 Criteria of Permissibility

The COP-1 presented in Section 7 is obviously valid in the case of isotropic SRF.

Example 5: In R^1 the covariance function

$$c_x(r) = \frac{c_x(0)}{\cosh\left(\frac{r}{a}\right)} \tag{16}$$

where $(a > 0)$ is associated with the spectral density

$$C_x(\omega) = \frac{ac_x(0)}{2\cosh\left(\frac{\pi a \omega}{2}\right)} > 0$$

for all ω. Thus, by reference to COP-1, the permissibility of the covariance (16) is assured.

For further illustration, certain important covariance functions are plotted in Fig. 2.2; the spectral densities are plotted in Fig. 2.3.

In addition, the following *third criterion of permissibility* (*COP-3* Christakos, 1984b), despite the fact that it is only sufficient and deals with a rather specific class of models, is very convenient for it imposes conditions directly on the covariance and not on the spectral density function: A continuous function $c_x(r)$, where $r = |\mathbf{h}| \in R^n$, is a permissible covariance (in other words, it is a nonnegative-definite function) if at the origin it holds true that

$$c'_x(r)\Big|_{r=0} = \frac{dc_x(r)}{dr}\Big|_{r=0} < 0 \tag{17}$$

if at infinity it is true that

$$\lim_{r \to \infty} \frac{c_x(r)}{r^{(1-n)/2}} = 0 \tag{18}$$

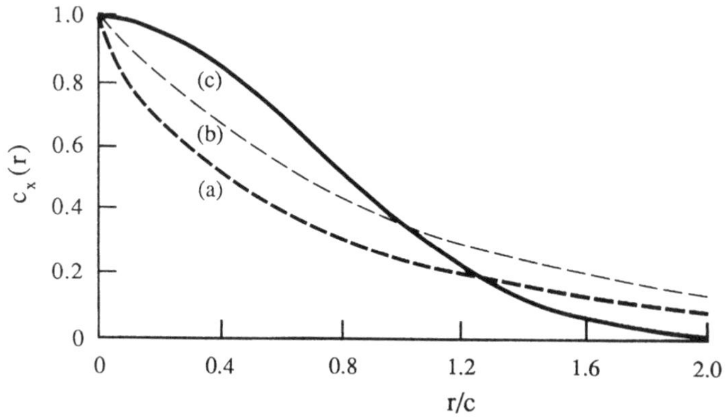

Figure 2.2 The covariances (a) $c_x(r) = [K_{1/3}(r/c)/G(1/3)]\sqrt[3]{4r/c}$; (b) $c_x(r) = \exp[-r/c]$; and (c) $c_x(r) = \exp[-(r/c)^2]$

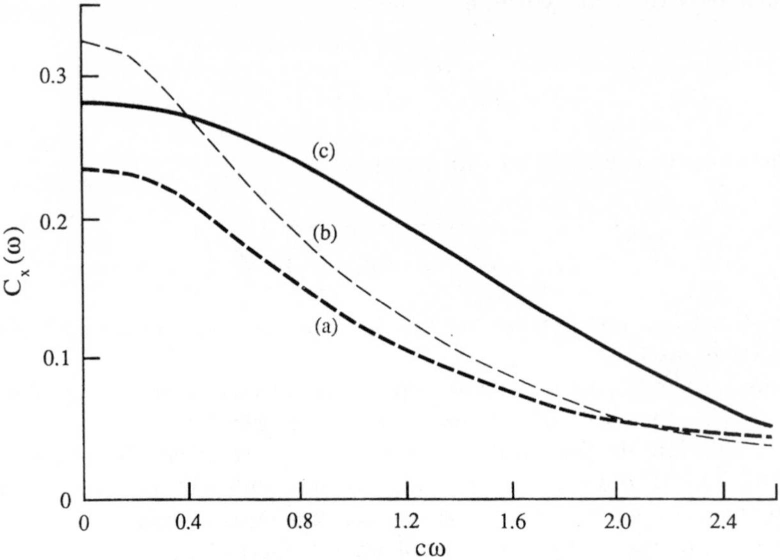

Figure 2.3 The spectral densities of the covariances of Fig. 2.2

and if

$$c_x''(r) = \frac{d^2 c_x(r)}{dr^2} \geq 0 \qquad \text{in} \quad R^1 \tag{19}$$

$$\int_r^\infty \frac{u}{\sqrt{u^2 - r^2}}\, dc_x''(r) \geq 0 \qquad \text{in} \quad R^2 \tag{20}$$

$$c_x''(r) - r c_x'''(r) \geq 0 \qquad \text{in} \quad R^3 \tag{21}$$

Example 6: In R^3, the exponential model

$$c_x(r) = \exp\left[-\frac{r}{a}\right], \qquad a > 0$$

gives

$$c_x'(r)\big|_{r=0} = -\frac{1}{a} < 0$$

$$\lim_{r\to\infty} \frac{c_x(r)}{r^{-1}} = \lim_{r\to\infty} r \exp\left[-\frac{r}{a}\right] = 0$$

and

$$c_x''(r) - r c_x'''(r) = \frac{r+a}{a^3} \exp\left[-\frac{r}{a}\right] > 0$$

Therefore, according to COP-3 the above model is a permissible covariance model in R^3 and, of course, the same applies in R^2 and in R^1.

8.4 Certain Classes of Isotropic Correlation Functions

Let $\rho_x(r)$ be an isotropic correlation function. Three important classes of correlation functions will be considered:

(a) The class I_n of isotropic, in general, correlation functions in R^n.
(b) The class $I_{n,O}$ of isotropic correlation functions in R^n that are everywhere continuous except, perhaps, at the origin.
(c) The class $I_{n,c}$ of isotropic correlation functions in R^n, which are continuous everywhere. Clearly, $I_{n,c} \subset I_{n,O} \subset I_n$.

By means of the above classification, a number of interesting properties can be derived (Matern, 1960; Schoenberg, 1938; Veneziano, 1980).

Property 1: Assume that $\rho_x(r) \in I_n$; then,

$$\rho_x(r) \geq -\frac{1}{n} \tag{22}$$

for all r. If $\rho_x(r) \in I_{n,O}$, then

$$\rho_x(r) \geq \operatorname*{Inf}_u \left[\alpha!\left(\frac{2}{u}\right)^{\alpha} J_\alpha(u) \right] \tag{23}$$

where $\alpha = (n-2)/2$, and $J_\alpha(u)$ is the Bessel function of the first kind.

Property 2: If $\rho_x(r) \in I_n$ with $n > 1$, and $\rho_x(r^*) = 1$ for some $r^* > 0$, then $\rho_x(r) = 1$ for all r.

Property 3: Let $\rho_x(r) \in I_{n,c}$ with $n > 1$; then it can be written as

$$\rho_x(r) = a + (1-a)\rho_x^*(r) \tag{24}$$

where $0 \leq a \leq 1$, and $\rho_x^*(r)$ is a continuous correlation function such that $\lim_{r \to \infty} \rho_x^*(r) = 0$. If $\rho_x(r) \in I_{n,O}$, then $\rho_x(r)$ can be decomposed into three parts; more precisely, a delta function component is added to the right hand-side of Eq. (24).

Property 4: This is the so-called *Schoenberg's conjecture*. The class I_n-$I_{n,O}$ is empty for all $n > 1$.

Example 1: For $n = 2$, 3 and 4, Eq. (23) gives $\rho_x(r) \geq -0.403$, $\rho_x(r) \geq -0.218$ and $\rho_x(r) \geq -0.133$, respectively.

9. Scales of Spatial Correlation

The concept of a scale of spatial correlation provides a measure of the extent of spatial correlations (long versus short correlations in space, etc.). First, assume that we are dealing with isotropic SRF. Then, interesting scales of spatial correlation are introduced as follows:

Definition 1: The *correlation radius* r_c, is defined as

$$r_c = \frac{1}{c_x(0)} \int_0^{\infty} c_x(r)\, dr \tag{1}$$

This is the distance over which significant correlations prevail. At distance r_c the value of the covariance is approximately 50% of the value of the corresponding variance. In time series analysis, the counterpart of the correlation radius is called *correlation time.*

Example 1: For the Gaussian covariance, $r_c = (\sqrt{\pi}/2)a$; for the exponential covariance, $r_c = a$; and for the spherical covariance, $r_c = 3a/8$.

Definition 2: The *range* ε is another scale of spatial correlation defined as the distance beyond which the covariance can be considered approximately equal to zero. Practically, the range ε is the distance for which the value of the covariance is less or equal to approximately 5% of the variance.

Example 2: For the Gaussian covariance, $\varepsilon = \sqrt{3}\, a$; for the exponential covariance, $\varepsilon = 3a$; and for the spherical covariance, $\varepsilon = a$. On the basis of the values of the correlation radius r_c and the range ε we conclude that the exponential covariance has the longest correlations and the spherical the shortest.

With anisotropic but homogeneous covariances, both scales of spatial correlation, r_c and ε, vary with direction in space. Consider, for example, the anisotropic Gaussian covariance of Eq. (31), Section 7 above. The coefficients a_1, a_2, and a_3 characterize the scales of spatial correlation along the directions h_1, h_2, and h_3, respectively. With covariances of the form of Eq. (30), Section 7, the coefficient a characterizes the correlation radius r_c in the direction perpendicular to the plane defined by $c_1h_1 + c_sh_2 + c_3h_3 = 0$. On that plane, as well as on all planes parallel to it, spatial correlation occurs up to infinity.

10. Relationships between the Spatial and the Frequency Domains—The Uncertainty Principle

We saw in previous sections that the covariance function and the spectral density function form a Fourier transform pair. As a consequence, an isotropic covariance $c_x(r)$ is uniquely determined by means of the spectral density $C_x(\omega)$, and vice versa. Another noticeable feature is the inverse relationship between the *widths* of the two functions. More precisely, a wide $c_x(r)$ (which implies a long correlated SRF) corresponds to a narrow $C_x(\omega)$; conversely, a narrow $c_x(r)$ (short correlated SRF) corresponds to a wide $C_x(\omega)$.

This relationship between $c_x(r)$ and $C_x(\omega)$ can be expressed quantitatively by means of the so-called *uncertainty principle* (the term has been borrowed from the famous Heisenberg principle of quantum mechanics; Messiah, 1965). The distance r can be viewed as a random variable with probability density

$$f_r(r) = \frac{c_x(r)}{2r_c c_x(0)}$$

where r_c is the correlation radius defined by Eq. (1), Section 9 above. In R^1, for example, the mean and the variance of the random variable r are given by

$$m_r = \frac{1}{2\pi C_x(0)} \int_{R^1} r c_x(r)\, dr \tag{1}$$

and

$$\sigma_r^2 = \frac{1}{2\pi C_x(0)} \int_{R^1} (r - m_r)^2 c_x(r)\, dr \tag{2}$$

respectively. Similarly, ω can be considered as a random variable too with probability density $f_\omega(\omega) = C_x(\omega)/c_x(0)$. Its mean and variance are, respectively,

$$m_\omega = \frac{1}{c_x(0)} \int_{R^1} \omega C_x(\omega)\, d\omega \tag{3}$$

and

$$\sigma_\omega^2 = \frac{1}{c_x(0)} \int_{R^1} (\omega - m_\omega)^2 C_x(\omega)\, d\omega \tag{4}$$

On the basis of Eqs. (1) through (4), and by using some well-known Fourier transform properties, it can be shown that

$$\sigma_r \sigma_\omega \geq 1 \tag{5}$$

which is the aforementioned uncertainty principle. The σ_r and σ_ω are considered as the *uncertainties* (*widths*) of the covariance and the spectral density, respectively. The equality in (5) occurs when the covariance function $c_x(r)$ is Gaussian (see example below).

Example 1: In R^1, the Gaussian covariance and its spectral density are

$$c_x(r) = c_x(0) \exp\left[-\frac{r^2}{a^2}\right]$$

and

$$C_x(\omega) = \frac{c_x(0)a}{(4\pi)^{1/2}} \exp\left[-\frac{a^2\omega^2}{4}\right]$$

respectively. The spatial uncertainty is $\sigma_r = a/\sqrt{2}$, and the frequency uncertainty is $\sigma_\omega = \sqrt{2}/a$; hence, in this case $\sigma_r\sigma_\omega = 1$.

When the SRF of interest is anisotropic in $R^n(n \geq 2)$, the correlation radii differ along various directions in space. This means that there exist n uncertainty principles, namely,

$$\sigma_{r_i}\sigma_{w_i} \geq 1 \tag{6}$$

for all $i = 1, \ldots, n$; r_i are distances, and w_i are frequencies along directions $s_1, \ldots, s_n$ in space.

11. Spatial Random Fields with Homogeneous Increments

11.1 An Extension of the Hypothesis of Homogeneity

There are several important RP $X(s)$, like the Wiener or the Poisson processes, which, while nonstationary themselves (their variances increase continuously with s), the increments of $X(s)$ defined by

$$Y_h(s) = X(s+h) - X(s) \tag{1}$$

where h is a fixed real number, are stationary RP. Situations like this have led to relevant *extensions* of the hypotheses of stationarity (in R^1) and homogeneity (in R^n). These extensions were originally developed by Kolmogorov (1941), Ito (1954), Gel'fand (1955), Yaglom and Pinsker (1953), Yaglom (1955, 1957, 1986). In geosciences they have been elaborated in a more practical context by Gandin (1963) and others. In geostatistics, similar results have been introduced by Matheron (1965, 1973), David (1977), and Journel and Huijbregts (1978).

In view of these extensions, it seems quite appropriate to study in more detail the *incremental* SRF

$$Y_{\mathbf{h}}(\mathbf{s}) = X(\mathbf{s}+\mathbf{h}) - X(\mathbf{s}) \tag{2}$$

(It is assumed that $X(\mathbf{s})$ and $Y_{\mathbf{h}}(\mathbf{s})$ are continuous in the m.s.s.) Naturally, the statistical moments of order up to two of the $Y_{\mathbf{h}}(\mathbf{s})$ will be its mean

$$m_{Y_{\mathbf{h}}}(\mathbf{s}) = E[Y_{\mathbf{h}}(\mathbf{s})] = E[X(\mathbf{s}+\mathbf{h}) - X(\mathbf{s})] = m_x(\mathbf{s}+\mathbf{h}) - m_x(\mathbf{s}) \tag{3}$$

and its covariance function

$$\begin{aligned} c_{Y_{\mathbf{h}}}(\mathbf{s}_i, \mathbf{s}_j) &= E[Y_{\mathbf{h}_i}(\mathbf{s}_i)\, Y_{\mathbf{h}_j}(\mathbf{s}_j)] \\ &= E[[X(\mathbf{s}_i+\mathbf{h}_i) - X(\mathbf{s}_i)] \times [X(\mathbf{s}_j+\mathbf{h}_j) - X(\mathbf{s}_j)]] \\ &= D_x(\mathbf{s}_i, \mathbf{s}_j; \mathbf{h}_i, \mathbf{h}_j) \end{aligned} \tag{4}$$

Equations (3) and (4) establish duality principles between the SRF $X(\mathbf{s})$ and $Y_{\mathbf{h}}(\mathbf{s})$. The function $D_x(\mathbf{s}_i, \mathbf{s}_j; \mathbf{h}_i, \mathbf{h}_j)$ is called the *structure function* of the SRF $X(\mathbf{s})$ and is related to the covariance function by

$$\begin{aligned} D_x(\mathbf{s}_i, \mathbf{s}_j; \mathbf{h}_i, \mathbf{h}_j) = {} & c_x(\mathbf{s}_i+\mathbf{h}_i, \mathbf{s}_j+\mathbf{h}_j) - c_x(\mathbf{s}_i, \mathbf{s}_j+\mathbf{h}_j) \\ & - c_x(\mathbf{s}_i+\mathbf{h}_i, \mathbf{s}_j) + c_x(\mathbf{s}_i, \mathbf{s}_j) \end{aligned} \tag{5}$$

Let us define the function

$$\gamma_x(\mathbf{s}_i, \mathbf{s}_j; \mathbf{h}_i, \mathbf{h}_j) = \tfrac{1}{2} D_x(\mathbf{s}_i, \mathbf{s}_j; \mathbf{h}_i, \mathbf{h}_j) \tag{6}$$

If we assume $\mathbf{s}_i = \mathbf{s}_j = \mathbf{s}$ and $\mathbf{h}_i = \mathbf{h}_j = \mathbf{h}$, Eq. (6) implies

$$\gamma_x(\mathbf{s}, \mathbf{h}) = \tfrac{1}{2} E[X(\mathbf{s}+\mathbf{h}) - X(\mathbf{s})]^2 \tag{7}$$

which is the usual form of the *geostatistical semivariogram* of $X(\mathbf{s})$, where a constant mean value is assumed. If the mean is not constant the operator $E[\cdot]^2$ must be replaced by $\mathrm{Var}[\cdot]$. After some manipulations we find that

$$\begin{aligned} 2\gamma_x(\mathbf{s}_i, \mathbf{s}_j; \mathbf{h}_i, \mathbf{h}_j) = {} & \gamma_x(\mathbf{s}_i+\mathbf{h}_i, \mathbf{s}_j) + \gamma_x(\mathbf{s}_i, \mathbf{s}_j+\mathbf{h}_j) \\ & - \gamma_x(\mathbf{s}_i+\mathbf{h}_i, \mathbf{s}_j+\mathbf{h}_j) - \gamma_x(\mathbf{s}_i, \mathbf{s}_j) \end{aligned} \tag{8}$$

Hence, the general form (6) of the semivariogram is uniquely determined by the simpler form (7), which is the form to be used herein. Next, by further exploring the duality principles between $X(\mathbf{s})$ and $Y_{\mathbf{h}}(\mathbf{s})$ we establish the definition below.

Definition 1: An SRF $X(\mathbf{s})$ will be called an *SRF with homogeneous increments* $Y_{\mathbf{h}}(\mathbf{s})$, given by Eq. (2) above, if the mean value and the semivariogram depend only on the vector difference $\mathbf{h}$, that is,

$$m_{Y_{\mathbf{h}}}(\mathbf{s}) = E[Y_{\mathbf{h}}(\mathbf{s})] = m_x(\mathbf{h}) \tag{9}$$

and

$$\gamma_x(\mathbf{s}, \mathbf{h}) = \tfrac{1}{2} E[Y_{\mathbf{h}}(\mathbf{s})]^2 = \tfrac{1}{2} c_{Y_{\mathbf{h}}}(\mathbf{O}) = \gamma_x(\mathbf{h}) \tag{10}$$

for any vector difference $\mathbf{h}$.

By definition the $\gamma_x(\mathbf{h})$ is an even, nonnegative function such that $\gamma_x(\mathbf{O}) = 0$. However, we shall see in subsequent sections that it is possible that $\gamma_x(\mathbf{h}) \neq 0$ as $\mathbf{h} \rightarrow \mathbf{O}$, and then we talk about the already mentioned nugget-effect phenomenon. On the strength of Definition 1, Eq. (8) yields the interesting formula

$$\begin{aligned} 2\gamma_x(\mathbf{h}_i, \mathbf{h}_j) &= E[Y_{\mathbf{h}_i}(\mathbf{s}) Y_{\mathbf{h}_j}(\mathbf{s})] \\ &= E\{[X(\mathbf{s}+\mathbf{h}_i) - X(\mathbf{s})][X(\mathbf{s}+\mathbf{h}_j) - X(\mathbf{s})]\} \\ &= \gamma_x(\mathbf{h}_i) + \gamma_x(\mathbf{h}_j) - \gamma_x(\mathbf{h}_i - \mathbf{h}_j) \end{aligned} \tag{11}$$

Remark 1: In the case of a homogeneous SRF $X(\mathbf{s})$

$$m_x(\mathbf{h}) = 0 \tag{12}$$

and

$$\gamma_x(\mathbf{h}) = c_x(\mathbf{O}) - c_x(\mathbf{h}) \tag{13}$$

for all $\mathbf{h}$ (see Fig. 2.4).

Furthermore, Eq. (11) becomes

$$2\gamma_x(\mathbf{h}_i, \mathbf{h}_j) = c_x(\mathbf{O}) + c_x(\mathbf{h}_i - \mathbf{h}_j) - c_x(\mathbf{h}_i) - c_x(\mathbf{h}_j) \tag{14}$$

for all $\mathbf{h}_i$, $\mathbf{h}_j$.

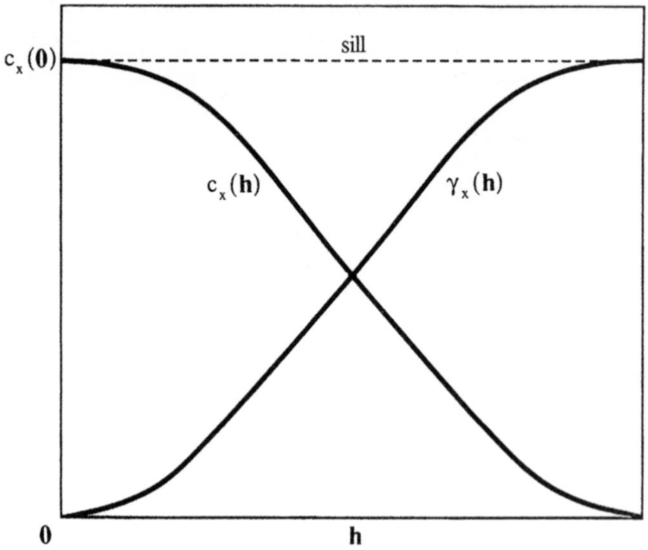

Figure 2.4 Spatial semivariogram $\gamma_x(\mathbf{h})$ and covariance $c_x(\mathbf{h})$ functions for a homogeneous SRF

Proposition 1: Let $X(\mathbf{s})$ be a nonhomogeneous SRF with homogeneous increments. Then

$$m_x(\mathbf{h}_i+\mathbf{h}_j)=m_x(\mathbf{h}_i)+m_x(\mathbf{h}_j) \tag{15}$$

and

$$\gamma_x(\mathbf{h}_i+\mathbf{h}_j)\leq \gamma_x(\mathbf{h}_i)+\gamma_x(\mathbf{h}_j)+2\sqrt{\gamma_x(\mathbf{h}_i)\gamma_x(\mathbf{h}_j)} \tag{16}$$

for all $\mathbf{h}_i$, $\mathbf{h}_j$.

Proof: Equation (15) is an immediate consequence of Eq. (9). To prove Eq. (16), let

$$E[X(\mathbf{s}+\mathbf{h}_i+\mathbf{h}_j)-X(\mathbf{s}+\mathbf{h}_j)]^2=2\gamma_x(\mathbf{h}_i)$$
$$E[X(\mathbf{s}+\mathbf{h}_j)-X(\mathbf{s})]^2=2\gamma_x(\mathbf{h}_j)$$

and

$$\begin{aligned}E\{[X(\mathbf{s}+\mathbf{h}_i+\mathbf{h}_j)-X(\mathbf{s}+\mathbf{h}_j)][X(\mathbf{s}+\mathbf{h}_j)-X(\mathbf{s})]\}\\ =\tfrac{1}{2}\{E[X(\mathbf{s}+\mathbf{h}_i+\mathbf{h}_j)-X(\mathbf{s})]^2\\ -E[X(\mathbf{s}+\mathbf{h}_i+\mathbf{h}_j)-X(\mathbf{s}+\mathbf{h}_j)]^2-E[X(\mathbf{s}+\mathbf{h}_j)-X(\mathbf{s})]^2\}\\ =\gamma_x(\mathbf{h}_i+\mathbf{h}_j)-\gamma_x(\mathbf{h}_i)-\gamma_x(\mathbf{h}_j)\end{aligned}$$

Then by applying Schwartz's inequality we obtain

$$\gamma_x(\mathbf{h}_i+\mathbf{h}_j)-\gamma_x(\mathbf{h}_i)-\gamma_x(\mathbf{h}_j)\leq\sqrt{2\gamma_x(\mathbf{h}_i)}\sqrt{2\gamma_x(\mathbf{h}_j)}$$

that is, Eq. (16). □

The $m_x(\mathbf{h})$ is continuous in $\mathbf{h}$ and the linearity condition (15) implies

$$m_x(\mathbf{h})=E[X(\mathbf{s}+\mathbf{h})-X(\mathbf{s})]=\mathbf{a}\cdot\mathbf{h} \tag{17}$$

where $\mathbf{a}$ is a vector coefficient in R^n. In geostatistics the $m_x(\mathbf{h})$ is called the *drift* of the SRF $X(\mathbf{s})$. If $\mathbf{a}=\mathbf{O}$ we say that the SRF has no drift.

Herein, unless stated otherwise, an SRF $X(\mathbf{s})$ will be considered as an SRF with homogeneous increments if Eq. (10) still holds true and, in addition, $m_x(\mathbf{h})=0$ is valid instead of Eq. (9) for any $\mathbf{h}\in R^n$. Equivalently, if the SRF $X(\mathbf{s})$ is differentiable, it will be termed a nonhomogeneous SRF with homogeneous increments if all its partial derivatives

$$Y_i(\mathbf{s})=\frac{\partial X(\mathbf{s})}{\partial s_i} \tag{18}$$

$i=1,2,\ldots,n$ are zero-mean homogeneous SRF. It is important to stress the fact that while the SRF $Y_{\mathbf{h}}(\mathbf{s})$ always exist, the SRF $Y_i(\mathbf{s})$ exist only when the $X(\mathbf{s})$ is differentiable. Moreover, it is obvious that any homogeneous SRF $X(\mathbf{s})$ is an SRF with homogeneous increments.

Corollary 1: It is valid that

$$\gamma_x(2^m \mathbf{h}) \leq 4^m \gamma_x(\mathbf{h}) \tag{19}$$

and for any $\mathbf{h}^* > \mathbf{0}$ there exists a constant b such that

$$\frac{\gamma_x(\mathbf{h})}{\mathbf{h}^2} \leq b \tag{20}$$

for all $\mathbf{h} > \mathbf{h}^*$.

Proof: Let in Eq. (16) $\mathbf{h}_i = \mathbf{h}_j = \mathbf{h}$, then $\gamma_x(2\mathbf{h}) \leq 4\gamma_x(\mathbf{h})$ and by induction we obtain Eq. (19). Furthermore by substituting $\mathbf{h}$ for $\mathbf{h}/2^m$ in (19) we find

$$\frac{\gamma_x(\mathbf{h})}{\mathbf{h}^2} \leq \frac{\gamma_x(\mathbf{h}/2^m)}{(\mathbf{h}/2^m)^2}$$

for any m. But any $\mathbf{h} > \mathbf{h}^*$ satisfies $\mathbf{h} \in [2^m \mathbf{h}^*, 2^{m+1}\mathbf{h}^*]$, which implies $(\mathbf{h}/2^m) \in [\mathbf{h}^*, 2\mathbf{h}^*]$. Hence, by letting b be the maximum of $\gamma_x(\mathbf{h})/\mathbf{h}^2$ in the interval $[\mathbf{h}^*, 2\mathbf{h}^*]$, we obtain Eq. (20). □

Example 1: Let $X(s)$ be a nonstationary RP with stationary increments

$$Y_h(s) = \Delta_s^1 X(s) = X(s+h) - X(s) \tag{21}$$

The corresponding covariance of the increments $Y_h(s)$ will be

$$\begin{aligned} c_{Y_h}(s-s') &= E[\Delta_s^1 X(s)\, \Delta_{s'}^1 X(s')] \\ &= c_x(s+h, s'+h) - c_x(s+h, s') - c_x(s, s'+h) + c_x(s, s') \\ &= \Delta_{s'}^1 \Delta_s^1 c_x(s, s') \qquad (22) \\ &= \gamma_x(s-s'+h) + \gamma_x(s-s'-h) - 2\gamma_x(s-s') \qquad (23) \end{aligned}$$

Thus, if the nonstationary RP and its stationary increments are related by Eq. (21), the corresponding covariances are related by Eq. (22). Now let $s = s'$; then Eq. (22) can be written

$$\begin{aligned} c_{Y_h}(0) &= \Delta_s^1 \Delta_s^1 c_x(s, s) \\ &= c_x(s+h, s+h) - c_x(s, s+h) - c_x(s+h, s) + c_x(s, s) = 2\gamma_x(h) \end{aligned} \tag{24}$$

These considerations imply that the semivariogram $\gamma_x(h)$ characterizes completely the correlation structure of the nonstationary $X(s)$. In conclusion, if we know $c_x(s, s')$ we can determine $c_{Y_h}(s-s')$ and $\gamma_x(h)$, but the converse is not generally true. In the special case of a stationary $X(s)$, Eq. (24) is written

$$c_{Y_h}(0) = 2c_x(0) - 2c_x(h) = 2\gamma_x(h)$$

or

$$\gamma_x(h) = c_x(0) - c_x(h) = \tfrac{1}{2} c_{Y_h}(0) \tag{25}$$

Remark 2: Based on the results of Section 5 above, it follows that if in Example 1 the derivative $\partial^2 c_x(s, s')/(\partial s\, \partial s')$ exists and is finite, it is the covariance of the m.s. derivative of $X(s)$, $Y(s) = dX(s)/ds$. Then

$$\frac{\partial^2}{\partial s\, \partial s'} c_x(s, s') = E[Y(s)\, Y(s')] = c_Y(s - s') \tag{26}$$

Note that $Y_h(s)$ always exist, while the $Y(s)$ exists only when $X(s)$ is differentiable.

11.2 Spectral Characteristics of Spatial Random Fields with Homogeneous Increments

The gist here is the existence of spectral moments corresponding to the semivariogram $\gamma_x(\mathbf{h})$ of the SRF $X(\mathbf{s})$ in R^n.

Assume that the $X(\mathbf{s})$ is a differentiable SRF with homogeneous increments. According to definition (18) the $Y_i(\mathbf{s}) = \partial X(\mathbf{s})/\partial s_i$, $i = 1, 2, \ldots, n$ are homogeneous SRF. If $c_{Y_i}(\mathbf{h})$ is the covariance of $Y_i(\mathbf{s})$, the theory of Section 7 indicates that

$$c_{Y_i}(\mathbf{h}) = \int_{R^n} \exp[i(\mathbf{w} \cdot \mathbf{h})]\, dQ_{Y_i}(\mathbf{w})$$

where $Q_{Y_i}(\mathbf{w})$ are positive summable measures in R^n without atom at origin.

Define the covariance

$$c_Y(\mathbf{h}) = \sum_{i=1}^{n} c_{Y_i}(\mathbf{h}) = \int_{R^n} \exp[i(\mathbf{w} \cdot \mathbf{h})]\, dQ_Y(\mathbf{w})$$

where

$$Q_Y(\mathbf{w}) = \sum_{i=1}^{n} Q_{Y_i}(\mathbf{w})$$

is also a positive summable measure without atom at origin. Note that

$$c_Y(\mathbf{h} = \mathbf{s} - \mathbf{s}') = \sum_{i=1}^{n} c_{Y_i}(\mathbf{h}) = \sum_{i=1}^{n} \frac{\partial^2 c_x(\mathbf{s}, \mathbf{s}')}{\partial s_i\, \partial s_i'}$$

The above considerations imply that the $\gamma_x(\mathbf{h})$ satisfies the relation

$$\nabla_{\mathbf{h}}^2 \gamma_x(\mathbf{h}) = c_Y(\mathbf{h}) \tag{27}$$

where

$$\nabla_{\mathbf{h}}^2 = \sum_{i=1}^{n} \frac{\partial^2}{\partial h_i^2}$$

is the Laplacian operator, and it admits the spectral representation

$$\gamma_x(\mathbf{h}) = \int_{R^n} \frac{[1-\cos(\mathbf{w}\cdot\mathbf{h})]}{\mathbf{w}^2}\, dQ_Y(\mathbf{w}) + a$$
$$= \int_{R^n} [1-\cos(\mathbf{w}\cdot\mathbf{h})]\, dQ_x(\mathbf{w}) + a \tag{28}$$

where a is an arbitrary constant. Since

$$|1-\cos(\mathbf{w}\cdot\mathbf{h})| \leq \frac{(\mathbf{w}\cdot\mathbf{h})^2}{2}$$

it follows that

$$|\gamma_x(\mathbf{h})| \leq \tfrac{1}{2}\int_{R^n} dQ_Y(\mathbf{w})|\mathbf{h}|^2 = \alpha|\mathbf{h}|^2$$

where $\alpha < \infty$, or by applying the dominated convergence theorem and using the fact that an SRF with homogeneous increments of first order has also homogeneous increments of higher orders,

$$\lim_{|\mathbf{h}|\to\infty} \frac{\gamma_x(\mathbf{h})}{|\mathbf{h}|^2} = 0 \tag{29}$$

which is the existence condition for the integrals (28) (Eq. (29) is a special case of Eq. (60), Section 3 of Chapter 3). If the SRF $X(\mathbf{s})$ is not differentiable and possesses a drift, the $\gamma_x(\mathbf{h})$ admits the spectral representation

$$\gamma_x(\mathbf{h}) = \int_{R^n} \frac{1-\cos(\mathbf{w}\cdot\mathbf{h})}{\mathbf{w}^2}\, dQ_Y(\mathbf{w}) + b\mathbf{h}^2 + a$$
$$= \int_{R^n} [1-\cos(\mathbf{w}\cdot\mathbf{h})]\, dQ_X(\mathbf{w}) + b\mathbf{h}^2 + a \tag{30}$$

or

$$\gamma_x(\mathbf{h}) = \int_{R^n} [1-\cos(\mathbf{w}\cdot\mathbf{h})] K_x(\mathbf{w})\, d\mathbf{w} + b\mathbf{h}^2 + a \tag{31}$$

where a and b are arbitrary coefficients; the measure $Q_Y(\mathbf{w})$ now satisfies

$$\int_{R^n} \frac{dQ_Y(\mathbf{w})}{1+\mathbf{w}^2} = \int_{R^n} \frac{\mathbf{w}^2\, dQ_x(\mathbf{w})}{1+\mathbf{w}^2} < \infty \tag{32}$$

and

$$C_Y(\mathbf{w}) = \mathbf{w}^2 K_x(\mathbf{w}) \tag{33}$$

where the $K_x(\mathbf{w})$ and $C_Y(\mathbf{w})$ are the corresponding spectral functions [assuming that the $Q_x(\mathbf{w})$ and $Q_Y(\mathbf{w})$ are differentiable].

Conversely, every function $\gamma_x(\mathbf{h})$ of the form (30) or (31) is the semivariogram of an SRF $X(\mathbf{s})$ with homogeneous increments. The inversion of Eq. (31) yields (for simplicity we assume that $a=b=0$)

$$\mathbf{w}^2 K_x(\mathbf{w}) = \frac{1}{(2\pi)^n} \int_{R^n} \mathbf{w} \cdot \nabla \gamma_x(\mathbf{h}) \sin(\mathbf{w} \cdot \mathbf{h}) \, d\mathbf{h} \tag{34}$$

To illustrate some important aspects of the preceding analysis, a few examples are discussed below.

Example 2: Let $X(s)$ be an m.s. differentiable, nonstationary RP with stationary increments. That is, if

$$Y(s) = \frac{dX(s)}{ds} \tag{35}$$

exist, then the $Y(s)$ is a stationary RP. Equation (35) implies

$$\begin{aligned} X(s) - X(0) &= \int_0^s Y(t)\, dt = \int_{R^1} \left[\int_0^s \exp[itw]\, dt \right] d\aleph_Y(w) \\ &= \int_{R^1} \frac{\exp[isw]-1}{iw}\, d\aleph_Y(w) \end{aligned} \tag{36}$$

where

$$Y(t) = \int_{R^1} \exp[itw]\, d\aleph_Y(w)$$

is the spectral representation of $Y(s)$, or

$$X(s) = \int_{R^1} \frac{\exp[isw]-1}{iw}\, d\aleph_Y(w) + X(0) \tag{37}$$

On the other hand, we can also write

$$X(s) = \int_{R^1} \exp[isw]\, d\aleph_x(w)$$

which assigns the measure

$$d\aleph_x(w) = \frac{d\aleph_Y(w)}{iw}$$

to the RP $X(s)$. Let

$$c_Y(h) = \int_{R^1} \exp[ihw]\, dQ_Y(w)$$

Then,

$$\gamma_x(h) = \tfrac{1}{2}E[X(h) - X(0)]^2$$

$$= \frac{1}{2}\int_{R^1} \frac{\{\exp[ihw]-1\}\{\exp[-ihw]-1\}}{(iw)^2} E|d\aleph_Y(w)|^2$$

$$= \int_{R^1} \frac{1-\cos wh}{w^2} dQ_Y(w)$$

where

$$dQ_Y(w) = w^2\, dQ_x(w)$$

These measures are summable, viz.,

$$\int_{R^1} dQ_Y(w) = \int_{R^1} w^2\, dQ_x(w) < \infty$$

Hence,

$$\gamma_x(h) = \int_{R^1} (1 - \cos wh)\, dQ_x(w)$$

If the functions $Q_Y(w)$ and $Q_x(w)$ are differentiable it holds true that

$$\int_{R^1} C_Y(w)\, dw = \int_{R^1} w^2 K_x(w)\, dw < \infty$$

with $C_Y(w) = w^2 K_x(w)$, where $C_Y(w)$ and $K_x(w)$ are the corresponding spectral functions. [Note that the last equation relates the spectral density function of the stationary random process $Y(s)$ with the spectral density of the nonstationary process $X(s)$.] In this case

$$\gamma_x(h) = \int_{R^1} (1 - \cos wh) K_x(w)\, dw \tag{38}$$

For the existence of the integral in Eq. (38) it is required that

$$\lim_{|w|\to 0} [w^3 K_x(w)] = 0$$

and

$$\lim_{|w|\to\infty} [wK_x(w)] = 0$$

It is interesting to compare the last two requirements with those for the existence of the covariance (12) of Section 7 ($n = 1$), viz.,

$$\lim_{|w|\to 0} [wC_x(w)] = 0 \qquad \text{and} \qquad \lim_{|w|\to\infty} [wC_x(w)] = 0$$

We notice immediately that the requirements imposed on K_x and C_x for the convergence at infinity are the same, but the requirement for the convergence at zero is less severe for the semivariogram (38) than for the covariance (12), Section 7. Hence, there may be situations where the semivariogram exists and the covariance does not. By inverting Eq. (38) we obtain

$$K_x(w) = \frac{1}{\pi w} \int_0^\infty \sin(wh) \frac{d\gamma_x(h)}{dh}\, dh \tag{39}$$

and

$$K_x(w) = \frac{1}{\pi w^2} \int_0^\infty \cos(wh) \frac{d^2\gamma_x(h)}{dh^2}\, dh \tag{40}$$

For the integral (39) to exist we must have

$$\lim_{h\to 0} \left[h^2 \frac{d\gamma_x(h)}{dh} \right] = 0 \qquad \text{and} \quad \lim_{h\to\infty} \left[\frac{d\gamma_x(h)}{dh} \right] = 0$$

On the other hand, the existence of the integral (40) requires that

$$\lim_{h\to 0} \left[h \frac{d^2\gamma_x(h)}{dh^2} \right] = 0 \qquad \text{and} \quad \lim_{h\to\infty} \left[\frac{d^2\gamma_x(h)}{dh^2} \right] = 0$$

Now assume that the point $w = 0$ is a jump discontinuity of $\aleph_Y(w)$; that is,

$$\lim_{\varepsilon\to 0} [\aleph_Y(\varepsilon) - \aleph_Y(-\varepsilon)] = a \neq 0$$

where a is a random variable, in general. Then we find that

$$\lim_{\varepsilon\to 0} [Q_Y(\varepsilon) - Q_Y(-\varepsilon)] = E|a|^2 > 0$$

In other words, the point $w = 0$ is a jump discontinuity of $Q_Y(w)$, as well; and

$$\begin{aligned} X(s) &= \int_{R^1} \exp[isw]\, d\aleph_x(w) + a \lim_{w\to 0} \frac{\exp[isw] - 1}{iw} \\ &= \int_{R^1} \exp[isw]\, d\aleph_x(w) + as \end{aligned}$$

and

$$\begin{aligned} \gamma_x(h) &= \int_{R^1} (1 - \cos wh) K_x(w)\, dw + 2a^2 \lim_{w\to 0} \frac{1 - \cos wh}{w^2} \\ &= \int_{R^1} (1 - \cos wh) K_x(w)\, dw + a^2 h^2 \end{aligned} \tag{41}$$

Since $|\cos z - 1| \leq \frac{1}{2}z^2$, Eq. (38) gives

$$|\gamma_x(h)| \leq bh^2 \tag{42}$$

where

$$b = \frac{1}{2}\int_{R^1} w^2 K_x(w)\, dw$$

Note that by applying the dominated convergence theorem Eq. (42) yields [see also Eq. (29)]

$$\lim_{|h|\to\infty} \frac{\gamma_x(h)}{|h|^2} = 0$$

Now, taking the Laplace transform of (42) we obtain

$$\int_{R^1} \left(\frac{1}{t} - \frac{t}{t^2+w^2}\right) K_x(w)\, dw \leq b\frac{2}{t^3}$$

or

$$\int_{R^1} \frac{t^2 w^2}{t^2+w^2} K_x(w)\, dw \leq 2b$$

and since

$$\lim_{t\to\infty} \frac{t^2}{t^2+w^2} = 1$$

we get

$$\int_{R^1} w^2 K_x(w)\, dw < \infty$$

Consequently, Eq. (42) implies

$$\int_{R^1} \frac{w^2}{1+w^2} K_x(w)\, dw < \infty \tag{43}$$

An interesting point here is that the condition (43) for the existence of the spectral representation (41) is equivalent to the expression (42), which involves the semivariogram function itself. Spectral representations like the ones above may be obtained for any nondifferentiable RP. In the latter case, however, the spectral measure may be not summable (i.e., the integral

$$\int_{R^1} C_Y(w)\, dw = \int_{R^1} w^2 K_x(w)\, dw$$

may become infinite) and one needs the condition

$$\int_0^a w^2 K_x(w)\, dw + \int_a^\infty K_x(w)\, dw < \infty \tag{44}$$

for any $a > 0$. As a consequence, unlike the case of homogeneous RP in which it is required that

$$\int_{-\infty}^{\infty} C_x(w)\, dw < \infty$$

here it is necessary only that near zero frequency the integral

$$\int w^2 K_x(w)\, dw < \infty$$

while at infinity

$$\int K_x(w)\, dw < \infty$$

Within the framework of spectral analysis one may also define the so-called spectral variogram function.

Definition 3: The *spectral semivariogram function* $\Gamma_x(\mathbf{w})$ of a nonhomogeneous, in general, but with homogeneous increments SRF $X(\mathbf{s})$ is defined as the Fourier transform of the semivariogram function $\gamma_x(\mathbf{h})$. For nonhomogeneous SRF, $\Gamma_x(\mathbf{w})$ will exist only in the sense of generalized functions (Schwartz, 1950–51; see, also, tables of Fourier transforms of generalized functions, Gelfand and Shilov, 1964).

For homogeneous SRF, the following proposition can be proven (Christakos, 1984b).

Proposition 2: If $X(\mathbf{s})$ is a homogeneous SRF with variance $c_x(\mathbf{O})$, the relationships below are valid.

$$\Gamma_x(\mathbf{w}) = c_x(\mathbf{O})\, \delta(\mathbf{w}) - C_x(\mathbf{w}) \tag{45}$$

and

$$\int_{R^n} \Gamma_x(\mathbf{w})\, d\mathbf{w} = \gamma_x(\mathbf{O}) \tag{46}$$

Example 3: Let $X(s)$ be a differentiable and nonstationary RP with stationary increments and a semivariogram of the form

$$\gamma_x(h) = ar^m \tag{47}$$

where $r = |h|$, $a > 0$, and $0 \leq m < 2$. From Eqs. (27) and (33) and by using tables of Fourier transforms of generalized functions (e.g., Gelfand and Shilov, 1964) we get

$$K_x(w) = \frac{1}{w^2}\, \text{F.T.}\left[\frac{d^2\gamma_x(r)}{dr^2}\right]$$

$$= a\frac{G(m+1)}{2\pi} \sin\frac{m\pi}{2} |w|^{-m-1} \tag{48}$$

where G is the gamma function. Moreover, using the same tables it can be found that

$$\Gamma_x(w) = -K_x(w) \tag{49}$$

Assuming now that the RP $X(s)$ is Gaussian we get

$$X(s) = \int_0^s \frac{(s-u)^{(m-1)/2}}{G\left(\frac{m+1}{2}\right)}\, dW_0(s) \tag{50}$$

where $W_0(s)$ is a Wiener process with

$$\gamma_{w_0}(r) = ar \tag{51}$$

(For definition of the Wiener process, see Chapter 3.)

Example 4: In Fig. 2.5 the unidimensional semivariogram functions

$$\gamma_x(r) = \sqrt[3]{\left(\frac{r}{c}\right)^2} \tag{52}$$

(special case of Eq. (47)) and

$$\gamma_x(r) = c_x(0)\left[1 - \frac{\sqrt[3]{\frac{4r}{c}}}{G\left(\frac{1}{3}\right)} K_{1/3}\left(\frac{r}{c}\right)\right] \tag{53}$$

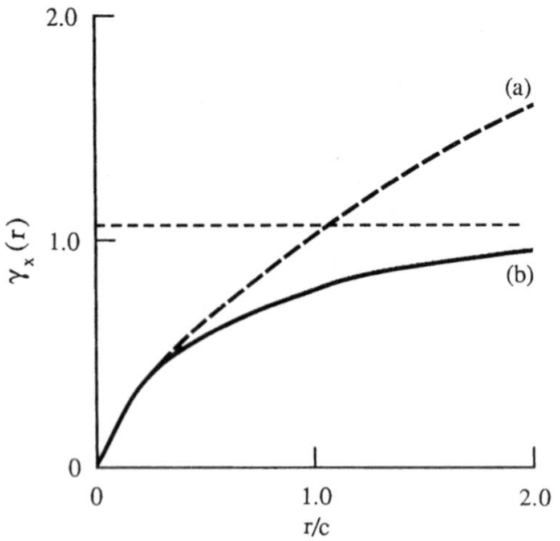

Figure 2.5 The semivariograms of (a) Eq. (52) and (b) Eq. (53)

where $K_{1/3}$ is a Bessel function of the second kind (Gradshteyn and Ryzhik, 1965), are plotted; their spectral densities are plotted in Fig. 2.6. Note that the second semivariogram is a special case of

$$\gamma_x(r) = c_x(0)\left[1 - \frac{2^{1-m}\left(\frac{r}{c}\right)^m}{G(m)} K_m\left(\frac{r}{c}\right)\right] \tag{54}$$

where $m > -1/2$.

In most cases we will consider *isotropic* semivariogram and spectral semivariogram functions. (Obviously, this does not necessarily imply isotropic SRF.) Working with n-dimensional spherical coordinates, Eq. (31) yields (assume $a = b = 0$)

$$\gamma_x(r) = \int_0^\infty [1 - \Lambda_n(\omega r)] K_x(\omega)\, d\omega \tag{55}$$

where

$$\Lambda_n(x) = \frac{2^{(n-2)/2} G(n/2)}{x^{(n-2)/2}} J_{(n-2)/2}(x)$$

and

$$\int_0^\infty \frac{\omega^2 K_x(\omega)\, d\omega}{1+\omega^2} < \infty, \qquad K_x(\omega) \geq 0$$

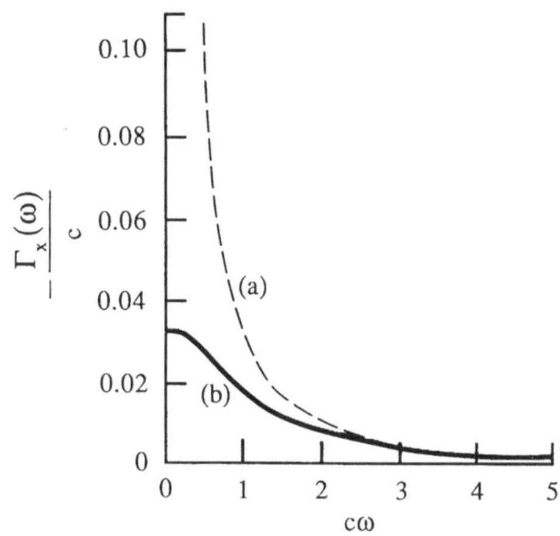

Figure 2.6 The spectral densities corresponding to the semivariograms of (a) Eq. (52) and (b) Eq. (53)

Finally, for isotropic SRF, using Eq. (4) of Section 8 and Eq. (13) above one obtains the following useful expressions:

$$\gamma_x(r) = 2\pi \int_0^\infty [1 - J_0(\omega r)] C_x(\omega)\omega \, d\omega \quad \text{in} \quad R^2 \tag{56}$$

and

$$\gamma_x(r) = 4\pi \int_0^\infty \left[1 - \frac{\sin(\omega r)}{\omega r}\right] C_x(\omega)\omega^2 \, d\omega \quad \text{in} \quad R^3 \tag{57}$$

where $r = |\mathbf{h}|$ and $\omega = |\mathbf{w}|$.

11.3 Criteria of Permissibility for Spatial Semivariograms

In the case that the SRF considered is homogeneous, the nonnegative-definiteness condition that a semivariogram $\gamma_x(\mathbf{h})$ should satisfy is readily obtained from COP-1 of Section 7.3 by simply making use of the relation introduced by Eq. (13) above. More precisely, assuming isotropic functions, the spectral domain equivalent of Eq. (13) is (see also Proposition 2 above)

$$\Gamma_x(\omega) = c_x(0)\, \delta(\omega) - C_x(\omega)$$

On the basis of this observation, the following *fourth criterion of permissibility* (*COP-4*) is, essentially, the equivalent of the COP-1 in terms of semivariogram functions: Let $X(\mathbf{s})$ be a homogeneous SRF in R^n. A continuous and isotropic function $\gamma_x(r)$ is a permissible semivariogram function if and only if

$$-\Gamma_x(\omega) \geq 0 \tag{58}$$

for all $\omega \geq 0^+$ [actually, $c_x(0)\, \delta(\omega) - \Gamma_x(\omega) \geq 0$ for all ω].

The semivariogram $\gamma_x(\mathbf{h})$ of a nonhomogeneous SRF $X(\mathbf{s})$ with homogeneous increments $Y_\mathbf{h}(\mathbf{s})$ is a so-called *conditionally nonnegative-definite* function; that is,

$$-\sum_{i=1}^{m} \sum_{j=1}^{m} q_i q_j \gamma_x(\mathbf{s}_i - \mathbf{s}_j) \geq 0 \tag{59}$$

for all nonnegative integers m, all points $\mathbf{s}_i, \mathbf{s}_j \in R^n$, and all real or complex numbers $q_1, q_2, \ldots, q_m$ satisfying the condition

$$\sum_{i=1}^{m} q_i = 0 \tag{60}$$

The conditional nonnegative-definiteness condition (59) is a consequence of the usual nonnegative-definiteness property of the ordinary covariance $c_x(\mathbf{s}_i, \mathbf{s}_j)$, in combination with the fact that the variance of the homogeneous increments $Y_\mathbf{h}(\mathbf{s})$ is finite.

Just as with the usual case of nonnegative-definiteness, the conditional nonnegative-definiteness property (59) is not usually applicable in practical situations. This is not a problem, however, since one can use the following equivalent but much more comprehensive *fifth criterion of permissibility* (*COP-5*; Christakos, 1984b): Let $X(\mathbf{s})$ be a nonhomogeneous SRF with homogeneous increments in R^n. A continuous and isotropic function $\gamma_x(r)$ is called a permissible semivariogram function if and only if it decreases slower than r^2 as $r \to \infty$; that is,

$$\lim_{r \to \infty} \frac{\gamma_x(r)}{r^2} = 0 \tag{61}$$

and the corresponding spectral function $K_x(\omega)$ exists (in the sense of Schwartz, 1950–51), includes no atom at origin, and is such that the $\omega^2 K_x(\omega)$ is nonnegative and bounded on R^n. In other words,

$$\omega^2 K_x(\omega) \geq 0 \tag{62}$$

for all ω.

Example 5: Consider Example 3 above. Clearly,

$$\lim_{r \to \infty} \frac{\gamma_x(r)}{r^2} = 0 \qquad \text{and} \qquad \omega^2 K_x(\omega) \geq 0$$

Therefore the function (47) is a permissible semivariogram model.

Example 6: The function $\gamma_x(r) = r[1 + \cos(ar)]$ is not a permissible semivariogram function in R^1, for the

$$\omega^2 K_x(\omega) = 2 + \frac{\omega^2}{(\omega - a)^2} + \frac{\omega^2}{(\omega + a)^2}$$

is not a bounded (finite) measure. The fact that the present function is not a permissible semivariogram model can be verified by using Eq. (19) above for $m = 1$, $a = 1$, and $r = \pi$. As Eq. (19) shows, a semivariogram should satisfy $\gamma_x(2\pi) < 4\gamma_x(\pi)$. However, this is not the case for the present function, which gives $\gamma_x(2\pi) = 4\pi$ and $\gamma_x(\pi) = 0$.

COP-5 is equivalent to the following criterion, which can be established with the help of representation (55): Any function of the form (55) where $K_x(\omega)$ satisfies the associated restrictions is a permissible semivariogram function.

11.4 The Geometry of Spatial Random Fields with Homogeneous Increments

We consider *continuous* semivariogram functions. By analogy with the results of Section 4 above, the following propositions can be proven.

Proposition 3: A continuous in the m.s.s. SRF implies a continuous semivariogram function. Conversely, a continuous semivariogram function implies an m.s. continuous SRF. Also, an SRF is differentiable in the m.s.s. if and only if the corresponding semivariogram is twice differentiable.

Proof: Indeed, if the SRF $X(\mathbf{s})$ is m.s. continuous at point $\mathbf{s}$, then

$$\lim_{\mathbf{h}\to\mathbf{O}} E[X(\mathbf{s}+\mathbf{h})-X(\mathbf{s})]^2=0$$

$$\Rightarrow \lim_{\mathbf{h}\to\mathbf{O}} 2\gamma_x(\mathbf{s}+\mathbf{h},\mathbf{s})=0=2\gamma_x(\mathbf{O}) \tag{63}$$

From the last equation it is obvious that the converse is also true. The proof for the differentiability can be established in an analogous setting. □

Let $X(\mathbf{s})$ be a nonhomogeneous, in general, SRF but with homogeneous increments. Assuming that the $X(\mathbf{s})$ is m.s. differentiable; that is, the

$$X_{(i)}(\mathbf{s})=\frac{\partial X(\mathbf{s})}{\partial s_i}=\underset{h\to 0}{\text{l.i.m.}}\,\frac{X(\mathbf{s}+h\boldsymbol{\varepsilon}_i)-X(\mathbf{s})}{h}$$

exist for all $i=1,2,\ldots,n$. By definition they will be homogeneous and such that (see Section 4 above)

$$\lim_{h\to 0} E\left[\frac{X(\mathbf{s}+h\boldsymbol{\varepsilon}_i)-X(\mathbf{s})}{h}-X_{(i)}(\mathbf{s})\right]^2=0$$

or

$$\lim_{h\to 0}\left[\frac{2}{h^2}\gamma_x(h\boldsymbol{\varepsilon}_i)\right]+c_{x_{(i)}}(\mathbf{O})-2\lim_{h\to 0} E\left[\frac{X(\mathbf{s}+h\boldsymbol{\varepsilon}_i)-X(\mathbf{s})}{\mathrm{h}}X_{(i)}(\mathbf{s})\right]=0$$

or

$$\lim_{h\to 0}\left[\frac{2}{h^2}\gamma_x(h\boldsymbol{\varepsilon}_i)\right]=c_{x_{(i)}}(\mathbf{O})$$

The last equation implies

$$\gamma_x(h\boldsymbol{\varepsilon}_i)=\frac{h^2}{2}c_{x_{(i)}}(\mathbf{O}) \qquad \text{as} \quad h\to 0 \tag{64}$$

and

$$\left.\frac{\partial\gamma_x(h\boldsymbol{\varepsilon}_i)}{\partial h}\right|_{h=0}=0 \tag{65}$$

for all $i=1,2,\ldots,n$. The generalization of the analysis above leads to the following useful corollary.

Corollary 2: The SRF $X(\mathbf{s})$ with homogeneous increments is m.s. differentiable of order ν (that is, the $\partial^\nu X(\mathbf{s})/\partial s_i^\nu$ exist in the m.s.s. for all $i=1,2,\ldots,n$) if and only if

$$\left.\frac{\partial^{2\nu-1}\gamma_x(\mathbf{h})}{\partial h_i^{2\nu-1}}\right|_{\mathbf{h}=\mathbf{O}}=0 \tag{66}$$

for all $i=1,2,\ldots,n$.

Analogous expressions may be derived in the frequency domain: Taking Eq. (31) into consideration we find that the $\partial^\nu X(\mathbf{s})/\partial s_i^\nu$ exist in the m.s.s. for all $i=1,2,\ldots,n$ if and only if

$$\left.\frac{\partial^{2\nu}\gamma_x(\mathbf{h})}{\partial h_i^{2\nu}}\right|_{\mathbf{h}=\mathbf{O}}=(-1)^{\nu+1}\int_{R^n} w_i^{2\nu}K_x(\mathbf{w})\,d\mathbf{w}<\infty \tag{67}$$

for all $i=1,2,\ldots,n$. Similar expressions may be obtained in terms of the spectral semivariogram functions $\Gamma_x(\mathbf{w})$.

Example 7: Let $X(s)$ be a random process with stationary increments, and assume that the derivative

$$Y(s)=\frac{dX(s)}{ds}=\underset{h\to 0}{\text{l.i.m.}}\frac{X(s+h)-X(s)}{h}$$

exists in the m.s.s. Then, by definition it will be stationary and such that

$$\lim_{h\to 0}E\left[\frac{X(s+h)-X(s)}{h}-Y(s)\right]^2=0$$

Consequently,

$$\lim_{h\to 0}\left[\frac{2}{h^2}\gamma_x(h)\right]+c_x(0)-2\lim_{h\to 0}E\left[\frac{X(s+h)-X(s)}{h}Y(s)\right]=0$$

or

$$\lim_{h\to 0}\left[\frac{2}{h^2}\gamma_x(h)\right]+c_Y(0)-2c_Y(0)=0$$

or

$$\gamma_x(h)=\frac{c_Y(0)}{2}h^2 \quad \text{as} \quad h\to 0 \tag{68}$$

The last equation implies that the derivative of the semivariogram must be zero at origin, viz.

$$\left.\frac{d\gamma_x(h)}{dh}\right|_{h=0}=0$$

Consider the Taylor expansion of a semivariogram around the origin in R^n, viz.

$$\gamma_x(\mathbf{h}) = \gamma_x(\mathbf{O}) + \sum_{i_1=1}^{n} a_{i_1} h_{i_1} + \frac{1}{2!} \sum_{i_1, i_2=1}^{n} a_{i_1 i_2} h_{i_1} h_{i_2} + \cdots$$

$$+ \frac{1}{(q-1)!} \sum_{i_1, \ldots, i_{q-1}=1}^{n} a_{i_1 \ldots i_{q-1}} h_{i_1} \cdots h_{i_{q-1}} + \cdots \tag{69}$$

where

$$a_{i_1 \ldots i_{q-1}} = \left. \frac{\partial^{q-1} \gamma_x(\mathbf{h})}{\partial h_{i_1} \cdots \partial h_{i_{q-1}}} \right|_{\mathbf{h}=\mathbf{O}}$$

Expansion (69), when possible, may provide useful information regarding the geometry of $X(\mathbf{s})$. In particular, (i) since by definition the $X(\mathbf{s})$ is m.s. continuous if and only if the $\gamma_x(\mathbf{h})$ is continuous at origin, Eq. (69) shows that this can happen if and only if $\gamma_x(\mathbf{O}) = 0$; (ii) according to Corollary 2, the $X(\mathbf{s})$ is differentiable in the m.s.s. if and only if $a_{i_1} = 0$ in expansion (69), etc.

Remark 3: It is instructive to examine the expansion (69) in the special case of an isotropic semivariogram $\gamma_x(r)$, $r = |\mathbf{h}|$. Taking Eq. (69) into consideration, the latter can, in general, be expanded around the origin as (Matheron, 1965)

$$\gamma_x(r) = \gamma_x(0) + \sum_k a_{2k} r^{2k} + \sum_\lambda b_\lambda r^\lambda + \sum_\mu c_{2\mu} r^{2\mu} \log(r) \tag{70}$$

where a_{2k}, b_λ, and c_{2k} are suitable coefficients, $k, \mu = 1, 2, \ldots$, and λ are real numbers different from even integers. Let λ^* and μ^* be the lower λ and μ values, respectively. If $\lambda^* < 2\mu^*$, the ν-th derivative of $X(\mathbf{s})$ exists in the m.s.s. if and only if $\lambda > 2\nu$; while if $2\mu^* < \lambda^*$, only the $\mu^* - 1$ m.s. derivative of $X(\mathbf{s})$ exists (and not the μ^* m.s. derivative). Let us consider a few examples.

Example 8: Assume that a nonhomogeneous SRF $X(\mathbf{s})$ with homogeneous increments is characterized by the semivariogram

$$\gamma_x(\mathbf{h}) = |\mathbf{h}|^m \tag{71}$$

where $0 < m < 2$. The semivariogram (71) is continuous but it is not twice differentiable. Moreover,

$$\left. \frac{d\gamma_x(\mathbf{h})}{d\mathbf{h}} \right|_{\mathbf{h}=\mathbf{O}} \neq 0$$

Therefore the underlying SRF $X(\mathbf{s})$ is m.s. continuous but it is not m.s. differentiable.

Example 9: The isotropic semivariogram

$$\gamma_x(r) = 1 - \exp\left[-\frac{r}{a}\right] \tag{72}$$

where $r = |\mathbf{h}| \in R^n$ and $a > 0$ is continuous at $r = 0$, but

$$\left.\frac{d\gamma_x(r)}{dr}\right|_{r=0} \neq 0$$

Hence the corresponding SRF $X(\mathbf{s})$ is m.s. continuous but not m.s. differentiable.

Finally, let us consider the sample function continuity and differentiability of a nonhomogeneous, in general, SRF in terms of its semivariogram function: By using previous results, we immediately find that if

$$\gamma_x(\mathbf{h}) \leq \frac{c|\boldsymbol{h}|^{2n}}{|\log|\mathbf{h}||^{1+\beta}} \tag{73}$$

where c is a positive constant and $\beta > 2$, then the SRF $X(\mathbf{s})$, $\mathbf{s} \in R^n$ is *a.s. continuous* over any compact set $C \subset R^n$. Sufficient conditions for the sample function continuity of the m.s. derivatives of $X(\mathbf{s})$ can be derived by replacing $\gamma_x(\mathbf{h})$ in (73) by the semivariogram function of the derivative. In the case of a real-valued *Gaussian* SRF $X(\mathbf{s})$, Eq. (73) leads to the following simpler condition

$$\gamma_x(\mathbf{h}) \leq \frac{c}{|\log|\mathbf{h}||^{1+\beta}} \tag{74}$$

for all $\mathbf{h} \in C$, where now $\beta > 0$.

11.5 Vector Spatial Random Fields with Homogeneous Increments

Several of the results above can be extended to vector SRF. However, one should be aware of certain differences.

Let $\mathbf{X}(\mathbf{s}) = [X_1(\mathbf{s}), X_2(\mathbf{s}), \ldots, X_k(\mathbf{s})]^{\mathrm{T}}$ be a vector SRF. The corresponding semivariogram matrix is

$$\mathbf{G_X} = [\gamma_{x_p x_{p'}}(\mathbf{h})] \tag{75}$$

where the component cross-semivariograms are given by

$$\gamma_{x_p x_{p'}}(\mathbf{h}) = \tfrac{1}{2}E\{[X_p(\mathbf{s}+\mathbf{h}) - X_p(\mathbf{s})][X_{p'}(\mathbf{s}+\mathbf{h}) - X_{p'}(\mathbf{s})]\} \tag{76}$$

Similar relations may be derived in terms of spatial cross-covariances. If the vector SRF is homogeneous, the cross-covariances are related to the cross-semivariograms by

$$2\gamma_{x_p x_{p'}}(\mathbf{h}) = 2c_{x_p x_{p'}}(\mathbf{O}) - c_{x_p x_{p'}}(\mathbf{h}) - c_{x_{p'} x_p}(\mathbf{h}) \tag{77}$$

Notice that the cross-covariance is in general nonsymmetric, and the $\gamma_{x_p x_{p'}}(\mathbf{h})$ and $c_{x_p x_{p'}}(\mathbf{h})$ are no longer equivalent spatial correlation functions. Moreover, while $\gamma_{x_p x_p}(\mathbf{h}) \geq 0$, the $\gamma_{x_p x_{p'}}(\mathbf{h})$, $(p \neq p')$ may take negative values, as well.

Under certain conditions, the isotropic semivariograms $\gamma_{x_p x_{p'}}(r)$ may be represented by

$$\gamma_{x_p x_{p'}}(r) = \int_0^\infty [1 - \Lambda_n(\omega r)] K_{x_p x_{p'}}(\omega)\, d\omega \tag{78}$$

where $\Lambda_n(x)$ is defined in Eq. (55), and $K_{x_p x_{p'}}(\omega)$ are cross-spectral density functions such that the trace

$$K_x(\omega) = \sum_{p=1}^{k} K_{x_p x_p}(\omega)$$

of the nonnegative-definite matrix

$$\mathbf{\Phi_X} = [K_{x_p x_{p'}}(\omega)]$$

satisfies

$$\int_0^\infty \frac{\omega^2 K_x(\omega)\, d\omega}{1+\omega^2} < \infty, \qquad K_x(\omega) \geq 0 \tag{79}$$

The above representation leads to the following *sixth criterion of permissibility* (*COP-6*): For $\mathbf{G_X}$ of Eq. (75) to be a matrix of permissible semivariograms, the matrix $\mathbf{\Phi_X}$ must be nonnegative-definite and its trace must satisfy the conditions of Eq. (79).

12. On the Ergodicity Hypotheses of Spatial Random Fields

We have seen above that ergodicity is a working hypothesis (the term *working* is used here to distinguish ergodicity from auxiliary hypotheses, such as homogeneity, etc.) that is needed because in most circumstances in nature we have only one available sequence of measurements. Then, an SRF $X(\mathbf{s})$ will be called an *ergodic SRF* if its mean and/or covariance (semivariogram) coincide with the corresponding *spatial averages* calculated over the single available realization. In this case we are talking about ergodicity in the mean and/or the covariance (semivariogram), respectively. Obviously, for ergodicity to hold it is necessary that one of the auxiliary hypotheses holds too. For example, ergodicity in the mean requires that the SRF has a constant mean (i.e., it is homogeneous in the mean).

More specifically the following types of ergodicity can be established:

(i) *Ergodicity in the mean* requires that the underlying SRF $X(\mathbf{s})$ has constant mean $E[X(\mathbf{s})] = m_x$, and that

$$\lim_{L\to\infty} m_x^* = \lim_{L\to\infty} \frac{1}{(2L)^n} \int_{U_L} X(\mathbf{s})\, d\mathbf{s} = m_x \tag{1}$$

where the domain of integration U_L is the n-dimensional cube $U_L = \{\mathbf{s}: |s_i| < L,\ i = 1, 2, \ldots, n\}$; m_x^* is a random variable depending on U_L but not on $\mathbf{s}$. Clearly

$$E[m_x^*] = m_x \tag{2}$$

Among the various forms of stochastic convergence discussed in Section 4, we will consider first the m.s. convergence. Then, some results will be presented with regard to the a.s. convergence, as well. In the case of m.s. convergence, Eq. (1) leads to the condition

$$\lim_{L\to\infty} E[m_x^* - m_x]^2 = \lim_{L\to\infty} \frac{1}{(2L)^{2n}} \int_{U_L} \int_{U_L} c_x(\mathbf{s}, \mathbf{s}')\, d\mathbf{s}\, d\mathbf{s}' = 0 \tag{3}$$

If the SRF $X(\mathbf{s})$ is homogeneous, Eq. (3) becomes

$$\lim_{L\to\infty} \frac{1}{(2L)^n} \int_{U_L} c_x(\mathbf{h})\, d\mathbf{h} = 0 \tag{4}$$

In the spectral domain, condition (4) is equivalent to the requirement that the spectral density $C_x(\mathbf{w})$ be continuous at the origin $\mathbf{w} = \mathbf{O}$. Equation (4) can be also written in terms of the semivariogram, namely,

$$\lim_{L\to\infty} \frac{1}{(2L)^n} \int_{U_L} \gamma_x(\mathbf{h})\, d\mathbf{h} = c_x(\mathbf{O}) \tag{5}$$

A mathematically sufficient but not necessary condition for (5) to hold is $\lim_{u\to\infty} \gamma_x(u\mathbf{e}) = c_x(\mathbf{O})$ for all $\mathbf{e} \neq \mathbf{O}$. In the case of isotropic SRF the sufficient condition reduces to

$$\lim_{r\to\infty} \gamma_x(r) = c_x(0) \tag{6}$$

Example 1: The semivariogram

$$\gamma_x(r) = c_x(0)\{1 - \exp[-ar]\}$$

satisfies (6) and, hence, it corresponds to an SRF that is ergodic in the mean. On the contrary the semivariogram

$$\gamma_x(r) = \frac{c_x(0)}{2}\{1 - \exp[-ar]\}$$

does not satisfy (5) and (6) and the underlying SRF is not ergodic in the mean.

(ii) *Ergodicity in the covariance* or *in the semivariogram* involves higher-order moments. The former also requires that the SRF is homogeneous, while for the latter it is necessary that the SRF has homogeneous increments. More specifically, a homogeneous SRF is ergodic in the covariance if

$$\lim_{L\to\infty} c_x^*(\mathbf{h}) = \lim_{L\to\infty} \frac{1}{(2L)^n} \int_{U_L} [X(\mathbf{s}) - m_x^*] \times [X(\mathbf{s}+\mathbf{h}) - m_x^*]\, d\mathbf{s} = c_x(\mathbf{h}) \tag{7}$$

Note that $E[c_x^*(\mathbf{h})] = c_x(\mathbf{h})$. In the light of m.s. convergence, definition (7) leads to the condition

$$\lim_{L\to\infty} E[c_x^*(\mathbf{h}) - c_x(\mathbf{h})]^2 = 0 \tag{8}$$

Equation (8) may also be expressed in the more tractable form

$$\lim_{L\to\infty} \frac{1}{(2L)^n} \int_{U_L} c_z(\mathbf{h})\, d\mathbf{h} = 0 \tag{9}$$

where

$$c_z(\mathbf{h}) = E\{[Z_{\mathbf{s}'}(\mathbf{s}+\mathbf{h}) - c_x(\mathbf{s}')][Z_{\mathbf{s}'}(\mathbf{s}) - c_x(\mathbf{s}')]\}$$

is the covariance of

$$Z_{\mathbf{s}'}(\mathbf{h}) = [X(\mathbf{h}+\mathbf{s}') - m_x^*][X(\mathbf{h}) - m_x^*]$$

In the special case of Gaussian homogeneous SRF, condition (9) reduces to

$$\lim_{L\to\infty} \frac{1}{(2L)^n} \int_{U_L} c_x(\mathbf{h})^2\, d\mathbf{h} = 0 \tag{10}$$

In the spectral domain, condition (10) is equivalent to the requirement that the spectral density $C_x(\mathbf{w})$ is continuous for all $\mathbf{w} \in R^n$. Again, Eq. (10) can be expressed in terms of the semivariogram, viz.,

$$\lim_{L\to\infty} \frac{1}{(2L)^n} \int_{U_L} \gamma_x(\mathbf{h})[2c_x(\mathbf{O}) - \gamma_x(\mathbf{h})]\, d\mathbf{h} = c_x^2(\mathbf{O}) \tag{11}$$

Now consider the case where the SRF $X(\mathbf{s})$ is nonhomogeneous but its increments $Y_{\mathbf{h}}(\mathbf{s}) = X(\mathbf{s}+\mathbf{h}) - X(\mathbf{s})$ are Gaussian homogeneous SRF with covariance

$$\begin{aligned} c_{Y_{\mathbf{h}}}(\mathbf{u} = \mathbf{s}-\mathbf{s}') &= E[Y_{\mathbf{h}}(\mathbf{s})\, Y_{\mathbf{h}}(\mathbf{s}')] \\ &= E\{[X(\mathbf{s}+\mathbf{h}) - X(\mathbf{s})][X(\mathbf{s}'+\mathbf{h}) - X(\mathbf{s}')]\} \\ &= \gamma_x(\mathbf{u}+\mathbf{h}) + \gamma_x(\mathbf{u}-\mathbf{h}) - 2\gamma_x(\mathbf{u}) \end{aligned}$$

By applying condition (11) we immediately find that the condition for ergodicity in the semivariogram can be written

$$\lim_{L\to\infty} \frac{1}{(2L)^n} \int_{U_L} [\gamma_x(\mathbf{u}+\mathbf{h}) + \gamma_x(\mathbf{u}-\mathbf{h}) - 2\gamma_x(\mathbf{u})]^2 \, d\mathbf{u} = 0 \tag{12}$$

An equivalent condition may be derived as follows: Consider the spatial integral

$$\gamma_x^*(\mathbf{h}) = \frac{1}{2(2L)^n} \int_{U_L} [X(\mathbf{s}+\mathbf{h}) - X(\mathbf{s})]^2 \, d\mathbf{s} \tag{13}$$

where $E[\gamma_x^*(\mathbf{h})] = \gamma_x(\mathbf{h})$. Ergodicity in the semivariogram implies

$$\begin{aligned} &\lim_{L\to\infty} E[\gamma_x^*(\mathbf{h}) - \gamma_x(\mathbf{h})]^2 \\ &\quad = \lim_{L\to\infty} \frac{1}{4(2L)^{2n}} \int_{U_L} \int_{U_L} E\{[X(\mathbf{s}+\mathbf{h}) - X(\mathbf{s})] \\ &\quad\quad \times [X(\mathbf{s}'+\mathbf{h}) - X(\mathbf{s}')]\}^2 \, d\mathbf{s} \, d\mathbf{s}' - \gamma_x(\mathbf{h})^2 = 0 \end{aligned} \tag{14}$$

Since the $Y_{\mathbf{h}}(\mathbf{s}) = X(\mathbf{s}+\mathbf{h}) - X(\mathbf{s})$ are Gaussian homogeneous SRF,

$$\begin{aligned} E[Y_{\mathbf{h}}^2(\mathbf{s})\, Y_{\mathbf{h}}^2(\mathbf{s}')] &= E[Y_{\mathbf{h}}^2(\mathbf{s})] E[Y_{\mathbf{h}}^2(\mathbf{s}')] + 2\{E[Y_{\mathbf{h}}(\mathbf{s})\, Y_{\mathbf{h}}(\mathbf{s}')]\}^2 \\ &= 4\gamma_x^2(\mathbf{h}) + 2[\gamma_x(\mathbf{s}-\mathbf{s}'+\mathbf{h}) + \gamma_x(\mathbf{s}-\mathbf{s}'-\mathbf{h}) - 2\gamma_x(\mathbf{s}-\mathbf{s}')]^2 \end{aligned}$$

and Eq. (14) yields

$$\lim_{L\to\infty} \frac{1}{2(2L)^{2n}} \int_{U_L} \int_{U_L} \{\gamma_x(\mathbf{s}-\mathbf{s}'+\mathbf{h}) + \gamma_x(\mathbf{s}-\mathbf{s}'-\mathbf{h}) - 2\gamma_x(\mathbf{s}-\mathbf{s}')\}^2 \, d\mathbf{s} \, d\mathbf{s}' = 0 \tag{15}$$

Example 2: The semivariogram $\gamma_x(\mathbf{h}) = a|\mathbf{h}|$ satisfies the above conditions and, hence, the underlying SRF is ergodic in the semivariogram.

An interesting aspect of ergodicity emerges in cases where we are only interested in the behavior of the covariance or the semivariogram near the origin, $\mathbf{h} \to \mathbf{O}$. It then suffices to verify some sort of ergodicity only near the origin. Practically, this means that the SRF is only ergodic in volumes significantly small as compared with the scales of spatial correlation. This sort of ergodicity has been assigned the names of *quasi-ergodicity* (Monin and Yaglom, 1971), and *microergodicity* (Matheron, 1978). For quasi-ergodicity to make sense, the volume V must satisfy certain conditions. For example, in the three-dimensional isotropic case, we usually require that

$$r_c \ll \sqrt[3]{V} \ll L \tag{16}$$

where r_c is the correlation radius and L is the transverse dimension of the volume V. Here microergodicity will be examined with the help of the results of Section 11.

Consider a nonhomogeneous SRF $X(\mathbf{s})$ that possesses homogeneous Gaussian increments with an isotropic semivariogram $\gamma_x(r)$, $r=|\mathbf{h}|$. According to the preceding theory, the $\gamma_x(r)$ expansion near the origin established by Eq. (70) of Section 11 is characterized by the term with the lowest degree; that is, $\gamma_x(r)=cr^{\lambda^*}$ as $r\to 0$, where $\lambda^*\leq 2$. In order that the actual semivariogram $\gamma_x(r)$ can be determined accurately in terms of the local integral $\gamma_x^*(r)$, which is a random variable, it is required that the m.s.

$$\underset{r\to 0}{\text{l.i.m.}}\frac{\gamma_x^*(r)}{\gamma_x(r)}=1 \tag{17}$$

After some manipulations the m.s. convergence condition (17) entails

$$\lim_{r\to 0} E\left[\frac{\gamma_x^*(r)}{\gamma_x(r)}-1\right]^2=\lim_{r\to 0}\text{Var}\left[\frac{\gamma_x^*(r)}{\gamma_x(r)}\right]$$
$$=\lim_{r\to 0}\frac{\text{Var}[\gamma_x^*(r)]}{\gamma_x^2(r)}=0$$

For small r,

$$\frac{\text{Var}[\gamma_x^*(r)]}{\gamma_x^2(r)}=\alpha r^{4-2\lambda^*}+\beta r^n$$

and, hence, Eq. (17) holds only if $\lambda^*<2$ and not for $\lambda^*=2$. The latter means that in order that microergodicity in the semivariogram be a sound assumption, the SRF must not be differentiable in the m.s.s.

Example 3: The semivariogram

$$\gamma_x(r)=c_0\left\{1-\exp\left[-\frac{r^2}{a^2}\right]\right\}$$

which corresponds to an m.s.-differentiable SRF is not microergodic, while the aforementioned semivariogram $\gamma_x(r)=ar^m$, $0<m<2$ is microergodic. This fact emphasizes the attractive properties of polynomial-type spatial correlation functions, regarding stochastic inferences of nonhomogeneous SRF. Such correlation functions will play an important role in subsequent investigations of this treatise.

To establish the ergodicity conditions above in the case of *a.s. convergence* it is convenient to assume that the domain U_L is an n-dimensional rectangular $\{\mathbf{s}: |s_i|<L_i, i=1,2,\ldots,n\}$ with $V=\prod_{i=1}^n s_i$. After some manipulations and assuming homogeneous SRF, the a.s. convergence corresponding

to Eq. (1) leads to the condition

$$\lim_{V\to\infty} \frac{2^n}{V} \int_0^{L_1} \int_0^{L_2} \cdots \int_0^{L_n} \left(1-\frac{h_1}{L_1}\right)\left(1-\frac{h_2}{L_2}\right)\cdots\left(1-\frac{h_n}{L_n}\right)$$
$$\times \gamma_x(h_1, h_2, \ldots, h_n)\, dh_1\, dh_2 \ldots dh_n = c_x(\mathbf{O}) \tag{18}$$

(Condition (18) also holds in the m.s. sense).

Example 4: The random process with a semivariogram of the form

$$\gamma_x(h) = c_x(0)\left\{1-\exp\left[-\frac{h}{a}\right]\right\}$$

satisfies condition (18) for it holds

$$\lim_{L\to\infty} \frac{2}{L} \int_0^{L} \left(1-\frac{h}{L}\right) \gamma_x(h)\, dh$$
$$= \lim_{L\to\infty} \frac{2c_x(0)}{L} \int_0^{L} \left(1-\frac{h}{L}\right)\left\{1-\exp\left[-\frac{h}{a}\right]\right\} dh = c_x(0)$$

Moreover, it is valid that $[\gamma_x^*(r)/r^\lambda] \xrightarrow{\text{a.s.}} b$ as $r \to 0$, where $b < \infty$ and $\lambda < 2$ (this is also true in the m.s. sense). Lastly, note that on the basis of Eq. (5) of Section 4, the m.s. and the a.s. convergence imply convergence in P and F as well.

Remark 1: It must be noted, however, that in many practical situations it may be difficult or even impossible to verify ergodicity by means of the above conditions. In these situations, while ergodicity may be a nonverifiable hypothesis, it can still be considered as a falsifiable hypothesis on the basis of the successes it leads to (in the sense of Chapter 1). This, in fact, explains the characterization "working hypothesis" given earlier. The matter will be discussed in more detail in Chapter 7.

13. Information and Entropy of Spatial Random Fields

Consider the SRF $X(\mathbf{s})$, and let $f_x(\chi_1, \ldots \chi_m)$ be the multivariate probability density function of the vector of random variables $\mathbf{X} = [x_1, \ldots, x_m]^T$ of $X(\mathbf{s})$, defined at points $\mathbf{s}_1, \ldots, \mathbf{s}_m$.

There are several definitions of the concept of information. Here, we will give the traditional definition (Shannon, 1948). Continuous random variables are considered, but similar definitions are valid for discrete random variables, as well.

Definition 1: The *information* contained in the vector of random variables $\mathbf{X}=[x_1,\ldots,x_m]^T$ about the SRF $X(\mathbf{s})$, is given by

$$\mathrm{Inf}[x_1,\ldots,x_m]=-\log[f_x(\chi_1,\ldots\chi_m)] \tag{1}$$

where the logarithm can be taken with arbitrary base $b>1$.

When the logarithm is taken to the base $b=2$, the unit of the entropy scale is called a "bit" (binary digit); when the natural logarithm to base e is taken, the unit is called a "nit." According to Eq. (1), the more informative is the random vector **X**, the less probable it is to occur.

Closely related to the concept of information is the notion of entropy, defined as follows.

Definition 2: The quantity

$$\begin{aligned}\varepsilon(\mathbf{X}) &= E\{\mathrm{Inf}[x_1,\ldots,x_m]\}=E\{-\log[f_x(\chi_1,\ldots,\chi_m)]\}\\ &= -\underbrace{\int\int\cdots\int}_{m\text{ times}}\log[f_x(\chi_1,\ldots,\chi_m)]f_x(\chi_1,\ldots,\chi_m)\,d\chi_1\ldots d\chi_m\end{aligned} \tag{2}$$

is called the *entropy* of the random vector **X**.

The $\varepsilon(\mathbf{X})$ provides a measure of the amount of uncertainty in the probability density $f_x(\chi_1,\ldots,\chi_m)$ *a priori*, before experimentation.

Example 1: The entropy of the Gaussian random vector **X** is given by

$$\varepsilon(\mathbf{X})=\log[(2\pi e)^m|\mathbf{C}_\mathbf{X}|]^{1/2} \tag{3}$$

Remark 1: Let $f_x(\chi)$ and $f_y(\psi)$ be the univariate probability densities of the SRF $X(\mathbf{s})$ and $Y(\mathbf{s})$, respectively. An interesting inequality can be established as follows.

$$\int_{R^1} f_x(\chi)\log\left[\frac{f_x(\chi)}{f_y(\psi)}\right]d\chi\geq 0 \tag{4}$$

The equality holds if and only if $f_x(\chi)\equiv f_y(\psi)$.

Probability densities are assigned by maximizing the entropy function (2), subject to constraints provided by the existing information and incorporated using Lagrange multipliers. This is the so-called *principle of maximum entropy* (Jaynes, 1957). Constraints usually concern statistical moments.

Example 2: The uniform probability density is the unique density possessing the maximum entropy among all the densities of continuous random variables with possible values within the same bounded region; the univariate exponential density possesses the maximum entropy among all the densities with the same mean value; and the univariate Gaussian density possesses the maximum entropy among all the densities with the same mean value and variance. Moreover, the multivariate Gaussian density possesses the maximum entropy among all the densities with the same covariance matrix.

Definition 3: The *conditional entropy* of a random variable x given that $y = \psi$ is given by

$$\begin{aligned} \varepsilon(x \mid y) &= E\{-\log[f_{x|y}(\chi \mid \psi)]\} \\ &= -\int_{R^1} \log[f_{x|y}(\chi|\psi)] f_{x|y}(\chi \mid \psi)\, d\chi \end{aligned} \tag{5}$$

The entropy of Eq. (5) is a random variable and, hence, one can define the *average conditional entropy* of x with respect to y as follows:

$$\begin{aligned} \overline{\varepsilon_y}(x) = E_y[\varepsilon(x \mid y)] &= \int_{R^1} \varepsilon(x \mid y) f_y(\psi)\, d\psi \\ &= -\int_{R^1}\int_{R^1} \log[f_{x|y}(\chi \mid \psi)] f_{x,y}(\chi, \psi)\, d\chi\, d\psi \end{aligned} \tag{6}$$

On the basis of the above definitions several interesting results can be derived. First, it holds true that

$$\varepsilon(x) \geq \overline{\varepsilon_y}(x) \tag{7}$$

(equality holds if and only if the random variables x and y are independent). An obvious extension of Eq. (7) is as follows:

$$\varepsilon(x) \geq \overline{\varepsilon_y}(x) \geq \overline{\varepsilon_{z,y}}(x) \geq \overline{\varepsilon_{w,z,y}}(x)$$

The entropy of x and y satisfies the relationships

$$\varepsilon(x, y) = \varepsilon(x) + \overline{\varepsilon_x}(y) = \varepsilon(y) + \overline{\varepsilon_y}(x) \tag{8}$$

and

$$\varepsilon(x, y) \leq \varepsilon(x) + \varepsilon(y) \tag{9}$$

(Again, equality holds if and only if the random variables x and y are independent.)

Definition 4: The amount of information on a random variable x in another variable y is defined by

$$\bar{I}_y(x) = \varepsilon(x) - \overline{\varepsilon_y}(x) \tag{10}$$

The physical meaning of Definition 4 is that while observing a natural process y one receives valuable information about another process x depending on y. This information is measured by means of Eq. (10), which expresses the intuitively justifiable concept that a process y provides information about another process x whenever it changes the probability law of x from its prior form $f_x(\chi)$ to the new, posterior form $f_{x|y}(\chi \mid \psi)$.

It is to be remarked that

$$\bar{I}_y(x) = \bar{I}_x(y) = \varepsilon(x) + \varepsilon(y) - \varepsilon(x, y) \tag{11}$$

In other words, the amount of information in y on x is equal to the amount of information in x on y. Hence, Eq. (11) measures the mutual information between the random variables x and y and is indifferent to which of the two is taken as source. Clearly,

$$\bar{I}_y(x) = \bar{I}_x(y) = \int_{R^1} \int_{R^1} \log\left[\frac{f_{x,y}(\chi, \psi)}{f_x(\chi) f_y(\psi)}\right] \times f_{x,y}(\chi, \psi)\, d\chi\, d\psi \geq 0 \tag{12}$$

where the equality holds if the x and y are independent random variables.

Remark 2: A problem with considerable practical consequences is as follows. Let $X(\mathbf{s})$ be an SRF modeling an observable natural process, and $Y(\mathbf{s})$ be an SRF modeling an unobservable process. Suppose that $Z(\mathbf{s}) = g[X(\mathbf{s})]$ is a transformation of $X(\mathbf{s})$. How does the amount of information about $Y(\mathbf{s})$ contained in $X(\mathbf{s})$ compare to that contained in $Z(\mathbf{s})$? With respect to this problem, it can be shown that

(a) No transformation $g[\cdot]$ can increase the amount of information about $Y(\mathbf{s})$.

(b) Transformations $g[\cdot]$ that are one-to-one mappings contain the same amount of information about $Y(\mathbf{s})$ as that contained in $X(\mathbf{s})$. The same happens with sufficient transformations (a transformation $g[\cdot]$ is called sufficient if for any other transformation $h[\cdot]$ such that the equations $g[\chi] = \zeta$ and $h[\chi] = \zeta$ have unique solution with respect to χ, the conditional density of $W(\mathbf{s}) = h[X(\mathbf{s})]$ at $Y(\mathbf{s}) = \psi$, $Z(\mathbf{s}) = \zeta$ is independent of ψ).

(c) Any other transformation leads to loss of information.

Interesting applications of the entropy function can be found in areas such as the modeling, estimation, and sampling of natural processes (see Chapters 7, 9, and 10).

On the basis of the analysis above, the so-called sysketogram function is defined in Section 9.3 of Chapter 10.

3

The Intrinsic Spatial Random Field Model

"All is flux, nothing is stationary."

Heracleitus

1. Introduction

This chapter is concerned with the most general case of nonhomogeneously distributed spatial processes, where certain of the spatial variability assumptions of Chapter 2 are not fulfilled in reality. Experimental data have shown, however, that such processes, while nonhomogeneous themselves, can be transformed to homogeneous spatial processes by means of some linear transformation. These experimental findings constitute the original and compelling motivation leading to the representation of spatial processes with complex nonhomogeneous characteristics by means of spatial random fields (SRF) with homogeneous increments of order ν or, equivalently, intrinsic SRF of order ν (ISRF-ν). These classes of SRF can be viewed as an extension of the theory of ordinary (nongeneralized) homogeneous SRF (see Yaglom and Pinsker, 1953; Yaglom, 1955; Pinsker, 1955). This view makes no reference to the mathematical theory of generalized (nonordinary) functions (or distributions, Schwartz, 1950–51) and is the one currently applied in geostatistical studies (Matheron, 1973). Geostatistics is also responsible for a number important results in the context of the correlation analysis of ISRF-ν.

The present treatise, however, favors another point of view. The ISRF-ν paradigm can arise within the context of the theory of random distributions or generalized random functions originally developed by Ito (1954), Gel'fand (1955), and Gel'fand and Vilenkin (1964). A unified theory of SRF will be developed that will include as special cases all classes of SRF considered in Chapter 2 as well as several other classes. The accommodation of the concept of generalized SRF (GSRF) strengthens the theoretical support of the ISRF model and further generalizes its results. In addition, it provides the means for extending the theory in the context of spatiotemporal random fields (see Chapter 5).

Aside from strengthening the theoretical background of the SRF model, practical considerations require a retreat to discrete notions of the ISRF concepts and definitions; this task is carried out by the second part of this chapter. There the results bear more particularly on a variety of issues: The power of the GSRF structure lies in its mathematical "smoothness" (in the mean square sense), which enables it to handle problems that cannot be handled by ordinary SRF; such problems are related to the practical significance of the mathematical notion of point SRF, the representation of complex, nonhomogeneously distributed natural processes, stochastic inferences of nonordinary correlation functions, etc. Also, under certain circumstances, the homogeneous increments of order ν constitute a filter that removes unimportant quantities and emphasizes the specific properties of interest, leading to a truer picture of the facts. On this view, every feature specification entered in the ISRF representation of a natural process is really an instruction that some particular spatial pattern is not to apply. ISRF representations comprise conditions on what type of nonhomogeneous spatial trends and correlations may be marked.

The GSRF capacity for capturing essential features of the natural process paves the way for establishing intriguing connections with stochastic differential equations representing, for example, flow in natural soil formations and pollutant transport in atmosphere, as well as stochastic difference equations modeling, for example, soil profiles.

Finally, it must be remarked that the results of this chapter get considerable play in the remainder of this book.

2. Generalized Spatial Random Fields

2.1 Definition and Basic Properties

Just as with the SRF of Chapter 2, the random fields to be considered in this chapter will be termed continuous-parameter or discrete-parameter according to whether the argument **s** takes discrete or continuous values. Let us commence with the notion of generalized SRF due to Ito (1954) and Gel'fand (1955). Let Q be some specified linear space of elements q and let $\mathscr{H}_{\mathscr{Q}} = L_2(\Omega, F, P)$ be the Hilbert space of all random variables $x(q)$ on Q: (i) endowed with the scalar product

$$(x(q_1), x(q_2)) = E[x(q_1)x(q_2)] = \iint \chi_1\chi_2 \, dF_x(\chi_1, \chi_2) \quad (1)$$

where $F_x(\chi_1, \chi_2)$ denotes the joint probability distribution of the random variables $x(q_1), x(q_2)$ with $\|x(q)\|^2 = E|x(q)|^2 < \infty$; and (ii) satisfying the linearity condition

$$x\left(\sum_{i=1}^{N} \lambda_i q_i\right) = \sum_{i=1}^{N} \lambda_i x(q_i) \tag{2}$$

for all $q_i \in Q$ and all (real or complex) numbers λ_i $(i = 1, 2, \ldots, N)$. The elements $q \in Q$ are in R^n; that is, $q = q(\mathbf{s})$, $\mathbf{s} \in R^n$. Various kinds of spaces Q may be suitable for the purpose of this study, such as the well-established Schwartz spaces (Schwartz, 1950–51): K, the space of infinite differentiable functions in R^n, which vanish, together with their derivatives, outside a finite support; and S, the space of infinitely differentiable functions, which, together with their derivatives of all orders, approach zero more rapidly than any power of $1/|\mathbf{s}|$ as $|\mathbf{s}| \to \infty$.

Definition 1: Let $L_2(\Omega, F, P)$ be the aforementioned Hilbert space of random variables in R^n. A *generalized SRF (GSRF)*, $X(q)$, is the random mapping

$$X: Q \to L_2(\Omega, F, P) \tag{3}$$

Herein it will be assumed that all GSRF considered are *continuous*; that is,

$$E|X(q_n) - X(q)|^2 \to 0 \qquad \text{when} \quad q_n \xrightarrow[n\to\infty]{} q \tag{4}$$

The $q_n \xrightarrow[n\to\infty]{} q$ means that all functions $q_n(\mathbf{s})$ and $q(\mathbf{s})$ vanish outside a compact support and all the partial derivatives of $q_n(\mathbf{s})$ converge to the corresponding partial derivatives of $q(\mathbf{s})$ on this support. The set of all continuous GSRF will be denoted by $\mathcal{G}$.

The *second-order* characteristics of the GSRF $X(q)$ include the

$$m_x(q) = E[X(q)] = \int \chi \, dF_x(\chi) \tag{5}$$

which is called the *mean value*, and the

$$c_x(q_1, q_2) = E[(X(q_1) - m_x(q_1))(X(q_2) - m_x(q_2))] \tag{6}$$

which is called the (*centered*) *covariance functional.* Both the mean and the covariance functional will be assumed to be real-valued and continuous in the sense that

$$m_x(q_n) \to m_x(q) \qquad \text{when} \quad q_n \xrightarrow[n\to\infty]{} q \tag{7}$$

and

$$c_x(q_n, q'_n) \to c_x(q, q') \qquad \text{when} \quad q_n \xrightarrow[n\to\infty]{} q,\ q'_n \xrightarrow[n\to\infty]{} q' \tag{8}$$

Notice that the functional $m_x(q)$ is linear, since

$$m_x\left(\sum_{i=1}^{N} \lambda_i q_i\right) = E\left[X\left(\sum_{i=1}^{N} \lambda_i q_i\right)\right] = E\left[\sum_{i=1}^{N} \lambda_i X(q_i)\right] = \sum_{i=1}^{N} \lambda_i m_x(q_i)$$

and therefore the $m_x(q)$ is a *generalized function* on Q. On the other hand, due to the linearity of $X(q)$, the covariance functional $c_x(q_1, q_2)$ is a *bilinear functional* on Q. Moreover, as

$$c_x(q, q) = E[|X(q) - m_x(q)|^2] \geq 0$$

the $c_x(q_1, q_2)$ is a *nonnegative-definite* bilinear functional. The above functionals $\in Q'$, where Q' is the *dual* space of Q.

2.2 Representations of Generalized Spatial Random Fields and Their Physical Significance

Let $X(\mathbf{s})$ be an ordinary SRF in the sense of Chapter 2 above (herein these fields will be denoted in brief by OSRF). To this OSRF we can associate a GSRF by means of the *continuous linear functional* (*CLF*)

$$X(q) = \langle q(\mathbf{s}), X(\mathbf{s})\rangle = \int_U q(\mathbf{s}) X(\mathbf{s})\, d\mathbf{s} \qquad (9)$$

where $U \subseteq R^n$; for simplicity the symbol U will usually be dropped. Clearly, the space $\mathbf{Y}$ of all OSRF may be considered as a subset of the space of all GSRF, that is, $\mathbf{Y} \subset \mathcal{G}$. Moreover, since a GSRF $X(q)$ cannot be assigned values at isolated points $\mathbf{s}$, we introduce the *convoluted SRF* (*CSRF*)

$$Y_q(\mathbf{s}) = \langle q(\mathbf{s}'), S_{\mathbf{s}} X(\mathbf{s}')\rangle = \int q(\mathbf{s}') S_{\mathbf{s}} X(\mathbf{s}')\, d\mathbf{s}' \qquad (10)$$

where $S_{\mathbf{s}} X(\mathbf{s}') = X(\mathbf{s}' + \mathbf{s})$ is the shift operator. Equation (10) yields $Y_q(\mathbf{0}) = X(q)$ for all $q \in Q$; also, taking into account the relation $S_{\mathbf{s}} X(q) = \langle q(\mathbf{s}'), S_{\mathbf{s}} X(\mathbf{s}')\rangle = \langle S_{\mathbf{s}} q(\mathbf{s}'), X(\mathbf{s}')\rangle = X(S_{-\mathbf{s}} q)$, we find that

$$Y_q(\mathbf{s}) = S_{\mathbf{s}} X(q) = X(S_{-\mathbf{s}} q) \qquad (11)$$

Let us now comment on the physical significance of representations (9) and (10).

(a) Quite often in practice one realizes that concepts such as "the value of the natural process X at point $\mathbf{s}$ in space, viz., $X(\mathbf{s})$" are purely mathematical, and that what one actually observes and measures is "the value of the natural process X averaged over some neighborhood U of the point $\mathbf{s}$ in space, viz., Eq. (9) or (10)" (e.g., the concept of porosity). Function $q(\mathbf{s})$ may be associated with a measuring device, an instrument's window, etc. Hence, under certain circumstances, it may be more realistic to develop mathematical models in terms of $X(q)$ and

$Y_q(\mathbf{s})$ rather than in terms of $X(\mathbf{s})$. In fact, most of the important instrumentation and scale results obtained within the context of the multiphase transport theory by Cushman and others (see, e.g., Cushman, 1984) can be derived and generalized further in terms of the GSRF theory above.

(b) On the other hand, the proper choice of $q(\mathbf{s})$ may assure that the derived SRF $X(q)$ and $Y_q(\mathbf{s})$ possess certain desirable properties. For example, since $X(q)$ and $Y_q(\mathbf{s})$ have smoother geometrical characteristics than the OSRF $X(\mathbf{s})$ (m.s. continuity, differentiability, etc.), they may provide better representations of spatial variability. (For a detailed discussion see below.)

(c) Equation (10) defines a *filter* whose input is the OSRF $X(\mathbf{s})$ and output the CSRF $Y_q(\mathbf{s})$. Depending on the choice of the function $q(\mathbf{s})$ one may develop a filter that removes noise and other useless quantities and emphasizes only the properties of interest, (e.g., a filter that yields maps containing high-frequency information—maps that enhance details of the spatial pattern of the natural process; see Chapter 9).

(d) Analysis in terms of $X(q)$ and $Y_q(\mathbf{s})$ can solve problems not otherwise tractable (e.g., the study of space nonhomogeneous SRF; Section 3 and 4). It may also lead to the development of *basic and adjoint differential models* of flow and pollutant transport, which capture essential features of the phenomena they describe and simplify considerably the solution of the related problems (Section 5).

It is easily shown that the means and covariances of $X(q)$ and $Y_q(\mathbf{s})$ are linearly related to those of the corresponding OSRF, particularly

$$m_x(q) = \langle q(\mathbf{s}), m_x(\mathbf{s})\rangle \tag{12}$$

$$m_Y(\mathbf{s}) = \langle q(\mathbf{s}'), m_{S_s X}(\mathbf{s}')\rangle = q(\mathbf{s}) * m_X(\mathbf{s}) \tag{13}$$

where $*$ denotes convolution and

$$c_x(q_1, q_2) = \langle\langle c_x(\mathbf{s}, \mathbf{s}'), q_1(\mathbf{s})\rangle, q_2(\mathbf{s}')\rangle \tag{14}$$

$$c_Y(\mathbf{s}, \mathbf{s}') = \langle\langle c_x(S_s X(\mathbf{s}''), S_{s'} X(\mathbf{s}''')), q_1(\mathbf{s}'')\rangle, q_2(\mathbf{s}''')\rangle \tag{15}$$

If we set $\mathbf{s} = \mathbf{s}'$ we find that the corresponding mean values and covariance functions are, respectively,

$$\begin{aligned} m_x(q) &= m_Y(\mathbf{O}) \\ c_x(q_1, q_2) &= c_Y(\mathbf{O}, \mathbf{O}) \end{aligned} \tag{16}$$

We have already seen that the covariance functional of the GSRF $X(q)$ is a nonnegative-definite bilinear functional in the sense that

$$c_x(q, q) = E[|X(q) - m_x(q)|^2] \geq 0 \tag{17}$$

for all $q \in Q$. Conversely, every continuous nonnegative-definite bilinear functional $c_x(q_1, q_2)$ in Q is a covariance functional of some GSRF $X(q)$. On the other hand, the covariance function $c_Y(\mathbf{s}, \mathbf{s}')$ is also a nonnegative-definite function in the ordinary sense. The fields $X(q)$ and $Y_q(\mathbf{s})$ are always *differentiable*, even when $X(\mathbf{s})$ is not. Indeed, choose $Q = K$ and let

$$X^{(\boldsymbol{\rho})}(q) = \langle q(\mathbf{s}), X^{(\boldsymbol{\rho})}(\mathbf{s})\rangle \tag{18}$$

where $\boldsymbol{\rho} = (\rho_1, \rho_2, \ldots, \rho_n)$ is a multi-index of nonnegative integers; the superscript $\boldsymbol{\rho}$ denotes partial differentiation of the order ρ in space; that is,

$$X^{(\boldsymbol{\rho})}(\mathbf{s}) = \frac{\partial^{|\boldsymbol{\rho}|} X(\mathbf{s})}{\partial s_1^{\rho_1} \ldots \partial s_n^{\rho_n}}, \qquad \text{where} \quad \rho = |\boldsymbol{\rho}| = \sum_{i=1}^{n} \rho_i$$

By applying integration by parts, Eq. (18) can be written

$$X^{(\boldsymbol{\rho})}(q) = (-1)^{\rho} X(q^{(\boldsymbol{\rho})}) \tag{19}$$

For the CSRF it is valid that

$$Y_q^{(\boldsymbol{\rho})}(\mathbf{s}) = (-1)^{\rho} S_{\mathbf{s}} X(q^{(\boldsymbol{\rho})}) \tag{20}$$

Since the mean $m_x(q)$ is a generalized function, it has a specific form, viz. (Gel'fand and Vilenkin, 1964)

$$m_x(q) = \left\langle \sum_{|\boldsymbol{\rho}| \leq \nu} q^{(\boldsymbol{\rho})}(\mathbf{s}), f_\rho(\mathbf{s}) \right\rangle = \left\langle \sum_{|\boldsymbol{\rho}| \leq \nu} (-1)^{\rho} f_\rho^{(\boldsymbol{\rho})}(\mathbf{s}), q(\mathbf{s}) \right\rangle \tag{21}$$

where ν is a suitable number, $q(\mathbf{s}) \in K$ and $f_\rho(\mathbf{s})$ are continuous functions in R^n (i.e., $f_\rho(\mathbf{s}) \in C$), only a finite number of which are different from zero on any given finite support of K. Working along the same lines it can be shown without any difficulty that

$$m_Y(\mathbf{s}) = \left\langle \sum_{|\boldsymbol{\rho}| \leq \nu} (-1)^{\rho} S_{\mathbf{s}} f_\rho^{(\boldsymbol{\rho})}(\mathbf{s}'), q(\mathbf{s}') \right\rangle \tag{22}$$

where for subsequent use let us set $g_\rho(\mathbf{s}) = f_\rho^{(\boldsymbol{\rho})}(\mathbf{s})$.

2.3 Homogeneous Generalized Spatial Random Fields

A *spatially homogeneous* (in the wide sense) GSRF $X(q)$ is a GSRF whose mean value $m_x(q)$ and covariance functional $c_x(q_1, q_2)$ are invariant with respect to any shift of the parameters, that is,

$$m_x(q) = m_x(S_{\mathbf{h}} q) \tag{23}$$

$$c_x(q_1, q_2) = c_x(S_{\mathbf{h}} q_1, S_{\mathbf{h}} q_2) \tag{24}$$

for any $\mathbf{h} \in R^n$, $q_1, q_2 \in Q$. The space of all homogeneous GSRF will be denoted by $\mathcal{G}_0 \subset \mathcal{G}$. The proof of the following proposition for n-dimensional homogeneous SRF is in no way different from the corresponding proof in terms of the unidimensional stationary random processes given in Ito (1954).

Proposition 1: Let $X(q)$ be a homogeneous GSRF on K. Then there exists one and only one generalized functional $c_x(q_1, q_2) \in K'$ satisfying (24).

Similarly, the CSRF $Y_q(\mathbf{s})$ will be termed a *homogeneous* CSRF if

$$m_Y(\mathbf{s}) = \text{constant} \tag{25}$$

$$c_Y(\mathbf{s}, \mathbf{s}') = c_Y(\mathbf{h} = \mathbf{s} - \mathbf{s}') \tag{26}$$

for any $\mathbf{h} \in R^n$. When the $X(q)$ is homogeneous, the $c_x(q_1, q_2)$ is a *translation invariant, nonnegative-definite bilinear functional* on Q, while the $c_Y(\mathbf{s}, \mathbf{s}')$ is a nonnegative-definite function. The former implies that the $c_x(q_1, q_2)$ can be written in the form

$$c_x(q_1, q_2) = \langle c_0, q_1 * \check{q}_2 \rangle \tag{27}$$

where $*$ denotes convolution and $\check{}$ denotes inversion [i.e., $\check{q}_2(\mathbf{h}) = q_2(-\mathbf{h})$]; the c_0 is a nonnegative-definite generalized function, which is the Fourier transform of some *positive tempered measure* $\phi(\mathbf{w})$ in R^n, that is

$$\int \frac{d\phi(\mathbf{w})}{(1 + |\mathbf{w}|^2)^p} < \infty$$

for some $p > 0$. Moreover, it is true that

$$c_x(q_1, q_2) = \langle c_0, q_1 * \check{q}_2 \rangle = \langle \phi, \tilde{q}_1 \overline{\tilde{q}_2} \rangle$$

which implies that

$$c_x(q_1, q_2) = \int \tilde{q}_1(\mathbf{w}) \overline{\tilde{q}_2(\mathbf{w})}\, d\phi(\mathbf{w}) \tag{28}$$

where $\tilde{q}_1(\mathbf{w})$ and $\overline{\tilde{q}_2(\mathbf{w})}$ are the Fourier transforms of $q_1(\mathbf{s})$ and $q_2(\mathbf{s})$, respectively [the line above $\tilde{q}_2(\mathbf{w})$ means "the complex conjugate of."] The $\phi(\mathbf{w})$ will be called the *spectral measure* of the GSRF $X(q)$.

Taking into consideration Eqs. (21) and (23) it follows that the functions $f_\rho(\mathbf{s})$ are constants; as a consequence,

$$\begin{aligned} g_\rho(\mathbf{s}) &= f_\rho^{(\rho)}(\mathbf{s}) = 0 \qquad \text{for all} \quad \rho \geq 1 \\ &= f_0^{(0)}(\mathbf{s}) = m \qquad \text{for} \quad \rho = 0 \end{aligned}$$

and the $m_x(q)$ will have the form

$$m_x(q) = m \langle q(\mathbf{s}), 1 \rangle \tag{29}$$

where m is a constant; moreover, the space of derivatives $q^{(\rho)}(\mathbf{s})$ in Eq. (21) coincides with the linear subspace $K_\rho \subset K$, which consists of functions $q(\mathbf{s}) \in K$ satisfying

$$\langle q^{(\rho)}(\mathbf{s}), \mathbf{s}^{\rho} \rangle = 0 \tag{30}$$

In this case, hence, it is convenient to restrict ourselves to $q(\mathbf{s}) \in K_\rho$. In the same context the homogeneity of the OSRF $X(\mathbf{s})$ implies that the covariance functional $c_x(q_1, q_2)$ is such that

$$\begin{aligned} c_x(q_1, q_2) &= \iint c_x(\mathbf{h}) q_1(\mathbf{s}) q_2(\mathbf{s}+\mathbf{h})\, d\mathbf{s}\, d\mathbf{h} \\ &= \int c_x(\mathbf{h}) \left[\int q_1(\mathbf{s}) q_2(\mathbf{s}+\mathbf{h})\, d\mathbf{s} \right] d\mathbf{h} \\ &= \langle c_x(\mathbf{h}), q_1 * \check{q}_2(\mathbf{h}) \rangle = c_x(q_1 * \check{q}_2) = c_x(q) \end{aligned} \tag{31}$$

for all $q_1, q_2 \in K_\rho$ and

$$q(\mathbf{h}) = \int q_1(\mathbf{s}) q_2(\mathbf{s}+\mathbf{h})\, d\mathbf{s} = q_1 * \check{q}_2(\mathbf{h}) \in K_\rho$$

Equation (31) is the variance of the CSRF $Y_q(\mathbf{s}) = X(S_{-\mathbf{s}} q)$, while its covariance can be written

$$\begin{aligned} c_{S_{\mathbf{h}}X}(q) = c_{Y_q}(\mathbf{h}) &= \langle c_x(\mathbf{s}+\mathbf{h}), q_1 * \check{q}_2(\mathbf{s}) \rangle \\ &= \int c_x(\mathbf{s}+\mathbf{h}) q(\mathbf{s})\, d\mathbf{s} \end{aligned} \tag{32}$$

The latter depends only on $\mathbf{h}$ and, therefore, the CSRF $Y_q(\mathbf{s}) = X(S_{-\mathbf{s}} q)$ is homogeneous too.

Example 1: As we saw above, depending on the choice of the space Q, the GSRF $X(q)$ can be differentiable as many times as we wish, even if the corresponding OSRF $X(\mathbf{s})$ is not differentiable. We reconsider this interesting property of GRSF in the light of the results on homogeneity. Let $X(\mathbf{s})$ be a homogeneous OSRF with mean $m_x = \text{constant}$ and covariance $c_x(\mathbf{h})$, and let $X(q)$ be the corresponding homogeneous GSRF, $X(q) = \langle q(\mathbf{s}), X(\mathbf{s}) \rangle$, with mean

$$\begin{aligned} m_{x_q} &= \int q(\mathbf{s}) m_x(\mathbf{s})\, d\mathbf{s} = m_{x_q}(\mathbf{h}) \\ &= \int q(\mathbf{s}) m_x(\mathbf{s}+\mathbf{h})\, d\mathbf{s} = m_x \langle q(\mathbf{s}), 1 \rangle \end{aligned}$$

and covariance

$$c_{x_q}(\mathbf{s}, \mathbf{s}') = c_{x_q}(\mathbf{s}, \mathbf{s}'+\mathbf{h})$$
$$= \iint q(\mathbf{s})q(\mathbf{s}')c_x(\mathbf{s}'-\mathbf{s}+\mathbf{h})\, d\mathbf{s}\, d\mathbf{s}'$$

The derivatives $X^{(\boldsymbol{\rho})}(q) = (-1)^{\rho} X(q^{(\boldsymbol{\rho})})$ always exist, assuming that we choose $q(\mathbf{s}) \in K_\rho$. By expanding the covariance around the origin $\mathbf{h} = \mathbf{O}$,

$$c_x(\mathbf{s}+\mathbf{h}) = c_x(\mathbf{s}) + \sum_{i_1=1}^{n} a_{i_1} h_{i_1} + \frac{1}{2!}$$
$$\times \sum_{i_1, i_2=1}^{n} a_{i_1 i_2} h_{i_1} h_{i_2} + \cdots$$

where $a_{i_1 \cdots i_{k-1}} = c_x^{(k-1)}(\mathbf{s})$; then by choosing $q(\mathbf{s})$ such that $c_x(q^{(1)}) = 0$, inserting the above expansion in (32) and differentiating we find that

$$\frac{\partial c_{S_\mathbf{h} X}(q)}{\partial h_i}\Big|_{\mathbf{h}=\mathbf{0}} = 0, \qquad \text{for all} \quad i = 1, 2, \ldots, n$$

which, since $X(q)$ is m.s. differentiable, is an expected result. It is also an intuitively justified result, in the sense that the averaging applied by $q(\mathbf{s})$ on $X(\mathbf{s})$ should lead to a field $X(q)$, which possesses a smoother spatial variability.

It would be interesting to investigate the conditions under which the SRF $Y_q(\mathbf{s}) = X(S_{-\mathbf{s}} q)$ is *zero-mean homogeneous*, even when the associated OSRF $X(\mathbf{s})$ is nonhomogeneous. As will be shown in the next section, this can happen under certain conditions concerning the choice of the functions $q(\mathbf{s})$ as well as the form of the functions $g_\rho(\mathbf{s})$.

3. Spatial Random Fields with Space Homogeneous Increments or Intrinsic Spatial Random Fields

3.1 Basic Notions

The previous results on homogeneous SRF lead us naturally to the concept of SRF with space homogeneous increments of order ν, in the ordinary or in the generalized sense. Let us begin with the following definition.

Definition 1: A CSRF $Y_q(\mathbf{s}) = X(S_{-\mathbf{s}} q)$ will be called a *CSRF of order* ν*(CSRF-*ν*)* if q belongs to the space

$$Q_\nu = \{q \in Q: \langle q(\mathbf{s}), g_\rho(\mathbf{s})\rangle = 0 \qquad \text{for all} \quad \rho \leq \nu\} \tag{1}$$

where the functions $g_\rho(\mathbf{s})$ belong to the space

$$C_\nu = \{g_\rho(\mathbf{s}) \in C: \langle q(\mathbf{s}), g_\rho(\mathbf{s})\rangle = 0$$
$$\Rightarrow \langle q(\mathbf{s}), S_{\mathbf{h}} g_\rho(\mathbf{s})\rangle = 0 \qquad \text{for all} \quad \rho \leq \nu\} \tag{2}$$

In this case the space Q_ν will be termed an *admissible space of order* ν (*AS*-ν).

Equation (1) assures a zero mean value for the SRF $Y_q(\mathbf{s})$ at the origin $\mathbf{s} = \mathbf{O}$, while the closeness of C_ν to translation [Eq. (2)] is necessary in order that the statistical properties of $X(q)$ remain unaffected by a shift $S_{\mathbf{h}}$ of the origin. Functions $g_\rho(\mathbf{s})$ that satisfy the conditions (1) and (2) are of the general form $g_\rho(\mathbf{s}) = \mathbf{s}^{\boldsymbol{\rho}} \exp[\boldsymbol{\alpha} \cdot \mathbf{s}]$, where $\boldsymbol{\alpha}$ is a (real or complex) vector. Herein, due to convenient invariance and linearity properties, it will be assumed that the $g_\rho(\mathbf{s})$ are *pure polynomials*, viz.,

$$g_\rho(\mathbf{s}) = \mathbf{s}^{\boldsymbol{\rho}} = s_1^{\rho_1} s_2^{\rho_2} \ldots s_n^{\rho_n} \tag{3}$$

where $\rho = |\boldsymbol{\rho}| = \sum_{i=1}^n \rho_i$. In connection with the notion of AS-ν, the properties below are valid.

$$\forall\, q(\mathbf{s}) \in Q_\nu \Rightarrow q(\mathbf{s}) \in Q_{\nu-1} \tag{4}$$

$$\forall\, q(\mathbf{s}) \in Q_\nu \Rightarrow q^{(\nu+1)}(\mathbf{s}) \in Q_\nu \tag{5}$$

$$\forall\, q(\mathbf{s}) \in Q_\nu \Rightarrow S_{\mathbf{h}} q(\mathbf{s}) \in Q_\nu \tag{6}$$

and

$$\forall\, q^*(\mathbf{s}) \in S \Rightarrow q(\mathbf{s}) = \delta(\mathbf{s}) - q^*(\mathbf{s}) \in Q_\nu \tag{7}$$

In view of the preceding theory, the validity of Eqs. (4) and (5) is obvious. For Eq. (6) note that, by definition, for any $g_\rho(\mathbf{s})$ of the form (3) and $q \in Q_\nu$ it holds true that

$$\int q(\mathbf{s}) \mathbf{s}^{\boldsymbol{\rho}} \, d\mathbf{s} = 0$$

Next, by applying the shift operator $S_{\mathbf{h}}$ we get

$$\int q(\mathbf{s})(\mathbf{s}+\mathbf{h})^{\boldsymbol{\rho}} \, d\mathbf{s} = \int q(\mathbf{s}) \left[\sum_{|\mathbf{k}|=0}^{\rho} C_{\rho!}^{|\mathbf{k}|!} \mathbf{s}^{\mathbf{k}} \mathbf{h}^{\boldsymbol{\rho}-\mathbf{k}} \right] d\mathbf{s}$$
$$= \sum_{|\mathbf{k}|=0}^{\rho} C_{\rho!}^{|\mathbf{k}|!} \mathbf{h}^{\boldsymbol{\rho}-\mathbf{k}}$$
$$\times \int q(\mathbf{s}) \mathbf{s}^{\mathbf{k}} \, d\mathbf{s} = 0$$

Hence $S_{\mathbf{h}} q(\mathbf{s}) \in Q_\nu$. Alternatively, the proof of (6) is a straightforward consequence of the following:

$$\begin{aligned}\langle S_{\mathbf{h}} q(\mathbf{s}), g_\rho(\mathbf{s})\rangle &= \langle q(\mathbf{s}), S_{\mathbf{h}} g_\rho(\mathbf{s})\rangle \\ &= \langle q(\mathbf{s}), \sum_{\zeta=0}^{\rho} C_{\rho!}^{\zeta!} g_\zeta(\mathbf{s}) g_{\rho-\zeta}(\mathbf{h})\rangle \\ &= \sum_{\zeta=0}^{\rho} C_{\rho!}^{\zeta!} g_{\rho-\zeta}(\mathbf{h}) \langle q(\mathbf{s}), g_\zeta(\mathbf{s})\rangle = 0\end{aligned}$$

Finally, to prove Eq. (7) assume that the $q^*(\mathbf{s}) \in S$ is defined by its Fourier transform

$$\tilde{q}^*(\mathbf{w}) = \left[1 + \frac{\alpha|\mathbf{w}|^2}{2!} + \cdots + \frac{\alpha^{\nu'}|\mathbf{w}|^{2\nu'}}{\nu'!}\right] \exp[-\alpha|\mathbf{w}|^2]$$

where $\alpha > 0$ and $2\nu' > \nu$. Then

$$\begin{aligned}&\int g_\rho(\mathbf{s})[\delta(\mathbf{s}) - q^*(\mathbf{s})]\, d\mathbf{s} \\ &\quad = \int g_\rho(\mathbf{s}) \left\{ \int [1 - \tilde{q}^*(\mathbf{w})] \exp[-i(\mathbf{w} \cdot \mathbf{s})]\, d\mathbf{w} \right\} d\mathbf{s} \\ &\quad = \int \delta^{(\rho)}(\mathbf{w})[1 - \tilde{q}^*(\mathbf{w})]\, d\mathbf{w} \\ &\quad = (-1)^\rho \int \delta(\mathbf{w})[1 - \tilde{q}^*(\mathbf{w})]^{(\rho)}\, d\mathbf{w} \\ &\quad = (-1)^\rho [1 - \tilde{q}^*(\mathbf{w})]^{(\rho)}|_{\mathbf{w}=\mathbf{O}} = 0\end{aligned}$$

for all ρ up to $2\nu' > \nu$, by definition of $\tilde{q}^*(\mathbf{w})$. The last equation implies Eq. (7).

Definition 2: A *GSRF* $X(q)$ with *homogeneous increments of order* ν (*GSRF*-ν) is a linear mapping

$$X\colon Q_\nu \to L_2(\Omega, F, P) \tag{8}$$

where the corresponding CSRF $Y_q(\mathbf{s})$ is zero-mean homogeneous for all $q \in Q_\nu$ and all $\mathbf{s} \in R^n$.

Remark 1: In view of Definition 2 all partial derivatives of order $\nu+1$ of a GSRF-ν, namely,

$$Y(q) = D^{(\nu+1)} X(q) \tag{9}$$

where $D^{(\nu+1)} X(q) = X^{(\nu+1)}(q) = (-1)^{\nu+1} X(q^{(\nu+1)})$ are zero-mean homogeneous GSRF. A similar result was first proven by Ito (1954) for stationary random processes. In particular, he showed that any generalized stationary random process $Y(q)$ whose spectral measure $Q_Y(w)$ satisfies the condition

$$\int_{R^1} \frac{dQ_Y(w)}{(1+w^2)^{\nu+1}} < \infty$$

coincides with the $\nu+1$ derivative of some ordinary random process $X(s)$. On the basis of the foregoing, an ordinary or intrinsic SRF-ν will be defined as follows.

Definition 3: An OSRF $X(\mathbf{s})$ is called an *ordinary or intrinsic SRF of order* ν (*OSRF-ν or ISRF-ν*) if for all $q \in Q_\nu$ the corresponding CSRF-ν $Y_q(\mathbf{s})$ is zero-mean homogeneous.

In fact, the zero mean condition is imposed for convenience and does not restrict generality. In connection with Definition 3, if the derivatives

$$D^{(\nu+1)}X(\mathbf{s}) = X^{(\nu+1)}(\mathbf{s}) = Y(\mathbf{s}) \tag{10}$$

exist and are zero-mean homogeneous SRF, the $X(\mathbf{s})$ is an ISRF-ν. (Note that unlike GSRF-ν, which are always differentiable, there exist ISRF-ν that are not differentiable; they can, however, be expressed as the sum of an infinitely differentiable ISRF-ν and a homogeneous SRF; Proposition 4 in Section 3.2.) To further illustrate this point consider the proposition below.

Proposition 1: Let $Y(s)$ be a zero-mean continuous homogeneous SRF and let $X_\nu(s)$ be a differentiable SRF, both in R^1 satisfying

$$D^{(\nu+1)}X_\nu(s) = Y(s) \tag{11}$$

Then the $X_\nu(s)$ is an ISRF-ν.

Proof: First note that the unidimensional stochastic differential equation (11) yields

$$X_\nu(s) = \int_0^s \frac{(s-u)^\nu}{\nu!} Y(u)\, du \tag{12}$$

For $\nu=0$, Eq. (12) yields $X_0(s) = \int_0^s Y(u)\, du$, which is an ISRF-0 since for $q \in Q_0$, the

$$Y_q(s) = \langle q(s'), S_s X_0(s')\rangle = \left\langle q(s'), S_s\left[\left[\int_0^{s'} Y(u)\, du\right]\right]\right\rangle$$

is a homogeneous CSRF-0. For $\nu > 0$ Eq. (12) can be written in the recursive form

$$X_\nu(s) = \int_0^s X_{\nu-1}(u)\, du \tag{13}$$

where $X_{\nu-1}$ is an ISRF-$(\nu-1)$. To continue the proof we will apply induction: If $X_{\nu-1}$ is an ISRF-$(\nu-1)$, it will be shown that the SRF defined by Eq. (13) is an ISRF-ν. Let us choose $q \in Q_\nu$ and set

$$\begin{aligned} Y_q(s) &= \langle q(s'), S_s X_\nu(s') \rangle \\ &= \left\langle q(s'), S_s \int_0^{s'} X_{\nu-1}(u)\, du \right\rangle \end{aligned}$$

Next define q^* so that

$$Y_{q^*}(s) = Y_q(s) \tag{14}$$

In other words,

$$\langle q^*(s'), S_s X_{\nu-1}(s') \rangle = \left\langle q(s'), S_s \int_0^{s'} X_{\nu-1}(u)\, du \right\rangle$$

The q^* satisfies

$$\begin{aligned} \langle q^*(s'), g_{\rho \le \nu-1}(s') \rangle &= \left\langle q(s'), \int_0^{s'} g_{\rho \le \nu-1}(u)\, du \right\rangle \\ &= \langle q(s'), g_{\rho \le \nu}(s') \rangle = 0 \end{aligned}$$

which implies that $q^* \in Q_{\nu-1}$. This result, in combination with the fact that the $X_{\nu-1}$ is an ISRF-$(\nu-1)$, implies that the $Y_{q^*}(s)$ is a zero-mean homogeneous CSRF-$(\nu-1)$. But then, due to Eq. (14), $Y_q(s)$ is also a zero-mean homogeneous SRF and, finally, since $q \in Q_\nu$, the $X_\nu(s)$ is an ISRF-ν. For completeness, note that the

$$Y(q) = \langle q(s), Y(s) \rangle = \langle q(s), X_\nu^{(\nu+1)}(s) \rangle = Y_q^{(\nu+1)}(0)$$

is a homogeneous generalized field. □

Clearly, if $X(\mathbf{s})$ is an ISRF-ν, it is also an ISRF-μ for all $\mu > \nu$. By convention, the *homogeneous* SRF is considered as an *ISRF*-(-1).

Example 1: The SRF with homogeneous increments defined in Section 11 of Chapter 2 is an ISRF-0. In addition it is an ISRF of any order $\nu > 0$. The following proposition describes a certain case of ISRF-ν that can be characterized by means of an analytic expression in **s** having random variables as parameters.

Proposition 2: The SRF

$$X(\mathbf{s}) = \sum_{\rho = |\boldsymbol{\rho}| \le \nu} \beta_\rho g_\rho(\mathbf{s}) \tag{15}$$

where β_ρ are random variables in $L_2(\Omega, F, P)$, is an ISRF-ν.

Proof: Let $q \in Q_\nu$:

$$\langle q(\mathbf{s}'), S_{\mathbf{s}} g_\rho(\mathbf{s}')\rangle = 0 \qquad \text{for all} \quad \rho \leq \nu$$

Then it is true that

$$Y_q(\mathbf{s}) = \langle q(\mathbf{s}'), S_{\mathbf{s}} X(\mathbf{s}')\rangle$$
$$= \sum_{\rho=0}^{\nu} \beta_\rho \langle q(\mathbf{s}'), S_{\mathbf{s}} g_\rho(\mathbf{s}')\rangle = 0 \qquad \square$$

According to Eq. (15) the ISRF-ν is completely characterized by the joint distribution of the random variables $\beta_\rho (\rho \leq \nu)$. The realizations of the field (15) have a polynomial form and, therefore, issues such as continuity and differentiability can be studied in the usual sense.

3.2 The Generalized Representation Set

By definition, to a given generalized $X(q)$ one may assign various ordinary $X(\mathbf{s})$, all having the same CSRF $Y_q(\mathbf{s})$. More precisely, the following definition makes sense.

Definition 4: The set of all ISRF-ν $\mathbf{X}_q = \{X^a(\mathbf{s}), a = 1, 2, \ldots\}$ that have the same CSRF-ν $Y_q(\mathbf{s})$ will be called the *generalized representation set of order* ν (*GRS-*ν). Each member of the GRS-ν $\mathbf{X}_q$ will be called a *representation* of $X(q)$. In other words, the correspondences below are valid.

$$\begin{array}{ccl} X(q) & \leftrightarrow & \{X^a(\mathbf{s}), \quad a = 1, 2, \ldots\} \\ \Updownarrow & & \Updownarrow \\ X(S_{-\mathbf{s}} q) & = & Y_q(\mathbf{s}) \end{array} \tag{16}$$

Proposition 3: Let $X^0(\mathbf{s})$ be a representation of $X(q)$ in the sense of Definition 4. For the SRF $X^a(\mathbf{s})$ to be another representation it is necessary and sufficient that it can be expressed by

$$X^a(\mathbf{s}) = X^0(\mathbf{s}) + \sum_{\rho=|\boldsymbol{\rho}| \leq \nu} c_\rho g_\rho(\mathbf{s}) \tag{17}$$

where the c_ρ, $\rho \leq \nu$ are random variables in $L_2(\Omega, F, P)$ such that

$$c_\rho = \langle \eta_\rho(\mathbf{s}), X^a(\mathbf{s})\rangle \tag{18}$$

and the $\eta_\rho(\mathbf{s})$ satisfy the

$$\langle \eta_\rho(\mathbf{s}), g_{\rho'}(\mathbf{s})\rangle = \begin{cases} 1 & \text{if} \quad \rho = \rho' \\ 0 & \text{otherwise} \end{cases} \tag{19}$$

Proof: If $X^0(\mathbf{s})$ is any specific representation of $X(q)$, then for all $q \in Q_\nu$ Eq. (17) gives

$$\langle q(\mathbf{s}), X^a(\mathbf{s})\rangle = \langle q(\mathbf{s}), X^0(\mathbf{s})\rangle + \sum_{\rho=|\boldsymbol{\rho}|\leq\nu} c_\rho \langle q(\mathbf{s}), g_\rho(\mathbf{s})\rangle$$

and, since $\langle q(\mathbf{s}), g_\rho(\mathbf{s})\rangle = 0$, we obtain $X^a(\mathbf{s}) = X^0(\mathbf{s})$. Hence, $X^a(\mathbf{s})$ is a representation of $X(q)$ too. Conversely, if $X^0(\mathbf{s})$ is a representation of $X(q)$ we shall show that any other representation $X^a(\mathbf{s})$, $a \neq 0$ will be of the form (17). If $q \in Q_\nu$, then

$$\left\langle \delta_{\mathbf{s}}(\mathbf{s}') - \sum_{\rho'=|\boldsymbol{\rho}'|\leq\nu} g_{\rho'}(\mathbf{s})\eta_{\rho'}(\mathbf{s}'), g_\rho(\mathbf{s}')\right\rangle$$
$$= \langle \delta_{\mathbf{s}}(\mathbf{s}'), g_\rho(\mathbf{s}')\rangle - \sum_{\rho'=|\boldsymbol{\rho}'|\leq\nu} g_{\rho'}(\mathbf{s})\langle \eta_{\rho'}(\mathbf{s}'), g_\rho(\mathbf{s}')\rangle$$
$$= g_\rho(\mathbf{s}) - g_\rho(\mathbf{s}) = 0$$

for all $\rho \leq \nu$, due to Eq. (19). [Here, $\delta_{\mathbf{s}}(\mathbf{s}')$ means $S_{-\mathbf{s}}\,\delta(\mathbf{s}')$.] Therefore we have

$$q^*_{\mathbf{s}}(\mathbf{s}') = \delta_{\mathbf{s}}(\mathbf{s}') - \sum_{\rho=|\boldsymbol{\rho}|\leq\nu} g_\rho(\mathbf{s})\eta_\rho(\mathbf{s}') \in Q_\nu \tag{20}$$

for all $\rho \leq \nu$, $q^*_{\mathbf{s}}(\mathbf{s}') = S_{\mathbf{s}} q^*(\mathbf{s}')$. Moreover, $\langle q(\mathbf{s}), q^*_{\mathbf{s}}(\mathbf{s}')\rangle = q(\mathbf{s})$ and, thus,

$$X(q^*_{\mathbf{s}}) = X^0(\mathbf{s}) = \langle q^*_{\mathbf{s}}(\mathbf{s}'), X^0(\mathbf{s}')\rangle \tag{21}$$

or

$$X^0(\mathbf{s}) - \sum_{\rho=|\boldsymbol{\rho}|\leq\nu} g_\rho(\mathbf{s})\langle \eta_\rho(\mathbf{s}'), X^0(\mathbf{s}')\rangle = X^0(\mathbf{s})$$

or

$$\langle \eta_\rho(\mathbf{s}'), X^0(\mathbf{s}')\rangle = 0 \tag{22}$$

for all $\rho \leq \nu$. If $X^a(\mathbf{s})$ is another representation,

$$\langle q^*_{\mathbf{s}}(\mathbf{s}'), X^a(\mathbf{s}')\rangle = \langle q^*_{\mathbf{s}}(\mathbf{s}'), X^0(\mathbf{s}')\rangle$$

or

$$X^a(\mathbf{s}) - \sum_{\rho=|\boldsymbol{\rho}|\leq\nu} g_\rho(\mathbf{s})\langle \eta_\rho(\mathbf{s}'), X^a(\mathbf{s}')\rangle$$
$$= X^0(\mathbf{s}) - \sum_{\rho=|\boldsymbol{\rho}|\leq\nu} g_\rho(\mathbf{s})\langle \eta_\rho(\mathbf{s}'), X^0(\mathbf{s}')\rangle$$

or, due to Eq. (22),

$$X^a(\mathbf{s}) = X^0(\mathbf{s}) + \sum_{\rho=|\boldsymbol{\rho}|\leq\nu} g_\rho(\mathbf{s})\langle \eta_\rho(\mathbf{s}'), X^a(\mathbf{s}')\rangle$$

and the validity of Eq. (17) is evident. □

Proposition 4: A continuous but not necessarily differentiable ISRF-ν $X(\mathbf{s})$ can be expressed as the sum of an infinitely differentiable ISRF-ν $X^*(\mathbf{s})$ and a homogeneous SRF $Y_q(\mathbf{s})$, viz.,

$$X(\mathbf{s}) = X^*(\mathbf{s}) + Y_q(\mathbf{s}) \tag{23}$$

Proof: We saw above [Eq. (7)] that if we define $q^*(\mathbf{s}) \in S$, then

$$q(\mathbf{s}) = \delta(\mathbf{s}) - q^*(\mathbf{s}) \in Q_\nu$$

By Definition 3,

$$Y_q(\mathbf{s}) = \langle \delta(\mathbf{s}') - q^*(\mathbf{s}'), S_\mathbf{s} X(\mathbf{s}') \rangle$$

is homogeneous and

$$Y_q(\mathbf{s}) = X(\mathbf{s}) - \langle q^*(\mathbf{s}'), S_\mathbf{s} X(\mathbf{s}') \rangle = X(\mathbf{s}) - X^*(\mathbf{s}) \tag{24}$$

For all $q \in Q_\nu$,

$$\begin{aligned}\langle q(\mathbf{s}), X^*(\mathbf{s}) \rangle &= \langle q(\mathbf{s}), \langle q^*(\mathbf{s}'), S_\mathbf{s} X(\mathbf{s}') \rangle \rangle \\ &= \langle q^*(\mathbf{s}), X(S_{-\mathbf{s}} q(\mathbf{s}')) \rangle\end{aligned}$$

that is, the $X^*(\mathbf{s})$ is an ISRF-ν. Particularly, it is a representation of the GSR-ν

$$X^*(q) = \langle q^*(\mathbf{s}), X[S_{-\mathbf{s}} q(\mathbf{s}')] \rangle$$

and, since $q^*(\mathbf{s}) \in S$, the $X^*(\mathbf{s})$ is an infinitely differentiable ISRF-ν. □

3.3 The Correlation Structure

Turning to the correlation structure of the ISRF-ν, we observe that in view of Eq. (9) above the generalized field $X^{(\nu)}(q)$ has a constant mean given by

$$E[X^{(\nu)}(q)] = m_x^{(\nu)}(q) = (-1)^\nu m_x(q^{(\nu)}) = a\langle q(\mathbf{s}), 1 \rangle = a\tilde{q}(\mathbf{O}) \tag{25}$$

The covariance $c_x(q_1^{(\nu+1)}, q_2^{(\nu+1)})$ can be written

$$\begin{aligned}c_x(q_1^{(\nu+1)}, q_2^{(\nu+1)}) &= E[X(q_1^{(\nu+1)}) X(q_2^{(\nu+1)})] \\ &= E[X^{(\nu+1)}(q_1) X^{(\nu+1)}(q_2)] = c_Y(q_1, q_2)\end{aligned} \tag{26}$$

which is a translation-invariant bilinear functional. This entails that the mean value and the covariance functional of a GSRF-ν $X(q)$ have as follows:

$$\begin{aligned}m_x(q) &= \sum_{0 \le |\mathbf{\rho}| \le \nu} a_\mathbf{\rho} \langle \mathbf{s}^{|\mathbf{\rho}|}, q(\mathbf{s}) \rangle \\ &= \sum_{\rho_1} \sum_{\rho_2} \cdots \sum_{\rho_n} a_{\rho_1 \rho_2 \ldots \rho_n} \langle s_1^{\rho_1} s_2^{\rho_2} \ldots s_n^{\rho_n}, q(\mathbf{s}) \rangle\end{aligned} \tag{27}$$

where $a_{\boldsymbol{\rho}}$ are suitable coefficients, $0 \leq \rho = |\boldsymbol{\rho}| = \sum_{i=1}^{n} \rho_i \leq \nu$, and

$$c_x(q_1, q_2) = \int_{\mathfrak{R}} \tilde{q}_1(\mathbf{w}) \overline{\tilde{q}_2(\mathbf{w})} \, d\phi(\mathbf{w}) + \sum_{|\mathbf{k}|=\nu+1} \sum_{|\boldsymbol{\lambda}|=\nu+1} \alpha_{k\lambda} \beta_k \bar{\eta}_\lambda \tag{28}$$

where $\mathfrak{R} = R^n - \{\mathbf{O}\}$, ϕ is a positive-tempered measure such that

$$\int_{0<|\mathbf{w}|<1} |\mathbf{w}|^{2\nu} \, d\phi(\mathbf{w}) < \infty; \qquad \beta_k = \langle q_1(\mathbf{s}), \mathbf{s}^{\mathbf{k}} \rangle$$

and

$$\eta_\lambda = \langle q_2(\mathbf{s}), \mathbf{s}^{\boldsymbol{\lambda}} \rangle$$

satisfying $\beta_k = \eta_\lambda = 0$ for $|\mathbf{k}| \leq \nu$, $|\boldsymbol{\lambda}| \leq \nu$; the $\alpha_{k\lambda}$ are numbers such that for any set of complex numbers $\{c_i, |\mathbf{i}| = \nu\}$ the

$$\sum_{|\mathbf{k}|=\nu} \sum_{|\boldsymbol{\lambda}|=\nu} c_k \bar{c}_\lambda \alpha_{k\lambda}$$

is positive-definite.

Conversely, any bilinear functional of the form (28) is the covariance functional of a GSRF-ν. Next we define the important concept of generalized spatial covariance first introduced by Matheron (1973).

Definition 5: Let $X(\mathbf{s})$ be a continuous ISRF-ν, and let $Y_q(\mathbf{s})$ be the associated CSRF-ν. A continuous function $k_x(\mathbf{h})$, $\mathbf{h} = \mathbf{s} - \mathbf{s}'$ in R^n is called a *generalized spatial covariance of order* ν (*GSC*-ν) if and only if

$$(X(q_1), X(q_2)) = \langle k_x(\mathbf{s} - \mathbf{s}'), q_1(\mathbf{s}) q_2(\mathbf{s}') \rangle \geq 0 \tag{29}$$

for all $q_1, q_2 \in Q_\nu$.

In practice it is usually assumed that the GSC-ν is *isotropic*, that is,

$$k_x(\mathbf{h}) = k_x(r) \tag{30}$$

where $r = |\mathbf{h}|$. Let us now return to Eq. (26). It is true that

$$\begin{aligned} c_x(q_1^{(\nu+1)}, q_2^{(\nu+1)}) &= \langle c_x(\mathbf{s}, \mathbf{s}'), q_1^{(\nu+1)}(\mathbf{s}) q_2^{(\nu+1)}(\mathbf{s}') \rangle \\ &= \langle D^{(2\nu+2)} c_x(\mathbf{s}, \mathbf{s}'), q_1(\mathbf{s}) q_2(\mathbf{s}') \rangle \\ &= \langle c_Y(\mathbf{s} - \mathbf{s}'), q_1(\mathbf{s}) q_2(\mathbf{s}') \rangle \end{aligned}$$

That is,

$$D^{(2\nu+2)} c_x(\mathbf{s}, \mathbf{s}') = c_Y(\mathbf{s} - \mathbf{s}') \tag{31}$$

The *deterministic* differential equation (31) is associated with the *stochastic* differential equation (10) and it may be solved with respect to $c_x(\mathbf{s}, \mathbf{s}')$. For illustration let us consider the unidimensional case: According to Proposition 1, if $X(s)$ is a differentiable ISRF-ν in R^1 such that $D^{(\nu+1)}X(s) = Y(s)$, then the $Y(s)$ is a homogeneous SRF. We saw above that any generalized stationary random process $Y(q)$ whose spectral measure $Q_Y(w)$ satisfies the condition

$$\int_{R^1} \frac{dQ_Y(w)}{(1+w^2)^{\nu+1}} < \infty \tag{32}$$

(tempered measure), coincides with the $\nu+1$ derivative of some ordinary nonstationary, in general, random process $X(s)$. In this case the latter process is an ISRF-ν and can be represented by

$$X(s) = \int_{R^1} \left[\exp[iws] - \frac{1}{1+w^{\nu+1}} \sum_{\rho=0}^{\nu} \frac{(iws)^\rho}{\rho!} \right] d\tilde{X}(w) \tag{33}$$

where $i = \sqrt{-1}$, and the $\tilde{X}(w)$ is such that the spectral measure $d\Theta = E|d\tilde{X}|^2$ satisfies

$$\int_{-\infty}^{\varepsilon} d\Theta(w) < \infty, \qquad \int_{-\varepsilon}^{\varepsilon} w^{2\nu+2}\, d\Theta(w) < \infty$$

and

$$\int_{-\varepsilon}^{\infty} d\Theta(w) < \infty$$

for all $\varepsilon > 0$. The covariances of $X(s)$ and $Y(s)$ are related by $D^{(2\nu+2)}c_x(s, s') = c_Y(h)$, where $h = s - s'$. The solution of the last differential equation is

$$c_x(s, s') = k_x(h) + p_\nu(s, s') \tag{34}$$

where

$$k_x(h) = (-1)^{\nu+1} \int_0^h \frac{(h-u)^{2\nu+1}}{(2\nu+1)!} c_Y(u)\, du \tag{35}$$

and

$$p_\nu(s, s') = \sum_{\rho=0}^{\nu} a_\rho(s') s^\rho + \sum_{\rho=0}^{\nu} a_\rho(s) s'^\rho \tag{36}$$

Solving for $c_Y(h)$ we obtain

$$c_Y(h) = (-1)^{\nu+1} D^{(2\nu+2)} k_x(h) \tag{37}$$

To throw additional light on the nature of the GSC-ν, let us discuss a few examples.

Example 2: Take the random process in R^1 defined by $X(s) = a_0 + a_1 s$, where a_0, a_1 are uncorrelated random variables. This process is a unidimensional ISRF-0 with linear drift and GSC-0

$$k_x(h) = -\gamma_x(h) = -\tfrac{1}{2}E[|a_1|^2]h^2$$

Note that the $k_x(h)$ is here unbounded, for

$$E[|a_1|^2] \neq 0$$

Example 3: The Wiener process $W_0(s)$ in R^1 is a zero-mean Gaussian process with stationary independent increments (once more recall that the term "stationary" is the one-dimensional equivalent of the term "homogeneous," which is used in more than one dimension) satisfying

$$c_{w_0}(s, s') = E[W_0(s)W_0(s')] = \sigma^2 \quad \min(s, s') \tag{38}$$

and

$$\text{Var}[W_0(s+h) - W_0(s)] = \sigma^2|h| \tag{39}$$

where $|h| = |s - s'|$ and σ^2 is a parameter (for simplicity one may assume $\sigma^2 = 1$). The $W_0(s)$ is an ISRF-0, and Eq. (38) can be written

$$c_{w_0}(s, s') = -\frac{\sigma^2}{2}|h| + \frac{\sigma^2}{2}(s + s')$$

which is Eq. (34) with

$$k_{w_0}(h) = -\frac{\sigma^2}{2}|h|$$

and

$$p_0(s, s') = a_0(s') + a_0(s) = \frac{\sigma^2}{2}s' + \frac{\sigma^2}{2}s$$

The covariance of $e(s) = dW_0(s)/ds$ is written

$$E[e(s)e(s')] = \frac{d^2}{ds\,ds'}E[W_0(s)W_0(s')]$$

$$= \frac{d^2}{ds\,ds'}\sigma^2 \quad \min(s, s') = \sigma^2\delta(s - s') \tag{40}$$

Due to the presence of the delta function, the derivative above exists in the generalized sense. In the stochastic context this implies that the $e(s)$ is a generalized stationary RP, but not an ordinary stationary RP. Since covariance (40) is a function of the difference $s - s'$, $e(s)$ is a homogeneous process (with zero mean). Moreover, the spectral density corresponding to (40) is equal to σ^2 and consequently the $e(s)$ is a *white-noise* process. The latter is also a Gaussian process, for the increments of $W_0(s)$ are Gaussian. Consequently, if $W_0(s)$ is a Wiener process with covariance (38), its derivative is a Gaussian white-noise process with covariance (40). Now consider the random process

$$W_\nu(s)=\int_0^s \frac{(s-u)^{\nu-1}}{(\nu-1)!}\,W_0(u)\,du \tag{41}$$

This is an ISRF-ν, and by differentiating with respect to s we get

$$\frac{d^\nu}{ds^\nu}\,W_\nu(s)=W_0(s) \qquad \text{and} \qquad \frac{d^{\nu+1}}{ds^{\nu+1}}\,W_\nu(s)=e(s)$$

Clearly, the corresponding covariances satisfy

$$\frac{\partial^{2\nu}}{\partial s^\nu\,\partial s'^\nu}\,c_{w_\nu}(s,s')=c_{w_0}(s-s')=c_{w_0}(h)$$

and

$$\frac{\partial^{2(\nu+1)}}{\partial s^{\nu+1}\,\partial s'^{\nu+1}}\,c_{w_\nu}(s,s')=c_e(s-s')=c_e(h)=\sigma^2\,\delta(h) \tag{42}$$

where $h=s-s'$. By integrating (42) first $\nu+1$ times with respect to s and then $\nu+1$ times with respect to s' we obtain an expression of the form (34), namely,

$$c_{w_\nu}(s,s')=k_{w_\nu}(h)+p_\nu(s,s')$$

where

$$k_{w_\nu}(h)=(-1)^{\nu+1}\int_0^h \frac{(h-u)^{2\nu+1}}{(2\nu+1)!}\,\delta(u)\,du=(-1)^{\nu+1}\frac{|h|^{2\nu+1}}{(2\nu+1)!} \tag{43}$$

On the strength of Definition 5, a continuous function $k_x(\mathbf{h})$ in R^n is a GSC-ν if and only if it is a *conditionally nonnegative-definite function*, namely,

$$\langle k_x(\mathbf{h}),\, q_1(\mathbf{s})q_2(\mathbf{s}')\rangle\geq 0 \tag{44}$$

for all $q_1, q_2\in Q_\nu$. In this case, the function $k_x(\mathbf{h})$ will be called a *permissible* GSC-ν. Assume now the following representation of the ISRF-ν $X(\mathbf{s})$, $X(q_{\mathbf{s}}^*)=X^0(\mathbf{s})=\langle q_{\mathbf{s}}^*(\mathbf{s}'),\,X^0(\mathbf{s}')\rangle$.

The covariance is written

$$\begin{aligned} c_x(\mathbf{s},\mathbf{s}') &= E[\langle q_{\mathbf{s}}^*(\mathbf{s}''),\,X^0(\mathbf{s}'')\rangle\langle q_{\mathbf{s}'}^*(\mathbf{s}''),\,X^0(\mathbf{s}'')\rangle]\\ &= E[Y_{q_{\mathbf{s}}^*}Y_{q_{\mathbf{s}'}^*}]=\langle q_{\mathbf{s}}^*(\mathbf{s}'')q_{\mathbf{s}'}^*(\mathbf{s}'''),\,k_x(\mathbf{s}''-\mathbf{s}''')\rangle \end{aligned}$$

with

$$q_{\mathbf{s}}^*(\mathbf{s}'')=\delta_{\mathbf{s}}(\mathbf{s}'')-\sum_{\rho=|\boldsymbol{\rho}|\leq\nu} g_\rho(\mathbf{s})\,\eta_\rho(\mathbf{s}'')\in Q_\nu$$

More precisely,

$$\begin{aligned} c_x(\mathbf{s}, \mathbf{s}') = &\langle \delta_{\mathbf{s}}(\mathbf{s}'')\delta_{\mathbf{s}'}(\mathbf{s}'''), k_x(\mathbf{s}''-\mathbf{s}''')\rangle \\ &- \sum_{\rho=|\boldsymbol{\rho}|\leq\nu} g_\rho(\mathbf{s})\langle \eta_\rho(\mathbf{s}''), k_x(\mathbf{s}''-\mathbf{s}')\rangle \\ &- \sum_{\rho=|\boldsymbol{\rho}|\leq\nu} g_\rho(\mathbf{s}')\langle \eta_\rho(\mathbf{s}''), k_x(\mathbf{s}-\mathbf{s}'')\rangle \\ &+ \sum_{\rho=|\boldsymbol{\rho}|\leq\nu}\ \sum_{\rho'=|\boldsymbol{\rho}'|\leq\nu} g_\rho(\mathbf{s})g_{\rho'}(\mathbf{s}') \\ &\times \langle \eta_\rho(\mathbf{s}'')\eta_{\rho'}(\mathbf{s}'''), k_x(\mathbf{s}''-\mathbf{s}''')\rangle \end{aligned}$$

or

$$\begin{aligned} c_x(\mathbf{s}, \mathbf{s}') = k_x(\mathbf{s}-\mathbf{s}') &- \sum_{\rho=|\boldsymbol{\rho}|\leq\nu} \alpha_\rho(\mathbf{s}')g_\rho(\mathbf{s}) \\ &- \sum_{\rho=|\boldsymbol{\rho}|\leq\nu} \alpha_\rho(\mathbf{s})g_\rho(\mathbf{s}') + \sum_{\rho=|\boldsymbol{\rho}|\leq\nu} \\ &\times \sum_{\rho'=|\boldsymbol{\rho}'|\leq\nu} c_{\rho\rho'}g_\rho(\mathbf{s})g_{\rho'}(\mathbf{s}') \end{aligned}$$

for appropriate coefficients $\alpha_\rho(\cdot)$ and $c_{\rho\rho'}$. These results lead to the following proposition.

Proposition 5: Assume that $X(\mathbf{s})$ is an ISRF-ν in R^n. Then its covariance function can be expressed by

$$c_x(\mathbf{s}, \mathbf{s}') = k_x(\mathbf{h}) + p_\nu(\mathbf{s}, \mathbf{s}') \tag{45}$$

where $\mathbf{h} = \mathbf{s} - \mathbf{s}'$, $k_x(\mathbf{h})$ is the associated GSC-ν and $p_\nu(\mathbf{s}, \mathbf{s}')$ is a polynomial with variable coefficients, of degree ν in $\mathbf{s}, \mathbf{s}'$.

On the basis of Proposition 5, the following corollary is straightforward.

Corollary 1: Let $X(q)$ be a GSRF-ν in R^n; then it is valid that

$$\begin{aligned} c_x(q_1, q_2) &= \langle c_x(\mathbf{s}, \mathbf{s}'), q_1(\mathbf{s})q_2(\mathbf{s}')\rangle \\ &= \langle k_x(\mathbf{h}), q_1(\mathbf{s})q_2(\mathbf{s}')\rangle \end{aligned} \tag{46}$$

Let $X(\mathbf{s})$ be a differentiable ISRF-ν in R^n. By definition all

$$Y_a(\mathbf{s}) = D_a^{(\nu+1)}X(\mathbf{s}) = \frac{\partial^{\nu+1}}{\partial s_1^{\nu_1}\ldots\partial s_n^{\nu_n}}X(\mathbf{s}) \tag{47}$$

are zero-mean homogeneous SRF for any

$$a \in A_{\nu+1} = \left\{a = (\nu_1, \nu_2, \ldots, \nu_n): \sum_{i=1}^{n} \nu_i = \nu+1\right\}$$

Consider the homogeneous linear differential operator of order $\nu+1$

$$L_{\nu+1}[X(\mathbf{s})] = \sum_{|\boldsymbol{\rho}|=\nu+1} \beta_{\boldsymbol{\rho}} D_a^{(\rho)}X(\mathbf{s}) = Y(\mathbf{s}) \tag{48}$$

where $\beta_{\boldsymbol{\rho}}=\beta_{\rho_1\ldots\rho_n}$ are coefficients and $Y(\mathbf{s})$ is a zero-mean homogeneous SRF. To the operator (48) we associate the auxiliary homogeneous polynomial of order ν

$$p_\nu(\mathbf{s})=\sum_{|\boldsymbol{\rho}|=\nu} b_{\boldsymbol{\rho}}\mathbf{s}^{\boldsymbol{\rho}}=\sum_{\rho_1}\cdots\sum_{\rho_n} b_{\rho_1\ldots\rho_n}s_1^{\rho_1}\ldots s_n^{\rho_n} \tag{49}$$

such that $L_{\nu+1}[p_\nu(\mathbf{s})]=0$. We usually assume that $\beta_{\boldsymbol{\rho}}=1$ for all $\boldsymbol{\rho}$. An interesting case of the operator (48) arises by setting

$$Y_i(\mathbf{s})=\frac{\partial^{\nu+1}}{\partial s_i^{\nu+1}}X(\mathbf{s}) \tag{50}$$

These are, by definition, homogeneous SRF for all $i=1,2,\ldots,n$. The corresponding homogeneous linear differential operator of order $\nu+1$ is

$$L_{\nu+1}[X(\mathbf{s})]=\nabla^{\nu+1}X(\mathbf{s})=Y(\mathbf{s}) \tag{51}$$

The covariance of each $Y_a(\mathbf{s})$ in Eq. (47) can be written

$$\begin{aligned} c_{Y_a}(\mathbf{h}) &= E[D_a^{(\nu+1)}X(\mathbf{s})\, D_a^{(\nu+1)}X(\mathbf{s}')] \\ &= D_a^{(2\nu+2)}c_x(\mathbf{s},\mathbf{s}') \end{aligned} \tag{52}$$

and, naturally, that of $Y(\mathbf{s})$ becomes

$$c_Y(\mathbf{h})=\sum_{a\in A_{\nu+1}} c_{Y_a}(\mathbf{h})=\sum_{a\in A_{\nu+1}} D_a^{(2\nu+2)}c_x(\mathbf{s},\mathbf{s}') \tag{53}$$

In the case of Eq. (50), the corresponding covariances are given by

$$c_{Y_i}(\mathbf{h})=\frac{\partial^{2\nu+2}}{\partial s_i^{\nu+1}\,\partial s_i'^{\nu+1}}c_x(\mathbf{s},\mathbf{s}') \tag{54}$$

and

$$c_Y(\mathbf{h})=\sum_{i=1}^{n}\sum_{j=1}^{n}\frac{\partial^{2\nu+2}}{\partial s_i^{\nu+1}\,\partial s_j'^{\nu+1}}c_x(\mathbf{s},\mathbf{s}') \tag{55}$$

In view of Eq. (45), Eq. (55) yields

$$\begin{aligned} c_Y(\mathbf{h}) &= (-1)^{\nu+1}\sum_{i=1}^{n}\frac{\partial^{2\nu+2}}{\partial h_i^{2\nu+2}}k_x(\mathbf{h}) \\ &= (-1)^{\nu+1}\,\nabla^{2\nu+2}k_x(\mathbf{h}) \end{aligned} \tag{56}$$

Example 4: If $X(\mathbf{s})$ is an ISRF-1, Eq. (51) gives the usual *Laplace* equation $\nabla^2X(\mathbf{s})=\Delta X(\mathbf{s})=Y(\mathbf{s})$. The covariances of the fields involved are related by $c_Y(\mathbf{h})=\Delta_{\mathbf{s}}\,\Delta_{\mathbf{s}'}c_x(\mathbf{s},\mathbf{s}')=\Delta^2k_x(\mathbf{h})$.

The spectral representations of the covariances $c_{Y_a}(\mathbf{h})$ are written

$$c_{Y_a}(\mathbf{h})=\int \exp[i(\mathbf{w}\cdot\mathbf{h})]\,dQ_{Y_a}(\mathbf{w})$$

where by Bochner's theorem the Q_{Y_a} are positive summable measures in R^n, without atom at the origin. Naturally,

$$c_Y(\mathbf{h}) = \sum_{a \in A_{\nu+1}} c_{Y_a}(\mathbf{h}) = \int \exp[i(\mathbf{w} \cdot \mathbf{h})]\, dQ_Y(\mathbf{w}) \tag{57}$$

where $Q_Y = \sum_{a \in A_{\nu+1}} Q_{Y_a}$ is also a positive summable measure in R^n, without atom at the origin. Clearly, the spectral representation (57) remains valid for the case of the covariance (55); furthermore, by combining Eqs. (56) and (57) we see that the measure Q_Y is the Fourier transform of $(-1)^{\nu+1}\Delta^{\nu+1}k_x(\mathbf{h})$, and that

$$k_x(\mathbf{h}) = \int \frac{\exp[i(\mathbf{w} \cdot \mathbf{h})]}{|\mathbf{w}|^{2\nu+2}}\, dQ_Y(\mathbf{w}) = \int \frac{\cos(\mathbf{w} \cdot \mathbf{h})}{|\mathbf{w}|^{2\nu+2}}\, dQ_Y(\mathbf{w}) \tag{58}$$

taking into account that $k_x(\mathbf{h})$ is a real function. But this integral is not convergent near zero. We can, however, take care of this difficulty by writing

$$k_x(\mathbf{h}) = \int \frac{\cos(\mathbf{w} \cdot \mathbf{h}) - p_\nu(\mathbf{w} \cdot \mathbf{h})}{|\mathbf{w}|^{2\nu+2}}\, dQ_Y(\mathbf{w})$$

The existence of the polynomial

$$p_\nu(\mathbf{w} \cdot \mathbf{h}) = \sum_{\rho = |\boldsymbol{\rho}| \le \nu} (-1)^\rho \frac{(\mathbf{w} \cdot \mathbf{h})^{2\rho}}{(2\rho)!}$$

does not cause any problem, since it is filtered out by the $\Delta^{\nu+1}$ operator. In general, GSC-ν $k_x(\mathbf{h})$ of the form

$$k_x(\mathbf{h}) = \int \frac{\cos(\mathbf{w} \cdot \mathbf{h}) - p_\nu(\mathbf{w} \cdot \mathbf{h})}{|\mathbf{w}|^{2\nu+2}}\, dQ_Y(\mathbf{w}) + p_{2\nu}(\mathbf{h}) < \infty \tag{59}$$

where $p_{2\nu}(\mathbf{h})$ is an even polynomial of degree $\le 2\nu$, provide solutions to the partial differential equation (56). In other words, if one GSC-ν is known, all the others can be derived by adding even polynomials of degree $\le 2\nu$. These covariances constitute the *class* $\mathcal{F}_\nu^X$ of the GSC-ν. A GSC-ν is also a GSC-μ for all $\mu \ge \nu$. Then, from the obvious inequality

$$|\cos(\mathbf{w} \cdot \mathbf{h}) - p_\nu(\mathbf{w} \cdot \mathbf{h})| \le \frac{(\mathbf{w} \cdot \mathbf{h})^{2\nu+2}}{(2\nu+2)!}$$

it follows that

$$|k_x(\mathbf{h})| \le \frac{\int dQ_Y(\mathbf{w})}{(2\nu+2)!} |\mathbf{h}|^{2\nu+2} = \alpha |\mathbf{h}|^{2\nu+2}$$

where $\alpha < \infty$, or

$$\lim_{|\mathbf{h}| \to \infty} \frac{k_x(\mathbf{h})}{|\mathbf{h}|^{2\nu+2}} = 0 \tag{60}$$

which assures the existence of the integral (59). Moreover the uniqueness of the spectral representation (59) together with Eq. (56) implies that $k_x(\mathbf{h})$ is $2\nu+2$ times differentiable [hence, the $X(\mathbf{s})$ is $\nu+1$ times differentiable] if and only if Eq. (59) is valid. Since

$$(-1)^{\nu+1}[\cos(\mathbf{w}\cdot\mathbf{h})-p_\nu(\mathbf{w}\cdot\mathbf{h})]\geq 0$$

it follows that

$$(-1)^{\nu+1}k_x(\mathbf{h})\geq 0 \tag{61}$$

Representations (58) and (59) associate with $k_x(\mathbf{h})$ the *spectral measure*

$$K_x(\mathbf{w})\,d\mathbf{w}=\frac{dQ_Y(\mathbf{w})}{|\mathbf{w}|^{2\nu+2}}$$

which is the same for the various elements $k_x(\mathbf{h})\in\mathscr{F}_\nu^X$. The latter fact emphasizes the importance of the spectral measure in stochastic analysis. For example, the conditions for differentiability can be generalized as follows:

Corollary 2: The GSC-ν $k_x(\mathbf{h})$ is 2λ times differentiable, which implies that the ISRF-ν $X(\mathbf{s})$ is λ times differentiable, if and only if

$$\int\frac{dQ_Y(\mathbf{w})}{(1+|\mathbf{w}|^2)^{\nu+1-\lambda}}<\infty \tag{62}$$

For $\lambda=\nu+1$, Eq. (62) implies

$$\int dQ_Y(\mathbf{w})=\int|\mathbf{w}|^{2\nu+2}K_x(\mathbf{w})\,d\mathbf{w}<\infty$$

The reader may find it interesting to compare this condition with the condition imposed by the existence of $\nabla^{\nu+1}X(\mathbf{s})$ in the case of homogeneous SRF (see remark 4, Section 7 of Chapter 2). When $\lambda=\nu=0$, Eq. (62) gives

$$\int\frac{dQ_Y(\mathbf{w})}{1+|\mathbf{w}|^2}=\int\frac{|\mathbf{w}|^2}{1+|\mathbf{w}|^2}K_x(\mathbf{w})\,d\mathbf{w}<\infty$$

(compare with the relevant condition for the spectral representation of SRF with homogeneous increments; see also Eq. (32), Section 11 of Chapter 2).

To proceed further we note that, since the GSC-ν $k_x(\mathbf{h})$ satisfies Eq. (60) it is a tempered generalized function and, therefore, $k_x(\mathbf{h})$ admits a Fourier transform in the sense of generalized functions. Hence, the following definition makes sense.

Definition 6: The *generalized spectral density function of order* ν, $K_x(\mathbf{w})$, is defined in the sense of generalized functions as the n-fold Fourier transform of the GSC-ν $k_x(\mathbf{h})$.

In view of the last definition we can write

$$|\mathbf{w}|^{2\nu+2}K_x(\mathbf{w}) = C_Y(\mathbf{w}) \tag{63}$$

where $C_Y(\mathbf{w}) = dQ_Y(\mathbf{w})/d\mathbf{w}$ is the spectral density function [Fourier transform of the covariance $c_Y(\mathbf{h})$]. Equation (63) is the spectral equivalent of the differential form (56). The analysis above can be extended to the case where the ISRF-ν is not in general differentiable. More precisely, by using Proposition 4 the following proposition, which is a generalization of the results obtained in Section 11 of Chapter 2, can be proven (Matheron, 1973).

Proposition 6: A continuous and symmetric function $k_x(\mathbf{h})$ on R^n is a GSC-ν; that is, it is a conditionally nonnegative-definite function in the sense of Definition 6 if and only if it admits the representation

$$k_x(\mathbf{h}) = \int \frac{\cos(\mathbf{w}\cdot\mathbf{h}) - 1_B(\mathbf{w})p_\nu(\mathbf{w}\cdot\mathbf{h})}{|\mathbf{w}|^{2\nu+2}}\, dQ_Y(\mathbf{w}) + p_{2\nu}(\mathbf{h}) \tag{64}$$

where $1_B(\mathbf{w})$ is the indicator of a neighborhood of the origin $\mathbf{w} = \mathbf{O}$ and Q_Y is a positive measure not necessarily summable, but without atom at the origin, and such that

$$\int \frac{dQ_Y(\mathbf{w})}{(1+|\mathbf{w}|^2)^{\nu+1}} < \infty \tag{65}$$

[tempered measure, see also Eq. (32); Eq. (65) is (62) for $\lambda = 0$]. Notice that it is always assumed that the ISRF-ν is without drift ($E[X]=0$); in the case that there is a drift, $p_{2\nu}(\mathbf{h})$ should be replaced by an even polynomial of degree $\leq 2\nu+2$.

Example 5: If $Y(s)$ is a zero-mean white-noise random process with covariance

$$c_Y(r) = \delta(r) \tag{66}$$

Eq. (35) gives

$$k_x(r) = (-1)^{\nu+1}\frac{r^{2\nu+1}}{(2\nu+1)!} \tag{67}$$

[Compare with Eq. (43) above.]

Example 6: Suppose that

$$c_Y(r) = \exp[-r] \tag{68}$$

The corresponding unidimensional GSC-ν from Eq. (35) is

$$k_x(r) = \frac{(-1)^{\nu+1}}{(2\nu+1)!}\,\frac{\gamma(2\nu+2, -r)}{\exp[r]} \tag{69}$$

where $\gamma(\cdot, \cdot)$ denotes the incomplete gamma function. For example, if $\nu=0$, Eq. (69) becomes

$$k_x(r)=\frac{-1-(r-1)\exp[r]}{\exp[r]}$$

Finally, the following proposition is due to Matheron (1973).

Proposition 7: Let $X(\mathbf{s})$ be a zero-mean ISRF-ν. If the GSC-ν of $X(\mathbf{s})$ satisfy inequalities of the form $|k_x(\mathbf{h})| \leq a+b|\mathbf{h}|^{2\nu}$, where a and b are constants, then $X(\mathbf{s})$ is the *restriction* of a continuous ISRF-$(\nu-1)$; that is, $X(\mathbf{s})$ can be considered to be a ISRF-$(\nu-1)$.

3.4 Permissibility Criteria for Generalized Spatial Covariances of Order ν

In view of Definition 5, a function can serve as a permissible GSC-ν if and only if it is a conditionally nonnegative-definite function. However, the application of condition (44) is not possible in practical situations. It is thus necessary to establish alternative criteria capable of testing, in an analytically and computationally tractable way, whether a function is permissible as a GSC-ν.

Such a criterion—the *seventh criterion of permissibility* (*COP-7*; Christakos, 1984b)—which can be derived with the help of Proposition 6 above, is as follows: A continuous, symmetric, and isotropic function $k_x(r)$, $r=|\mathbf{h}|$ in R^n is a permissible GSC-ν if and only if it decreases slower than $r^{2\nu+2}$ as $r\to\infty$; that is,

$$\lim_{r\to\infty}\frac{k_x(r)}{r^{2\nu+2}}=0 \tag{70}$$

and the corresponding generalized spectral function $K_x(\omega)$ exists (in the sense of generalized functions), includes no atom at the origin, and is such that the $\omega^{2\nu+2}K_x(\omega)$ is a nonnegative measure on R^n; that is,

$$\omega^{2\nu+2}K_x(\omega)\geq 0 \tag{71}$$

Remark 2: Clearly, an isotropic function that is a permissible GSC-ν in n dimensions is also permissible in any dimension $m<n$. In view of the foregoing analysis, several classes of GSC-ν may be constructed as follows:

(i) Assume a model for the stationary unidimensional covariance $c_Y(r)$.
(ii) Apply Eq. (35) to derive a GSC-ν $k_x(r)$ in R^1.
(iii) Use space transformations (Chapter 6) to derive the isotropic GSC-ν model in the space of interest R^n.

The following examples give some insight regarding this matter.

Table 3.1 Permissibility Conditions for Model (72)

$\nu=0$: $a_0, c_0 \geq 0$
$\nu=1$: $a_0, c_0, c_1 \geq 0$
$\nu=2$: $a_0, c_0, c_2 \geq 0$, and $c_1 \geq -\sqrt{\frac{20(n+3)}{3(n+1)}\, c_0 c_2}$

Example 7: Covariance (67) can be generalized to R^n (Delfiner, 1976)

$$k_x(r) = a_0\,\delta(r) + \sum_{\rho=0}^{\nu} (-1)^{\rho+1} c_\rho r^{2\rho+1} \tag{72}$$

where the coefficients a_0 and c_ρ satisfy certain permissibility conditions derived by means of COP-7; see Table 3.1.

At this point, it is interesting to note that a unidimensional ISRF-ν admits a GSC-ν of the polynomial form (72) if and only if it can be expressed as

$$X(s) = b_0 W(s) + b_1 \int_0^s W(v)\, dv + \cdots + b_\nu \int_0^s \frac{(s-v)^{\nu-1}}{(\nu-1)!} W(v)\, dv \tag{73}$$

where b_ρ, $\rho = 0, 1, \ldots, \nu$ are real coefficients and $W(s)$ is a unidimensional ISRF-0 with GSC-0 $k_w(h) = -|h|$, say a Wiener process.

Example 8: Consider the model

$$k_x(r) = \frac{(-1)^{\nu+1}}{(2\nu+1)!} \frac{\gamma(2\nu+2, -br)}{b^{2\nu+2} \exp[br]} \tag{74}$$

where $b > 0$. Equation (74) is obtained from the homogeneous unidimensional covariance $c_Y(r) = \exp[-br]$, $b > 0$ (see also Example 6 above). After some manipulations Eq. (74) may also be written as

$$k_x(r) = \frac{(-1)^{\nu+1}}{b^{2\nu+2} \exp[br]} \left[1 - \exp(br) \sum_{i=0}^{2\nu+1} \frac{(-br)^i}{i!} \right] \tag{75}$$

Equation (74) is a permissible model in R^1. In R^n the coefficient b may need to satisfy some additional constraints, which again may arise from COP-7.

3.5 Cross-Generalized Spatial Covariances of Orders ν and ν'

Let $X_p(\mathbf{s})$ and $X_{p'}(\mathbf{s})$ be two ISRF of orders ν and ν', respectively. The $Y_p(\mathbf{s}) = X_p^{(\nu+1)}(\mathbf{s})$ and $Y_{p'}(\mathbf{s}) = X_{p'}^{(\nu'+1)}(\mathbf{s})$ are, by definition, homogeneous

SRF. Consider the matrix

$$\mathbf{K_X} = [k_{x_p x_{p'}}(\mathbf{h})] \tag{76}$$

where $k_{x_p x_{p'}}(\mathbf{h})$, $p, p' = 1, \ldots, k$, and $\mathbf{h} = \mathbf{s} - \mathbf{s}' \in R^n$ are *cross-GSC-*(ν, ν'). The $k_{x_p x_{p'}}(\mathbf{h})$ are continuous functions in R^n satisfying

$$E[Y_{q_p}(\mathbf{s}) Y_{q_{p'}}(\mathbf{s}')] = \langle k_{x_p x_{p'}}(\mathbf{h}), q_p(\mathbf{s}) q_{p'}(\mathbf{s}')\rangle \geq 0 \tag{77}$$

for all $q_p \in Q_\nu$, $q_{p'} \in Q_{\nu'}$. Here $Y_{q_p}(\mathbf{s})$ and $Y_{q_{p'}}(\mathbf{s}')$ are the corresponding CSRF-ν and ν', respectively. The generalized and ordinary covariances are related by

$$\begin{aligned} D^{(\nu+1,\nu'+1)} c_{x_p x_{p'}}(\mathbf{s}, \mathbf{s}') &= (-1)^{\nu+\nu'} D^{(\nu+1,\nu'+1)} k_{x_p x_{p'}}(\mathbf{h}) \\ &= c_{Y_p Y_{p'}}(\mathbf{h}) \end{aligned} \tag{78}$$

Example 9: Consider the processes $W_\nu(s)$ and $W_{\nu'}(s)$, defined by Eq. (41) for $\nu' \neq \nu$. The cross-covariances are written

$$\frac{\partial^{\nu+\nu'}}{\partial s^\nu \, \partial s'^{\nu'}} c_{w_\nu w_{\nu'}}(s, s') = c_{w_0}(s - s') \tag{79}$$

Working along similar lines we find that the cross-GSC-ν, ν' $(\nu \geq \nu')$ of $W_\nu(s)$ and $W_{\nu'}(s)$ is written

$$k_{w_\nu w_{\nu'}}(h) = (-1)^{\nu'+1} \frac{|h|^{\nu+\nu'+1}}{(\nu+\nu'+1)!} \tag{80}$$

Under certain conditions (see analysis of previous sections), the GSC-ν $k_{x_p x_{p'}}(\mathbf{h})$ may be represented by

$$k_{x_p x_{p'}}(\mathbf{h}) = \int [\cos(\mathbf{w} \cdot \mathbf{h}) - p_\nu(\mathbf{w} \cdot \mathbf{h})] K_{x_p x_{p'}}(\mathbf{w}) \, d\mathbf{w} \tag{81}$$

where $K_{x_p x_{p'}}(\mathbf{w})$ are cross-spectral functions such that the trace

$$K_x(\mathbf{w}) = \sum_{p=1}^{k} K_{x_p x_{p'}}(\mathbf{w})$$

of the nonnegative-definite matrix $\tilde{\mathbf{K}}_\mathbf{X} = [K_{x_p x_{p'}}(\mathbf{w})]$ satisfies

$$\int \frac{|\mathbf{w}|^{2\nu+2} K_x(\mathbf{w}) \, d\mathbf{w}}{(1+\mathbf{w}^2)^{\nu+1}} < \infty, \qquad K_x(\mathbf{w}) \geq 0 \tag{82}$$

The above representation leads to the following *eighth criterion of permissibility* (*COP-8*): For $\mathbf{K_X}$ of Eq. (76) to be a matrix of permissible GSC-ν, the matrix $\tilde{\mathbf{K}}_\mathbf{X}$ must be nonnegative-definite and its trace must satisfy the conditions of Eq. (82).

4. Discrete Linear Representations of Spatial Random Fields

4.1 Discrete Intrinsic Spatial Random Fields of Order ν

Since in practice the data are discretely distributed in space, the matter calls for a *discrete linear representation*, suitable for practical applications. Within this context the notion of spatial polynomials is relevant. More precisely, in R^n a polynomial of the form

$$p_\nu(\mathbf{s}) = \sum_{0\le|\boldsymbol{\rho}|\le\nu} b_{\boldsymbol{\rho}} g_{\boldsymbol{\rho}}(\mathbf{s}) = \sum_{0\le|\boldsymbol{\rho}|\le\nu} b_{\boldsymbol{\rho}} \mathbf{s}^{\boldsymbol{\rho}}$$
$$= \sum_{\rho_1} \cdots \sum_{\rho_n} b_{\rho_1\ldots\rho_n} s_1^{\rho_1} \ldots s_n^{\rho_n} \tag{1}$$

will be called the *spatial polynomial of order* ν $(SP\text{-}\nu)$.

Example 1: In one dimension and for $\nu = 2$

$$p_2(\mathbf{s}) = \sum_{0\le\rho_1+\rho_2\le 2} b_{\rho_1,\rho_2} s_1^{\rho_1} s_2^{\rho_2}$$
$$= b_{00} + b_{10}s_1 + b_{01}s_2 + b_{20}s_1^2 + b_{02}s_2^2 + b_{11}s_1 s_2 \tag{2}$$

Consider now the discrete SRF $X(\mathbf{s}_i)$, $\mathbf{s}_i \in R^n$, $i = 1, 2, \ldots, m$. Let $q \in Q_m$; that is, q is a real measure on R^n with finite support such that

$$q(\mathbf{s}) = \sum_{i=1}^{m} q(\mathbf{s}_i)\, \delta(\mathbf{s}_i - \mathbf{s}) = \sum_{i=1}^{m} q_i\, \delta_i(\mathbf{s}) \tag{3}$$

The corresponding discrete GSRF and CSRF are written, respectively,

$$X(q) = \left\langle \sum_{i=1}^{m} q_i\, \delta_i(\mathbf{s}), X(\mathbf{s}) \right\rangle = \sum_{i=1}^{m} q_i X(\mathbf{s}_i) \tag{4}$$

and

$$Y_q(\mathbf{s}) = \left\langle \sum_{i=1}^{m} q_i\, \delta_i(\mathbf{s}'), S_{\mathbf{s}} X(\mathbf{s}') \right\rangle = \sum_{i=1}^{m} q_i S_{\mathbf{s}} X(\mathbf{s}_i) \tag{5}$$

Equipped with the above notions we can now formulate definitions and results analogous to those obtained in Section 3.

Definition 1: The discrete SRF $Y_q(\mathbf{s})$ of Eq. (5) will be called a *spatial increment of order* ν $(SI\text{-}\nu)$ if

$$\sum_{i=1}^{m} q_i g_\rho(\mathbf{s}_i) = 0 \tag{6}$$

for all $\rho = |\boldsymbol{\rho}| \le \nu$. In this case the coefficients $\{q_i\}$, $i = 1, 2, \ldots, m$ will be called *admissible coefficients of order* ν $(AC\text{-}\nu)$.

In other words, condition (6) assigns weights q_i to points $\mathbf{s}_i$, $i = 1, 2, \ldots, m$ so that monomials of degree up to ν in the coordinates of $\mathbf{s}_i$ are canceled.

Table 3.2 Conditions (6) in Two Dimensions

$\nu = 0$:			
$\sum_{i=1}^{m} q_i = 0$			
$\nu = 1$:			
$\sum_{i=1}^{m} q_i = 0$;	$\sum_{i=1}^{m} q_i s_{i1} = 0$	and	$\sum_{i=1}^{m} q_i s_{i2} = 0$
$\nu = 2$:			
$\sum_{i=1}^{m} q_i = 0$;	$\sum_{i=1}^{m} q_i s_{i1} = 0$	and	$\sum_{i=1}^{m} q_i s_{i2} = 0$
$\sum_{i=1}^{m} q_i s_{i1}^2 = 0$;	$\sum_{i=1}^{m} q_i s_{i2}^2 = 0$	and	$\sum_{i=1}^{m} q_i s_{i1} s_{i2} = 0$

The number of conditions is equal to that of monomials of degree $\rho \leq \nu$. In Table 3.2 conditions (6) are presented for the common in practice two-dimensional case $[\mathbf{s}_i = (s_{i1}, s_{i2}) \in R^2]$.

Remark 1: Since by Definition 1, $\sum_{i=1}^{m} q_i = 0$ $(\rho = 0)$, the SI-ν $Y_q(\mathbf{s})$ can be viewed as a high-pass filter, so that the resulting pattern enhances details of the original SRF $X(\mathbf{s})$.

Definition 2: A discrete *ISRF-ν* is an ISRF $X(\mathbf{s})$ for which the corresponding SI-ν $Y_q(\mathbf{s})$ is a zero-mean homogeneous SRF.

Example 2: See Fig. 3.1, where $\mathbf{s}_i = (s_{i1}, s_{i2}) \in R^2$, $i = 1, 2, \ldots, 13$. Let

$$\begin{aligned} Y_q(\mathbf{s}_1) = \sum_{i=1}^{13} q_i X(\mathbf{s}_i) &= 20X(\mathbf{s}_1) - 8X(\mathbf{s}_2) - 8X(\mathbf{s}_3) \\ &- 8X(\mathbf{s}_4) - 8X(\mathbf{s}_5) + 2X(\mathbf{s}_6) + 2X(\mathbf{s}_7) + 2X(\mathbf{s}_8) \\ &+ 2X(\mathbf{s}_9) + X(\mathbf{s}_{10}) + X(\mathbf{s}_{11}) + X(\mathbf{s}_{12}) + X(\mathbf{s}_{13}) \end{aligned} \tag{7}$$

By taking into account the geometry of Fig. 3.1 we find

$$\sum_{i=1}^{13} q_i = 20 - 4 \times 8 + 4 \times 2 + 4 \times 1 = 0$$

and

$$\begin{aligned} \sum_{i=1}^{13} q_i s_{i1} = 20 s_{11} &- 8(s_{11} + \Delta s) - 8 s_{11} - 8(s_{11} - \Delta s) \\ &- 8 s_{11} + 2(s_{11} + \Delta s) + 2(s_{11} + \Delta s) + 2(s_{11} - \Delta s) \\ &+ 2(s_{11} - \Delta s) + s_{11} + 2\Delta s + s_{11} \\ &+ s_{11} - 2\Delta s + s_{11} = 0 \end{aligned}$$

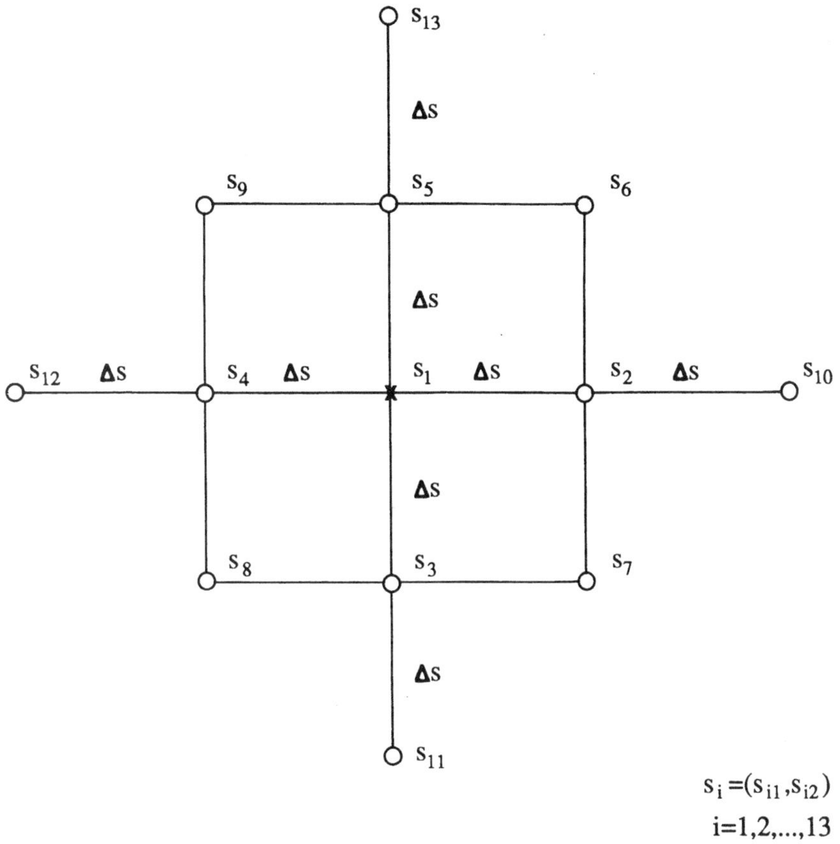

Figure 3.1 The R^2 case

Similarly,

$$\sum_{i=1}^{13} q_i s_{i2} = 0, \qquad \sum_{i=1}^{13} q_i s_{i1}^2 = 0, \qquad \sum_{i=1}^{13} q_i s_{i2}^2 = 0$$

and

$$\sum_{i=1}^{13} q_i s_{i1} s_{i2} = 0$$

Hence, the conditions of Definition 1 are satisfied and the $Y_q(\mathbf{s})$ is an SI-2. If $Y_q(\mathbf{s})$ is a zero-mean homogeneous SRF, the corresponding $X(\mathbf{s})$ is an ISRF-2. Conversely, if $X(\mathbf{s})$ is an ISRF-2, $Y_q(\mathbf{s})$ is a zero-mean SI-2.

Proposition 1: Any discrete unidimensional process $X(s)$ represented by the difference equation

$$\Delta_s^{\nu+1} X(s) = Y_q(s) \tag{8}$$

where

$$\Delta_s^{\nu+1} X(s) = \sum_{i=0}^{\nu+1} (-1)^i C_{\nu+1}^i X(s+\nu+1-i) \tag{9}$$

is the finite difference operator of order $\nu+1$ in s, $C_{\nu+1}^i = \binom{\nu+1}{i}$, and $Y_q(s)$ is homogeneous, is an ISRF-ν. [The argument in Eq. (9) is, in general, of the form $s-(i-\nu-1)\ \Delta s$, where Δs is the unit step; nevertheless, for simplicity and without loss of generality, it is assumed that $\Delta s = 1$.]

Proof: According to Definition 2 it suffices to show that

$$S_{\nu,\rho} = \sum_{i=0}^{\nu+1} (-1)^i C_{\nu+1}^i i^\rho = 0 \tag{10}$$

for all $\rho \leq \nu$. Indeed, Eq. (10) is valid (Riordan, 1983), and this completes the proof. □

Remark 2: Solutions of the difference equation (8) are

$$X(s) = \sum_{i=0}^{s-\nu-1} C_{s-i-1}^{\nu} Y_q(i) = \sum_s^{\nu+1} Y_q(s) \tag{11}$$

where $\sum_s^{\nu+1}$ denotes the summation of order $\nu+1$ in s, $s \geq \nu+1$ and $X(s \leq \nu) = 0$. All discrete ISRF-ν $X(s)$ admit representations of the form (11).

Example 3: Consider the ISRF-ν

$$X_\nu(s) = \int_0^s \frac{(s-u)^\nu}{\nu!} Y(u)\, du \tag{12}$$

where $Y(s)$ is a homogeneous field (see Proposition 1, Section 3 above). By choosing

$$q(s) = \sum_{i=0}^{\nu+1} (-1)^i C_{\nu+1}^i\ \delta_{\nu+1-i}(s) = \Delta_s^{\nu+1}\ \delta(s) \tag{13}$$

the SI-ν $Y_q(s)$ may be expressed by Eq. (8) above, viz.,

$$Y_q(s) = \Delta_s^{\nu+1} X_\nu(s) \tag{14}$$

which is the discrete analog of the CSRF-ν. Furthermore, assuming that $X_\nu(s)$ is differentiable, the generalized field

$$\begin{aligned} Y(q) &= \sum_{i=0}^{\nu+1} (-1)^i C_{\nu+1}^i Y(s_i) \\ &= Y_q^{(\nu+1)}(0) = \sum_{i=0}^{\nu+1} (-1)^i C_{\nu+1}^i X_\nu^{(\nu+1)}(s_i) \end{aligned} \tag{15}$$

is homogeneous too. It is interesting to compare representations (11) and (12): All discrete-parameter ISRF-ν admit a representation of the form (11), while for a continuous-parameter ISRF-ν to be represented by (12) it is necessary that it is $\nu+1$ times differentiable in s.

4.2 The Generalized Spatial Covariance of Order ν

Just as for continuously distributed SRF, the discrete version of the conditional nonnegative-definiteness property of the GSC-ν states that a function $k_x(\mathbf{h})$ in R^n is a *GSC-ν* if and only if for all pairs of AC-ν $\{q_i\}$ and $\{q_i'\}$,

$$E[X(q)X(q')] = E[Y_q(\mathbf{O})\,Y_{q'}(\mathbf{O})] = E\left[\sum_{i=1}^{m} q_i X(\mathbf{s}_i) \sum_{i=1}^{m} q_i' X(\mathbf{s}_i')\right] = \sum_{i=1}^{m}\sum_{j=1}^{m} q_i q_j' k_x(\mathbf{h}_{ij}) \geq 0 \tag{16}$$

or equivalently,

$$E[X(q)]^2 = E[Y_q(\mathbf{O})]^2 = E\left[\sum_{i=1}^{m} q_i X(\mathbf{s}_i)\right]^2 = \sum_{i=1}^{m}\sum_{j=1}^{m} q_i q_j k_x(\mathbf{h}_{ij}) \geq 0 \tag{17}$$

where $\mathbf{h}_{ij} = \mathbf{s}_i - \mathbf{s}_j$.

Example 4: Consider, again, Proposition 1 above. The corresponding homogeneous covariance will be

$$\begin{aligned} c_{Y_q}(s-s') &= E[\Delta_s^{\nu+1} X(s)\, \Delta_{s'}^{\nu+1} X(s')] \\ &= \sum_{i=0}^{\nu+1}\sum_{i'=0}^{\nu+1} (-1)^{i+i'} C_{\nu+1}^{i} C_{\nu+1}^{i'} \\ &\quad \times c_x(s-i+\nu+1,\, s'-i'+\nu+1) \\ &= \Delta_{s'}^{\nu+1}\, \Delta_s^{\nu+1} c_x(s, s'). \end{aligned} \tag{18}$$

Since $X(s)$ is an ISRF-ν, there exists a GSC-ν satisfying (18). The latter, taking Eq. (8) into account, becomes

$$\begin{aligned} c_{Y_q}(h) &= (-1)^{\nu+1} \sum_{i=0}^{2\nu+2} (-1)^i C_{2\nu+2}^i k_x(h+\nu+1-i) \\ &= (-1)^{\nu+1}\, \Delta_h^{2(\nu+1)} k_x(h-\nu-1) \end{aligned} \tag{19}$$

where $h = s - s'$. If $k_x(\cdot)$ is known, Eq. (19) for $h=0$ allows us to calculate the variance of any SI-ν $Y_q(s)$. Moreover, by inverting the last equation, the GSC-ν can be expressed in terms of the ordinary homogeneous covariance as follows

$$k_x(h) = (-1)^{\nu+1} \sum_{i=0}^{h-\nu-1} C_{h+\nu-i}^{1+2\nu} c_{Y_q}(i) - \tfrac{1}{2}(-1)^{\nu+1} C_{h+\nu}^{1+2\nu} c_{Y_q}(0) \tag{20}$$

Remark 3: Notice the difference between the continuous formula (35), Section 3 and the discrete (20): Assuming that the $Y(s)$ is a white-noise process with $c_Y(h) = \delta(h)$, Eq. (35), Section 3 gives

$$k_x(h) = (-1)^{\nu+1} \frac{|h|^{2\nu+1}}{(2\nu+1)!} \tag{21}$$

while, with the same assumptions regarding $Y_q(s)$, Eq. (20) gives ($h > 0$)

$$k_x(h) = (-1)^{\nu+1} \frac{h(h^2-1)\ldots(h^2-\nu^2)}{2(2\nu+1)!} \tag{22}$$

Example 5: Consider now Eq. (12) above. The corresponding GSC-ν satisfy the following interesting formulas

$$k_{x_\nu}(h) = -\int_0^h du_2 \int_0^{u_2} du_1\, k_{x_{\nu-1}}(u_1) \tag{23}$$

and

$$k_{x_\nu}(h) = (-1)^\nu \int_0^h du_{2\nu} \ldots \int_0^{u_2} du_1\, k_{x_0}(u_1)$$

$$= (-1)^\nu \int_0^h \frac{(h-u)^{2\nu-1}}{(2\nu-1)!} k_{x_0}(u)\, du \tag{24}$$

For example, if $k_{x_0}(h) = -|h|$, Eq. (24) yields Eq. (21) above.

The proposition that follows is useful in constructing SI-ν in practice. It also plays a central role in deriving the system of spatial estimation equations (see Chapter 9 later).

Proposition 2: Assume that $X(\mathbf{s})$ is an SRF and let

$$\hat{X}(\mathbf{s}_0) = \sum_{i=1}^{m} \lambda_i X(\mathbf{s}_i) \tag{25}$$

be the linear estimator of $X(\mathbf{s})$ at the location $\mathbf{s}_0$ so that

$$E[\hat{X}(\mathbf{s}_0) - X(\mathbf{s}_0)] = 0 \tag{26}$$

and

$$E[X(\mathbf{s}_0)] = \sum_{|\rho| \leq \nu} \eta_\rho \mathbf{s}_0^\rho \tag{27}$$

where $\eta_{\mathbf{\rho}}$ are suitable coefficients. Then, the difference

$$Y_q(\mathbf{s}_0) = \hat{X}(\mathbf{s}_0) - X(\mathbf{s}_0) = \sum_{i=0}^{m} \lambda_i X(\mathbf{s}_i) \tag{28}$$

where $\lambda_0 = -1$, is a SI-ν.

Proof: From Eqs. (25), (26) we immediately obtain

$$\sum_{i=0}^{m} \lambda_i E[X(\mathbf{s}_i)] = 0$$

and by inserting Eq. (27),

$$\sum_{i=0}^{m} \lambda_i \left[\sum_{|\mathbf{\rho}| \le \nu} \eta_{\mathbf{\rho}} \mathbf{s}_i^{\mathbf{\rho}} \right] = 0$$

or

$$\sum_{|\mathbf{\rho}| \le \nu} \eta_{\mathbf{\rho}} \left[\sum_{\mathrm{i}=0}^{\mathrm{m}} \lambda_{\mathrm{i}} \mathbf{s}_i^{\mathbf{\rho}} \right] = 0$$

or

$$\sum_{i=0}^{m} \lambda_i \mathbf{s}_i^{\mathbf{\rho}} = 0 \qquad \text{for all} \quad |\mathbf{\rho}| \le \nu$$

Then, according to Definition 1 above the SRF (28) is an SI-ν. □

Remark 4: Obviously, if the $Y_q(\mathbf{s})$ of Eq. (28) is homogeneous, the $X(\mathbf{s})$ is by definition an ISRF-ν. Conversely, if $X(\mathbf{s})$ is an ISRF-ν, the $Y_q(\mathbf{s})$ of Eq. (28) is a homogeneous SI-ν.

Remark 5: There are several linear estimators of the form (25). A convenient choice may be as follows. Assume that the ISRF-ν $X(\mathbf{s})$ is of the form (15) of Section 3 above. We can fit an estimator

$$\hat{X}(\mathbf{s}_0) = \sum_{|\mathbf{\rho}| \le \nu} \hat{\beta}_\rho \mathbf{s}_0^{\mathbf{\rho}} \tag{29}$$

by least squares, where

$$\hat{\beta}_\rho = \sum_{i=1}^{m} b_{i,\rho} X(\mathbf{s}_i) \tag{30}$$

are estimates of β_ρ of Eq. (15) of Section 3 for suitable coefficients $b_{i,\rho}$. This yields

$$\hat{X}(\mathbf{s}_0) = \sum_{i=1}^{m} \lambda_i X(\mathbf{s}_i)$$

where

$$\lambda_i = \sum_{|\mathbf{\rho}| \le \nu} b_{i,\rho} \mathbf{s}_0^{\mathbf{\rho}}$$

and the SRF

$$Y_q(\mathbf{s}_0)=\hat{X}(\mathbf{s}_0)-X(\mathbf{s}_0)=\sum_{i=0}^{m}\lambda_i X(\mathbf{s}_i)$$
$$=\sum_{i=1}^{m}\left[\sum_{|\boldsymbol{\rho}|\leq\nu} b_{i,\rho}\mathbf{s}_0^{\boldsymbol{\rho}}\right]X(\mathbf{s}_i)-X(\mathbf{s}_0) \tag{31}$$

is a SI-ν. Note that

$$\sum_{i=0}^{m}\lambda_i\mathbf{s}_i^{\boldsymbol{\rho}}=0 \qquad (\lambda_0=-1, |\boldsymbol{\rho}|\leq\nu)$$

or

$$\mathbf{s}_0^{\boldsymbol{\rho}'}-\sum_{i=1}^{m}\left[\sum_{|\boldsymbol{\rho}|\leq\nu} b_{i,\rho}\mathbf{s}_0^{\boldsymbol{\rho}}\right]\mathbf{s}_0^{\boldsymbol{\rho}'}=0 \tag{32}$$

for all $|\boldsymbol{\rho}'|\leq\nu$. In summary: First find coefficients $b_{i,\rho}(i=1,\ldots,m$ and $\rho=0,\ldots,\nu)$ satisfying Eq. (32); then use Eq. (31) to construct SI-ν.

Remark 6: The above results may be used to provide an instructive proof of Proposition 5, Section 3 above. The field $Y_q(\mathbf{s}_0)=\hat{X}(\mathbf{s}_0)-X(\mathbf{s}_0)$, where

$$\hat{X}(\mathbf{s}_0)=\sum_{|\boldsymbol{\rho}|\leq\nu}\hat{\beta}_\rho\mathbf{s}_0^{\boldsymbol{\rho}}$$

is a least squares estimate of the ISRF-ν $X(\mathbf{s}_0)$, is a SI-ν. The corresponding covariance is

$$\begin{aligned} c_x(\mathbf{s}_0,\mathbf{s}_0')=E[X(\mathbf{s}_0)X(\mathbf{s}_0')]&=E[Y_q(\mathbf{s}_0)Y_q(\mathbf{s}_0')] \\ &-\sum_{|\boldsymbol{\rho}|\leq\nu}E[Y_q(\mathbf{s}_0)\hat{\beta}_\rho]\mathbf{s}_0'^{\boldsymbol{\rho}} \\ &-\sum_{|\boldsymbol{\rho}|\leq\nu}E[Y_q(\mathbf{s}_0')\hat{\beta}_\rho]\mathbf{s}_0^{\boldsymbol{\rho}} \\ &+\sum_{|\boldsymbol{\rho}|\leq\nu}\sum_{|\boldsymbol{\rho}'|\leq\nu}E[\hat{\beta}_\rho\hat{\beta}_{\rho'}]\mathbf{s}_0^{\boldsymbol{\rho}}\mathbf{s}_0'^{\boldsymbol{\rho}'} \end{aligned}$$

On the other hand,

$$\begin{aligned} E[Y_q(\mathbf{s}_0)Y_q(\mathbf{s}_0')]&=\sum_{i=0}^{m}\sum_{j=0}^{m}\lambda_i\lambda_j' k_x(\mathbf{s}_i-\mathbf{s}_j') \\ &=k_x(\mathbf{s}_0-\mathbf{s}_0')-\sum_{|\boldsymbol{\rho}|\leq\nu}\left[\sum_{j=1}^{m}b_{j,\rho}k_x(\mathbf{s}_0-\mathbf{s}_j')\right]\mathbf{s}_0^{\boldsymbol{\rho}} \\ &-\sum_{|\boldsymbol{\rho}'|\leq\nu}\left[\sum_{i=1}^{m}b_{i,\rho'}k_x(\mathbf{s}_i-\mathbf{s}_0')\right]\mathbf{s}_0'^{\boldsymbol{\rho}'} \\ &+\sum_{|\boldsymbol{\rho}|\leq\nu}\sum_{|\boldsymbol{\rho}'|\leq\nu}\left[\sum_{i=1}^{m}\sum_{j=1}^{m}b_{i,\rho}b_{i,\rho'}k_x(\mathbf{s}_i-\mathbf{s}_j')\right]\mathbf{s}_0^{\boldsymbol{\rho}}\mathbf{s}_0'^{\boldsymbol{\rho}'} \end{aligned}$$

By combining the last two equations we obtain Eq. (45) of Section 3 above.

4.3 The Generalized Representation Set of Order ν

In Section 3.2 we demonstrated that, given a GSRF-ν $X(q)$ one may assign various ISRF-ν $X(\mathbf{s})$, all having the same CSRF-ν $Y_q(\mathbf{s}) = X(S_{-\mathbf{s}}q)$. In fact we saw that there can be defined a whole set of $X(\mathbf{s})$, which was termed GRS-ν and was denoted by $\mathbf{X}_q$. The subscript q means that while the members $X(\mathbf{s})$ of the GRS-ν are functions of $\mathbf{s}$, the set $\mathbf{X}_q$ itself is a function of the AC-ν $\{q_i\}$ chosen. Here is an example of such a set.

Example 6: Let $X_1(s)$ be an ISRF-2 in R^1 and let

$$Y_1(s) = X_1(s+3h) - 3X_1(s+2h) + 3X_1(s+h) - X_1(s) \tag{33}$$

be a zero mean stationary process. The latter is a SI-2 since

$$\sum_{i=1}^{4} q_i = 1 - 3 + 3 - 1 = 0$$

$$\sum_{i=1}^{4} q_i s_i = s + 3h - 3(s+2h) + 3(s+h) - s = 0$$

and

$$\sum_{i=1}^{4} q_i s_i^2 = (s+3h)^2 - 3(s+2h)^2 + 3(s+h)^2 - s^2 = 0$$

Now assume that $X_2(s)$ is another random process defined by

$$X_2(s) = X_1(s) + a_2 s^2 + a_1 s + a_0 \tag{34}$$

where the coefficients a_0, a_1, and a_2 are all random variables. After some manipulations involving the last two equations, the incremental process

$$Y_2(s) = X_2(s+3h) - 3X_2(s+2h) + 3X_2(s+h) - X_2(s)$$

leads to the equality $Y_2(s) = Y_1(s)$. In other words, we see here that if the ISRF-2 $X_1(s)$ is a representation $\in \mathbf{X}_q$, the $X_2(s)$ given by Eq. (34) is also an ISRF-2, sharing the same SI-2 with $X_1(s)$, that is, $X_2(s) \in \mathbf{X}_q$. In fact, $X_1(s)$ and $X_2(s)$ are two different solutions of Eq. (33).

To fix ideas, Proposition 3 of Section 3 will be now reconsidered in the light of the discretely distributed SRF.

Proposition 3: Assume that the discrete ISRF-ν $X^0(\mathbf{s})$ is a representation from a set $\mathbf{X}_q$. The SRF $X^a(\mathbf{s})$ is another representation of $\mathbf{X}_q$ if and only if it can be expressed as

$$X^a(\mathbf{s}) = X^0(\mathbf{s}) + \sum_{|\boldsymbol{\rho}| \leq \nu} a_\rho \mathbf{s}^{\boldsymbol{\rho}} \tag{35}$$

where the coefficients a_ρ are random variables in $L_2(\Omega, F, P)$.

Proof: First, let $X^0(\mathbf{s}) \in \mathbf{X}_q$. We will show that any other representation, say $X^a(\mathbf{s}) \in \mathbf{X}_q$, will be of the form (35). Both ISRF-ν $X^0(\mathbf{s})$ and $X^a(\mathbf{s})$ have the same SI-ν so that

$$\sum_{i=1}^{m} q_i[X^0(\mathbf{s}_i) - X^a(\mathbf{s}_i)] = 0 \qquad \text{for all AC-}\nu \quad \{q_i\}$$

Then $X^b(\mathbf{s}) = X^0(\mathbf{s}) - X^a(\mathbf{s})$ is an ISRF-ν with SI-ν

$$Y_q^b(\mathbf{s}_0) = \sum_{i=1}^{m} q_i X^b(\mathbf{s}_i)$$

By letting

$$\lambda_i = \sum_{|\boldsymbol{\rho}| \le \nu} b_{i,\rho} \mathbf{s}_0^{\boldsymbol{\rho}}$$

as in Remark 5 above, we get

$$\begin{aligned} X^b(\mathbf{s}_0) &= \sum_{i=1}^{m} \lambda_i X^b(\mathbf{s}_i) \\ &= \sum_{|\boldsymbol{\rho}| \le \nu} \left[\sum_{i=1}^{m} b_{i,\rho} X^b(\mathbf{s}_i) \right] \mathbf{s}_0^{\boldsymbol{\rho}} \\ &= \sum_{|\boldsymbol{\rho}| \le \nu} a_\rho \mathbf{s}_0^{\boldsymbol{\rho}} \end{aligned}$$

where

$$a_\rho = \sum_{i=1}^{m} b_{i,\rho} X^b(\mathbf{s}_i)$$

Thus,

$$X^0(\mathbf{s}) - X^a(\mathbf{s}) = \sum_{|\boldsymbol{\rho}| \le \nu} a_\rho \mathbf{s}_0^{\boldsymbol{\rho}}$$

which is Eq. (35). Conversely, assume that $X^a(\mathbf{s})$ is given by Eq. (35), where $X^0(\mathbf{s})$ is a representation from the GRS-ν $\mathbf{X}_q$ and a_ρ, $\rho = 0, 1, \ldots, \nu$ are random variables. It will be shown that $X^a(\mathbf{s})$ is another representation from $\mathbf{X}_q$. Indeed, (35) is written

$$\sum_{i=1}^{m} q_i X^a(\mathbf{s}_i) = \sum_{i=1}^{m} q_i X^0(\mathbf{s}_i) + \sum_{i=1}^{m} q_i \left[\sum_{|\boldsymbol{\rho}| \le \nu} a_\rho \mathbf{s}_0^{\boldsymbol{\rho}} \right]$$

But the AC-ν $\{q_i\}$ $(i = 1, \ldots, m)$ cancel out all polynomials of degree $\rho \le \nu$, and so

$$\sum_{i=1}^{m} q_i X^a(\mathbf{s}_i) = \sum_{i=1}^{m} q_i X^0(\mathbf{s}_i) = Y_q(\mathbf{s}) \qquad \square$$

According to the analysis above, there should exist several covariances $\{c^a(\mathbf{s}, \mathbf{s}'), a = 1, 2, \ldots\}$ corresponding to the members of the set $\mathbf{X}_q$. Let $X^0(\mathbf{s})$ be a representation of the $\mathbf{X}_q$ and $c^0(\mathbf{s}, \mathbf{s}')$ be the corresponding covariance; also, let $X^a(\mathbf{s})$ be another representation from $\mathbf{X}_q$, so that Eq. (35) is valid. Then, if $c^a(\mathbf{s}, \mathbf{s}')$ is the covariance of $X^a(\mathbf{s})$,

$$c^a(\mathbf{s}, \mathbf{s}') = c^0(\mathbf{s}, \mathbf{s}') + \sum_{|\boldsymbol{\rho}| \leq \nu} E[a_\rho X^0(\mathbf{s}')]\mathbf{s}^{\boldsymbol{\rho}} + \sum_{|\boldsymbol{\zeta}| \leq \nu} E[a_\zeta X^0(\mathbf{s})]\mathbf{s}'^{\boldsymbol{\zeta}} + \sum_{|\boldsymbol{\rho}| \leq \nu} \sum_{|\boldsymbol{\zeta}| \leq \nu} E[a_\rho a_\zeta]\mathbf{s}^{\boldsymbol{\rho}}\mathbf{s}'^{\boldsymbol{\zeta}}$$

Hence, we can write, in more concise form,

$$c^a(\mathbf{s}, \mathbf{s}') = c^0(\mathbf{s}, \mathbf{s}') + \sum_{|\boldsymbol{\rho}| \leq \nu} \alpha_\rho(\mathbf{s}')\mathbf{s}^{\boldsymbol{\rho}} + \sum_{|\boldsymbol{\zeta}| \leq \nu} \beta_\zeta(\mathbf{s})\mathbf{s}'^{\boldsymbol{\zeta}} + \sum_{|\boldsymbol{\rho}| \leq \nu} \sum_{|\boldsymbol{\zeta}| \leq \nu} c_{\rho\zeta}\mathbf{s}^{\boldsymbol{\rho}}\mathbf{s}'^{\boldsymbol{\zeta}} \tag{36}$$

where $\alpha_\rho(\mathbf{s}')$, $\beta_\zeta(\mathbf{s})$, and $c_{\rho\zeta}$ are suitable coefficients.

4.4 Cross-Generalized Spatial Covariances of Orders ν and ν'

In Example 9, Section 3 the determination of the unidimensional cross-GSC-ν, ν' $c_{w_\nu w_{\nu'}}(s, s')$ of the ISRF-ν $W_\nu(s)$ and ISRF-ν' $W_{\nu'}(s)$ was discussed. Here we will examine these results in a discrete context.

Let $X_1(\mathbf{s})$ and $X_2(\mathbf{s})$ be two ISRF of order ν and ν', respectively. Let $Y_{q_1}(\mathbf{s})$ and $Y_{q_2}(\mathbf{s})$ be the corresponding SI-ν and SI-ν'. A function $k_{x_1x_2}(\mathbf{h})$, $\mathbf{h} \in R^n$ will be called a *cross-generalized spatial covariance of order* ν, ν' (*CGSC-*ν, ν') if and only if

$$E[Y_{q_1}(\mathbf{O})Y_{q_2}(\mathbf{O})] = E\left[\sum_{i=1}^{m_1} q_{1,i}X_1(\mathbf{s}_i)\sum_{j=1}^{m_2} q_{2,j}X_2(\mathbf{s}_j)\right] = \sum_{i=1}^{m_1}\sum_{j=1}^{m_2} q_{1,i}q_{2,j}k_{x_1x_2}(\mathbf{s}_i - \mathbf{s}_j) \tag{37}$$

for all AC-ν $\{q_{1,i}\}$, $i = 1, 2, \ldots, m_1$ and AC-ν' $\{q_{2,j}\}$, $j = 1, 2, \ldots, m_2$.

Example 7: The setting is the same as in Example 5 above. Here, though, we seek the determination of the cross-GSC-ν, ν' $k_{x_\nu x_{\nu'}}(h)$ for any two ISRF

$X_\nu(s)$ and $X_{\nu'}(s)$ of orders ν and ν', respectively. Assume that $\nu \geq \nu'$ so that $X_{\nu'}(s)$ can be considered as an ISRF-ν, also. Then by letting

$$P_{\nu\nu'} = E[\Delta_s^{\nu+1} X_\nu(s)\, \Delta_s^{\nu'+1} X_{\nu'}(s)]$$

$$= \sum_{i=0}^{\nu+1} \sum_{j=0}^{\nu'+1} (-1)^{i+j} C_{\nu+1}^{i} C_{\nu'+1}^{j} k_{x_\nu x_{\nu'}}(i-j) \tag{38}$$

and by working along similar lines as in Example 5 we obtain

$$\frac{d^{\nu+\nu'}}{dh^{\nu+\nu'}} k_{x_\nu, x_{\nu'}}(h) = (-1)^\nu k_{x_0}(h) \tag{39}$$

For example, if $k_{x_0}(h) = -|h|$, Eq. (39) gives

$$k_{x_\nu x_{\nu'}}(h) = (-1)^{\nu+1} \frac{|h|^{\nu+\nu'+1}}{(\nu+\nu'+1)!} \tag{40}$$

5. Stochastic Differential and Difference Equations

5.1 Basic Equations

Stochastic differential equations (SDE) over space are mathematical descriptions of physical systems expressed under the general form

$$L[X(\mathbf{s})] = Y(\mathbf{s}) \tag{1}$$

where $X(\mathbf{s})$ is the unknown SRF, $L[\cdot]$ is a given operator, and $Y(\mathbf{s})$ is a known SRF—also called a forcing function. The notion of randomness enters the SDE models in three basic ways:

(a) random initial conditions;
(b) random forcing function; and
(c) random coefficient of the operator $L[\cdot]$.

Forms (a), (b), and (c) are not mutually exclusive; in fact, many SDE are a mixture of these three forms.

While many of the basic theoretical problems in the study of SDE are essentially the same as those for classical (deterministic) differential equations (existence and uniqueness of solutions, stability, dependence of solutions on coefficients and initial conditions, etc.), there are considerable differences, as well. These differences naturally arise from the study of the

SRF described by the SDE. For example, the sense of an SDE will depend on whether the SRF is viewed as a collection of random variables (mean square sense) or as a family of realizations (sample function sense).

As a consequence, stochastic solutions to SDE may be obtained in more than one way:

(i) In terms of SRF representations (in the mean square or the sample function sense).

(ii) By determining the probability distribution functions or the characteristic functions of the SRF involved (in more complicated situations functionals need to be used).

(iii) By means of the deterministic differential equations that govern the corresponding statistical moments, or the deterministic algebraic equations relating the corresponding spectral functions.

When approach (iii) is used, one should first verify the existence and uniqueness of the SDE solutions (this should be done, even when the derivation of the solutions themselves is not a feasible objective). Another important aspect of the statistical moments approach (iii) is the so-called *closure* problem. More specifically, a closure problem arises when we have a hierarchy of N equations with $N+1$ statistical moments. It is then necessary to establish a suitable approximation technique of converting the infinite hierarchy of equations into a closed set.

There are several good references on the subject of SDE, including Syski (1967), Gihman and Skorokhod (1972), Arnold (1974), Friedman (1975, 1976), Da Prato and Tubaro (1987), Sobczyk (1991). Some examples of SDE related to earth sciences are given in Chapter 6. Here, as well as in the following sections, we will focus only on a limited number of SDE topics closely related to the random field models considered in the book.

Certain interesting results emerge from a fruitful interaction of the generalized SRF theory and the theory of SDE. For illustration purposes, let $X(\mathbf{s})$ be an ISRF-ν. By definition, all

$$Y_i(\mathbf{s}) = \frac{\partial^{\nu+1}}{\partial s_i^{\nu+1}} X(\mathbf{s})$$

are homogeneous SRF and so is the field defined by the SDE

$$Y(\mathbf{s}) = \sum_{i=1}^{n} Y_i(\mathbf{s}) = \nabla^{\nu+1} X(\mathbf{s}) \tag{2}$$

Here $L[\cdot] = \nabla^{\nu+1}[\cdot]$. Equation (2) is interpreted in the m.s.s. [under certain conditions—$Y(\mathbf{s})$ is a white-noise SRF, etc.—Eq. (2) may also be interpreted as an equation for the sample function].

In R^3 the solution to Eq. (2)—assuming that all stochastic conditions of its validity exist—is written

$$X(\mathbf{s}) = \frac{(-1)^{\nu+1/2} G\left(\frac{2-\nu}{2}\right)}{2^{\nu+1}\pi^{3/2} G\left(\frac{\nu+1}{2}\right)} \int |\mathbf{u}-\mathbf{s}|^{\nu-2} Y(\mathbf{u})\, d\mathbf{u}$$

with appropriate boundary conditions. Similarly, in R^2 the

$$X(\mathbf{s}) = \frac{1}{2^{\nu}\pi\left[\left(\frac{\nu-1}{2}\right)!\right]^2} \int |\mathbf{u}-\mathbf{s}|^{\nu-1} \log|\mathbf{u}-\mathbf{s}|\, Y(\mathbf{u})\, d\mathbf{u} \tag{3}$$

also satisfies (2). The corresponding covariance functions are related by (see also Section 3 above)

$$\nabla^{2\nu+2} k_x(\mathbf{h}) = \Delta^{\nu+1} k_x(\mathbf{h}) = (-1)^{\nu+1} c_Y(\mathbf{h}) \tag{4}$$

Equation (4) can be solved with respect to $k_x(\mathbf{h})$: In R^3

$$k_x(\mathbf{h}) = \frac{(-1)^{\nu+1} G\left(\frac{1-2\nu}{2}\right)}{2^{2\nu+2}\pi^{3/2} G(\nu+1)} \int |\mathbf{v}-\mathbf{h}|^{2\nu-1} c_Y(\mathbf{v})\, d\mathbf{v} \tag{5}$$

and in R^2

$$k_x(\mathbf{h}) = \frac{1}{2^{2\nu+1}\pi(\nu!)^2} \int |\mathbf{v}-\mathbf{h}|^{2\nu} \log|\mathbf{v}-\mathbf{h}|\, c_Y(\mathbf{v})\, d\mathbf{v} \tag{6}$$

Example 1: If the ISRF-1 $X(\mathbf{s})$ and the homogeneous SRF $Y(\mathbf{s})$ satisfy the *Poisson* SDE (which is used, e.g., in hydrology to model steady-state groundwater flow)

$$\nabla^2 X(\mathbf{s}) = \Delta X(\mathbf{s}) = Y(\mathbf{s}) \tag{7}$$

one may study $X(\mathbf{s})$ by means of $Y(\mathbf{s})$. More precisely, in R^3 we obtain

$$X(\mathbf{s}) = -\frac{1}{4\pi} \int \frac{Y(\mathbf{u})}{|\mathbf{u}-\mathbf{s}|}\, d\mathbf{u} \tag{8}$$

and in R^2

$$X(\mathbf{s}) = \frac{1}{2\pi} \int \log|\mathbf{u}-\mathbf{s}|\, Y(\mathbf{u})\, d\mathbf{u} \tag{9}$$

The covariances satisfy

$$\begin{aligned} \nabla^2_{\mathbf{s}} \nabla^2_{\mathbf{s}'} c_x(\mathbf{s}, \mathbf{s}') &= \sum_{i=1}^{n} \sum_{j=1}^{n} \frac{\partial^4 c_x(\mathbf{s}, \mathbf{s}')}{\partial s_i^2\, \partial s_j'^2} \\ &= \nabla^4_{\mathbf{h}} k_x(\mathbf{h}) = c_Y(\mathbf{h}) \end{aligned} \tag{10}$$

Given the homogeneous covariance $c_Y(\mathbf{h})$, the last equation can be solved with respect to $k_x(\mathbf{h})$. In R^3,

$$k_x(\mathbf{h}) = -\frac{1}{8\pi}\int |\mathbf{v}-\mathbf{h}| c_Y(\mathbf{v})\, d\mathbf{v} \tag{11}$$

and in R^2

$$k_x(\mathbf{h}) = \frac{1}{8\pi}\int (\mathbf{v}-\mathbf{h})^2 \log|\mathbf{v}-\mathbf{h}| c_Y(\mathbf{v})\, d\mathbf{v} \tag{12}$$

For example, in the isotropic three-dimensional circumstance with $c_Y(r) = \exp[-br]$, the GSC-1 is given by

$$k_x(r) = -\frac{4}{b^5 r} - \frac{r}{b^3} + \frac{4\exp[-br]}{b^5 r} + \frac{\exp[-br]}{b^4} \tag{13}$$

Other GSC-1 may be obtained by adding to Eq. (13) the polynomial $c_0 + c_1 r + c_2 r^2$ for properly chosen c_0, c_1, and c_2.

In the case of discrete-parameter ISRF, the semivariogram function is closely related to stochastic difference equation models.

Example 2: Consider the on-line difference model (Christakos, 1982)

$$X(s_i) = \alpha X(s_i - h) + W(s_i) \tag{14}$$

where α is a deterministic coefficient and $W(s_i)$ is an SRF with, usually, zero mean and variance σ_w^2. Physically, models of the form (14) are based on the ascertainment that many natural processes have a structure not so arbitrary but of some low-order recursive form. It is easily seen that for $\alpha = 1$, it holds that

$$\gamma_x(h) = \tfrac{1}{2}\sigma_w^2 \tag{15}$$

For more complex situations of spatial variability, it is possible to fit to the data difference equations of the form

$$X(\mathbf{s}_i) = \sum_{k=1}^{K} \alpha_k X(\mathbf{s}_{i-k}) + \beta W(\mathbf{s}_i) \tag{16}$$

where α_k and β are suitable coefficients. Equations (15) and (16) imply a connection between stochastic difference equations and the notion of spatial increments discussed above: If the coefficients α_k and β are chosen so that they constitute a set of admissible coefficients of some order ν (AC-ν), then the SRF $W(\mathbf{s}_i)$ is an SI-ν.

5.2 Adjoint Equations—The Air Pollution Problem

The generalized SRF formalism discussed in the previous sections provides simple solutions to stochastic partial differential equation (SPDE) problems, which are difficult to solve or cannot be solved in terms of ordinary SRF.

Consider an SPDE of the general form (1) above, with the appropriate initial conditions. Equation (1) generates an ordinary SRF $X(\mathbf{s})$; it will be called here the "basic SPDE." To Eq. (1) we associate the functional

$$\langle L[X(\mathbf{s})], X^*(\mathbf{s})\rangle = \langle Y(\mathbf{s}), X^*(\mathbf{s})\rangle \tag{17}$$

where $X^*(\mathbf{s})$ is an ordinary SRF whose meaning will become clear shortly.

Assume now that by using the functional operators introduced in previous sections Eq. (17) is transformed into the form

$$\langle L^*[X^*(\mathbf{s})], X(\mathbf{s})\rangle = F[X(\mathbf{s}), X^*(\mathbf{s})] + \langle Y(\mathbf{s}), X^*(\mathbf{s})\rangle \tag{18}$$

where $F[\cdot]$ is a suitable function of SRF. Let us next denote

$$L^*[X^*(\mathbf{s})] = q(\mathbf{s}) \tag{19}$$

Equation (19) will be termed the "adjoint SPDE." In view of Eq. (19), Eq. (18) becomes

$$X(q) = F[X(\mathbf{s}), X^*(\mathbf{s})] + \langle Y(\mathbf{s}), X^*(\mathbf{s})\rangle \tag{20}$$

where $X(q) = \langle X(\mathbf{s}), q(\mathbf{s})\rangle$ is a generalized SRF in the sense given in the preceding sections.

Therefore, the "basic SPDE" (1) has been transformed into the functional equation (20), where $X^*(\mathbf{s})$ is the solution of the "adjoint SPDE" (19). What are the practical implications of this transformation? To answer this question, it is necessary to consider the following facts:

(a) First, the physical interpretation of the generalized SRF $X(q)$ depends on the choice of the function $q(\mathbf{s})$. If we choose $q(\mathbf{s}) = \delta(\mathbf{s}-\mathbf{s}')$, then

$$X(q) = X(\mathbf{s}') \tag{21}$$

In other words, Eq. (21) gives the value of $X(\mathbf{s}')$ at point $\mathbf{s}'$; if we let

$$q(\mathbf{s}) = \begin{cases} \dfrac{1}{V} & \text{if} \quad \mathbf{s} \in V \\ 0 & \text{otherwise} \end{cases}$$

then

$$X(q) = \frac{1}{V}\int_V X(\mathbf{s})\, d\mathbf{s} \tag{22}$$

that is, the mean value of $X(\mathbf{s})$ within the volume V.

(b) A class of very important practical problems in earth sciences can be solved faster and more efficiently by means of the generalized SRF. For illustration purposes consider the example below (see also Marchuk, 1986, for the deterministic counterpart of the problem).

Example 3: Let the ordinary SRF $X(s)$ represent the concentration of aerosol substance transported in the atmosphere over a region A. For simplicity, assume that the phenomenon is spatially one dimensional (the three-dimensional, time-dependent version will be studied in Chapter 5) and is governed by the basic stochastic differential equation

$$\mu\frac{\partial^2 X(s)}{\partial s^2} - \upsilon X(s) = W\delta(s - s_0) \tag{23}$$

with initial conditions $X(\pm\infty) = 0$, where μ is the diffusion coefficient, υ is a quantity that has inverse time dimension, W is the capacity of the pollution source, and s_0 determines the location of the pollution source. The problem is to find a region $U \subset A$ so that for all $s_0 \in U$, the pollution over a specific area $G \subset A$ does not exceed a permissible level c, viz.,

$$X(s') < c \tag{24}$$

for all $s' \in G$.

To solve this problem in the "traditional" way, one needs to solve Eq. (23) many times for various possible $s_0 \in A$ and check the solution with respect to Eq. (24), to determine the region U. This is clearly not practical. The generalized SRF concept, on the other hand, offers a better alternative, as follows. First, by applying the procedure of Eqs. (18) and (19) we find that the adjoint SPDE is

$$\mu\frac{\partial^2 X^*(s)}{\partial s^2} - \upsilon X^*(s) = q(s) \tag{25}$$

with initial conditions $X^*(\pm\infty) = 0$. Next, from Eq. (20) we get

$$X(q) = WX^*(s_0) \tag{26}$$

By choosing $q(s) = \delta(s - s')$, we obtain [Eq. (21)] $X(q) = X(s')$, and the solution of Eq. (25) is given by

$$X^*(s, s') = \frac{1}{2\mu\sqrt{\frac{\upsilon}{\mu}}}\exp\left[-\sqrt{\frac{\upsilon}{\mu}}(s - s')\operatorname{sgn}(s - s')\right] \tag{27}$$

where $\operatorname{sgn}(s - s') > 0$ if $s \geq s'$, and <0 if $s \leq s'$ [assuming that all the stochastic conditions for the validity of solution (27) are satisfied]. By substituting

Eq. (27) into Eq. (26) and taking Eq. (24) into account we find

$$W\frac{1}{2\mu\sqrt{\frac{\upsilon}{\mu}}}\exp\left[-\sqrt{\frac{\upsilon}{\mu}}(s_0-s')\operatorname{sgn}(s_0-s')\right]<c \tag{28}$$

for all $s'\in G$, which determines the required region U. Moreover, for each given location $s_0\in U$ of the pollution source, spatial realizations of the aerosol concentration $X(s)$ can be found from Eq. (26), viz.,

$$X(s)=W\frac{1}{2\mu\sqrt{\frac{\upsilon}{\mu}}}\exp\left[-\sqrt{\frac{\upsilon}{\mu}}(s_0-s)\operatorname{sgn}(s_0-s)\right] \tag{29}$$

4

The Factorable Random Field Model

"One of the peculiar difficulties in Probability Theory is that its difficulties sometimes are not seen."

G. Boole

1. Introduction

This chapter introduces the class of factorable random fields (FRF). The factorability concept stems from the study of the orthogonality structure of the probability density functions involved and applies in a spatial, temporal, or a spatiotemporal context. The key role is played here by a probabilistic entity called the theta function of the RF on a probability space. Under certain integrability conditions, this function has an interesting orthogonality structure in a measurable space.

Hence attention is focused on fields that exhibit such a structure and are called factorable random fields. These concepts are defined precisely in Section 2, where their properties both in the space and the frequency domains are explored. In Section 3, it is proven that any strictly monotonic function of an FRF is itself factorable. How an FRF can arise is the topic of Section 4. It is shown that, under fairly broad conditions, random fields satisfying the Pearson differential equation may belong to the class of FRF. Also, SRF whose bivariate probability density function is of the so-called isofactorial form (these densities have been used in nonlinear geostatistics) constitute a special group of FRF. This being the case, the applicability of nonlinear geostatistics can be significantly extended by means of the theory of FRF. Section 5 introduces the on-line nonlinear state-nonlinear observation system, and its compatibility with the assumption of factorability is discussed.

Common features of the unidimensional FRF (also called factorable random processes, FRP) and the Markov and martingale processes are pointed out. The nonlinear state-nonlinear observation system is introduced and its compatibility with the assumption of factorability is discussed. Analysis shows that under the factorability assumption, on-line nonlinear state-nonlinear observation systems are equivalent to linear ones with coefficients determined in terms of theta functions. This is an interesting property of FRF that is directly linked to optimal real-time estimation problems, such as estimation of air pollution concentration at the receptor, stream flow forecasting, and soil parameter identification. This set of problems, which will be discussed in Chapter 9, is intended as nuclei around which new areas of applications are expected to crystallize.

2. The Theory of Factorable Random Fields

In earth sciences, the idea of representing natural processes in terms of mutually orthogonal functions has been used extensively. Good examples are the Fourier integrals and series, which depend on the orthogonality of the sinusoids. In this section, the same basic idea will be considered in a stochastic context, by means of the theory of factorable random fields.

Let (Ω, F, P) be a probability space and let (R, B) be a measurable space, where R is the real line and B is the σ-field of Borel sets on R. The random variables x_i, $i=1,2,\ldots,m$ are defined on (Ω, F, P) and take values in (R, B). Assume that $g(\chi_1,\ldots,\chi_m)$ is a function of $L_2(R^m, P_x)$; that is, it is a square integrable function with respect to a measure P_x with density

$$\frac{P_x(d\chi_1\ldots d\chi_m)}{d\chi_1\ldots d\chi_m}=f_x(\chi_1)\ldots f_x(\chi_m)$$

This means that

$$r_m=\underbrace{\iint\cdots\int}_{m\text{ times}} g^2(\chi_1,\ldots,\chi_m)f_x(\chi_1)\ldots f_x(\chi_m)\,d\chi_1\ldots d\chi_m<\infty \tag{1}$$

where $f_x(\chi_i)$, $i=1,2,\ldots,m$ are univariate probability densities. Let $\{p_{i,k}(\chi_i)\}$, $i=1,2,\ldots,m$ and $k=0,1,\ldots,$ be sets of complete polynomials of degree k in $L_2(R, P_i)$, which are orthogonal with respect to $P_i(d\chi_i)/d\chi_i=f_x(\chi_i)$, in the sense that

$$E[p_k(\chi_i)p_\lambda(\chi_i)]=\int p_k(\chi_i)p_\lambda(\chi_i)f_x(\chi_i)\,d\chi_i=0 \tag{2}$$

for all $k\neq\lambda=0,1,\ldots$ and $i=1,2,\ldots,m$. Then, it is possible to expand the function $g(\chi_1, \ldots, \chi_m)$ as follows:

$$g(\chi_1, \ldots, \chi_m) = \underbrace{\sum_{k_1=0}^{\infty} \cdots \sum_{k_m=0}^{\infty}}_{m \text{ times}} g_{k_1 \ldots k_m} p_{k_1}(\chi_1) \ldots p_{k_m}(\chi_m) \tag{3}$$

where $\{p_{k_1}(\chi_1), \ldots, p_{k_m}(\chi_m)\}$, $k_1, \ldots, k_m = 0, 1, \ldots,$ are complete sets of orthogonal polynomials. Thus, quite formally we have that

$$g_{k_1 \ldots k_m} = \int \cdots \int g(\chi_1, \ldots, \chi_m) p_{k_1}(\chi_1) \ldots p_{k_m}(\chi_m)$$
$$\times g_1(\chi_1) \ldots g_m(\chi_m)\, d\chi_1 \ldots d\chi_m \tag{4}$$

The coefficients $g_{k_1 \ldots k_m}$ determined by Eq. (4) have an interesting minimizing property: the minimum value of

$$Q = \int \cdots \int \left[g(\chi_1, \ldots, \chi_m) - \sum_{k_1=0}^{M_1} \cdots \sum_{k_m=0}^{M_m} \zeta_{k_1 \ldots k_m} p_{k_1}(\chi_1) \ldots p_{k_m}(\chi_m) \right]^2$$
$$\times f_x(\chi_1) \ldots f_x(\chi_m)\, d\chi_1 \ldots d\chi_m \tag{5}$$

is attained only for $\zeta_{k_1 \ldots k_m} = g_{k_1 \ldots k_m}$. The corresponding completeness relationship is given by

$$\sum_{k_1=0}^{\infty} \cdots \sum_{k_m=0}^{\infty} g^2_{k_1 \ldots k_m} = r_m \tag{6}$$

which assures that the series expansion of $g(\chi_1, \ldots, \chi_m)$ in L_2 converges to $g(\chi_1, \ldots, \chi_m)$ in L_2.

Let $L_2(\Omega, F, P)$ be the Hilbert space of the random variables $\{x_i\}$ at $\mathbf{s} = \mathbf{s}_1, \mathbf{s}_2, \ldots, \mathbf{s}_i, \ldots$. The SRF $X(\mathbf{s}), \mathbf{s} \in R^n$ is defined as a function on R^n with values in the Hilbert space $L_2(\Omega, F, P)$, viz.,

$$X: R^n \to L_2(\Omega, F, P) \tag{7}$$

(see also Chapter 2). The manipulations (1)–(7) motivate the following definition.

Definition 1: Let $f_x(\chi_i, \chi_j)$ be the bivariate probability density of the random variables $x_i = X(\mathbf{s}_i)$ and $x_j = X(\mathbf{s}_j)$. Let $\{\tilde{p}_{i,k}(\chi_i)\}$ and $\{\tilde{p}_{j,\lambda}(\chi_j)\}$, $k, \lambda = 0, 1, \ldots,$ be two complete sets of orthogonal polynomials in $L_2(R, f_i)$ and $L_2(R, f_j)$, respectively [where $f_i = f_x(\mathbf{s}_i)$ and $f_j = f_x(\mathbf{s}_j)$ are the corresponding univariate probability densities]. The SRF $X(\mathbf{s})$, $\mathbf{s} \in R^n$ will be called a *factorable random field (FRF)* if (a) the theta function

$$\theta(\chi_i, \chi_j) = \frac{f_x(\chi_i, \chi_j)}{f_x(\chi_i) f_x(\chi_j)} \tag{8}$$

is in $L_2(R^2, f_i f_j)$; that is, the integral

$$r_2 = \iint \theta^2(\chi_i, \chi_j) f_x(\chi_i) f_x(\chi_j)\, d\chi_i\, d\chi_j \tag{9}$$

is finite, and (b) the quantities

$$h_q(\chi_i) = \int \theta(\chi_i, \chi_j) f_x(\chi_j) \tilde{p}_q(\chi_j)\, d\chi_j \tag{10}$$

$$h_q(\chi_j) = \int \theta(\chi_i, \chi_j) f_x(\chi_i) \tilde{p}_q(\chi_i)\, d\chi_i \tag{11}$$

are polynomials of degrees up to q in χ_i and χ_j respectively.

The completeness of $\{\tilde{p}_k(\chi_i)\}$ and $\{\tilde{p}_\lambda(\chi_j)\}$ in $L_2(R, f_i)$ and $L_2(R, f_j)$, respectively, is of great consequence for the convergence of orthogonal expansions. This circumstance enhances the significance of the fact that any arbitrary orthogonal set in L_2 is capable of being completed to an orthogonal set, complete in L_2 (see, e.g., Alexits, 1981). It is convenient that the coefficients

$$\theta_{k\lambda} = \iint \theta(\chi_i, \chi_j) \tilde{p}_k(\chi_i) \tilde{p}_\lambda(\chi_j) f_x(\chi_i) f_x(\chi_j)\, d\chi_i\, d\chi_j \tag{12}$$

called the *factorability coefficients*, be chosen such that

$$\theta_{k\lambda} = \delta_{k\lambda} \theta_k \tag{13}$$

where

$$\theta_k = E[p_k(x_i) p_k(x_j)] \tag{14}$$

and where the polynomials $\{p_k(\chi_i)\}$ and $\{p_k(\chi_j)\}$ are subsets of the complete sets $\{\tilde{p}_k(\chi_i)\}$ and $\{\tilde{p}_\lambda(\chi_j)\}$, respectively. Then, condition (13) leads to the following proposition (Christakos, 1989).

Proposition 1: If $X(\mathbf{s})$ is an FRF, complete sets $\{p_k(\chi_i)\}$ and $\{p_\lambda(\chi_j)\}$ $k, \lambda = 0, 1, \ldots$, of orthogonal polynomials in $L_2(R, f_i)$ and $L_2(R, f_j)$, respectively, can be defined, such that

$$E[p_k(x_i) p_\lambda(x_j)] = 0 \qquad \text{when} \quad k \neq \lambda \tag{15}$$

and the theta function $\theta(\chi_i, \chi_j)$ can be expanded as the bilinear diagonal form

$$\theta(\chi_i, \chi_j) = \sum_{k=0}^{\infty} \theta_k p_k(\chi_i) p_k(\chi_j) \tag{16}$$

where

$$\sum_{k=0}^{\infty} \theta_k^2 = r_2 \tag{17}$$

Definition 2: Definition 1 can be naturally extended to a pair of SRF $X(\mathbf{s})$ and $Y(\mathbf{s})$, in the sense that, if the joint theta function

$$\theta(\chi_i, \psi_j) = \frac{f_{XY}(\chi_i, \psi_j)}{f_x(\chi_i) f_Y(\psi_j)} \tag{18}$$

is in $L_2(R^2, f_i, f_j)$, and the quantities

$$h_q(\chi_i) = \int \theta(\chi_i, \psi_j) f_Y(\psi_j) p_q(\psi_j)\, d\psi_j \tag{19}$$

$$h_q(\psi_j) = \int \theta(\chi_i, \psi_j) f_x(\chi_i) p_q(\chi_i)\, d\chi_i \tag{20}$$

are polynomials of degrees up to q in χ_i and ψ_j, respectively, then the processes $X(\mathbf{s})$ and $Y(\mathbf{s})$ are termed *jointly factorable random fields* (JFRF).

The corresponding factorability coefficients

$$\theta_{k\lambda} = \iint \theta(\chi_i, \psi_j) p_k(\chi_i) p_\lambda(\psi_j) f_x(\chi_i) f_Y(\psi_j)\, d\chi_i\, d\psi_j \tag{21}$$

are such that $\theta_{k\lambda} = \delta_{k\lambda}\theta_k$, where $\theta_k = E[p_k(x_i)p_k(y_j)]$.

Remark 1: Early investigations of bilinear expansions of bivariate densities include those of Lancaster (1958), Pearson (1901), and Barrett and Lampard (1955). Pearson introduced the so-called *tetrachoric series expansion* of the bivariate Gaussian density. Barrett and Lampard showed how such bilinear models can be generated by a physical process. Lancaster modified Pearson's contingency coefficient ϕ^2 by using the Radon–Nykodim derivative of the bivariate distribution with respect to the univariate distributions; and he showed that, if ϕ^2 is bounded, the bivariate distribution can be expanded in an eigenfunction expansion. In fact, in the light of the general theory of orthogonal expansions, ϕ^2 is related to r_2 by $r_2 = 1 + \phi^2$.

The polynomials $p_k(\chi_i)$ and $p_k(\chi_j)$ depend, in general, on the values $\mathbf{s} = \mathbf{s}_i$ and $\mathbf{s} = \mathbf{s}_j$. So do the theta functions $\theta(\chi_i, \chi_j)$ and the factorability coefficients θ_k. In the case that the SRF is homogeneous, the orthogonal polynomials are independent of $\mathbf{s}$, and the θ_k functions depend only on $\mathbf{h} = \mathbf{s}_i - \mathbf{s}_j$. It is easily shown that

$$p_0(\chi_i) = p_0(\chi_j) = 1 \tag{22}$$

$$p_1(\chi_i) = \frac{\chi_i - E[x_i]}{\sigma[x_i]} \tag{23}$$

and

$$p_1(\chi_j) = \frac{\chi_j - E[x_j]}{\sigma[x_j]} \tag{24}$$

In other words, the mean $E[\cdot]$ and the standard deviation $\sigma[\cdot]$ of the random variables may be derived from the orthogonal polynomials, and vice versa. The preceding results emphasize the importance of the coefficients θ_k in the study of FRF. Some additional properties are summarized below.

Proposition 2: For the factorability coefficients θ_k the following relations hold:

$$|\theta_k| \leq 1 \tag{25}$$

$$\theta_0 = 1 \tag{26}$$

$$\theta_1 = \frac{c_x(\mathbf{s}_i, \mathbf{s}_j)}{\sigma[x_i]\sigma[x_j]} \tag{27}$$

Moreover,

$$E[p_k(x_i)p_\lambda(x_j)] = \delta_{k\lambda}\theta_k \tag{28}$$

which is the factorability property of significant consequence in the nonlinear estimation context to be considered in Chapter 9, and

$$\theta_k = \frac{E[p_k(x_i)|\chi_j]}{p_k(\chi_j)} \tag{29}$$

Proof: See Christakos (1989).

Interesting results can be obtained in the frequency domain, as well.

Definition 3: The *spectral theta function* is defined as

$$\phi(\omega_i, \omega_j) = \iint \theta(\chi_i, \chi_j) \exp[i(\omega_i\chi_i + \omega_j\chi_j)]\, d\chi_i\, d\chi_j \tag{30}$$

which is the two-dimensional Fourier transform of $\theta(\chi_i, \chi_j)$ and $i = \sqrt{-1}$.

In the light of the above definition, the following proposition can be proven (Christakos, 1989).

Proposition 3: The spectral theta function can be expressed as

$$\phi(\omega_i, \omega_j) = 4\pi^2\, \delta(\omega_i)\, \delta(\omega_j) \sum_{k=0}^{\infty} \theta_k \eta_k\left(\frac{1}{\omega_i}\right) \eta_k\left(\frac{1}{\omega_j}\right) \tag{31}$$

where $\eta_k(\cdot)$ and $\eta_k(\cdot)$ are polynomials of degree k, and $\delta(\cdot)$ is the usual delta measure.

In most cases of practical importance, one deals with a *symmetric* theta function, in the sense that

$$\theta(\chi_i, \chi_j) = \theta(\chi_j, \chi_i) \tag{32}$$

This implies that the bivariate density is also symmetrical,

$$f_x(\chi_i, \chi_j) = f_x(\chi_j, \chi_i) \tag{33}$$

The univariate probability densities have the same functional form $f_x(\cdot)$ and so do the orthogonal polynomials $p_k(\cdot)$ Consequently, the *symmetric* theta function of an FRF $X(\mathbf{s})$ is an $L_2(R^2, f_i, f_j)$ function given by

$$\theta(\chi_i, \chi_j) = \sum_{k=0}^{\infty} \theta_k p_k(\chi_i) p_k(\chi_j) \tag{34}$$

where $\{p_k(\chi_i)\}$, $\{p_k(\chi_j)\}$ are complete sets so that Eq. (28) holds. Similarly, in the case of JFRF $X(\mathbf{s})$ and $Y(\mathbf{s})$, the symmetric theta function becomes

$$\theta(\chi_i, \psi_j) = \frac{f_{XY}(\chi_i, \psi_j)}{f_X(\chi_i) f_Y(\psi_j)} = \sum_{k=0}^{\infty} \theta_k p_k(\chi_i) p_k(\psi_j) \tag{35}$$

which belongs to the class of L_2 functions too.

For simplicity, unless stated otherwise, FRF with symmetric theta functions, zero mean, and unit variance is considered for the rest of this chapter. Since classification of SRF based on statistical regularity or memory does not make specific reference to the detailed form of the probability densities, an FRF may be coupled with properties from either one or both of these classifications (e.g., homogeneous and/or Gaussian properties). Proposition 4 below illustrates this fact in the unidimensional case.

Naturally, in one dimension we deal with *factorable random processes* (FRP). Before we proceed, let us recall some results regarding RP. Let χ_υ be the σ-field generated by $\{X(\upsilon), 0 \le \upsilon \le s\}$. First, suppose that $X(s)$ is a Markov process; that is,

$$E[X(s)|\chi_\upsilon] = E[X(s)|X(\upsilon)]$$

By definition of factorability, the conditional mean can be expressed as

$$E[X(s)|X(\upsilon)] = \theta_1 X(s)$$

where $\theta_1 = \rho \le 1$ is the correlation coefficient between state $X(s)$ and state $X(\upsilon)$. On the basis of the last two equations the following proposition can be proven (Christakos, 1989).

Proposition 4: Let (Ω, F, P) be a probability space and let $X(s)$, $s \in T$ be an FRP defined on it such that $E|X(s)| < \infty$. If $X(s)$ is a Markov FRP with respect to χ_υ, then it is also a *martingale* process (with respect to χ_υ); that is, it is valid that

$$E[X(s)|\chi_\upsilon] = \theta_1 X(s) \tag{36}$$

where $\upsilon \le s$. Note that the $X(s)$ is a supermartingale if $\theta_1 X(\upsilon) \le X(\upsilon)$, and it is a submartingale if $\theta_1 X(\upsilon) > X(\upsilon)$.

Based on Proposition 4, the proof of the following corollary is straightforward.

Corollary 1: If $X(s)$ is a Markov FRP with respect to $\chi_v = \sigma\{X(v), 0 \leq v \leq s\}$ and $E|X^k(s)| < \infty$, $p_k[X(s)]$ is a martingale with respect to χ_v; particularly,

$$E\{p_k[X(s)] | \chi_v\} = \theta_k p_k[X(s)] \tag{37}$$

Remark 2: It is interesting to compare the result of Proposition 4 with Jensen's theorem (e.g., Doob, 1960) of martingale theory, which under the same integrability conditions states that if $X(s)$ is a martingale process, then the $|X(s)|$ is a submartingale.

3. Nonlinear Transformations of Factorable Random Fields

In various applications of nonlinear systems of modeling and estimation, one may define an output process $Z(\mathbf{s})$, which is the nonlinear transformation of the input process $X(\mathbf{s})$. In other words,

$$Z(\mathbf{s}) = F[X(\mathbf{s})] \tag{1}$$

Let $z_s = F(x_s)$ and $z_{s'} = F(x_{s'})$. The corresponding univariate densities are

$$f_{z_s} = f[x_s = F^{-1}(z_s)] \frac{dF^{-1}(z_s)}{dz_s}$$

and

$$f_{z_{s'}} = f[x_{s'} = F^{-1}(z_{s'})] \frac{dF^{-1}(z_{s'})}{dz_{s'}}$$

The bivariate probability density is

$$f_x(z_s, z_{s'}) = f_x(x_s, x_{s'}) \frac{dF^{-1}(z_s)}{dz_s} \frac{dF^{-1}(z_{s'})}{dz_{s'}}$$

$$= f_{z_s} f_{z_{s'}} \sum_{k=0}^{\infty} \theta_k p_k[F^{-1}(z_s)] p_k[F^{-1}(z_{s'})]$$

which leads to the following proposition.

Proposition 1: If $X(\mathbf{s})$, $\mathbf{s} \in R^n$ is an FRF, the process $Z(\mathbf{s})$, $\mathbf{s} \in R^n$ defined by a transformation of the form (1), where $F[\cdot\}$ is a strictly monotonic function, is also an FRF. The theta function associated to $Z(\mathbf{s})$ has the same functional form as that of $X(\mathbf{s})$, particularly if $\theta(x_s, x_{s'})$ is a function in $L_2(R^2, f_{x_s} f_{x_{s'}})$ the function

$$\theta(z_s, z_{s'}) = \theta[x_s = F^{-1}(z_s), x_{s'} = F^{-1}(z_{s'})] \tag{2}$$

is in $L_2(R^2, f_{z_s} f_{z_{s'}})$.

4. Construction of Factorable Random Fields

Factorability restricts the class of SRF of interest to the ones with theta functions $\theta(\chi_i, \chi_j)$ satisfying Definition 1 of Section 2. That is, $\theta(\chi_i, \chi_j)$ should be a function in $L_2(R^2, f_i f_j)$ and complete sets $\{p_k(\chi_i)\}$, $\{p_k(\chi_j)\}$ of orthogonal polynomials in $L_2(R, f_i)$, $L_2(R, f_j)$, respectively, should be derived.

Essentially, constructing a factorable theta function means establishing an expansion of the form (34), Section 2. In this context a comprehensive approach to calculating $p_k(\chi)$, which are orthogonal with respect to $f_x(\chi)$ is available in the case that the density satisfies the *Pearson differential equation*

$$\frac{1}{f_x(\chi)} \frac{df_x(\chi)}{d\chi} = \frac{w(\chi)}{\upsilon(\chi)} \tag{1}$$

where $\chi \in [\chi_1, \chi_2]$ and

$$w(\chi) = w_1 \chi + w_0 \tag{2}$$

$$\upsilon(\chi) = \upsilon_2 \chi^2 + \upsilon_1 \chi + \upsilon_0 \tag{3}$$

The product $\upsilon(\chi) f_x(\chi)$ vanishes at both boundaries χ_1 and χ_2; that is,

$$\upsilon(\chi_1) f_x(\chi_1) = \upsilon(\chi_2) f_x(\chi_2) = 0 \tag{4}$$

For $f_x(\chi)$ to serve as a weighting function for sets of orthogonal polynomials $\{p_k(\chi)\}$, the statistical moments of any order must be bounded; that is,

$$E[x^n] < \infty \tag{5}$$

where $n < \infty$.

The density functions that satisfy the above restrictions are of significant theoretical and practical importance. These include the Gaussian, exponential, and Pearson (Type I) densities (Jackson, 1941). The corresponding orthogonal polynomials are given by

$$p_k(\chi) = \frac{1}{f_x(\chi)} \frac{d^k}{d\chi^k} [\upsilon^k(\chi) f_x(\chi)] \tag{6}$$

Let us first consider the case where the bivariate density is known, the $\theta(\chi_i, \chi_j)$ is an L_2-function and, therefore, the problem is to establish a series expansion of the form (34), Section 2.

Example 1: The given bivariate density is of the Gaussian form

$$f_x(\chi_i, \chi_j) = \frac{1}{2\pi\sqrt{1-\rho^2}} \exp\left[-\frac{\chi_i^2 + \chi_j^2 - 2\rho\chi_i\chi_j}{1-\rho^2} \right] \tag{7}$$

where $-\infty \leq \chi_i, \chi_j \leq \infty$, and ρ is the correlation coefficient. The univariate probability density is

$$f_x(\chi) = \frac{1}{\sqrt{2\pi}} \exp\left[-\frac{\chi^2}{2}\right] \tag{8}$$

The theta function of $L_2(R^2, f_x^2)$ is expressed by

$$\theta(\chi_i, \chi_j) = \frac{1}{\sqrt{1-\rho^2}} \exp\left[-\frac{(\rho^2+1)(\chi_i^2+\chi_j^2) - 4\rho\chi_i\chi_j}{2(1-\rho^2)}\right] \tag{9}$$

where $r_2 = 1/(1-\rho^2)$, $|\rho| < 1$. Since $f_x(\chi)$ given by Eq. (8) satisfies the conditions (1)–(5), the corresponding orthogonal polynomials can be found from Eq. (6): They are the Hermite polynomials $H_k(\cdot)$. In these circumstances, the expression (9) may be expanded as

$$\theta(\chi_i, \chi_j) = \sum_{k=0}^{\infty} \rho^k H_k(\chi_i) H_k(\chi_j) \tag{10}$$

Obviously, $\theta_k = \rho^k$.

Equation (10) is the so-called *Hermitean* model of nonlinear geostatistics (Journel and Huijbregts, 1978). In fact, all the *isofactorial* bivariate probability densities used in disjunctive kriging satisfy the factorability hypothesis (e.g., Armstrong and Matheron, 1986). Furthermore, the theory of FRF provides the means for constructing numerous other bivariate probability densities of significant importance in the context of nonlinear geostatistics (Christakos, 1986c).

An interesting situation of building FRF arises when the univariate probability density $f_x(\chi)$ is given and the factorability coefficients θ_k are prescribed. In this case, the polynomials $p_k(\chi)$ can be calculated as before and $\theta(\chi_i, \chi_j)$ can be constructed as in Eq. (34), Section 2, provided that the θ_k are such that the expansion (34), Section 2, is nowhere negative.

Example 2: The density

$$f_x(\chi) = \frac{4\sqrt{\pi}}{3} \sqrt{1-\chi^2} \tag{11}$$

where $-1 \leq \chi \leq 1$. Equation (11) satisfies conditions (1)–(5); thus, Eq. (6) gives

$$p_k(\chi) = \sqrt{\frac{2(k+1)!}{G(k+2)}}\, C_k^1(\chi) \tag{12}$$

where $G(\cdot)$ is the gamma function and $C_k^1(\cdot)$ are Gegenbauer polynomials. The theta function is written

$$\theta(\chi_i, \chi_j) = \sum_{k=0}^{\infty} \theta_k \frac{2(k+1)!}{G(k+2)} C_k^1(\chi_i) C_k^1(\chi_j) \tag{13}$$

where the θ_k assigned must be such that Eq. (13) is nowhere negative. For this, a proper choice is

$$\theta_k = \frac{2k!}{G(k+2)} C_k^1(\chi)$$

Example 3: The density

$$f_x(\chi) = \exp[-\chi] \tag{14}$$

where $0 \le \chi \le \infty$ is the weighting function for the Laguerre polynomials $L_k(\chi)$. The associated process is a stationary Markov process; thus, using some results by Karlin and McGregor (1960), the coefficients θ_k are prescribed as

$$\theta_k = \exp[-a_k \tau], \tag{15}$$

where a_k is a function of k and $\tau = t_i - t_j \ge 0$. For the θ_k of Eq. (15), Eq. (34) of Section 2 becomes

$$\theta(\chi_i, \chi_j) = \sum_{k=0}^{\infty} \exp[-a_k \tau] L_k(\chi_i) L_k(\chi_j) \tag{16}$$

and the corresponding FRP is a martingale.

Remark 1: The choice of factorability coefficients (15) produces a rich class of FRP. Equation (34), Section 2, after some manipulations, can be written as the integral equation

$$\int k(\chi_i, \chi_j) e_k(\chi_j)\, d\chi_j = \theta_k e_k(\chi_i) \tag{17}$$

where

$$k(\chi_i, \chi_j) = \frac{\theta(\chi_i, \chi_j)}{\sqrt{f_x(\chi_i) f_x(\chi_j)}}$$

is the kernel and

$$e_k(\cdot) = p_k(\cdot)\sqrt{f_x(\cdot)}$$

are the eigenfunctions. The θ_k play here the role of the corresponding eigenvalues, which must form a complete set and should not depend on $\tau = t_i - t_j$. If θ_k are given by (15), the process is Markov with transition probability density

$$q(\chi_i, \chi_j; \tau) = f_x(\chi_j) \sum_{k=0}^{\infty} \exp[-a_k \tau] p_k(\chi_i) p_k(\chi_j) \tag{18}$$

satisfying the well-known Kolmogorov equations (see, e.g., Gihman and Skorohod, 1974a). The class of theta functions corresponding to (18) is of the general form

$$\theta(\chi_i, \chi_j) = \sum_{k=0}^{\infty} \exp[-a_k \tau] p_k(\chi_i) p_k(\chi_j) \tag{19}$$

and Eq. (16) is a special case of (19).

For a given univariate density $f_x(\chi)$ there may be more than one possible factorable theta function.

Example 4: Starting from the Gaussian univariate density (8) one may derive the theta function

$$\theta(\chi_i, \chi_j) = \sum_{k=0}^{\infty} \rho^{2k} H_{2k}(\chi_i) H_{2k}(\chi_j) \tag{20}$$

where the Hermite polynomials are of even order. The associated FRP is different from that generated by Eq. (10), since the bivariate density that corresponds to (20) is no longer Gaussian, but it is given by

$$f_x(\chi_i, \chi_j) = \frac{1}{1-\rho} \exp\left[-\frac{\chi_i + \chi_j}{1-\rho}\right] I_0\left(\frac{2\sqrt{\rho \chi_i \chi_j}}{1-\rho}\right) \tag{21}$$

Furthermore, by comparing Eqs. (10) and (20), one finds that, Eq. (20) $= \frac{1}{2}$ [Eq. (10), with correlation coefficient ρ] $+ \frac{1}{2}$ [Eq. (10), with correlation coefficient $-\rho$].

A short list of FRF together with the associated univariate probability densities and orthogonal polynomials is given in Table 4.1 Moreover, by

Table 4.1 FRF with the Associated Univariate Probability Densities and Orthogonal Polynomials

UPD	Weighting function	$p_k(\chi)$
Normal	$\exp\left[-\frac{\chi^2}{2}\right]$	Hermite
Gamma/chi-squared	$\chi^k \exp[-\chi], \chi > 0$	Generalized Laguerre
Negative exponential/Rayleigh	$\exp[-\chi], \chi > 0$	Laguerre
Uniform	$1, -1 < \chi < 1$	Legendre
	$\frac{(1-\chi^2)^k}{\sqrt{1-\chi^2}}, -1 < \chi < 1$	Gegenbauer
	$\frac{1}{\sqrt{1-\chi^2}}, -1 < \chi < 1$	Chebyshev
	$\chi^{k \log \chi}, \chi > 0$	Stieltjes-Wigert
	$(1-\chi)^k (1+\chi)^\lambda, -1 < \chi < 1$	Jacobi

using Proposition 1 of Section 3, various factorable processes may be constructed. For instance, we may take any of the examples of this section and apply a monotonic transformation: The resulting process will be an FRP too. Several other examples are given in Christakos (1986c).

5. The Nonlinear State–Nonlinear Observation System

Nonlinear phenomena occur quite frequently in earth sciences. Examples include the forecasting of block grade distributions (Matheron, 1975), streamflow estimation and flood forecasting (Mein *et al.*, 1974; Bras and Georkakakos, 1980), soil parameter identification (Christakos, 1985), and herring-bone cloud formations (Infeld and Rowlands, 1990).

Typically, nonlinear characteristics are represented through

(a) direct transformations, if the nonlinearities can be expressed by explicit functions that are analytically tractable;
(b) power series expansions, such as Taylor series and polynomial series; and
(c) integral transformations, such as Fourier and Laplace transforms.

In addition to these methods, the results of the preceding sections can be used to study nonlinear characteristics of stochastic systems by means of FRP. Let us consider the on-line *nonlinear state-nonlinear observation system* (*NSNOS*) of the scalar form

$$X(s) = G[X(s-1), s-1] + W(s-1) \tag{1}$$

$$Y(s) = B[X(s), s] + V(s) \tag{2}$$

where $X(s)$ and $Y(s)$ are FRP, and the $W(s)$, $V(s)$ are white noises; $W(s)$ and $V(s)$ are assumed to be independent, and $W(s)$ is independent of the initial condition $X(0)$.

Results regarding NSNOS are derived by making the following assumptions (Christakos 1989).

Assumption 1: The processes $X(s)$ and $Y(s)$ are FRP in the sense defined in the previous sections.

Assumption 2: The state $G[X(s-1), s-1]$ and the observation $B[X(s), s]$ are $L_2(R, f_x(\chi))$ functions, in the sense defined above.

Proposition 1: Retain the assumptions of NSNOS above. If $G_{k,s-1}$ and $B_{k,s}$, $k = 0, 1, \ldots$ are expansion coefficients of the L_2 functions $G[X(s-1), s-1]$ and $B[X(s), s]$, respectively, then

$$G_{k,s-1} = \delta_{1k}\theta_1^x = \delta_{1k}\frac{G[X(s-1), s-1]}{X(s-1)} \tag{3}$$

and

$$B_{k,s} = \delta_{1k}\theta_1^{xy} = \delta_{1k}\frac{B[X(s), s]}{X(s)} \tag{4}$$

where

$$\theta_1^x = E[X(s)X(s-1)], \qquad \theta_1^{xy} = E[X(s)Y(s)]$$

$$G_{1,s-1} = E\{G[X(s-1), s-1]X(s-1)\}$$

and

$$B_{1,s} = E\{B[X(s), s]X(s)\}$$

Equation (1) can be used to evaluate the univariate density $f_x(\chi_s)$ in terms of $f_x(\chi_{s-1})$. By applying Bayes rule, we have

$$f_x(\chi_s) = \int f_x(\chi_s|\chi_{s-1})f_x(\chi_{s-1})\, d\chi_{s-1} \tag{5}$$

The $f_x(\chi_s|\chi_{s-1})$ can be computed from Eq. (1), namely,

$$f_x(\chi_s|\chi_{s-1}) = f_w(\omega_{s-1}) \qquad \text{at} \quad W(s-1) = X(s) - G[X(s-1), s-1]$$

Equation (5) then gives

$$\int \chi_s f_x(\chi_s|\chi_{s-1})\, d\chi_s = G(\chi_{s-1}, s-1) \tag{6}$$

Also, on the strength of Eqs. (28) of Section 2 above and

$$f_x(\chi_s|\chi_{s-1}) = \theta(\chi_s, \chi_{s-1})f_x(\chi_s)$$

the left-hand side of (6) yields

$$\int \chi_s f_x(\chi_s|\chi_{s-1})\, d\chi_s = \sum_{k=0}^{\infty} \theta_k^x p_k(\chi_{s-1})$$
$$\times \int \chi_s p_k(\chi_s) f_x(\chi_s)\, d\chi_s = \theta_1^x \chi_{s-1} = G(\chi_{s-1}, s-1)$$

where Proposition 1 has been used. The state model (1) is compatible with the assumption of an FRP.

Since Eq. (1) generates a Markov process with respect to the σ-field

$$Z_{s-1} = \sigma\{X(u), 0 \le u \le s-1\}$$

by Proposition 4 of Section 2, we have

$$E[X(s)|Z_{s-1}] = \theta_1^x \chi_{s-1} \tag{7}$$

thus leading to the corollary below.

Corollary 1: If $X(s)$ is an FRP, Eq. (1) generates a martingale process with respect to the σ-field

$$Z_{s-1} = \sigma\{X(u), \qquad 0 \leq u \leq s-1\}$$

The following corollary is a direct consequence of Proposition 1.

Corollary 2: Retain Assumptions 1 and 2. The NSNOS (1) and (2) reduces to the linear state-linear observation system (LSLOS)

$$X(s) = a_{s-1}X(s-1) + W(s-1) \tag{8}$$

$$Y(s) = b_s X(s) + V(s) \tag{9}$$

where $a_{s-1} = \theta_1^x = G_{1,s-1}$, and $b_s = \theta_1^{xy} = B_{1,s}$.

Corollary 2 is particularly important within the estimation context, as we see in Chapter 9.

5

The Spatiotemporal Random Field Model

"The views of space and time which I wish to lay before you have sprung from the soil of experimental physics, and therein lies their strength. They are radical. Henceforth space by itself, and time by itself, are doomed to fade away into mere shadows, and only a kind of union of the two will preserve an independent reality."

H. Minkowski

1. Introduction

This chapter is devoted to the study of spatiotemporal natural processes, that is, processes that develop simultaneously in space and in time. In Section 2 we discuss the emergence of spatiotemporal natural processes in various branches of applied physical sciences and address the fundamental hypotheses and problems regarding the quantitative description of such processes. Several practical issues of spatiotemporal data analysis and processing are presented and the variety of potential applications is reviewed. The latter is followed by a critical discussion of the inadequacies of previous works on the subject.

To proceed with the rigorous mathematical modeling of natural processes that change in space and time, one must elaborate on a theory of *spatiotemporal random field* (S/TRF). This theory is presented in Section 3 through 6. The preceding mathematical results then act as the theoretical support for the discrete parameter representations, as well as the optimal space-time estimation and simulation methods discussed in a more practical context in the second part of the present chapter (section 7), and in Chapters 8 and 9, respectively.

The presentation in this chapter is intended to be sufficiently general to allow the application of the analysis to space-time problems not mentioned in this book.

2. Spatiotemporal Natural Processes—A Review

Spatiotemporal processes, that is, processes that develop simultaneously in space and time, occur in nearly all the areas of applied sciences, such as hydrogeology (e.g., water vapor concentrations, soil moisture content, and precipitation data consisting of long time series at various locations in space); environmental engineering (e.g., concentrations of pollutants in environmental media—water/air/soil/biota); climate predictions and meteorology (e.g., variations of atmospheric temperature, density, moisture content, and velocity); and oil reservoir engineering (e.g., porosities, permeabilities, and fluid saturations during the production phase).

In this context, important issues include the assessment of the spatiotemporal variability of the earth's surface temperature and the prediction of extreme conditions; the assessment of space-time trends in runoff on the basis of a spatially and temporally sparse database; the estimation of the soil moisture content at unmeasured locations in space and instants in time; the reconstruction of the whole field of a climate parameter using all the space-time data efficiently; the study of the transport of pollutants through porous media; the elucidation of the spatiotemporal distribution of rainfall for satellite remote-sensing studies; the optimal sampling design of meteorological observations; and the simulation of oil reservoir characteristics as a function of spatial position and production time.

The issues above are parts of the general problem of *analysis and processing of data from space-time physical phenomena.* In all these situations, the spatiotemporal pattern of change of the natural processes involved possesses a certain structure at the macroscopic level and a purely random character at the microscopic level. The latter implies a significant amount of uncertainty in spatiotemporal variation. Moreover, this variation is, in general, space nonhomogeneous and time nonstationary (there may exist complex trends in space, time-varying correlation structures, significant space-time cross-effects, etc.). Practicing scientists and engineers need to have a working understanding and quantitative assessment of this uncertainty and its implications. For example, spatiotemporal variability plays an extremely substantial role in the understanding, modeling, and prediction of surficial processes in space-time. It is also very important in improving our basic knowledge regarding the climatological influences on the hydrogeology of a region. If neglected, spatiotemporal-parameter variability of water management models may adversely influence management decisions.

Typically, space–time data analysis and processing problems have been handled under some convenient but rather simplistic assumptions. In *hydrogeology and water resources research*, common statistical methods of analysis create artificial decompositions of hydrologic processes—one in space and one in time—and study them separately (Rodriguez-Iturbe and Mejia, 1974; Delhome, 1977); or focus on time averages (monthly, seasonal, annual) of the hydrologic parameters; or make additional assumptions, like space homogeneity and weak time dependency (e.g., Bras and Rodriguez-Iturbe, 1985). The multivariate analysis concept that has been used in a number of hydrologic problems (e.g., Quenouille, 1957; Fiering, 1964) accounts for the vector formulation of the scalar time series model, where the component time series are correlated to each other. Variability in space is not taken into consideration and the modeling of the combined evolution of these series in space and in time is clearly not an issue addressed by multivariate analysis. Similar decomposition has been applied in some recent studies on the assessment of Ireland's wind power resource (e.g., Haslett and Raftery, 1989). Moreover, classical statistics and time-series methods have failed to provide a conceptual framework determining the correlation structure of the spatiotemporal heterogeneity of soil–water properties from local to global scales.

In *environmental research* the existing models (e.g., Bennett and Chorley, 1978; Gilbert, 1987) either apply traditional methods of classical statistics, which are incapable to capture important features of the space–time structure or have been designed to handle problems that are significantly different in nature than those arising in the spatiotemporal data analysis and processing context considered above. In particular, the class of classical statistics models does not determine any law of change of the environmental parameters, and the relative distances of the sample locations/instances over space–time do not enter the analysis of the correlation structure. Under certain circumstances, stationarity may be impossible to define (Granger, 1975).

The second class of environmental models available concern either specific space–time interaction systems where the input/output physical parameters are treated at each spatial location as separate time series, or the description of the system's transfer function by means of some special space–time patterns. These models do not provide an adequate quantitative assessment of spatiotemporal variability in general, and they do not account for the space nonhomogeneous and/or time nonstationary characteristics of the environmental parameters in particular. In some recent environmental studies the spatio-chronological order of the data is not properly considered, and arbitrary but not well justified decompositions of the correlation functions are assumed. Moreover, optimal reconstruction schemes general enough to cover the majority of applications have not been developed (see, e.g., comments made in Shinn and Lynn (1979), also by Bilonick (1985), and by Rouhani and Hall (1989) in a geostatistical framework). Space-time models based on the distributed parameter concept (Stavroulakis, 1983) are not in general appropriate for most environmental problems. These models are assumed to be governed by a differential equation of a particular form that does not represent adequately the majority of the spatiotemporal natural processes of interest; issues of stability, controllability, and observability involve serious difficulties.

In *reservoir characterization*, space-time data processing does not exist at present. Most of the techniques available exclusively account for the spatial variation of geological reservoir processes, when in reality these processes are simultaneously a function of spatial location and production time (Journel and Alabert 1988; Lake and Carrol, 1986). Also, current practices in data collection—with the exception of some oil sand deposits—do not account for time. One of the reasons that space-time models do not exist at present in reservoir characterization is because the need for detailed and advanced reservoir characterization has been recognized only recently.

The methods used for statistical *climate modeling and prediction* are usually somewhat primitive versions of the methods used for *weather analysis and prediction* (e.g., Gandin, 1963; Lindzen, 1989; Von Storch *et al.*, 1989). Many of them suffer the same limitations with the methods used in hydrology. For example, the basic ansatz of multivariate techniques such as "principal oscillation pattern" and "principal interaction pattern" (Von Storch *et al.*, 1989) is based on the arbitrary assumption that the space-time characteristics of a low-order system are the same as those of the full system. Also, important issues such as the characterization of spatiotemporal intermittency or spottiness in rainfall as it pertains to various notions of scaling as well as the physically observed features of clustering, growth, and decay of convective cells, and larger-scale spatiotemporal forms observed in mesoscale rainfall systems cannot be addressed by the existing statistical methods (see, e.g., NRC, 1991).

In *global warming* research, aspects of current interest are as follows:

(a) Many eminent authors claim that while one cannot assert that no warming occurred, the existing statistical analysis of earth's surface temperature data does not provide adequate assessments regarding temperature's space-time variability and it does not lead to convincing arguments supporting the concept that changes at the macroscopic level are due to greenhouse warming rather than to space-time natural variability (e.g., Lindzen, 1989).

(b) In water resources management the existence of a warming trend raises the question whether the global warming has been sufficient to translate into a corresponding change in the spatiotemporal structure of runoff series. Again, current statistical analyses of runoff series are subject to serious questions given that they are based on observations relating to a spatially and temporally sparse database and they assume no model about the underlying spatiotemporal evolution of the runoff series.

Clearly, the temperature data (a) and the rainfall series (b) studies above are typical examples of analyses where the theoretical models used are incapable of providing adequate representations of the spatiotemporal variability and, hence, they cannot give satisfactory answers to crucial questions concerning climate and water resources problems.

The main reasons for such—clearly inadequate from various viewpoints—analyses of spatiotemporal data should be attributed to the following facts:

(i) the importance of spatiotemporal variability in the study of space–time phenomena was not fully appreciated until recently; and

(ii) most of the theoretical tools and mathematical techniques of data processing available have been designed to operate exclusively in time (time series methods; e.g., Grenander and Rosenblatt, 1957) or exclusively in space (random fields, geostatistics; e.g., Matheron, 1973; Ivanov and Leonenko, 1989).

Undoubtedly, the literature on the subject of applied space–time data analysis and processing is very limited and most aspects of importance in the analysis, modeling, and estimation of spatiotemporal parameters have not been studied adequately.

In view of the foregoing considerations, the following conclusions are drawn:

(a) Any modeling assumption should reflect adequately the macroscopic and microscopic evolution characteristics of the underlying processes over space and time. Spatial and temporal scales of variability should be intimately connected. This is a requisite for the understanding and prediction of spatiotemporal processes in hydrogeology, climate modeling, and environmental pollution monitoring and control.

(b) Due to the random character in the variability of the data at the microscopic level, these processes must naturally be described stochastically; the concept of randomness should be viewed as an intrinsic part of the space–time evolution, and not only as a statistical description of possible states.

(c) The proper model should be capable to assess quantitatively any space-nonhomogeneous/time-nonstationary variability features and to provide efficient solutions to practical problems, such as space–time estimation.

Taking these issues into account, it seems quite reasonable that the concept of an S/TRF is the appropriate stochastic model for spatiotemporal processes. Within the framework of the S/TRF model (see Fig. 5.1), space and time form a combined process having simultaneous and interrelated effects on the evolution of the natural variable it represents. Suitable methodological hypotheses and operational tools assure that the mathematical concept of S/TRF is compatible with the physics of the variate it describes and, thus, it is applicable in practice. Finally, conclusions regarding the spatiotemporal variability (trends in space, periodicities in time, nonhomogeneous/nonstationary correlations, etc.) can be established in terms of duality principles that relate the mathematical notions and the physical behavior of the process they model (see Chapter 1, the methodology of the stochastic research program). Here, stochastic spatiotemporal correlation functions provide the means for structural inferences.

In general, the objectives of spatiotemporal data analysis and processing include two important aspects:

(a) To assess quantitatively the spatiotemporal variability of the natural processes of interest (degree of regularity, continuity, nonhomogeneous spatial features, nonstationary characteristics, etc.).

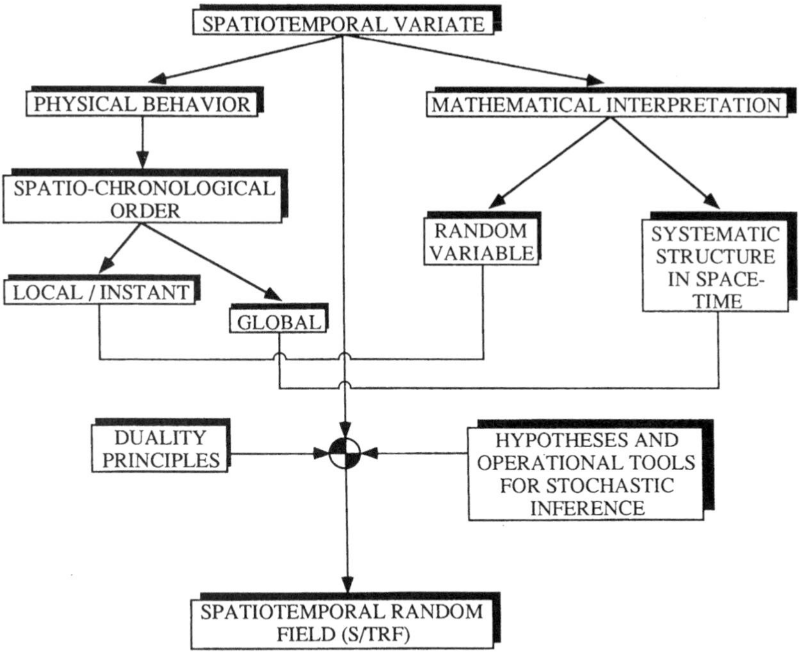

Figure 5.1 Modeling spatiotemporal natural processes by means of S/TRF

(b) To provide efficient and computationally attractive procedures for deriving optimal (in a well-defined mathematical sense) and physically meaningful estimation maps of the natural process, at unknown points in space and/or instants in time, based on fragmentary space–time data.

Of course, the outcome of space–time data analysis and processing may be not an end in itself. Several important consequences will emerge in the context of earth sciences. More specifically,

(i) A deeper understanding of the physics of the space–time processes will be obtained. For instance, knowledge about the spatiotemporal variability of the various climate parameters will improve our basic understanding of how the global climate actually functions.

(ii) The predictive capabilities of many computer-based differential equation models in hydrology and environmental research are limited because the parameters of the models are difficult to determine. Much of this difficulty may stem from (a) the spatiotemporal variability of the media, and (b) identifiable differences in initial physical assumptions. It is, hence, of significant importance to understand how (a) and (b) influence the outcomes of modeling.

(iii) Space–time data analysis and processing will provide the necessary means for solving important problems in various areas of water resources. Information about the spatiotemporal-parameter variability of a water resource system will allow the detailed simulation of the system and will influence considerably management decisions. The assessment of the spatiotemporal variability of pollutant concentrations will provide the knowledge needed to monitor and control environmental pollution. Storm runoff depends on the space–time structure of rainfall and other parameters. Random space–time rainfall patterns can be used in evaluating strategies for satellite remote-sensing of rainfall and for development of procedures for converting radiant intensity received by an instrument from its field of view into rainfall amount. S/TRF simulations of the anticipated effects on surface temperature due to the increase of carbon dioxide in the atmosphere over a specific time period will provide valuable insight into the study of global warming issues. In connection with this, the possible effects of the coupled increase of precipitation and temperature on the hydrology of a particular region can be determined; then, conclusions could be derived about the incorporation of climatic changes into the planning of future earth systems, and the modification of the operating rules of existing water resource systems.

An S/TRF is termed continuous parameter or discrete parameter according to whether its space–time arguments of an S/TRF take continuous or discrete sets of values. The first part of this chapter (Sections 3–6) treats continuous parameter S/TRF and certainly constitutes a legitimate mathematical activity on its own merit—but this has not been our only purpose. The first part will act as the theoretical support to the discrete parameter spatiotemporal variability models developed in a more practical context in Section 7.

3. Ordinary Spatiotemporal Random Fields

3.1 The Basic Space–Time Notions

Let $\mathbf{s}=(s_1, s_2, \ldots, s_n)\in R^n$ (R^n is the Euclidean space of dimensionality $n \geq 1$) with

$$|\mathbf{s}| = \sqrt{\sum_{i=1}^{n} \mathbf{s}_i^2} \qquad \text{and} \quad t \in T$$

$$(T \subseteq R^1_{+,0} = \{t \in R^1 \colon t \geq 0\})$$

In the Cartesian product $R^n \times T$ let $(\mathbf{s}, t) \in R^n \times T$ denote space–time coordinates, such that $|(\mathbf{s}, t)|^2 = |\mathbf{s}|^2 + t^2$. We also define

$$(\mathbf{s}, t)^{\boldsymbol{\alpha},\beta} = \mathbf{s}^{\boldsymbol{\alpha}} t^{\beta} = s_1^{\alpha_1} s_2^{\alpha_2} \ldots s_n^{\alpha_n} t^{\beta}$$

where β is a nonnegative integer and $\boldsymbol{\alpha} = (\alpha_1, \alpha_2, \ldots, \alpha_n)$ is a multi-index of nonnegative integers such that $|\boldsymbol{\alpha}| = \sum_{i=1}^{n} \alpha_i$ and $\boldsymbol{\alpha}! = \alpha_1!\alpha_2! \ldots \alpha_n!$. Let (R^n, Φ, ϕ) and (T, Ψ, ψ) denote the corresponding measured spaces in R^n and T, respectively.

In the Cartesian product $R^n \times T$ let $\mathfrak{J}$ be the set of all $A \times B$ with $A \in R^n$ and $B \in T$. For these sets we define the spatiotemporal measure $\mu(A \times B) = \phi(A)\psi(B)$, which is countably additive on $\mathfrak{J}$. On $R^n \times T$ let $\Phi \otimes \Psi$ be the σ-field generated by the set $\mathfrak{J}$. Then $\Phi \otimes \Psi$ is termed the product σ-field on $R^n \times T$. The measure μ on $\Phi \otimes \Psi$ will be called a product measure $\phi \times \psi$. The spatiotemporal measure μ can be expanded to a countably additive measure on the product σ-field $\Phi \otimes \Psi$ generated by $\mathfrak{J}$ but, in general, such an expansion is not unique. Here, however, we will assume that the necessary conditions under which this extension is unique are satisfied and, therefore, it can be written in terms of iterated integrals in $R^n \times T$. Let (R^n, Φ, ϕ) and (T, Ψ, ψ) be σ-finite measured spaces, and let $X(\mathbf{s}, t)$ be an ordinary spatiotemporal function from $R^n \times T$ into $[0, \infty]$. We can view $X(\mathbf{s}, t)$ either as a random field in an $n+1$-dimensional space, or as a time-dependent random field in an n-dimensional space. The latter view is more convenient for nonrelativistic problems. The $X(\mathbf{s}, t)$ is measurable for $\Phi \otimes \Psi$; that is,

$$X(\mathbf{s}, t) \in L_1(R^n \times T, \quad \Phi \otimes \Psi, \quad \mu = \phi \times \psi)$$

Then

$$\int_{R^n \times T} X(\mathbf{s}, t)\, d\mu(\mathbf{s}, t) = \int_{R^n} \int_T X(\mathbf{s}, t)\, d\phi(\mathbf{s})\, d\psi(t)$$

$$= \int_T \int_{R^n} X(\mathbf{s}, t)\, d\psi(t)\, d\phi(\mathbf{s})$$

(Naturally, the support of the $X(\mathbf{s}, t)$ is the closure of the set of all $(\mathbf{s}, t) \in R^n \times T$ that satisfy $X(\mathbf{s}, t) \neq 0$.) To prove that

$$X(\mathbf{s}, t) \in L_1(R^n \times T, \quad \Phi \otimes \Psi, \quad \mu = \phi \times \psi)$$

we can prove that $X(\mathbf{s}, t)$ is $\Phi \otimes \Psi$-measurable and such that

$$\int_{R^n} \int_T |X(\mathbf{s}, t)|\, d\phi(\mathbf{s})\, d\psi(t) < \infty$$

or

$$\int_T \int_{R^n} |X(\mathbf{s}, t)|\, d\psi(t)\, d\phi(\mathbf{s}) < \infty$$

(For simplicity, in the following the symbol $R^n T$ under the integrals will usually be omitted.)

We define some spaces of spatiotemporal function $X(\mathbf{s}, t)$ in $R^n \times T$, which are useful within the framework of the present study: the space $\boldsymbol{C}$ of all real and continuous functions in space-time with compact support; the space $\boldsymbol{K}$ of all real, continuous and infinitely differentiable functions in space and time with compact support; the space $\boldsymbol{S}$ of all real, continuous, and infinitely differentiable functions which, together with their derivatives of all orders, approach zero more rapidly than any power of $1/|(\mathbf{s}, t)|$ as $|(\mathbf{s}, t)| \to \infty$. Notice that $\boldsymbol{S} \supset \boldsymbol{K}$, as all functions in $\boldsymbol{K}$ vanish identically outside a finite support, whereas those in $\boldsymbol{S}$ merely decrease rapidly at infinity. Spaces $\boldsymbol{K}$ and $\boldsymbol{S}$ are of particular importance in this study. The topology in $\boldsymbol{K}$ and $\boldsymbol{S}$ is in the sense of Schwartz (1950–51) where, in view of the aforementioned space-time considerations, the argument is now $(\mathbf{s}, t) \in R^n \times T$.

In the special case that the time-instant $t \in T$ is fixed, the symbol

$$\mathbf{s}_{/t} = (s_{1/t}, s_{2/t}, \ldots, s_{n/t})$$

may be used to denote the spatial position in R^n. Then

$$\mathbf{s}_{/t} \cdot \mathbf{s}'_{/t} = \sum_{i=1}^{n} s_{i/t} s'_{i/t}$$

is the scalar product of vectors $\mathbf{s}_{/t}$ and $\mathbf{s}'_{/t}$,

$$d(\mathbf{s}_{/t}, \mathbf{s}'_{/t}) = \sqrt{\sum_{i=1}^{n} (s_{i/t} - s'_{i/t})^2}$$

is the distance between $\mathbf{s}_{/t}$ and $\mathbf{s}'_{/t}$, and

$$|\mathbf{s}_{/t}| = \sqrt{\sum_{i=1}^{n} s_{i/t}^2}$$

is the length of the vector $\mathbf{s}_{/t}$. Similarly, the symbol $t_{/\mathbf{s}}$ may be used in the case where t varies while $\mathbf{s}$ is fixed.

By considering the space–time coordinates $(\mathbf{s}, t)$ as auxiliary variables for the geometrical description of the spatiotemporal distribution of a natural process, one may assume that the process is projected on the domain $(\mathbf{s}, t)$. Then, since the $(\mathbf{s}, t)$-domain is an auxiliary element, there may be constructed other domains on which natural processes could be projected. As we shall see in the next section, such a domain is, for example, the Fourier (or frequency) domain $(\mathbf{w}, \lambda)$, where $\mathbf{w}$ and λ are frequency characteristics. Another useful domain can be constructed in terms of the space transformation operator; this is the $(\mathbf{s}\cdot\boldsymbol{\theta}, \boldsymbol{\theta}, t)$-domain (see Chapter 6). The $(\mathbf{s}, t)$-, $(\mathbf{w}, \lambda)$-, and $(\mathbf{s}\cdot\boldsymbol{\theta}, \boldsymbol{\theta}, t)$-domains provide equivalent representations of the natural process defined on them. And while one's intuition is better adopted to the $(\mathbf{s}, t)$-domain, in many cases it is more convenient to work in the other two domains.

3.2 Definition of Ordinary Spatiotemporal Random Fields and Certain of Their Physical Applications

Let $\mathbf{z} = (\mathbf{s}, t)$. We denote by $\mathscr{H}_k = L_2(\Omega, F, P)$ the Hilbert space of all continuous-parameter random variables $x_1, \ldots, x_m$ defined at $\mathbf{z}_1, \ldots, \mathbf{z}_m$ and endowed with the scalar product

$$(x_1, x_2) = E[x_1 x_2] = \iint \chi_1 \chi_2 \, dF_x(\chi_1, \chi_2) \tag{1}$$

where $F_x(\chi_1, \chi_2)$ denotes the joint probability distribution of the random variables x_1 and x_2, while

$$\|x\|^2 = E|x|^2 = \int \chi^2 \, dF_x(\chi) < \infty \tag{2}$$

where $F_x(\chi)$ denotes the probability distribution of x. Usually $F_x(\chi)$ and $F_x(\chi_1, \chi_2)$ are assumed to be differentiable so that they can be replaced by the probability densities $f_x(\chi)$ and $f_x(\chi_1, \chi_2)$.

Definition 1: The *ordinary S/TRF (OS/TRF)* $X(\mathbf{s}, t)$ is defined as the function on the Cartesian product $R^n \times T$ with values in the Hilbert space $L_2(\Omega, F, P)$, viz.,

$$X: R^n \times T \to L_2(\Omega, F, P) \tag{3}$$

Just as for SRF (Chapter 2), an S/TRF $X(\mathbf{z}) = X(\mathbf{s}, t)$ is specified completely by means of all finite dimensional probability measures $\mu_x(B)$ associated with the families of random variables $x_1, \ldots, x_m$ at $\mathbf{z}_1, \ldots, \mathbf{z}_m$; viz.,

$$\mu_x(B) = \mu_{\mathbf{z}_1,\ldots,\mathbf{z}_m}(B) = P[(x_1, \ldots, x_m) \in B]$$

for every $B \in \mathfrak{F}^m$ ($\mathfrak{F}^m$ is a suitably chosen σ-field of subsets of R^m) and all $m = 1, 2, \ldots$. The corresponding probability density functions are written as

$$\begin{aligned} f_x(\chi_1, \ldots, \chi_m)\, d\chi_1 \ldots d\chi_m &= f_{\mathbf{z}_1,\ldots,\mathbf{z}_m}(\chi_1, \ldots, \chi_m)\, d\chi_1 \ldots d\chi_m \\ &= P[\chi_1 \leq x(\mathbf{z}_1) \leq \chi_1 + d\chi_1, \ldots; \chi_m \leq x(\mathbf{z}_m) \leq \chi_m + d\chi_m] \end{aligned} \tag{4}$$

for all m. All OS/TRF to be considered will be continuous in the mean square sense; that is, $E\,|X(\mathbf{s}', t') - X(\mathbf{s}, t)|^2 \to 0$, when $\mathbf{s}' \to \mathbf{s}$ and $t' \to t$. Moreover, OS/TRF are, in general, taken to represent *space-nonhomogeneous / time-nonstationary* natural processes (e.g., spatiotemporal history of soil shear stresses during an earthquake, oil reservoir porosity distribution in space–time during the production phase). The simplest examples of space-nonhomogeneous/time-nonstationary fields are those of the form

$$X(\mathbf{s}, t) = Y(\mathbf{s}, t) + p_{\nu,\mu}(\mathbf{s}, t)$$

where $Y(\mathbf{s}, t)$ is a space-homogeneous/time-stationary OS/TRF (Section 3.3 below) and $p_{\nu,\mu}(\mathbf{s}, t)$ is a polynomial of degree ν in $\mathbf{s}$ and μ in t, with random coefficients. The space of all continuous OS/TRF will be denoted by $\mathcal{K}$.

The stochastic model of Eq. (3) will be equipped with these analytical tools, which will allow it to describe adequately the manner in which the underlying natural process develops over space–time. This development is a reflection of a certain pattern of combined spatiotemporal correlations between values in the natural process. In the sequel we will consider *second-order* OS/TRF; that is, the analysis will be based on up to second-order statistical moments assumed to be continuous and finite. More precisely, an OS/TRF $X(\mathbf{s}, t)$ will be characterized in terms of its *spatiotemporal mean value*

$$m_x(\mathbf{s}, t) = E[X(\mathbf{s}, t)] = \int \chi f_x(\chi)\, d\chi \tag{5}$$

the *centered spatiotemporal covariance function*

$$\begin{aligned} c_x(\mathbf{s}, t; \mathbf{s}', t') &= E[(X(\mathbf{s}, t) - m_x(\mathbf{s}, t))(X(\mathbf{s}', t') - m_x(\mathbf{s}', t'))] \\ &= \iint (\chi_1 - m_1)(\chi_2 - m_2) f_x(\chi_1, \chi_2)\, d\chi_1\, d\chi_2 \end{aligned} \tag{6}$$

and the *spatiotemporal semivariogram* or *structure function*

$$\gamma_x(\mathbf{s}, t; \mathbf{s}', t') = \tfrac{1}{2}E[X(\mathbf{s}, t) - X(\mathbf{s}', t')]^2$$
$$= \tfrac{1}{2}\iint (\chi_1 - \chi_2)^2 f_x(\chi_1, \chi_2)\, d\chi_1\, d\chi_2 \tag{7}$$

In some cases of nonzero mean the operator $E[\cdot]$ may be replaced by $\mathrm{Var}[\cdot]$. When $\mathbf{s} = \mathbf{s}'$ and $t = t'$,

$$c_x(\mathbf{s}, t; \mathbf{s}, t) = E[X(\mathbf{s}, t) - m_x(\mathbf{s}, t)]^2 = V_x(\mathbf{s}, t)$$

where $V_x(\mathbf{s}, t)$ is the *spatiotemporal variance* of the S/TRF $X(\mathbf{s}, t)$ at point/instant $(\mathbf{s}, t)$. A continuous function $c_x(\mathbf{s}, t; \mathbf{s}', t')$ is the covariance function of an OS/TRF if and only if it satisfies the *nonnegative-definiteness condition*

$$\sum_{i=1}^{m} \sum_{j=1_i}^{k_i} \sum_{i'=1}^{m} \sum_{j'=1_{i'}}^{k_{i'}} q_{ij} q_{i'j'} c_x(\mathbf{s}_i, t_j; \mathbf{s}_{i'}, t_{j'}) \geq 0 \tag{8}$$

for all $m, k_i, k_{i'} (=1, 2, \ldots)$, all $(\mathbf{s}_i, t_j) \in R^n \times T$, and all numbers (real or complex) q_{ij}, $q_{i'j'}$; here k_i denotes the number of time instants t_j, $j = 1_i, 2_i, \ldots, k_i$ used, given that we are at the spatial position $\mathbf{s}_i$.

Remark 1: Interesting special cases of the spatiotemporal covariance $c_x(\mathbf{s}, t; \mathbf{s}', t')$ are the purely spatial covariance (fixed t), namely,

$$c_x(\mathbf{s}_{/t}, \mathbf{s}'_{/t}) = E[(X(\mathbf{s}_{/t}) - m_x(\mathbf{s}_{/t}))(X(\mathbf{s}'_{/t} - m_x(\mathbf{s}'_{/t}))]$$

and the purely temporal covariance (fixed $\mathbf{s}$), namely,

$$c_x(t_{/\mathbf{s}}, t'_{/\mathbf{s}}) = E[(X(t_{/\mathbf{s}}) - m_x(t_{/\mathbf{s}}))(X(t'_{/\mathbf{s}}) - m_x(t'_{/\mathbf{s}}))]$$

Instead of the centered covariance function one may also define the *noncentered* spatiotemporal covariance function

$$\sigma_x(\mathbf{s}, t; \mathbf{s}', t') = E[X(\mathbf{s}, t)X(\mathbf{s}', t')] = c_x(\mathbf{s}, t; \mathbf{s}', t') + m_x(\mathbf{s}, t)m_x(\mathbf{s}', t') \tag{9}$$

The other mode of second-order analysis is that in the *frequency domain.* The harmonic expansions of $X(\mathbf{s}, t)$ can be considered as an extension in the space–time context of the relevant results for SRF presented in Chapter 2. In particular

$$X(\mathbf{s}, t) = \iint \exp[i(\mathbf{w} \cdot \mathbf{s} + \lambda t)]\tilde{X}(\mathbf{w}, \lambda)\, d\mathbf{w}\, d\lambda \tag{10}$$

where $i = \sqrt{-1}$, and $\tilde{X}(\mathbf{w}, \lambda)$ is the so-called *spectral amplitude* of $X(\mathbf{s}, t)$. [For certain spatiotemporal RF, it may be more convenient to use the factor

$\exp[-i\lambda t]$ than $\exp[i\lambda t]$ in the harmonic expansions of $X(\mathbf{s}, t)$.] The corresponding *spectral density function* $C_x(\mathbf{w}, \lambda; \mathbf{w}', \lambda')$ is defined by

$$c_x(\mathbf{s}, t; \mathbf{s}', t') = \iiiint \exp[i(\mathbf{w}\cdot\mathbf{s}+\mathbf{w}'\cdot\mathbf{s}'+\lambda t+\lambda' t')] \times C_x(\mathbf{w}, \lambda; \mathbf{w}', \lambda')\, d\mathbf{w}\, d\lambda\, d\mathbf{w}'\, d\lambda' \tag{11}$$

where $C_x(\mathbf{w}, \lambda; \mathbf{w}', \lambda')$ is a positive summable function in $R^n \times T$. The $C_x(\mathbf{w}, \lambda; \mathbf{w}', \lambda')$ forms an $R^n \times T$-fold Fourier transform pair with the spatiotemporal covariance $c_x(\mathbf{s}, t; \mathbf{s}', t')$.

The example below discusses the emergence of the S/TRF model in the context of nonlinear wave phenomena. The latter play a very important role in applied sciences, including hydrodynamics, meteorology, and geophysics.

Example 1: Consider the stochastic Kortweg–de Vries (KdV) equation modeling dispersive waves in a nonlinear one-dimensional medium

$$\frac{\partial X(s,t)}{\partial t} - 6X(s,t)\frac{\partial X(s,t)}{\partial s} + \frac{\partial^3 X(s,t)}{\partial s^3} = Y(t) \tag{12}$$

where the S/TRF $X(s, t)$ represents the wave field, and the $Y(t)$ is a zero-mean stationary and Gaussian random process. Given suitable initial conditions the solution of Eq. (12) is (Orlowski and Sobczyk, 1989)

$$X(s,t) = V(t) - 2k^2 \operatorname{sech}^2[ks - 4k^3 t + kw(t)] \tag{13}$$

where k is a function of known wave parameters,

$$V(t) = \int_0^t Y(u)\, du$$

and

$$w(t) = 6\int_0^t V(u)\, du$$

The mean of the S/TRF $X(s, t)$, $m_x(s, t)$ is plotted in Fig. 5.2, where $Y(t)$ is assumed to be a white-noise process with covariance $c_Y(t, t') = 2\delta(\tau)$, $\tau = t - t'$.

Additional examples that offer an idea of the variety of applications of the S/TRF model are discussed below.

Example 2: In climate modeling, the OS/TRF $X(\mathbf{s}, t)$ may denote the space–time distribution of earth's surface temperature. Space–time correlation analysis can provide valuable information regarding the spatiotemporal patterns followed by temperature.

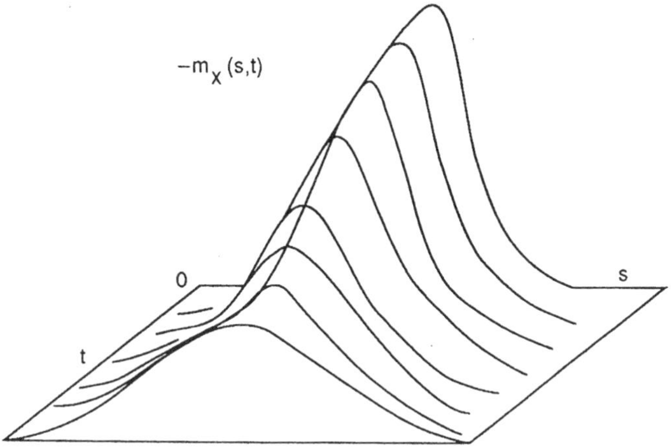

Figure 5.2 The mean value of the S/TRF $X(s, t)$ of Eq. (13)

Moreover, if $V(\mathbf{s})$ denotes the space sampled around point **s** during the time period $T(t)$ (say, $t \pm \delta t$), the $V(\mathbf{s}) \cdot T(t)$ will denote the space–time support of the sample, and the average temperature within $V(\mathbf{s}) \cdot T(t)$ will be

$$\bar{X}(\mathbf{s}, t) = \frac{1}{V(\mathbf{s})T(t)} \int_{V(\mathbf{s})} \int_{T(t)} S_{\mathbf{s},t} X(\mathbf{s}', t')\, d\mathbf{s}'\, dt' \tag{14}$$

On the basis of $\bar{X}(\mathbf{s}, t)$, maps of the averaged surface temperature over space and time may be constructed.

Example 3: In hydrology, the S/TRF $X(\mathbf{s}, t)$ may represent the precipitation intensity at location **s** at time t. Then the total streamflow for an area V over the time period T may be written as

$$S = \int_V \int_T f(\mathbf{s}, t) X(\mathbf{s}, t)\, d\mathbf{s}\, dt \tag{15}$$

where $f(\mathbf{s}, t)$ is a weighting function.

Similarly, if $X(\mathbf{s}, t)$ is the rainfall rate at location and time t, the average rainfall over an area V during the time period T is given by

$$R = \frac{1}{VT} \int_V \int_T X(\mathbf{s}, t)\, d\mathbf{s}\, dt \tag{16}$$

(e.g., Bell, 1987).

Moreover, weather processes such as dynamic, thermodynamic, and cloud microphysical processes operate over a variety of spatiotemporal scales, which are strongly related to each other; but they also interact with other variables like soil moisture, surface orography, and roughness discontinuities. The spatiotemporal structure of these processes must be explored, if we hope to understand the physical interactions between hydrology, weather, and climate, and to construct reliable predictions.

Example 4: In meteorology, $X(\mathbf{s}, t)$ could be the wind velocity at site $\mathbf{s}$ on day t. Then knowledge of the space–time correlation structure of the S/TRF $X(\mathbf{s}, t)$ can be translated into quite precise knowledge of the average available kinetic energy in the wind at a new site (e.g., Haslett and Raftery, 1989).

Example 5: In environmental studies, the S/TRF $X(\mathbf{s}, t)$ may represent the sulfate deposition at location $\mathbf{s}$ and time t. Then, analysis of the spatiotemporal covariance and semivariogram functions may provide valuable information regarding the patterns followed by the sulfate deposition and produce useful sulfate deposition maps (e.g., Bilonick, 1985).

S/TRF models can be used to simulate air-pollution processes from routine monitoring data. Consider, for example, data for sulfur dioxide (SO_2) and assume that the region is one-dimensional along, say, the east-west coordinate; SO_2 concentrations $X(\mathbf{s}, t)$ are described by the atmospheric diffusion equation

$$\frac{\partial X(s, t)}{\partial t} + V\frac{\partial X(s, t)}{\partial s} - \mu\frac{\partial^2 X(s, t)}{\partial s^2} = Y(s, t) \tag{17}$$

where V is the wind velocity, μ is a diffusion coefficient, and $Y(\mathbf{s}, t)$ is the rate of emission of SO_2 (e.g., Omatu and Seinfeld, 1981).

The quantity

$$\bar{X} = \sum_{i=1}^{k} \int_{D_i} \int_{T_i} \left(\frac{1}{T_i} + \beta_i\right) X(\mathbf{s}, t)\, d\mathbf{s}\, dt \tag{18}$$

[$X(\mathbf{s}, t)$ is the concentration of aerosol substance, β_i are coefficients that account for the fraction of the aerosol that gets into the soil] provides a global measure of the environmental pollution over the ecologically important (non-overlapping) zones D_i during the time periods T_i $(i = 1, \ldots, k)$. $\bar{X}$ accounts for the total amount of aerosol that settles on the surface of the earth, and is of considerable importance in making sound judgements regarding the pollution of soil and water. The effect of the latter on the environmental ecology can be significant within the framework of biocenosis.

3.3 Space-Homogeneous/Time-Stationary Ordinary Spatiotemporal Random Fields

An OS/TRF $X(\mathbf{s}, t)$, $(\mathbf{s}, t) \in R^n \times T$ will be called *space homogeneous/time stationary in the strict sense*, if all the multidimensional probability densities are invariant under the translation $\mathbf{z} \to \mathbf{z} + \delta\mathbf{z}$ (where, as before, $\mathbf{z} = (\mathbf{s}, t)$); viz.,

$$\begin{aligned} P[\chi_1 \le x(\mathbf{z}_1) \le \chi_1 + d\chi_1, \ldots, \chi_m \le x(\mathbf{z}_m) \le \chi_m + d\chi_m] \\ = P[\chi_1 \le x(\mathbf{z}_1 + \delta\mathbf{z}) \le \chi_1 + d\chi_1, \ldots, \chi_m \\ \le x(\mathbf{z}_m + \delta\mathbf{z}) \le \chi_m + d\chi_m] \end{aligned} \tag{19}$$

or

$$f_{\mathbf{z}_1,\ldots,\mathbf{z}_m}(\chi_1, \ldots, \chi_m) = f_{\mathbf{z}_1+\delta\mathbf{z},\ldots,\mathbf{z}_m+\delta\mathbf{z}}(\chi_1, \ldots, \chi_m) \tag{20}$$

for all $m = 1, 2, \ldots$. Space-homogeneous/time-stationary RF occur, for example, in the case of blackbody radiation within a large cavity maintained at a constant temperature.

An OS/TRF $X(\mathbf{s}, t)$ will be called *space homogeneous/time stationary in the wide sense*, if its mean and covariance do not change under a shift of the parameters; that is,

$$m_x(\mathbf{s}, t) = \text{constant} \tag{21}$$

and

$$c_x(\mathbf{s}, t; \mathbf{s}', t') = c_x(\mathbf{h}, \tau) \tag{22}$$

where $\mathbf{h} = \mathbf{s} - \mathbf{s}'$, $\tau = t - t'$. The spatial-homogeneity/time-stationarity property of the OS/TRF $X(\mathbf{s}, t)$, the latter considered as a function with values in the Hilbert space $L_2(\Omega, F, P)$, amounts to the fact that there exist in the closed linear subspace H spanned by the random variables in $L_2(\Omega, F, P)$ a group of unitary operators $U_{\mathbf{h},\tau}$ such that

$$U_{\mathbf{h},\tau} X(\mathbf{s}, t) = S_{\mathbf{h},\tau} X(\mathbf{s}, t) \tag{23}$$

where $\mathbf{s}, \mathbf{h} \in R^n$ and $t, \tau \in T$ [here, $S_{\mathbf{h},\tau} X(\mathbf{s}, t) = X(\mathbf{s} + \mathbf{h}, t + \tau)$ is the shift operator]. It is easily seen that in the case of space-homogeneous/time-stationary fields the covariance (6) and the semivariogram (7) are related by (assuming zero mean)

$$c_x(\mathbf{h}, \tau) = c_x(\mathbf{O}, 0) - \gamma_x(\mathbf{h}, \tau) \tag{24}$$

The set of all space-homogeneous/time-stationary ordinary fields will be denoted by $\mathscr{K}_0 \subset \mathscr{K}$.

The results of Chapter 2 on homogeneous SRF can be extended in the space–time context. In particular, the space-homogeneous/time-stationary RF $X(\mathbf{s}, t)$ admits the Fourier–Stieltjes representation

$$X(\mathbf{s}, t) = \iint \exp[i(\mathbf{w} \cdot \mathbf{s} + \lambda t)]\, d\aleph_x(\mathbf{w}, \lambda) \tag{25}$$

where $\aleph_x(\mathbf{w}, \lambda)$ is a random field such that

$$C_x(\mathbf{w}, \lambda)\, \delta(\mathbf{w} - \mathbf{w}')\, \delta(\lambda - \lambda') = E[d\aleph_x(\mathbf{w}, \lambda)\, \overline{d\aleph_x(\mathbf{w}', \lambda')}]$$

where $C_x(\mathbf{w}, \lambda)$ is the spectral function satisfying the spectral representation of the covariance $c_x(\mathbf{h}, \tau)$, viz.,

$$c_x(\mathbf{h}, \tau) = \int\int \exp[i(\mathbf{w} \cdot \mathbf{h} + \lambda\tau)]C_x(\mathbf{w}, \lambda)\, d\mathbf{w}\, d\lambda \tag{26}$$

and

$$C_x(\mathbf{w}, \lambda) = \frac{1}{(2\pi)^{n+1}} \int\int \exp[-i(\mathbf{w} \cdot \mathbf{h} + \lambda\tau)c_x(\mathbf{h}, \tau)\, d\mathbf{h}\, d\tau \tag{27}$$

Equations (26) and (27) are the space–time extensions of Khitchin's (1949) time-stationary covariance, spectral-density pair of functions. Since the covariance $c_x(\mathbf{h}, \tau)$ is a nonnegative-definite function, according to *Bochner's theorem*

$$C_x(\mathbf{w}, \lambda) \geq 0 \tag{28}$$

for all $\mathbf{w}$, λ. For real-valued even spatiotemporal covariances, $c_x(\mathbf{h}, \tau) = c_x(-\mathbf{h}, -\tau)$, and the exponential factor above can be replaced by $\cos(\mathbf{w} \cdot \mathbf{h} + \lambda\tau)$. When physically justified, the class of space-homogeneous/time-stationary RF largely simplifies computations.

Example 6: Consider the stochastic partial differential equation

$$a\frac{\partial^2 X(\mathbf{s}, t)}{\partial t^2} + b\left[\frac{\partial^4 X(\mathbf{s}, t)}{\partial s_1^4} + \frac{\partial^4 X(\mathbf{s}, t)}{\partial s_2^4}\right] + 2b\frac{\partial^4 X(\mathbf{s}, t)}{\partial s_1^2\, \partial s_2^2} = Y(\mathbf{s}, t) \tag{29}$$

In the light of the analysis above and assuming that a solution to Eq. (29) exists in the mean square sense, the latter can be written in terms of spectral representations as follows:

$$\tilde{X}(\mathbf{w}, \lambda) = \frac{\tilde{Y}(\mathbf{w}, \lambda)}{b\mathbf{w}^4 - a\lambda^2} \tag{30}$$

As a consequence, the correlation structure of $X(\mathbf{s}, t)$ can be expressed by

$$c_x(\mathbf{h}, \tau) = \int\int \exp[i(\mathbf{w} \cdot \mathbf{h} + \lambda\tau)]\frac{C_Y(\mathbf{w}, \lambda)}{[b\mathbf{w}^4 - a\lambda^2]^2}\, d\mathbf{w}\, d\lambda \tag{31}$$

The purely spatial spectral representations of a random field can be derived by means of $c_x(\mathbf{h}, \tau)$ and $C_x(\mathbf{w}, \lambda)$, as follows:

$$C_x(\mathbf{w}) = \frac{1}{(2\pi)^n} \int e^{-i\mathbf{w} \cdot \mathbf{h}} c_x(\mathbf{h}, 0)\, d\mathbf{h} \tag{32}$$

and

$$C_x(\mathbf{w}) = \int C_x(\mathbf{w}, \lambda)\, d\lambda \tag{33}$$

Similarly, the purely temporal spectral representations can be expressed as

$$C_x(\lambda) = \frac{1}{2\pi} \int e^{-i\lambda\tau} c_x(\mathbf{0}, \tau)\, d\tau \tag{34}$$

and

$$C_x(\lambda) = \int C_x(\mathbf{w}, \lambda)\, d\mathbf{w} \tag{35}$$

Example 7: A noteworthy use of the preceding analysis arises in the context of modeling spatiotemporal processes such as turbulent fluid flows and rainfall intensity at the ground surface. In these situations the variations of the S/TRF $X(\mathbf{s}, t)$ are determined by the variability of the SRF $Y(\mathbf{s})$, which travels with a constant velocity vector $\mathbf{v}$. As a consequence, the so-called *frozen field* is defined as $X(\mathbf{s}, t) = Y(\mathbf{s} - \mathbf{v}t)$. The corresponding covariances and spectral density functions are related by

$$c_x(\mathbf{h}, \tau) = c_Y(\mathbf{h} - \mathbf{v}\tau)$$

and

$$C_x(\mathbf{w}, \lambda) = C_Y(\mathbf{w})\, \delta(\lambda - \mathbf{w} \cdot \mathbf{v})$$

where $C_x(\mathbf{w}) = C_Y(\mathbf{w})$. Assuming that the SRF $Y(\mathbf{s})$ has the isotropic covariance in R^3

$$c_Y(r) = \sigma^2 \exp\left[-\frac{r^2}{a^2}\right]$$

where $r = |\mathbf{h}|$, one obtains the spectral density function

$$C_x(\omega) = \frac{\sigma^2 a^3}{8\sqrt{\pi^3}} \exp[-\tfrac{1}{4}(a\omega)^2]$$

where $\omega = |\mathbf{w}|$ and $v = |\mathbf{v}|$, and

$$C_x(\lambda) = \frac{\sigma^2 a}{2\sqrt{\pi}\, v} \exp\left[-\frac{a^2\lambda^2}{4v^2}\right]$$

The $c_x(\mathbf{h}, \tau)$ will be termed *space-time separable* if

$$c_x(\mathbf{h}, \tau) = c_x(\mathbf{h}) c_x(\tau) \tag{36}$$

Clearly, this implies $C_x(\mathbf{w}, \lambda) = C_x(\mathbf{w}) C_x(\lambda)$. When physically justified, separability is an extremely convenient property, from a mathematical point of view.

Example 8: In hydrology, the point-precipitation intensity at location $\mathbf{s}$ during the time t is considered as an S/TRF $X(\mathbf{s}, t)$ with separable covariance of the form

$$c_x(\mathbf{h}, \tau) = \sigma_x^2 \rho_x(\mathbf{h}) \rho_x(\tau)$$

where σ_x^2 is the point variance of $X(\mathbf{s}, t)$, $\rho_x(\mathbf{h})$ is the spatial correlation, and $\rho_x(\tau)$ is the temporal correlation. By taking advantage of covariance separability, comprehensive charts for rainfall-network design can be constructed (see Chapter 10; also Bras and Rodriguez-Iturbe, 1985).

In practice one usually makes an additional assumption, namely, that of *space-isotropic/time-stationary* RF: The covariance and spectral functions are

$$c_x(\mathbf{h}, \tau) = c_x(r, \tau) \tag{37}$$

and

$$C_x(\mathbf{w}, \lambda) = C_x(\omega, \lambda) \tag{38}$$

where $r = |\mathbf{h}|$ and $\omega = |\mathbf{w}|$. Using spherical coordinates in Eq. (26) above, the following *ninth criterion of permissibility* (*COP-9*) can be proven: In order that $c_x(r, \tau)$ be a covariance function of a space-isotropic/time-stationary RF, it is necessary and sufficient that this function admits a representation of the form

$$c_x(r, \tau) = (2\pi)^{n/2} \int_{-\infty}^{\infty} \int_0^{\infty} \frac{J_{(n-2)/2}(\omega r)}{(\omega r)^{(n-2)/2}} \times \exp[i\lambda\tau]\omega^{n-1} C_x(\omega, \lambda)\, d\omega\, d\lambda \tag{39}$$

where $C_x(\omega, \lambda) \geq 0$ on the half-plane (ω, λ), $\omega \in [0, \infty)$, $\lambda \in (-\infty, \infty)$. An analogous criterion is valid in terms of the corresponding space–time semivariogram defined in Eq. (24) above.

Example 9: A useful spatiotemporal semivariogram in $R^1 \times T$ is

$$\gamma_x(r, \tau) = c\left\{1 - \exp\left[-\sqrt{\frac{r^2}{a^2} + \frac{\tau^2}{b^2}}\right]\right\} \tag{40}$$

where $c > 0$. The corresponding spectral density is

$$C_x(\omega, \lambda) = \frac{abc}{2\pi\sqrt{(1 + a^2\omega^2 + b^2\lambda^2)^3}} > 0$$

The space-range of the semivariogram is defined as the r-value for which $\gamma_x(r, 0) = 0.95c$; viz., $r = \varepsilon_s = 3a$. Similarly the time-range is defined as the τ-value for which $\gamma_x(0, \tau) = 0.95c$; viz., $\tau = \varepsilon_t = 3b$.

Other combinations of spatial homogeneity and temporal stationarity, in the strict or the wide sense, are also possible. Thus, an S/TRF is called *time stationary in the strict sense* if its multivariate probability density does not depend on the absolute time, but only on the time differences. This is the case, for example, of the steady-state turbulent flow through a pipe. An S/TRF $X(\mathbf{s}, t)$ is called *time stationary in the wide sense* if

$$c_x(\mathbf{s}, t; \mathbf{s}', t') = c_x(\mathbf{s}, \mathbf{s}'; \tau) \tag{41}$$

where $\tau = t - t'$. An S/TRF is said to be *space homogeneous in the strict sense* if its multivariate probability density does not depend on the absolute positions in space, but only on the vector distances between these positions. Finally, the $X(\mathbf{s}, t)$ is called *space homogeneous in the wide sense* if

$$c_x(\mathbf{s}, t; \mathbf{s}', t') = c_x(\mathbf{h}; t, t') \tag{42}$$

where $\mathbf{h} = \mathbf{s} - \mathbf{s}'$. The space–time covariances satisfy relationships similar to those discussed for purely spatial RF (Chapter 2); for example, for a time-stationary S/TRF (in the wide sense) it is valid that

$$|c_x(\mathbf{s}, \mathbf{s}'; \tau)| \leq \sqrt{c_x(\mathbf{s}, \mathbf{s}; 0) c_x(\mathbf{s}', \mathbf{s}'; 0)}$$

Moreover, the sysketogram function can also be defined in the space–time context (see also Section 9.3 of Chapter 10).

4. Generalized Spatiotemporal Random Fields

4.1 Definition and Basic Properties

In dealing with space nonhomogeneous and/or time nonstationary natural processes it will be useful to introduce the notion of generalized S/TRF. The latter is an extension in the space–time context of the notion of random distribution due to Ito (1954) and Gel'fand (1955). Let Q be some specified linear space of elements q and let $\mathscr{H}_{\mathscr{G}} = L_2(\Omega, F, P)$ be the Hilbert space of all random variables $x(q)$ on Q endowed with the scalar product

$$(x(q_1), x(q_2)) = E[x(q_1)x(q_2)] = \iint \chi_1 \chi_2 \, dF(\chi_1, \chi_2) \tag{1}$$

where $F(\chi_1, \chi_2)$ denotes the joint probability distribution of the random variables $x(q_1), x(q_2)$ with $\|x(q)\|^2 = E|x(q)|^2 < \infty$, and satisfying the linearity condition

$$x\left(\sum_{i=1}^{N} \lambda_i q_i\right) = \sum_{i=1}^{N} \lambda_i x(q_i) \tag{2}$$

for all $q_i \in Q$ and all (real or complex) numbers λ_i $(i = 1, 2, \ldots, N)$. The elements $q \in Q$ are in $R^n \times T$; i.e. that is, $q = q(\mathbf{s}, t)$. Among the Q spaces suitable for the purpose of this study are the spaces $\boldsymbol{K}$ and $\boldsymbol{S}$ of Section 3.1.

Definition 1: A *generalized S/TRF (GS/TRF)* on Q, $X(q)$, is the random mapping

$$X: Q \rightarrow L_2(\Omega, F, P) \tag{3}$$

The GS/TRF considered will always be assumed continuous in the sense that $E|X(q_n)-X(q)|^2 \to 0$ when $q_n \xrightarrow[n\to\infty]{} q$. [The $q_n \xrightarrow[n\to\infty]{} q$ means that all functions $q_n(\mathbf{s}, t)$ and $q(\mathbf{s}, t)$ vanish outside a compact support and all the partial derivatives of $q_n(\mathbf{s}, t)$ converge to the corresponding partial derivatives of $q(\mathbf{s}, t)$ on this support.] The set of all continuous GS/TRF on Q will be denoted by $\mathcal{G}$.

The second-order characteristics of the GS/TRF are the *spatiotemporal mean value*

$$m_x(q) = E[X(q)] = \int \chi \, dF_x(\chi) \tag{4}$$

where $F_x(\chi)$ denotes the probability distribution of $X(q)$, and the

$$c_x(q_1, q_2) = E[(X(q_1) - m_x(q_1))(X(q_2) - m_x(q_2))] \tag{5}$$

which will be called the (*centered*) *spatiotemporal covariance functional* of the GS/TRF $X(q)$. Both the mean and the covariance functional will be assumed to be real-valued and continuous relative to the topology of Q, in the sense that $m_x(q_n) \to m_x(q)$ when $q_n \xrightarrow[n\to\infty]{} q$, and $c_x(q_n, q'_n) \to c_x(q, q')$ when $q_n \xrightarrow[n\to\infty]{} q$, $q'_n \xrightarrow[n\to\infty]{} q'$ for q, q', q_n, $q'_n \in Q$. Also, a useful second-order characteristic is the *spatiotemporal structure* or *semivariogram functional* defined by

$$\gamma_x(q_1, q_2) = \tfrac{1}{2} E[X(q_1) - X(q_2)]^2 \tag{6}$$

Finally, mathematically equivalent space-time second-order functionals may be constructed in the frequency domain by taking the Fourier transform of the covariance and the semivariogram functionals. The functional $m_x(q)$ is linear, for $m_x(\sum_{i=1}^N \lambda_i q_i) = \sum_{i=1}^N \lambda_i m_x(q_i)$. Hence, $m_x(q)$ is a *distribution* (or *generalized function*) on Q, in the sense of Schwartz (1950–51). Moreover, due to the linearity of $X(q)$, $c_x(q_1, q_2)$ is a *bilinear functional* on Q. These functionals $\in Q'$, where Q' is the *dual* space of Q.

4.2 Continuous Linear Functional Representations of Generalized Spatiotemporal Random Fields

In the sequel we will concentrate on GS/TRF that are of the *continuous linear functional form* (*CLF*)

$$X(q) = \langle q(\mathbf{s}, t), X(\mathbf{s}, t)\rangle = \iint q(\mathbf{s}, t) X(\mathbf{s}, t) \, d\mathbf{s} \, dt \tag{7}$$

where $q \in Q$ and $X(\mathbf{s}, t)$ is an OS/TRF in the sense of Definition 1, Section 3 above. As Eq. (7) shows, an ordinary S/TRF $X(\mathbf{s}, t)$ admits a linear extension that is the GS/TRF $X(q)$ defined by (7). Depending on the choice of the function q, the CLF (7) may admit a variety of physical interpretations. Let us consider the following example.

Example 1: Assume that $X(\mathbf{s}, t)$ represents the concentration of an aerosol substance in the atmosphere. By choosing $q(\mathbf{s}, t) = \delta(\mathbf{s} - \mathbf{s}^*)\, \delta(t - t^*)$, Eq. (7) gives the value of the substance at the point/instant $(\mathbf{s}, t)$. If one let

$$q(\mathbf{s}, t) = \begin{cases} 1, & \text{if} \quad \mathbf{s} \in V \quad \text{and} \quad t \in [t_1, t_2] \\ 0, & \text{otherwise} \end{cases}$$

Eq. (7) provides the total amount of substance in the volume V during the time period $[t_1, t_2]$.

Since a GS/TRF $X(q)$ cannot be assigned values at isolated points/instances $(\mathbf{s}, t)$ (unless q is a delta function), we introduce the following field.

Definition 2: A *convoluted S/TRF (CS/TRF)* is defined as the S/TRF

$$\begin{aligned} Y_q(\mathbf{s}, t) &= \langle q(\mathbf{s}', t'), S_{\mathbf{s},t} X(\mathbf{s}', t') \rangle \\ &= \iint q(\mathbf{s}', t') S_{\mathbf{s},t} X(\mathbf{s}', t')\, d\mathbf{s}'\, dt' \end{aligned} \tag{8}$$

We can now make the following observations: The CS/TRF (8) is characterized by $Y_q(\mathbf{O}, 0) = X(q)$ for all $q \in Q$. Also, since

$$\begin{aligned} S_{\mathbf{s},t} X(q) &= \langle q(\mathbf{s}', t'), S_{\mathbf{s},t} X(\mathbf{s}', t') \rangle \\ &= \langle S_{\mathbf{s},t} q(\mathbf{s}', t'), X(\mathbf{s}', t') \rangle = X(S_{-\mathbf{s},-t} q) \end{aligned}$$

it holds true that

$$Y_q(\mathbf{s}, t) = S_{\mathbf{s},t} X(q) = X(S_{-\mathbf{s},-t} q)$$

for all $q \in Q$ and all $(\mathbf{s}, t) \in R^n \times T$. The space $\mathscr{H}$ of OS/TRF may be considered as a subset of the space $\mathscr{G}$ of GS/TRF, viz., $\mathscr{H} \subset \mathscr{G}$. Moreover, the fields $X(q)$ and $Y_q(\mathbf{s}, t)$ have certain important properties, as follows.

Property i: The means and covariances of $X(q)$ and $Y_q(\mathbf{s}, t)$ are written

$$m_x(q) = E[X(q)] = \langle m_x(\mathbf{s}, t), q(\mathbf{s}, t) \rangle \tag{9}$$

$$m_Y(\mathbf{s}, t) = E[Y_q(\mathbf{s}, t)] = \langle m_{S_{\mathbf{s},t} X}(\mathbf{s}', t'), q(\mathbf{s}', t') \rangle \tag{10}$$

and

$$\begin{aligned} c_x(q_1, q_2) &= E[(X(q_1) - m_x(q_1))(X(q_2) - m_x(q_2))] \\ &= \langle \langle c_x(\mathbf{s}, t; \mathbf{s}', t'), q_1(\mathbf{s}, t) \rangle, q_2(\mathbf{s}', t) \rangle \end{aligned} \tag{11}$$

$$\begin{aligned} c_Y(\mathbf{s}, t; \mathbf{s}', t') &= E[(Y_{q_1}(\mathbf{s}, t) - m_Y(\mathbf{s}, t))(Y_{q_2}(\mathbf{s}', t') - m_Y(\mathbf{s}', t'))] \\ &= \langle \langle c_x(S_{\mathbf{s},t} X(\mathbf{s}'', t''), S_{\mathbf{s}',t'} X(\mathbf{s}''', t''')), q_1(\mathbf{s}'', t'') \rangle, q_2(\mathbf{s}''', t''') \rangle \end{aligned} \tag{12}$$

The means and covariances of the GS/TRF and CS/TRF are linearly related to those of the corresponding OS/TRF. From Eqs. (10) and (12) we find that the corresponding mean values and covariances functions are written, respectively,

$$m_x(q) = m_Y(\mathbf{O}, 0) \tag{13}$$

$$c_x(q_1, q_2) = c_Y(\mathbf{O}, 0; \mathbf{O}, 0) \tag{14}$$

Property ii: The covariance functional of the GS/TRF $X(q)$ is a *nonnegative-definite bilinear functional* in the sense that

$$c_x(q, q) = E[|X(q) - m_x(q)|^2] \geq 0 \tag{15}$$

for all $q \in Q$. Conversely, every continuous nonnegative-definite bilinear functional $c_x(q_1, q_2)$ in Q is a covariance functional of some GS/TRF $X(q)$. On the other hand, the covariance function $c_Y(\mathbf{s}, t; \mathbf{s}', t')$ is also a nonnegative-definite function in the ordinary sense defined earlier.

Property iii: The fields $X(q)$ and $Y_q(\mathbf{s}, t)$ are always *differentiable*, even when $X(\mathbf{s}, t)$ is not. To see this assume $Q = \boldsymbol{K}$ and let

$$\begin{aligned} X^{(\boldsymbol{\rho},\zeta)}(q) &= \langle q(\mathbf{s}, t), X^{(\boldsymbol{\rho},\zeta)}(\mathbf{s}, t)\rangle \\ &= \iint q(\mathbf{s}, t) X^{(\boldsymbol{\rho},\zeta)}(\mathbf{s}, t)\, d\mathbf{s}\, dt \end{aligned} \tag{16}$$

where ζ is a nonnegative integer and $\boldsymbol{\rho} = (\rho_1, \rho_2, \ldots, \rho_n)$ is a multi-index of nonnegative integers; in other words, the superscript $(\boldsymbol{\rho}, \zeta)$ denotes partial differentiation of the order ρ in space and differentiation of order ζ in time

$$X^{(\boldsymbol{\rho},\zeta)}(\mathbf{s}, t) = D^{(\boldsymbol{\rho},\zeta)} X(\mathbf{s}, t) = \frac{\partial^{|\boldsymbol{\rho}|}}{\partial s_1^{\rho_1} \ldots \partial s_n^{\rho_n}} \left[\frac{\partial^{\zeta}}{\partial t^{\zeta}} X(\mathbf{s}, t) \right] \tag{17}$$

where $\rho = |\boldsymbol{\rho}| = \sum_{i=1}^{n} \rho_i$. By applying integration by parts, Eq. (16) can be written

$$\begin{aligned} X^{(\boldsymbol{\rho},\zeta)}(q) &= (-1)^{\rho+\zeta} \iint q^{(\boldsymbol{\rho},\zeta)}(\mathbf{s}, t) X(\mathbf{s}, t)\, d\mathbf{s}\, dt \\ &= (-1)^{\rho+\zeta} \langle q^{(\boldsymbol{\rho},\zeta)}(\mathbf{s}, t), X(\mathbf{s}, t)\rangle = (-1)^{\rho+\zeta} X(q^{(\boldsymbol{\rho},\zeta)}) \end{aligned} \tag{18}$$

Similarly for the CS/TRF,

$$\begin{aligned} Y_q^{(\boldsymbol{\rho},\zeta)}(\mathbf{s}, t) &= (-1)^{\rho+\zeta} \iint q^{(\boldsymbol{\rho},\zeta)}(\mathbf{s}, t) S_{\mathbf{s},t} X(\mathbf{s}', t')\, d\mathbf{s}'\, dt' \\ &= (-1)^{\rho+\zeta} \langle q^{(\boldsymbol{\rho},\zeta)}(\mathbf{s}, t), S_{\mathbf{s},t} X(\mathbf{s}', t')\rangle \\ &= (-1)^{\rho+\zeta} S_{\mathbf{s},t} X(q^{(\boldsymbol{\rho},\zeta)}) \end{aligned} \tag{19}$$

Therefore, although there may exist no $X^{(\boldsymbol{\rho},\zeta)}(\mathbf{s}, t)$ as such, we can always obtain $X^{(\boldsymbol{\rho},\zeta)}(q)$ and $Y_q^{(\boldsymbol{\rho},\zeta)}(\mathbf{s}, t)$ in the sense defined above. This feature of the GS/TRF has several interesting consequences. For example, it leads to a more realistic evaluation of the *microscale* properties that are closely related to the space–time pattern of the random field: The behavior of the covariance or semivariogram near the space–time origin governs the most important geometrical characteristics of random fields, such as mean square continuity and differentiability.

Property iv: By applying the Riesz–Radon theorem in terms of generalized functions we find that the mean $m_x(q)$ can be written as

$$m_x(q) = \left\langle \sum_{\rho \le \nu} \sum_{\zeta \le \mu} q^{(\boldsymbol{\rho},\zeta)}(\mathbf{s}, t), f_{\rho,\zeta}(\mathbf{s}, t) \right\rangle \tag{20}$$

where ν and μ are nonnegative integers, $q(\mathbf{s}, t) \in \boldsymbol{K}$, and $f_{\rho,\zeta}(\mathbf{s}, t)$ are continuous functions in $R^n \times T$, only a finite number of which are different from zero on any given finite support $\boldsymbol{U}$ of $\boldsymbol{K}$. Integration by parts yields

$$m_x(q) = \left\langle \sum_{\rho \le \nu} \sum_{\zeta \le \mu} (-1)^{\rho+\zeta} f_{\rho,\zeta}^{(\boldsymbol{\rho},\zeta)}(\mathbf{s}, t), q(\mathbf{s}, t) \right\rangle \tag{21}$$

A similar expression may be derived for the mean $m_Y(\mathbf{s}, t)$ of $Y_q(\mathbf{s}, t)$, namely,

$$m_Y(\mathbf{s}, t) = \left\langle \sum_{\rho \le \nu} \sum_{\zeta \le \mu} (-1)^{\rho+\zeta} S_{\mathbf{s},t} f_{\rho,\zeta}^{(\boldsymbol{\rho},\zeta)}(\mathbf{s}', t'), q(\mathbf{s}', t') \right\rangle \tag{22}$$

For convenience in the subsequent analysis let us put

$$g_{\rho,\zeta}(\mathbf{s}, t) = f_{\rho,\zeta}^{(\boldsymbol{\rho},\zeta)}(\mathbf{s}, t) \tag{23}$$

Closely related to Property (iv) is the following section.

4.3 Space-Homogeneous/Time-Stationary Generalized Spatiotemporal Random Fields

A GS/TRF $X(q)$, $q(\mathbf{s}, t) \in Q$, $(\mathbf{s}, t) \in R^n \times T$ will be called *space homogeneous/time stationary in the wide sense* if its mean value $m_x(q)$ and covariance functional $c_x(q_1, q_2)$ are invariant with respect to any shift of the parameters; that is,

$$m_x(q) = m_x(S_{\mathbf{h},\tau} q) \tag{24}$$

$$c_x(q_1, q_2) = c_x(S_{\mathbf{h},\tau} q_1, S_{\mathbf{h},\tau} q_2) \tag{25}$$

for any $(\mathbf{h}, \tau) \in R^n \times T$. Clearly, when the $X(q)$ is space homogeneous/time stationary, the $c_x(q_1, q_2)$ is a translation-invariant, nonnegative-definite bilinear functional on Q, and the following proposition can be proven (Christakos, 1991c).

Proposition 1: If $X(q)$ is a space-homogeneous/time-stationary GS/TRF on Q, there exists one and only one generalized functional $c_x(q_1, q_2) \in Q'$ such that

$$\langle X(q_1), X(q_2)\rangle = c_x(q_1, q_2), \qquad q_1, q_2 \in Q$$

We shall denote by $\mathscr{G}_0$ the set of all space-homogeneous/time-stationary generalized fields. Note that $\mathscr{H}_0 \subset \mathscr{G}_0 \subset \mathscr{G}$. Similarly, the CS/TRF $Y_q(\mathbf{s}, t)$ is called *space homogeneous/time stationary* if

$$m_Y(\mathbf{s}, t) = \text{constant} \tag{26}$$

and

$$c_Y(\mathbf{s}, t; \mathbf{s}', t') = c_Y(\mathbf{h}, \tau) \tag{27}$$

where $\mathbf{h} = \mathbf{s} - \mathbf{s}'$, $\tau = t - t'$, for any $(\mathbf{h}, \tau) \in R^n \times T$. The $c_Y(\mathbf{s}, t; \mathbf{s}', t')$ is a nonnegative-definite function. In view of Eq. (22) and condition (24) it follows that the functions $f_{\rho,\zeta}(\mathbf{s}, t)$ are constants. Therefore,

$$\begin{aligned} g_{\rho,\zeta}(\mathbf{s}, t) &= f_{\rho,\zeta}^{(\rho,\zeta)}(\mathbf{s}, t) = 0 \qquad \text{for all} \quad \rho \geq 1, \zeta \geq 1 \\ &= f_{0,0}^{(0,0)}(\mathbf{s}, t) = m \qquad \text{for} \quad \rho = \zeta = 0 \end{aligned} \tag{28}$$

and the $m_x(q)$ will have the form

$$m_x(q) = m \iint q(\mathbf{s}, t)\, d\mathbf{s}\, dt = m\langle q(\mathbf{s}, t), 1\rangle \tag{29}$$

The $c_x(q_1, q_2) \in Q'$ can be expressed in terms of the corresponding $c_x(\mathbf{h}, \tau)$ as follows:

$$c_x(q_1, q_2) = \langle c_x(\mathbf{h}, \tau), q_1 * \check{q}_2(\mathbf{h}, \tau)\rangle = c_x(q_1 * \check{q}_2) \tag{30}$$

for all $q_1, q_2 \in Q$, where $*$ denotes convolution and $\vee$ denotes inversion [i.e., $\check{q}_2(\mathbf{h}, \tau) = q_2(-\mathbf{h}, -\tau)$].

Example 1: Let us define in $R^1 \times T$ a zero-mean *Wiener* S/TRF $W(s, t)$, $s \in [s_1, s_2]$, $t \in [0, \infty)$ as a *Gaussian* S/TRF with covariance function

$$c_x(s, t; s', t') = \min(s - s_1, s' - s_2) \min(t, t') \tag{31}$$

The $X(s, t) = \partial W(s, t)/\partial s\, \partial t$ will be zero-mean *white-noise* S/TRF with covariance function

$$c_x(h, \tau) = \delta(h, \tau) \tag{32}$$

where $h = s - s'$, $\tau = t - t'$ and the spatiotemporal delta function $\delta(h, \tau)$ is such as

$$\iint \delta(h, \tau)\, dh\, d\tau = \iint \delta(h)\, \delta(\tau)\, dh\, d\tau$$

The corresponding GS/TRF $X(q) = \langle X(s, t), q(s, t) \rangle$ has covariance

$$c_x(q_1, q_2) = \langle c_x(h, \tau), q_1 * \check{q}_2(h, \tau) \rangle$$
$$= \langle \delta, q_1 * \check{q}_2 \rangle = (q_1 * \check{q}_2)(0, 0) = \delta(q_1 * \check{q}_2) \tag{33}$$

The above results can be generalized to more than one spatial dimension. More specifically, one may define in $R^n \times T$ the so-called *Brownian sheet* $W(\mathbf{s}, t)$, which is a zero-mean Gaussian S/TRF with covariance.

$$c_w(\mathbf{s}, t; \mathbf{s}', t') = \min(s_1, s_1') \ldots \min(s_n, s_n') \min(t, t') \tag{34}$$

Brownian sheet has important applications in the context of stochastic partial differential equations.

Since $c_x(q_1, q_2)$ is a translation-invariant bilinear functional, it will have the form $c_x(q_1, q_2) = \langle c_0, q_1 * q_2 \rangle$, where c_0 is a nonnegative-definite generalized function that is the Fourier transform of some positive-tempered measure $\phi(\mathbf{w}, \lambda)$ in $R^n \times T$; that is,

$$c_0(q) = \iint \tilde{q}(\mathbf{w}, \lambda)\, d\phi(\mathbf{w}, \lambda)$$

and

$$\iint \frac{d\phi(\mathbf{w}, \lambda)}{(1 + |(\mathbf{w}, \lambda)|^2)^p} < \infty \qquad \text{for some} \quad p > 0$$

Moreover, in the light of the Fourier transform properties of generalized functions, it is valid that

$$c_x(q_1, q_2) = \langle c_0, q_1 * \check{q}_2 \rangle = \langle \phi, \tilde{q}_1 \tilde{q}_2 \rangle$$

which yields the following result (see, also, Christakos, 1991c).

Proposition 2: Let $X(q)$ be a GS/TRF in $R^n \times T$. The covariance functional is written

$$c_x(q_1, q_2) = \iint \tilde{q}_1(\mathbf{w}, \lambda) \tilde{q}_2(\mathbf{w}, \lambda)\, d\phi(\mathbf{w}, \lambda) \tag{35}$$

where $\tilde{q}_1(\mathbf{w}, \lambda)$ and $\tilde{q}_2(\mathbf{w}, \lambda)$ are the Fourier transform of the $q_1(\mathbf{s}, t)$ and $q_2(\mathbf{s}, t)$, respectively, and $\phi(\mathbf{w}, \lambda)$ is some positive-tempered measure in $R^n \times T$. In this case the $\phi(\mathbf{w}, \lambda)$ is called the *spectral measure* of the GS/TRF $X(q)$.

Example 2: Consider once more the Example 1 above. Since

$$c_x(q_1, q_2) = \langle c_0, q_1 * \check{q}_2 \rangle = \langle \delta, q_1 * \check{q}_2 \rangle = \langle \phi, \tilde{q}_1 \tilde{q}_2 \rangle$$

and the Fourier transform of $c_0 = \delta$ is $d\mathbf{w}\, d\lambda$ (Lebesgue measure), we conclude that the spectral measure of $X(q)$ is $d\phi(\mathbf{w}, \lambda) = d\mathbf{w}\, d\lambda$.

The space-homogeneous/time-stationary analysis leads to the fifth property.

Property v: The CS/TRF $Y_q(\mathbf{s}, t)$ can be zero-mean space homogeneous/time stationary even when the associated OS/TRF $X(\mathbf{s}, t)$ is space nonhomogeneous/time nonstationary. This can happen under certain conditions concerning the choice of the functions $q(\mathbf{s}, t)$ as well as the form of the functions $g_{\rho,\zeta}(\mathbf{s}, t)$. More specifically, we must define spaces

$$Q_{\nu/\mu} = \{q \in Q : \langle q(\mathbf{s}, t), g_{\rho,\zeta}(\mathbf{s}, t)\rangle = 0 \quad \text{for all} \quad \rho \leq \nu, \zeta \leq \mu\} \tag{36}$$

and

$$\mathscr{C}_{\nu/\mu} = \{g_{\rho,\zeta}(\mathbf{s}, t) \in \boldsymbol{C} : \langle q(\mathbf{s}, t), g_{\rho,\zeta}(\mathbf{s}, t)\rangle = 0$$
$$\Rightarrow \langle q(\mathbf{s}, t), \boldsymbol{S}_{\mathbf{h},\tau} g_{\rho,\zeta}(\mathbf{s}, t)\rangle = 0 \quad \text{for all} \quad \rho \leq \nu, \zeta \leq \mu\} \tag{37}$$

where $\boldsymbol{C}$ is the space of continuous functions in $R^n \times T$ with compact support. Equation (36) assures a zero-mean value for the CS/TRF $Y_q(\mathbf{s}, t)$ at $(\mathbf{s}, t) = (\mathbf{O}, 0)$, while the closeness of $\mathscr{C}_{\nu/\mu}$ to translation [Eq. (37)] is necessary in order that stochastic inference about $X(q)$ makes sense (i.e., in order that the stochastic correlation properties of $X(q)$ remain unaffected by a shift $\boldsymbol{S}_{\mathbf{h},\tau}$ of the space/time origin). Functions $g_{\rho,\zeta}(\mathbf{s}, t)$ that satisfy these conditions are of the form

$$g_{\rho,\zeta}(\mathbf{s}, t) = \mathbf{s}^{\boldsymbol{\rho}} t^{\zeta} \exp[\boldsymbol{\alpha} \cdot \mathbf{s} + \beta t] \tag{38}$$

where $\boldsymbol{\alpha}$ and β are (real or complex) vector and number, respectively. Indeed, suppose that $g_{\rho,\zeta}(\mathbf{s}, t)$ is of the form (38) and let

$$\iint q(\mathbf{s}, t)\mathbf{s}^{\boldsymbol{\rho}} t^{\zeta} \exp[\boldsymbol{\alpha} \cdot \mathbf{s} + \beta t]\, d\mathbf{s}\, dt = 0$$

Then, by applying a shift $\boldsymbol{S}_{\mathbf{h},\tau}$ we get

$$\iint q(\mathbf{s}, t)(\mathbf{s}+\mathbf{h})^{\boldsymbol{\rho}}(t+\tau)^{\zeta} \exp[\boldsymbol{\alpha} \cdot (\mathbf{s}+\mathbf{h}) + \beta(t+\tau)]\, d\mathbf{s}\, dt$$
$$= \iint q(\mathbf{s}, t)\left[\sum_{|\mathbf{k}|=0}^{\rho} C_{\rho!}^{|\mathbf{k}|!}\mathbf{s}^{\mathbf{k}}\mathbf{h}^{\boldsymbol{\rho}-\mathbf{k}} \sum_{m=0}^{\zeta} C_{\zeta!}^{m!} t^m \tau^{\zeta-m}\right]$$
$$\times \exp[\boldsymbol{\alpha} \cdot (\mathbf{s}+\mathbf{h}) + \beta(t+\tau)]\, d\mathbf{s}\, dt$$
$$= \sum_{|\mathbf{k}|=0}^{\rho} \sum_{m=0}^{\zeta} C_{\rho!}^{|\mathbf{k}|!} C_{\zeta!}^{m!} \mathbf{h}^{\boldsymbol{\rho}-\mathbf{k}} \tau^{\zeta-m} \exp[\boldsymbol{\alpha} \cdot \mathbf{h} + \beta\tau]$$
$$\times \iint q(\mathbf{s}, t)\mathbf{s}^{\mathbf{k}} t^m \exp[\boldsymbol{\alpha} \cdot \mathbf{s} + \beta t]\, d\mathbf{s}\, dt = 0$$

In other words, the conditions of Eq. (37) are fulfilled.

Example 3: Let us choose the function $q^*(\mathbf{s}, t) \in \boldsymbol{S}$ so that its Fourier transform $\tilde{q}^*(\mathbf{w}, \lambda)$ satisfies

$$[1-\tilde{q}^*(\mathbf{w}, \lambda)]^{(\rho,\zeta)}|_{(\mathbf{w},\lambda)=(\mathbf{O},0)}=0$$

for all ρ up to $2\nu' > \nu$ and all ζ up to $2\mu' > \mu$. (For example, a function with a Fourier transform

$$\tilde{q}^*(\mathbf{w}, \lambda)=\left[1+\frac{\alpha\beta}{4}|\mathbf{w}|^2\lambda^2+\cdots+\frac{\alpha^{\nu'}\beta^{\mu'}}{4\rho\zeta}|\mathbf{w}|^{2\nu'}\lambda^{2\mu'}\right]\exp[-(\alpha|\mathbf{w}|^2+\beta\lambda^2)]$$

where α, $\beta > 0$ and ν', μ' are integers satisfying $2\nu' > \nu$, $2\mu' > \mu$.) We will show that $q(\mathbf{s}, t) = \delta(\mathbf{s}, t) - q^*(\mathbf{s}, t) \in Q_{\nu/\mu}$. Indeed,

$$\begin{aligned}
&\iint g_{\rho,\zeta}(\mathbf{s}, t)[\delta(\mathbf{s}, t)-q^*(\mathbf{s}, t)]\, d\mathbf{s}\, dt \\
&\quad=\iint g_{\rho,\zeta}(\mathbf{s}, t)\left\{\iint [1-\tilde{q}^*(\mathbf{w}, \lambda)]\exp[-i(\mathbf{w}\cdot\mathbf{s}+\lambda t)]\, d\mathbf{w}\, d\lambda\right\} d\mathbf{s}\, dt \\
&\quad=\iint \delta^{(\rho,\zeta)}(w, \lambda)[1-\tilde{q}^*(\mathbf{w}, \lambda)]\, d\mathbf{w}\, d\lambda \\
&\quad=(-1)^{\rho+\zeta}\iint \delta(\mathbf{w}, \lambda)[1-\tilde{q}^*(\mathbf{w}, \lambda)]^{(\rho,\zeta)}\, d\mathbf{w}\, d\lambda \\
&\quad=(-1)^{\rho+\zeta}[1-\tilde{q}^*(\mathbf{w}, \lambda)]^{(\rho,\zeta)}|_{(\mathbf{w},\lambda)=(\mathbf{O},0)}=0
\end{aligned}$$

for all ρ up to $2\nu' > \nu$ and all ζ up to $2\mu' > \mu$, by definition of $\tilde{q}^*(\mathbf{w}, \lambda)$.

From a practical point of view, the modeling of spatiochronological variations and the estimation of spatiotemporal processes is easier and more efficiently carried out when the $g_{\rho,\zeta}(\mathbf{s}, t)$ are *pure polynomials*, viz.,

$$g_{\rho,\zeta}(\mathbf{s}, t)=\mathbf{s}^{\boldsymbol{\rho}}t^{\zeta}=s_1^{\rho_1}s_2^{\rho_2}\ldots s_n^{\rho_n}t^{\zeta} \tag{39}$$

where $\rho=|\boldsymbol{\rho}|=\sum_{i=1}^{n}\rho_i$. This is due mainly to convenient invariance and linearity properties that the latter satisfy. In conclusion, the "derived" fields $X(q)$ and $Y_q(\mathbf{s}, t)$ have a very convenient mathematical structure. From a physical viewpoint this means that even if $X(\mathbf{s}, t)$ represents an actual natural process which has, in general, very irregular, space-nonhomogeneous/time-nonstationary features, we can derive fields $X(q)$ and $Y_q(\mathbf{s}, t)$ that have regular, space-homogeneous/time stationary-features (see Fig. 5.3). Hence, analysis and processing are much easier.

We will close this section with a final remark. Just as for SRF (Chapter 2), a complete stochastic characterization of an S/TRF $X(\mathbf{s}, t)$ is given by its *characteristic functional* defined as follows:

$$\Phi(q)=E\{\exp[iX(q)]\} \tag{40}$$

which must be known for any $q(\mathbf{s}, t)$ such that the integral (7) exists for all possible realizations of $X(\mathbf{s}, t)$.

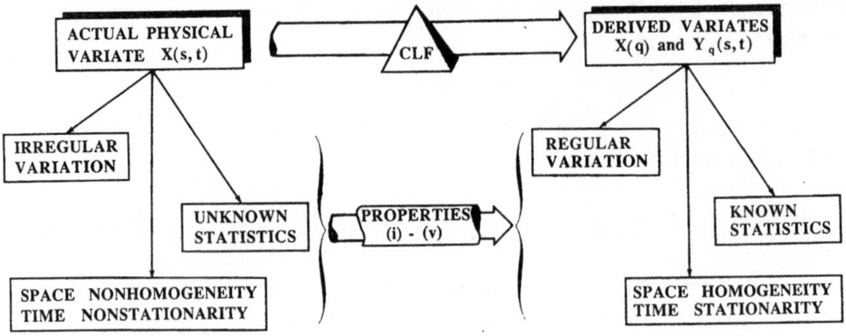

Figure 5.3 Linking actual and derived variates (processes) by means of Properties i through v

When the $\Phi(q)$ is available, one may derive the characteristic function, probability density, and the space–time moments of the S/TRF $X(\mathbf{s}, t)$. For example, it holds true that

$$m_x^{(k)}(\mathbf{s}, t) = E[X^k(\mathbf{s}, t)] = \frac{1}{i^k} \frac{\delta^k \Phi[q]}{\delta q^k(\mathbf{s}, t)}\bigg|_{q(\mathbf{s},t)=0} \tag{41}$$

where $\delta/\delta q$ denotes the functional derivative. Most of the characteristic functional properties of SRF discussed in Chapter 2 are valid for S/TRF, as well.

5. Spatiotemporal Random Fields of Order ν/μ (Ordinary and Generalized)

5.1 Random Fields with Space-Homogeneous/Time-Stationary Increments

We now come to what is, for our present concern, the most interesting aspect of S/TRF, namely, the concept of S/TRF with space-homogeneous/time-stationary increments of orders ν in space and μ in time, in the ordinary or in the generalized sense.

Definition 1: A CS/TRF $Y_q(\mathbf{s}, t)$ will be called a *CS/TRF of order* ν *in space/*μ *in time* (*CS/TRF-*ν/μ) if $q \in Q_{\nu/\mu}$. In this case the space $Q_{\nu/\mu}$ will be termed an *admissible space of order* ν/μ (*AS-*ν/μ).

Definition 2: Let $Q_{\nu/\mu}$ be an AS-ν/μ. A *GS/TRF* $X(q)$ *with space homogeneous of order* ν*/time stationary of order* μ *increments* (*GS/TRF-*ν/μ) is a linear mapping

$$X: Q_{\nu/\mu} \to L_2(\Omega, F, P) \tag{1}$$

where the corresponding CS/TRF $Y_q(\mathbf{s}, t)$ is zero-mean space-homogeneous/time-stationary for all $q \in Q_{\nu/\mu}$ and all $(\mathbf{h}, \tau) \in R^n \times T$.

The set of all continuous GS/TRF-ν/μ will be denoted by $\mathcal{G}_{\nu/\mu}$. The definition above is equivalent to the following one, which we unfold in an extended Ito–Gel'fand spirit:

Definition 3: A GS/TRF-ν/μ $X(q)$ is a GS/TRF for which all differential operators of the form

$$Y(q) = D^{(\nu+1,\mu+1)}X(q) \tag{2}$$

where $D^{(\nu+1,\mu+1)}X(q) = X^{(\nu+1,\mu+1)}(q) = (-1)^{\nu+\mu}X(q^{(\nu+1,\mu+1)})$ are zero-mean space-homogeneous/time-stationary generalized fields.

The OS/TRF associated with the space $\mathcal{G}_{\nu/\mu}$ will be defined as follows:

Definition 4: An OS/TRF $X(\mathbf{s}, t)$ is called an *OS/TRF of order* ν/μ (*OS/TRF-*ν/μ) if for all $q \in Q_{\nu/\mu}$ the corresponding CS/TRF $Y_q(\mathbf{s}, t)$ is zero-mean space-homogeneous/time-stationary.

The zero-mean condition is imposed for convenience and does not restrict generality. A flowchart summarizing the various steps involved in the construction of an S/TRF-ν/μ (ordinary or generalized) is presented in Fig. 5.4. In light of Definition 4, if the

$$Y(\mathbf{s}, t) = D^{(\nu+1,\mu+1)}X(\mathbf{s}, t) = X^{(\nu+1,\mu+1)}(\mathbf{s}, t) \tag{3}$$

exist and are zero-mean space-homogeneous/time-stationary fields, then the $X(\mathbf{s}, t)$ is an OS/TRF-ν/μ.

In connection with this, the following propositions can be proven (Christakos, 1991b and c).

Proposition 1: The solutions of the stochastic partial differential equation in $R^1 \times T$

$$D^{(\nu+1,\mu+1)}X_{\nu/\mu}(s, t) = Y(s, t) \tag{4}$$

where $Y(s, t)$ is zero-mean space-homogeneous/time-stationary, are OS/TRF ν/μ. Note that in this case the

$$Y(q) = \langle q(s, t), Y(s, t)\rangle = \langle q(s, t), X_{\nu/\mu}^{(\nu+1,\mu+1)}(s, t)\rangle = Y_q^{(\nu+1,\mu+1)}(0) \tag{5}$$

is a space-homogeneous/time-stationary generalized field.

Proposition 2: The OS/TRF

$$X(\mathbf{s}, t) = \sum_{\rho=0}^{\nu} \sum_{\zeta=0}^{\mu} \beta_{\rho,\zeta} g_{\rho,\zeta}(\mathbf{s}, t) \tag{6}$$

where $\beta_{\rho,\zeta}$ ($\rho \le \nu$ and $\zeta \le \mu$) are random variables in $\mathcal{H}_k = L_2(\Omega, F, P)$, is an OS/TRF ν/μ.

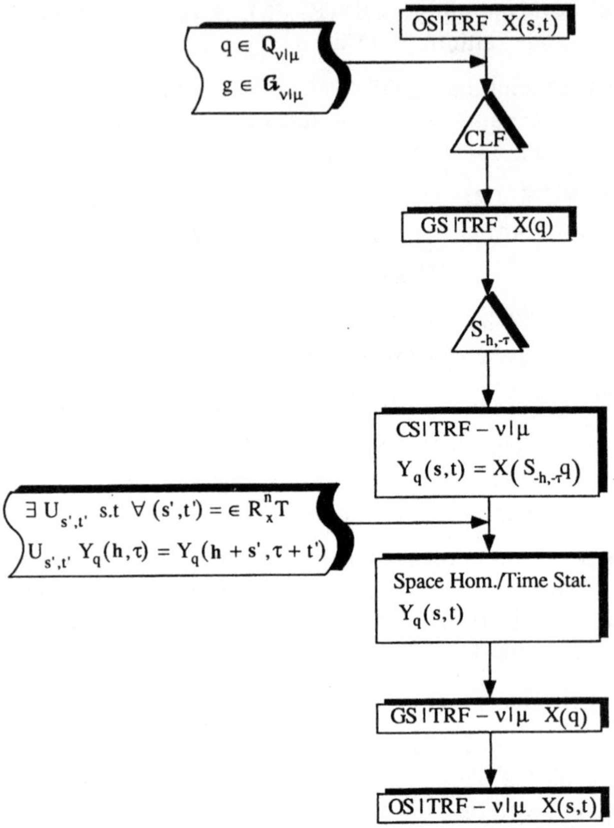

Figure 5.4 Flowchart depicting the various steps in constructing an $S/TRF\text{-}\nu/\mu$

In view of Eqs. (7) and (8) of Section 4 above, to each generalized $X(q)$ correspond various ordinary $X(\mathbf{s}, t)$, all having the same CS/TRF $Y_q(\mathbf{s}, t) = X(S_{-\mathbf{s},-t}q)$; that is, we can write

$$\begin{array}{ccc} X(q) & \leftrightarrow & \{X^a(\mathbf{s}, t),\ a = 1, 2, \ldots\} \\ \downarrow\uparrow & & \downarrow\uparrow \\ X(S_{-\mathbf{s},-t}q) = & & Y_q(\mathbf{s}, t) \end{array} \tag{7}$$

Hence,

Definition 5: The set

$$\mathscr{X}_q = \{X^a(\mathbf{s}, t), \quad a = 1, 2, \ldots\}$$

of all $OS/TRF\text{-}\nu/\mu$ that have the same $CS/TRF\text{-}\nu/\mu$ $Y_q(\mathbf{s}, t)$ in Q will be termed the *generalized representation set of order* ν/μ ($GRS\text{-}\nu/\mu$). Each member of the $GRS\text{-}\nu/\mu$ will be called a *representation* of the $X(q)$.

We can now state the proposition below (Christakos, 1991c).

Proposition 3: Let $X^0(\mathbf{s}, t)$ be a representation $\in \mathscr{X}_q$ of $X(q)$. The OS/TRF $X^a(\mathbf{s}, t)$ is another representation if and only if it can be expressed as

$$X^a(\mathbf{s}, t) = X^0(\mathbf{s}, t) + \sum_{\rho=|\boldsymbol{\rho}|\leq\nu} \sum_{\zeta\leq\mu} c_{\rho,\zeta} g_{\rho,\zeta}(\mathbf{s}, t) \tag{8}$$

where the $c_{\rho,\zeta}$, $\rho \leq \nu$, and $\zeta \leq \mu$ are random variables in $L_2(\Omega, F, P)$ such that

$$c_{\rho,\zeta} = \langle \eta_{\rho,\zeta}(\mathbf{s}, t), X^a(\mathbf{s}, t)\rangle \tag{9}$$

where the $\eta_{\rho,\zeta}(\mathbf{s}, t)$ satisfy the

$$\langle \eta_{\rho,\zeta}(\mathbf{s}, t), g_{\rho',\zeta'}(\mathbf{s}, t)\rangle = \begin{cases} 1 & \text{if} \quad \rho = \rho' \quad \text{and} \quad \zeta = \zeta' \\ 0 & \text{otherwise} \end{cases} \tag{10}$$

An OS/TRF-ν/μ is not always differentiable. It can, however, be expressed in terms of a differentiable OS/TRF-ν/μ as shown in the proposition below (Christakos, 1991b and c).

Proposition 4: Let $X(\mathbf{s}, t)$ be a continuous OS/TRF-ν/μ. Then it follows that

$$X(\mathbf{s}, t) = X^*(\mathbf{s}, t) + Y_q(\mathbf{s}, t) \tag{11}$$

where $X^*(\mathbf{s}, t)$ is an infinitely differentiable OS/TRF-ν/μ and $Y_q(\mathbf{s}, t)$ is a space-homogeneous/time-stationary random field.

5.2 The Correlation Structure of Spatiotemporal Random Fields of Order ν/μ

In this subsection we will study the spatiotemporal trend and correlation structure of a S/TRF-ν/μ. In view of the preceding results, the generalized field $X^{(\nu,\mu)}(q) = (-1)^{\nu+\mu} X(q^{(\nu,\mu)})$ has *constant mean*; that is, we can write

$$E[X^{(\nu,\mu)}(q)] = m_x^{(\nu,\mu)}(q) = (-1)^{\nu+\mu} m_x(q^{(\nu,\mu)}) = a\tilde{q}(\mathbf{O}, 0) \tag{12}$$

while the covariance functional is expressed as

$$\begin{aligned} c_x(q_1^{(\nu+1,\mu+1)}, q_2^{(\nu+1,\mu+1)}) &= E[X(q_1^{(\nu+1,\mu+1)})X(q_2^{(\nu+1,\mu+1)})] \\ &= E[X^{(\nu+1,\mu+1)}(q_1)X^{(\nu+1,\mu+1)}(q_2)] = c_Y(q_1, q_2) \end{aligned} \tag{13}$$

But as was shown above, the $c_Y(q_1, q_2)$ in Eq. (13) is a translation-invariant bilinear functional and, therefore, so is $c_x(q_1^{(\nu+1,\mu+1)}, q_2^{(\nu+1,\mu+1)})$. Taking into account the properties of bilinear functionals (for the relevant theory see Gel'fand and Vilenkin, 1964), Eqs. (12) and (13) lead to the proposition below.

Proposition 5: Let $X(q)$ be a GS/TRF-ν/μ in $R^n \times T$. Its mean value and covariance functional have the following forms

$$m_x(q) = \sum_{0 \leq |\boldsymbol{\rho}| \leq \nu} \sum_{0 \leq \zeta \leq \mu} a_{\boldsymbol{\rho},\zeta} \langle \mathbf{s}^{\boldsymbol{\rho}} t^{\zeta}, q(\mathbf{s}, t) \rangle$$

$$= \sum_{\rho_1} \sum_{\rho_2} \cdots \sum_{\rho_n} \sum_{\zeta} a_{\rho_1 \rho_2 \ldots \rho_n \zeta} \langle s_1^{\rho_1} s_2^{\rho_2} \ldots s_n^{\rho_n} t^{\zeta}, q(\mathbf{s}, t) \rangle \tag{14}$$

where $a_{\rho,\zeta}$ are suitable coefficients, $0 \leq \rho = |\boldsymbol{\rho}| = \sum_{i=1}^{n} \rho_i \leq \nu$; and

$$c_x(q_1, q_2) = \iint_{\mathfrak{R} \times \mathfrak{J}} \tilde{q}_1(\mathbf{w}, \lambda) \tilde{q}_2(\mathbf{w}, \lambda)\, d\phi_x(\mathbf{w}, \lambda)$$

$$+ G[\tilde{q}_1^{(\nu+1,\mu+1)}(\mathbf{O}, 0), \tilde{q}_2^{(\nu+1,\mu+1)}(\mathbf{O}, \mathrm{O})] \tag{15}$$

where $\mathfrak{R} = R^n - \{\mathbf{O}\}$ and $\mathfrak{J} = T - \{0\}$, ϕ_x is a certain positive-tempered measure, and G is some function in $\tilde{q}_1^{(\nu+1,\mu+1)}(\mathbf{O}, \mathrm{O})$ and $\tilde{q}_2^{(\nu+1,\mu+1)}(\mathbf{O}, \mathrm{O})$.

Example 1: Consider in $R^2 \times T$ the case $\nu = \mu = 1$. According to Proposition 5, the mean value of $X(q)$ will be

$$m_x(q) = \sum_{0 \leq \rho \leq 1} \sum_{0 \leq \zeta \leq 1} a_{\rho_1 \rho_2 \zeta} \langle s_1^{\rho_1} s_2^{\rho_2} t^{\zeta}, q(\mathbf{s}, t) \rangle$$

$$= a_{000} + a_{100} \langle s_1, q \rangle + a_{010} \langle s_2, q \rangle + a_{001} \langle t, q \rangle$$

$$+ a_{101} \langle s_1 t, q \rangle + a_{011} \langle s_2 t, q \rangle$$

The mean value of $Y(q)$ is

$$m_Y(q) = m_x^{(2,2)}(q) = D^{(2,2)} m_x(q) = \frac{\partial^4 m_x(q)}{\partial s_1^{\rho_1}\, \partial s_2^{\rho_2}\, \partial t^2}$$

where $\rho_1 + \rho_2 = 2$. It is valid that

$$\frac{\partial^4 m_x(q)}{\partial s_1^2\, \partial t^2} = \frac{\partial^4 m_x(q)}{\partial s_2^2\, \partial t^2} = \frac{\partial^4 m_x(q)}{\partial s_1\, \partial s_2\, \partial t^2} = 0$$

In other words, $Y(q)$ has zero mean as expected. Obviously,

$$m_x^{(1,1)}(q) = \frac{\partial^2 m_x(q)}{\partial s_1^{\rho_1}\, \partial s_2^{\rho_2}\, \partial t}$$

is constant for all possible combinations of ρ_1 and ρ_2 such that $\rho_1 + \rho_2 = 1$; that is, the $X^{(1,1)}(q)$ is a constant-mean, space-nonhomogeneous/time-nonstationary GS/TRF in general.

We proceed with the analysis of the spatiotemporal correlation structure of OS/TRF-ν/μ by introducing the following definition.

Definition 6: Consider a continuous OS/TRF-ν/μ $X(\mathbf{s}, t)$. A continuous and symmetric function $k_x(\mathbf{h}, \tau)$ in $R^n \times T$ is termed a *generalized spatiotemporal covariance of order* ν *in space and* μ *in time* (GS/TC-ν/μ) if and only if

$$(X(q_1), X(q_2)) = \langle k_x(\mathbf{h}, \tau), q_1(\mathbf{s}, t) q_2(\mathbf{s}', t')\rangle \geq 0 \tag{16}$$

where $\mathbf{h} = \mathbf{s} - \mathbf{s}'$ and $\tau = t - t'$, for all $q_1, q_2 \in Q_{\nu/\mu}$.

In other words, in order that a given function be a *permissible* model of some GS/TC-ν/μ it is necessary and sufficient that the condition (16) is satisfied. We saw above that with a particular GS/TRF-ν/μ $X(q)$ we can associate a GRS-ν/μ $\mathscr{X}_q$ whose elements are the corresponding OS/TRF-ν/μ $X(\mathbf{s}, t)$. Similarly, with a particular $X(q)$ we can associate a set of GS/TC-ν/μ satisfying Definition 6; this set will be called the *generalized spatiotemporal covariance representation set of order* ν/μ ($GS/TCRS$-ν/μ), and will be denoted by $\mathscr{F}^{\mathscr{X}}_{\nu/\mu}$. The concept of the GS/TC-ν/μ $k_x(\mathbf{h}, \tau)$ can be considered as the space–time extension of the purely spatial generalized covariance in the sense of Matheron (1973) (see also Chapter 3). We will see below that some interesting properties of $\mathscr{F}^{\mathscr{X}}_{\nu/\mu}$ may be obtained by assuming that the GS/TC-ν/μ is *space isotropic*, that is,

$$k_x(\mathbf{h}, \tau) = k_x(r, \tau) \tag{17}$$

where $r = |\mathbf{h}|$.

Let us now explore Eq. (13) a little more in light of Definition 6. We have

$$\begin{aligned} c_x(q_1^{(\nu+1,\mu+1)}, q_2^{(\nu+1,\mu+1)}) &= \langle c_x(\mathbf{s}, t; \mathbf{s}', t'), q_1^{(\nu+1,\mu+1)}(\mathbf{s}, t) q_2^{(\nu+1,\mu+1)}(\mathbf{s}', t')\rangle \\ &= \langle D^{(2\nu+2,2\mu+2)} c_x(\mathbf{s}, t; \mathbf{s}', t'), q_1(\mathbf{s}, t) q_2(\mathbf{s}', t')\rangle \\ &= \langle c_Y(\mathbf{s} - \mathbf{s}', t - t'), q_1(\mathbf{s}, t) q_2(\mathbf{s}', t')\rangle \end{aligned}$$

It is also true that

$$D^{(2\nu+2,2\mu+2)} c_x(\mathbf{s}, t; \mathbf{s}', t') = c_Y(\mathbf{s} - \mathbf{s}', t - t') \tag{18}$$

The above partial differential equation can be solved with respect to $c_x(\mathbf{s}, t; \mathbf{s}', t')$. For illustration consider first the $R^1 \times T$ case: According to Proposition 1 above, if $X(s, t)$ is a differentiable OS/TRF-ν/μ in $R^1 \times T$ such that $D^{(\nu+1,\mu+1)} X(s, t) = Y(s, t)$, the $Y(s, t)$ is space homogeneous/time stationary. The corresponding covariances of $X(s, t)$ and $Y(s, t)$ are related by $D^{(2\nu+2,2\mu+2)} c_x(s, t; s', t') = c_Y(r, \tau)$, where $r = s - s'$ and $\tau = t - t'$. The solution of this partial differential equation is

$$c_x(s, t; s', t') = k_x(r, \tau) + p_{\nu,\mu}(s, t; s', t') \tag{19}$$

where

$$k_x(r, \tau) = (-1)^{\nu+\mu} \int_0^r \int_0^\tau \frac{(r-u)^{2\nu+1} (\tau - v)^{2\mu+1}}{(2\nu+1)!(2\mu+1)!} c_Y(u, v)\, du\, dv \tag{20}$$

is the corresponding GS/TC-ν/μ and $p_{\nu,\mu}(s, t; s', t')$ is a polynomial of degree ν in s, s' and μ in t, t'. Equation (20) can be solved with respect to $c_Y(r, \tau)$, viz.,

$$c_Y(r, \tau) = (-1)^{\nu+\mu} D^{(2\nu+2, 2\mu+2)} k_x(r, \tau) \tag{21}$$

In $R^n \times T$ the analysis above leads to the following proposition (Christakos, 1991b and c).

Proposition 6: Let $X(\mathbf{s}, t)$ be an OS/TRF-ν/μ in $R^n \times T$. Its covariance function can be expressed in the following form

$$c_x(\mathbf{s}, t; \mathbf{s}', t') = k_x(\mathbf{h}, \tau) + p_{\nu,\mu}(\mathbf{s}, t; \mathbf{s}', t') \tag{22}$$

where $k_x(\mathbf{h}, \tau)$ ($\mathbf{h} = \mathbf{s} - \mathbf{s}'$ and $\tau = t - t'$) is the associated GS/TC-ν/μ and $p_{\nu,\mu}(\mathbf{s}, t; \mathbf{s}', t')$ is a polynomial with variable coefficients of degree ν in $\mathbf{s}$, $\mathbf{s}'$, and degree μ in t, t'.

Proposition 6 together with the definition of GS/TRF-ν/μ conclude the following result.

Corollary 1: If $X(q)$ is a GS/TRF-ν/μ in $R^n \times T$, then

$$\begin{aligned} c_x(q_1, q_2) &= \langle c_x(\mathbf{s}, t; \mathbf{s}', t'), q_1(\mathbf{s}, t) q_2(\mathbf{s}', t) \rangle \\ &= \langle k_x(\mathbf{h}, \tau), q_1(\mathbf{s}, t) q_2(\mathbf{s}', t') \rangle \end{aligned} \tag{23}$$

In view of Corollary 1, condition (16), satisfied by all GS/TC-ν/μ $k_x(\mathbf{h}, \tau)$, can also emerge from the fact that $c_x(q_1, q_2)$ is a nonnegative-definite bilinear functional in $Q_{\nu/\mu}$ that satisfies Eq. (23). A continuous and symmetric function $k_x(\mathbf{h}, \tau)$ in $R^n \times T$ is a permissible GS/TC-ν/μ if and only if

$$\langle k_x(\mathbf{h}, \tau), q(\mathbf{s}, t) q(\mathbf{s}', t') \rangle \geq 0 \tag{24}$$

for all $q \in Q_{\nu/\mu}$. We will also say that the $k_x(\mathbf{h}, \tau)$ is a *conditionally nonnegative-definite function* of order ν/μ.

Let $X(\mathbf{s}, t)$ be a differentiable OS/TRF-ν/μ. By definition the

$$Y_a(\mathbf{s}, t) = D_a^{(\nu+1, \mu+1)} X(\mathbf{s}, t) \tag{25}$$

is a zero-mean space-homogeneous/time-stationary random field for all $a \in A$, with

$$A = \left\{ a = (\boldsymbol{\nu}^*, \mu+1) = (\nu_1, \nu_2, \ldots, \nu_n, \mu+1) \colon \sum_{i=1}^{n} \nu_i = \nu + 1 \right\}$$

The spectral representation of the covariance of each $Y_a(\mathbf{s}, t)$ is written

$$c_{Y_a}(\mathbf{h}, \tau) = \iint \exp[i(\mathbf{w} \cdot \mathbf{h} + \lambda\tau)] \, d\phi_{Y_a}(\mathbf{w}, \lambda)$$

where $\phi_{Y_a}(\mathbf{w}, \lambda)$, $a \in A$ are positive summable measures in $R^n \times T$, without atom at the origin. We define the covariance

$$\begin{aligned} c_Y(\mathbf{h}, \tau) &= \sum_{a \in A} c_{Y_a}(\mathbf{h}, t) \\ &= E\left[\sum_{a \in A} D_a^{(\nu+1,\mu+1)} X(\mathbf{s}, t)\, D_a^{(\nu+1,\mu+1)} X(\mathbf{s}', t')\right] \\ &= \sum_{a \in A} D_a^{(2\nu+2,2\mu+2)} c_x(\mathbf{s}, t; \mathbf{s}', t') \\ &= \iint \exp[i(\mathbf{w}\cdot\mathbf{h} + \lambda\tau)]\, d\phi_Y(\mathbf{w}, \lambda) \end{aligned} \tag{26}$$

where $\phi_Y(\mathbf{w}, \lambda) = \sum_{a \in A} \phi_{Y_a}(\mathbf{w}, \lambda)$ is also a positive summable measure in $R^n \times T$, without atom at the origin.

A function $k_x(\mathbf{h}, \tau)$ is a permissible GS/TC-ν/μ in the sense of Definition 6 if and only if it admits the following spectral representation.

$$\begin{aligned} k_x(\mathbf{h}, \tau) = \iint & [\exp[i(\mathbf{w}\cdot\mathbf{h})] - p_{2\nu+1}[i(\mathbf{w}\cdot\mathbf{h})]] \\ & \times[\exp[i(\lambda\tau)] - p_{2\mu+1}(i\lambda\tau)] \\ & \times[\mathbf{w}^{2\nu+2}\lambda^{2\mu+2}]^{-1} \\ & \times d\phi_Y(\mathbf{w}, \lambda) + p_{2\nu,2\mu}(\mathbf{h}, \tau) \end{aligned} \tag{27}$$

where

$$p_{2\nu+1}[i(\mathbf{w}\cdot\mathbf{h})] = \sum_{\rho=0}^{2\nu+1} i^\rho \frac{(\mathbf{w}\cdot\mathbf{h})^\rho}{\rho!}$$

$$p_{2\mu+1}(i\lambda\tau) = \sum_{\zeta=0}^{2\mu+1} i^\zeta \frac{(\lambda\tau)^\zeta}{\zeta!}$$

and $p_{2\nu,2\mu}(\mathbf{h}, \tau)$ is an arbitrary polynomial of degree $\leq 2\nu$ in $\mathbf{h}$ and $\leq 2\mu$ in τ.

Note that since the GS/TC-ν/μ is a real-valued function, Eq. (27) can also be written in an imaginary part-free form, in terms of cosine and sine functions (Christakos, 1990b). A GS/TC-ν/μ is also a GS/TC-ν'/μ' for all $\nu' \geq \nu$ and $\mu' \geq \mu$. On the basis, now, of the obvious inequality

$$\begin{aligned} &|\exp[i(\mathbf{w}\cdot\mathbf{h})] - p_{2\nu+1}[i(\mathbf{w}\cdot\mathbf{h})]|\,|\exp[i(\lambda\tau)] - p_{2\mu+1}(i\lambda\tau)| \\ &\leq \frac{(\mathbf{w}\cdot\mathbf{h})^{2\nu+2}(\lambda\tau)^{2\mu+2}}{(2\nu+2)!(2\mu+2)!} \end{aligned}$$

it follows that

$$\begin{aligned} |k_x(\mathbf{h}, \tau)| &\leq \frac{\iint d\phi_Y(\mathbf{w}, \lambda)}{(2\nu+2)!(2\mu+2)!} |\mathbf{h}|^{2\nu+2}\tau^{2\mu+2} \\ &= \alpha|\mathbf{h}|^{2\nu+2}\tau^{2\mu+2} \end{aligned}$$

where $\alpha < \infty$, or

$$\lim_{\substack{|\mathbf{h}|\to\infty \\ \tau\to\infty}} \frac{k_x(\mathbf{h}, \tau)}{|\mathbf{h}|^{2\nu+2}\tau^{2\mu+2}} = 0 \tag{28}$$

which assures the existence of the integral (27). In view of the foregoing considerations, if $k_x(\mathbf{h}, \tau) \in \mathscr{F}^{\mathscr{X}}_{\nu/\mu}$, then $k_x(\mathbf{h}, \tau) + p_{2\nu,2\mu}(\mathbf{h}, \tau) \in \mathscr{F}^{\mathscr{X}}_{\nu/\mu}$ too.

Clearly, the GS/TC-ν/μ satisfies the relation

$$\nabla_{\mathbf{h}}^{2\nu+2} \frac{\partial^{2\mu+2}}{\partial\tau^{2\mu+2}} k_x(\mathbf{h}, \tau) = (-1)^{\nu+\mu} c_Y(\mathbf{h}, \tau) \tag{29}$$

In relation to Eq. (29), the measure $\phi_Y(\mathbf{w}, \lambda)$ is the Fourier transform of

$$(-1)^{\nu+\mu} \nabla_{\mathbf{h}}^{2\nu+2} \frac{\partial^{2\mu+2}}{\partial\tau^{2\mu+2}} k_x(\mathbf{h}, \tau)$$

Employing Proposition 4 it is not difficult to show that the representation (27) is in general true for any $X(\mathbf{s}, t)$, not necessarily differentiable.

In the case now where $\phi_Y(\mathbf{w}, \lambda)$ is differentiable, we can define the *generalized spectral density function of order* ν/μ $K_x(\mathbf{w}, \lambda)$ as the n-fold space/time Fourier transform of $k_x(\mathbf{h}, \tau)$. The lemma below is an immediate consequence of the preceding spectral analysis.

Lemma 1 (*Tenth criterion of permissibility*; *COP-10*): Let $X(\mathbf{s}, t)$ be a differentiable OS/TRF-ν/μ. A continuous function $k_x(\mathbf{h}, \tau)$ in $R^n \times T$ is a permissible GS/TC-ν/μ if and only if Eq. (28) holds true and the corresponding $K_x(\mathbf{w}, \lambda)$ exists (in the sense of generalized functions), includes no atom at origin, and is such that the $|\mathbf{w}|^{2\nu+2}\lambda^{2\mu+2} K_x(\mathbf{w}, \lambda)$ is a nonnegative measure.

It is noteworthy that if the space isotropic $c_Y(r, \tau)$, $r = |\mathbf{h}|$ is *space-time separable* [i.e., $c_Y(r, \tau) = c_Y(r) c_Y(\tau)$], then the $k_x(r, \tau)$ is separable too [i.e., $k_x(r, \tau) = k_x(r) k_x(\tau)$]. We shall be examining a series of cases of this type.

Example 2: Consider the stochastic partial differential equation (4), where $Y(s, t)$ is a zero-mean white-noise S/TRF in $R^1 \times T$ with covariance

$$c_Y(r, \tau) = \delta(r, \tau) = \delta(r)\,\delta(\tau) \tag{30}$$

Equation (20) gives

$$k_x(r, \tau) = (-1)^{\nu+\mu} \frac{r^{2\nu+1}\tau^{2\mu+1}}{(2\nu+1)!(2\mu+1)!} \tag{31}$$

A generalization of the covariance (31) in $R^n \times T$ is the isotropic GS/TC-ν/μ

$$k_x(r, \tau) = \sum_{\rho=0}^{\nu} \sum_{\zeta=0}^{\mu} (-1)^{\rho+\zeta} a_{\rho\zeta} r^{2\rho+1} \tau^{2\zeta+1} \tag{32}$$

where the coefficients $a_{\rho\zeta}$ should satisfy certain permissibility conditions so that the $k_x(r, \tau)$ is a conditionally nonnegative-definite function in the sense of Eq. (24); see also Lemma 1. More precisely the coefficients $a_{\rho\zeta}$ must be such that the following condition is satisfied:

$$\sum_{\rho=0}^{\nu} \sum_{\zeta=0}^{\mu} G((2\rho+n+1)/2)[(2\rho+1)!][(2\zeta+1)!]/\rho! \\ \times a_{\rho\zeta}\omega^{2(\nu-\rho)}\lambda^{2(\mu-\zeta)} \geq 0 \tag{33}$$

where $G(\cdot)$ is the gamma function for all $\omega \geq 0$ and $\lambda \geq 0$.

Based now on the observation that an OS/TRF-ν/μ of the form (4) can be assigned a GS/TC-ν/μ of the polynomial form (31), the proof of the following proposition is straightforward (see also Christakos, 1991b).

Proposition 7: Assume that an OS/TRF-ν/μ in $R^1 \times T$ can be expressed by

$$X(s, t) = \sum_{\rho=0}^{\nu} \sum_{\zeta=0}^{\mu} \alpha_{\rho\zeta} \int_0^s \int_0^t \frac{(s-u)^\rho (t-v)^\zeta}{\rho!\zeta!} Y(u, v)\, du\, dv \tag{34}$$

where $\alpha_{\rho\zeta}$, $\rho = 0, 1, \ldots, \nu$, and $\zeta = 0, 1, \ldots, \mu$ are suitable coefficients and $Y(s, t)$ is a zero-mean white-noise S/TRF in $R^1 \times T$. Then its GS/TC-ν/μ is of the form (32).

Example 3: Working along lines similar to those of Example 2 we find that if

$$c_Y(r, \tau) = a \exp[-br - c\tau] \tag{35}$$

the corresponding GS/TC-ν/μ in $R^1 \times T$ is

$$k_x(r, \tau) = \frac{a(-1)^{\nu+\mu}}{(2\nu+1)!(2\mu+1)!} \frac{\gamma(2\nu+2, -br)\gamma(2\mu+2, -c\tau)}{b^{2\nu+2}c^{2\mu+2} \exp[br + c\tau]} \tag{36}$$

where $\gamma(\cdot, \cdot)$ denotes the incomplete gamma function. After some manipulations Eq. (36) may also be written as

$$k_x(r, \tau) = \frac{a(-1)^{\nu+\mu}}{b^{2\nu+2}c^{2\mu+2} \exp[br + c\tau]} \left[1 - \exp(br) \sum_{i=0}^{2\nu+1} \frac{(-br)^i}{i!}\right] \\ \times \left[1 - \exp(c\tau) \sum_{i=0}^{2\mu+1} \frac{(-c\tau)^i}{i!}\right] \tag{37}$$

Consider, for instance, the case $\nu = \mu = 0$; then, Eq. (37) gives

$$k_x(r, \tau) = \frac{a[1 + (br-1)\exp(br)][1 + (c\tau - 1)\exp(c\tau)]}{b^2 c^2 \exp[br + c\tau]}$$

6. Stochastic Partial Differential Equations

6.1 Basic Equations

Stochastic differential equations over space–time have the general form

$$L[X(\mathbf{s}, t)] = Y(\mathbf{s}, t) \tag{1}$$

where $X(\mathbf{s}, t)$ is the unknown S/TRF, $L[\cdot]$ is a given operator, and $Y(\mathbf{s}, t)$ is a known S/TRF—also called a forcing function. Space–time stochastic differential equations can be classified in a fashion similar to the spatial stochastic differential equations discussed in Chapter 3. Despite significant progress over the last decade or so, much work remains to be done in the theory of stochastic partial differential equations (SPDE). A partial list of references was given in Section 5 of Chapter 3.

Just as in Chapter 3, our attention in this section will be focused on only a few specific SPDE topics that have certain interesting connections with the S/TRF models considered above. The relevance of such connections owe to the fact that a variety of natural processes are governed by such equations (e.g., flow through porous media, hydroclimatic systems, and transport and diffusion in the atmosphere).

Furthermore, these connections can be valuable tools in the improvement of existing physical models. For example, in flood prediction, an important problem is the quantitative precipitation forecasting. By studying the spatiotemporal residual series of model errors, it is possible to develop corrections to the model to account for persistent errors. In groundwater contaminant transport modeling, there often arise structural errors in model predictions due to complexities in the subsurface system that cannot reasonably be modeled deterministically. The ability to model stochastically the resulting space–time processes offers the potential of developing corrections to the model predictions to better reflect the true system.

We saw above that, by definition, a continuous-parameter OS/TRF-ν/μ obeys certain SPDE, and the corresponding covariances (ordinary and generalized) satisfy the corresponding deterministic differential equations: If $X(\mathbf{s}, t)$ is an OS/TRF-ν/μ, by definition, all

$$Y_i(\mathbf{s}, t) = \frac{\partial^{\nu+\mu+2}}{\partial s_i^{\nu+1}\, \partial t^{\mu+1}} X(\mathbf{s}, t) \tag{2}$$

are space-homogeneous/time-stationary RF. The field

$$Y(\mathbf{s}, t) = \sum_{i=1}^{n} Y_i(\mathbf{s}, t) = \nabla^{\nu+1} \frac{\partial^{\mu+1}}{\partial t^{\mu+1}} X(\mathbf{s}, t) \tag{3}$$

where $L[\cdot] = \nabla^{\nu+1}(\partial^{\mu+1}/\partial t^{\mu+1})[\cdot]$ is space homogeneous/time stationary, too. This observation leads to the following result.

Proposition 1: Let $Y(\mathbf{s}, t)$ be a space-homogeneous/time-stationary field. Then, there is one and only one OS/TRF-ν/μ $X(\mathbf{s}, t)$ with representations satisfying the differential equation

$$Y(\mathbf{s}, t) = \nabla^{\nu+1} \frac{\partial^{\mu+1}}{\partial t^{\mu+1}} X(\mathbf{s}, t) \tag{4}$$

An immediate consequence of Proposition 1 is the corollary below.

Corollary 1: If $X(\mathbf{s}, t)$ is an OS/TRF-ν/μ, then there exists an OS/TRF-$(\nu+2k)/(\mu+2\lambda)$ $Z(\mathbf{s}, t)$ such that

$$X(\mathbf{s}, t) = \nabla^{2k} \frac{\partial^{2\lambda}}{\partial t^{2\lambda}} Z(\mathbf{s}, t) \tag{5}$$

The covariances associated with expressions (2) and (3) are, respectively,

$$c_{Y_i}(\mathbf{h}, \tau) = \frac{\partial^{2\nu+2\mu+4}}{\partial s_i^{\nu+1}\, \partial s_i'^{\nu+1}\, \partial t^{\mu+1}\, \partial t'^{\mu+1}} c_x(\mathbf{s}, t; \mathbf{s}', t) \tag{6}$$

and

$$\begin{aligned} c_Y(\mathbf{h}, \tau) &= \frac{\partial^{2\mu+2}}{\partial t^{\mu+1}\, \partial t'^{\mu+1}} \sum_{i=1}^{n} \sum_{j=1}^{n} \frac{\partial^{2\nu+2}}{\partial s_i^{\nu+1}\, \partial s_j'^{\nu+1}} c_x(\mathbf{s}, t; \mathbf{s}', t') \\ &= \nabla_{\mathbf{s}}^{\nu+1} \nabla_{\mathbf{s}'}^{\nu+1} \frac{\partial^{2\mu+2}}{\partial t^{\mu+1}\, \partial t'^{\mu+1}} c_x(\mathbf{s}, t; \mathbf{s}', t') \\ &= (-1)^{\nu+\mu} \nabla^{2\nu+2} \frac{\partial^{2\mu+2}}{\partial \tau^{2\mu+2}} k_x(\mathbf{h}, \tau) \end{aligned} \tag{7}$$

Example 1: Suppose that the OS/TRF-1/1 $X(\mathbf{s}, t)$ and the space-homogeneous/time-stationary RF $Y(\mathbf{s}, t)$ are related by

$$\nabla_{\mathbf{s}}^2 \frac{\partial^2}{\partial t^2} X(\mathbf{s}, t) = Y(\mathbf{s}, t) \tag{8}$$

One may now study the stochastic properties of the field $X(\mathbf{s}, t)$ by means of $Y(\mathbf{s}, t)$; the corresponding covariance functions are related by

$$\begin{aligned} c_Y(\mathbf{h}, \tau) &= \sum_{i=1}^{n} \sum_{j=1}^{n} \frac{\partial^4}{\partial s_i^2\, \partial s_j'^2} \left[\frac{\partial^4 c_x(\mathbf{s}, t; \mathbf{s}', t')}{\partial t^2\, \partial t'^2} \right] \\ &= \nabla_{\mathbf{s}}^2 \nabla_{\mathbf{s}'}^2 \frac{\partial^4}{\partial t^2\, \partial t'^2} c_x(\mathbf{s}, t; \mathbf{s}', t') \\ &= \nabla_{\mathbf{h}}^4 \frac{\partial^4}{\partial \tau^4} k_x(\mathbf{h}, \tau) \end{aligned} \tag{9}$$

6.2 Adjoint Equations—The Air Pollution Problem

The results obtained in Section 5.2 of Chapter 3 can be easily extended to S/TRF models. In particular, the "basic SPDE" (1) can be transformed into the functional equation

$$X(q) = F[X(\mathbf{s}, t), X^*(\mathbf{s}, t)] + \langle Y(\mathbf{s}, t), X^*(\mathbf{s}, t)\rangle \tag{10}$$

where $X(q) = \langle X(\mathbf{s}, t), q(\mathbf{s}, t)\rangle$ is a generalized S/TRF in the sense given in the preceding sections, and $X^*(\mathbf{s}, t)$ is the solution of the "adjoint SPDE"

$$L^*[X^*(\mathbf{s}, t)] = q(\mathbf{s}, t) \tag{11}$$

Examples about the physical interpretation of the generalized S/TRF $X(q)$ were given in previous sections.

Example 2: Let $X(\mathbf{s}, t)$ denote the concentration of aerosol substance in the atmosphere in a domain of interest A within a time period T. Suppose that A is cylindrical with total surface $S = S_B + S_T + S_L$, where S_B, S_T, and S_L denote the base, top, and lateral surfaces of A. Substance transport and diffusion within A is governed by

$$\frac{\partial X(\mathbf{s}, t)}{\partial t} + \nabla\mathbf{V}\, X(\mathbf{s}, t) + \upsilon X(\mathbf{s}, t) - \frac{\partial}{\partial s_3}\nu\frac{\partial X(\mathbf{s}, t)}{\partial s_3} - \mu\,\nabla^2 X(\mathbf{s}, t) = W\delta(\mathbf{s}, \mathbf{s}_0) \tag{12}$$

with initial conditions

$$\begin{aligned} &X(\mathbf{s}, t) = 0 \qquad (\mathbf{s} \in S_L) \\ &\frac{\partial X(\mathbf{s}, t)}{\partial s_3} = \alpha X(\mathbf{s}, t) \qquad (\mathbf{s} \in S_B) \\ &\frac{\partial X(\mathbf{s}, t)}{\partial s_3} = 0 \qquad (\mathbf{s} \in S_T) \end{aligned} \tag{13}$$

Moreover, it is assumed that

$$X(\mathbf{s}, T) = X(\mathbf{s}, 0) \tag{14}$$

In the above equations, $\mathbf{V}$ is the velocity vector of air particles with components V_1, V_2, and V_3, along the horizontal directions s_1, s_2, and the vertical direction s_3; μ and ν are the horizontal and vertical diffusion coefficients, respectively; υ is a quantity that has an inverse time dimension, $\alpha \geq 0$ is a parameter determining the interaction of the impurities with the underlying surface; W is the intensity of the aerosol discharge, and $\mathbf{s}_0$ is the location of the aerosol source (e.g., industrial plant).

The problem is to find a region $U \subset A$ where a new industrial plant can be located, so that for all $\mathbf{s}_0 \in U$ the resulting pollution over a nearby populated area $D \subset A$ during the time period T does not exceed a permissible level c, imposed by global and local sanitary requirements, viz.,

$$X(\mathbf{s}) < c \tag{15}$$

for all $\mathbf{s} \in D$. It is assumed that all necessary information about the wind fields in the region is available.

Let us consider the functional

$$X(q) = \int_A \int_0^T q(\mathbf{s}, t) X(\mathbf{s}, t)\, d\mathbf{s}\, dt \tag{16}$$

where $q(\mathbf{s}, t) = 1/T + \beta\, \delta(s_3)$ ($\mathbf{s} \in A$), $= 0(\mathbf{s} \notin A)$; β is a coefficient discussed earlier in Example 5 of Section 3. The "adjoint SPDE" with respect to Eqs. (12) through (14) above is

$$-\frac{\partial X^*(\mathbf{s}, t)}{\partial t} + \nabla\, \mathbf{V} X^*(\mathbf{s}, t) + \upsilon X^*(\mathbf{s}, t)$$
$$-\frac{\partial}{\partial s_3} \nu \frac{\partial X^*(\mathbf{s}, t)}{\partial s_3} - \mu\, \nabla^2 X^*(\mathbf{s}, t) = q(\mathbf{s}, t) \tag{17}$$

and

$$\begin{aligned} &X^*(\mathbf{s}, t) = 0 \qquad (\mathbf{s} \in S_{\mathrm{L}}) \\ &\frac{\partial X^*(\mathbf{s}, t)}{\partial s_3} = \alpha X^*(\mathbf{s}, t) \qquad (\mathbf{s} \in S_{\mathrm{B}}) \\ &\frac{\partial X^*(\mathbf{s}, t)}{\partial s_3} = 0 \qquad (\mathbf{s} \in S_{\mathrm{T}}) \\ &X^*(\mathbf{s}, T) = X^*(\mathbf{s}, 0) \end{aligned} \tag{18}$$

After solving Eqs. (17) and (18) we substitute $X^*(\mathbf{s}, t)$ into Eq. (16) to find

$$X(q) = W \int_0^T X^*(\mathbf{s}, t)\, dt = X_q(\mathbf{s})$$

Finally, from the condition (15) we can find the locations $\mathbf{s}_0 : X_q(\mathbf{s}_0) < c$, which determines the region U.

7. Discrete Linear Representations of Spatiotemporal Random Fields

The key element in passing from abstract theory to a practical analysis of spatiotemporal data is the development of suitable *discrete linear representations* of the S/TRF model. This is necessary because real data are usually discretely distributed in space-time.

Let $X(\mathbf{s}_i, t_j)$, where $(\mathbf{s}_i, t_j) \in R^n \times T$, $i = 1, 2, \ldots, m$ and $j = 1, 2, \ldots, k$ be a discrete-parameter OS/TRF. Let $q \in Q = \mathcal{Q}_m$, where $\mathcal{Q}_m$ is the space of real measures on $R^n \times T$ with finite support and such that

$$q(\mathbf{s}, t) = \sum_{i=1}^{m} \sum_{j=1_i}^{p_i} q(\mathbf{s}_i, t_j)\, \delta(\mathbf{s}_i - \mathbf{s}, t_j - t)$$
$$= \sum_{i=1}^{m} \sum_{j=1_i}^{p_i} q_{ij}\, \delta_{ij}(\mathbf{s}, t) \tag{1}$$

where p_i denotes the number of time instances t_j $(j = 1_i, 2_i, \ldots, p_i)$ used, given that we are at the spatial position $\mathbf{s}_i$.

The corresponding discrete GS/TRF and CS/TRF are, respectively,

$$X(q) = \left\langle \sum_{i=1}^{m} \sum_{j=1_i}^{p_i} q_{ij}\, \delta_{ij}(\mathbf{s}, t), X(\mathbf{s}, t) \right\rangle$$
$$= \sum_{i=1}^{m} \sum_{j=1_i}^{p_i} q_{ij} X(\mathbf{s}_i, t_j) \tag{2}$$

and

$$Y_q(\mathbf{s}, t) = \left\langle \sum_{i=1}^{m} \sum_{j=1_i}^{p_i} q_{ij}\, \delta_{ij}(\mathbf{s}', t'), S_{\mathbf{s},t} X(\mathbf{s}', t') \right\rangle$$
$$= \sum_{i=1}^{m} \sum_{j=1_i}^{p_i} q_{ij} S_{\mathbf{s},t} X(\mathbf{s}_i, t_j) \tag{3}$$

Definition 1: The discrete S/TRF $Y_q(\mathbf{s}, t)$ of Eq. (3) will be called a *spatiotemporal increment of order ν in space and μ in time (S/TI-ν/μ)* on $Q_{\nu/\mu}$ if

$$\sum_{i=1}^{m} \sum_{j=1_i}^{p_i} q_{ij}\, \mathbf{s}_i^{\rho} t_j^{\zeta} = 0 \tag{4}$$

for all $\rho \leq \nu$ and $\zeta \leq \mu$. In this case the coefficients $\{q_{ij}\} \in Q_{\nu/\mu} \subset \mathcal{Q}_m$, $i = 1, 2, \ldots, m$ and $j = 1_i, 2_i, \ldots, p_i$ will be termed *admissible coefficients of order ν/μ (AC-ν/μ)*.

On the basis of the above definition the next follows naturally.

Definition 2: The discrete OS/TRF $X(\mathbf{s}, t)$ will be called a *OS/TRF-ν/μ* on $Q_{\nu/\mu}$ if the corresponding S/TI-ν/μ $Y_q(\mathbf{s}, t)$ is a zero-mean space-homogeneous/time-stationary RF.

A summary of continuous S/TRF-related notions and their discrete analogs are given in Table 5.1.

Example 1: Consider the case illustrated in Fig. 5.5, where $(\mathbf{s}, t) \in R^2 \times T$

Table 5.1 S/TRF-Related Notions and Their Discrete Analogs

$$q(\mathbf{s}, t) \in Q_{\nu/\mu}\colon q(\mathbf{s}, t) = \sum_{i=1}^{m} \sum_{j=1_i}^{l_i} q(\mathbf{s}_i, t_j)\, \delta(\mathbf{s}_i - \mathbf{s}, t_j - t)$$

Theory		Practice	
$X(q) = \langle q(\mathbf{s}, t), X(\mathbf{s}, t)\rangle$		$X(q) = \sum_{i=1}^{m} \sum_{j=1_i}^{l_i} q_{ij} X(\mathbf{s}_i, t_j)$	
$Y_q(\mathbf{s}, t) = X(S_{\mathbf{s},t} q)$		$Y_q(\mathbf{s}, t) = X(S_{\mathbf{s},t} q)$	
$= \langle q(\mathbf{s}', t'), S_{\mathbf{s},t} X(\mathbf{s}', t')\rangle$		$= \sum_{i=1}^{m} \sum_{j=1_i}^{l_i} q_{ij} X(\mathbf{s}_i + \mathbf{s}, t_j + t)$	
$m_x(q) = \langle q(s, t), m_x(\mathbf{s}, t)\rangle$		$m_x(q) = \sum_{i=1}^{m} \sum_{j=1_i}^{l_i} q_{ij} m_x(\mathbf{s}_i, t_j)$	
$\langle q(\mathbf{s}, t), g_{\rho\zeta}(\mathbf{s}, t)\rangle = 0$	$\rho \le \nu$, $\zeta \le \mu$	$\sum_{i=1}^{m} \sum_{j=1_i}^{l_i} q_{ij} g_{\rho\zeta}(\mathbf{s}_i, t_j) = 0$	$\rho \le \nu$, $\zeta \le \mu$
$m_x(q) = \sum_{0 \le \rho \le \nu} \sum_{0 \le \zeta \le \mu} \alpha_{\rho\zeta} \langle g_{\rho\zeta}(\mathbf{s}, t), q(\mathbf{s}, t)\rangle$		$m_x(q) = \sum_{0 \le \rho \le \nu} \sum_{0 \le \zeta \le \mu} b_{\rho\zeta} g_{\rho\zeta}(\mathbf{s}, t)$	
$\langle k_x(\mathbf{s} - \mathbf{s}', t - t'), q(\mathbf{s}, t) q(\mathbf{s}', t')\rangle \ge 0$		$\sum_{i=1}^{m} \sum_{j=1_i}^{l_i} \sum_{i'=1}^{m} \sum_{j'=1_{i'}}^{l_{i'}} q_{ij} q_{i'j'} k_x(\mathbf{s}_i - \mathbf{s}_{i'}, t_j - t_{j'}) \ge 0$	
$q \in Q_{\nu/\mu}$		$\{q_{ij}\}$	AC-ν/μ

and $\mathbf{s} = (s_1, s_2)$. Let

$$\begin{aligned}
Y_q(s_1, s_2, t) &= \sum_{i=1}^{5} \sum_{j=1}^{3} q_{ij} X(s_{i1}, s_{i2}, t_j) \\
&= X(s_1 + \Delta s, s_2, t + \Delta t) - 2X(s_1 + \Delta s, s_2, t) \\
&\quad + X(s_1 + \Delta s, s_2, t - \Delta t) + X(s_1 + \Delta s, s_2, t + \Delta t) \\
&\quad - 2X(s_1, s_2 + \Delta s, t) + X(s_1, s_2 + \Delta s, t - \Delta t) \\
&\quad + X(s_1 - \Delta s, s_2, t + \Delta t) - 2X(s_1 - \Delta s, s_2, t) \\
&\quad + X(s_1 - \Delta s, s_2, t - \Delta t) + X(s_1, s_2 - \Delta s, t + \Delta t) \\
&\quad - 2X(s_1, s_2 - \Delta s, t) + X(s_1, s_2 - \Delta s, t - \Delta t) \\
&\quad - 4[X(s_1, s_2, t + \Delta t) - 2X(s_1, s_2, t) + X(s_1, s_2, t - \Delta t)] \qquad (5)
\end{aligned}$$

It is easily shown that

$$\sum_{i=1}^{5} \sum_{j=1}^{3} q_{i1,i2,j} s_{i1}^{\rho_1} s_{i2}^{\rho_2} t_j^{\zeta} = 0$$

for all $\rho_1 + \rho_2 \le 1$ and $\zeta \le 1$. Therefore, the $Y_q(s_1, s_2, t)$ above is an S/TI-1/1. If in addition it is space homogeneous/time stationary, the corresponding

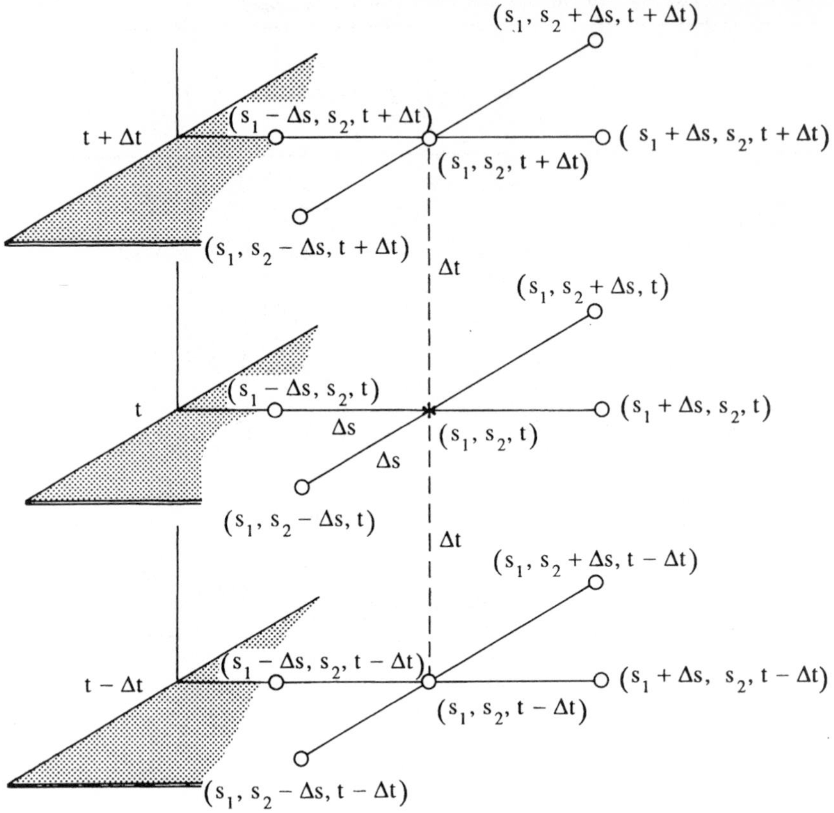

Figure 5.5 The $(\mathbf{s}, t) \in R^2 \times T$ case

$X(s_1, s_2, t)$ is an OS/TRF-1/1 with mean value

$$m_x(s_1, s_2) = \sum_{0 \le \rho_1 + \rho_2 \le 1} \sum_{0 \le \zeta \le 1} a_{\rho_1, \rho_2, \zeta} s_1^{\rho_1} s_2^{\rho_2} t^{\zeta}$$

$$= a_{000} + a_{100} s_1 + a_{010} s_2 + a_{001} t + a_{101} s_1 t + a_{011} s_2 t$$

Proposition 1: Any discrete space-nonhomogeneous/time-nonstationary RF in $R^1 \times T$ represented by the *spatiotemporal autoregressive model of order* $\nu+1$ *in space and* $\mu+1$ *in time, S/TAR* $(\nu+1, \mu+1)$

$$\Delta_{s,t}^{(\nu+1,\mu+1)} X(s, t) = Y_q(s, t) \tag{6}$$

where

$$\Delta_{s,t}^{(\nu+1,\mu+1)} X(s, t) = \sum_{\rho=0}^{\nu+1} \sum_{\zeta=0}^{\mu+1} (-1)^{\rho+\zeta} C_{\nu+1}^{\rho} C_{\mu+1}^{\zeta} \times X(s - \rho + \nu + 1, t - \zeta + \mu + 1) \tag{7}$$

is the finite difference of order $\nu+1$ in s and $\mu+1$ in t, $Y_q(s, t)$ is a space-homogeneous/time-stationary random field and $C_i^j=\binom{i}{j}$, is an OS/TRF-ν/μ.

Proof: See Christakos (1991c).

It is interesting to compare representations (4) of Section 5 and Eq. (7) above: All discrete-parameter OS/TRF-ν/μ admit a representation of the form (7), while for a continuous-parameter OS/TRF-ν/μ to be represented by (4) it is necessary that it is $\nu+1$ times differentiable in s and $\mu+1$ times in t.

Proposition 2: Let $X(\mathbf{s}, t)$ be an OS/TRF on $Q_{\nu/\mu}$ and let

$$\hat{X}(\mathbf{s}_0, t_0)=\sum_{i=1}^{m}\sum_{j=1}^{p_i}\lambda_{ij}X(\mathbf{s}_i, t_j) \tag{8}$$

be the linear estimator of $X(\mathbf{s}, t)$ at point/instant $(\mathbf{s}_0, t_0)$ such that

$$E[\hat{X}(\mathbf{s}_0, t_0)-X(\mathbf{s}_0, t_0)]=0 \tag{9}$$

and

$$E[X(\mathbf{s}_0, t_0)]=\sum_{\rho\leq\nu}\sum_{\zeta\leq\mu}\eta_{\rho\zeta}\mathbf{s}_0^{\rho}t_0^{\zeta} \tag{10}$$

where $\eta_{\rho\zeta}$ are suitable coefficients. Then the difference

$$\begin{aligned}Y_q(\mathbf{s}_0, t_0)&=\hat{X}(\mathbf{s}_0, t_0)-X(\mathbf{s}_0, t_0)\\&=\sum_{i=0}^{m}\sum_{j=0}^{p_i}\lambda_{ij}X(\mathbf{s}_i, t_j)\end{aligned} \tag{11}$$

where $\lambda_{00}=-1$ and $\lambda_{i0}=\lambda_{0j}=0$ $(i, j\neq 0)$, is an S/TI-ν/μ on $Q_{\nu/\mu}$.

Proof: See Christakos (1991c).

Notice that if the $Y_q(\mathbf{s}, t)$ of Eq. (11) is space homogeneous/time stationary, the $X(\mathbf{s}, t)$ is by definition an OS/TRF-ν/μ. Conversely, if $X(\mathbf{s}, t)$ is an OS/TRF-ν/μ, the $Y_q(\mathbf{s}, t)$ of Eq. (11) is a space-homogeneous/time-stationary S/TI-ν/μ.

In the discrete framework, Eq. (16) of Section 5 implies that a function $k_x(\mathbf{h}, \tau)$ in $R^n\times T$ is a *generalized spatiotemporal covariance of order ν in space and μ in time (GSTC-ν/μ)* if and only if for all AC-ν/μ $\{q_{ij}\}$

$$\begin{aligned}E[X(q)]^2=E[Y_q(\mathbf{O}, O)]^2&=E\left[\sum_{i=1}^{m}\sum_{j=1_i}^{p_i}q_{ij}X(\mathbf{s}_i, t_j)\right]^2\\&=\sum_{i=1}^{m}\sum_{j=1_i}^{p_i}\sum_{i'=1}^{m}\sum_{j'=1_{i'}}^{p_{i'}}q_{ij}q_{i'j'}k_x(\mathbf{h}_{ii'}, \tau_{jj'})\geq 0\end{aligned} \tag{12}$$

where $\mathbf{h}_{ii'}=\mathbf{s}_i-\mathbf{s}_{i'}$ and $\tau_{jj'}=t_j-t_{j'}$.

In practical applications where a finite number of discretely distributed data are available, it is convenient to use GS/TC-ν/μ $k_x(\mathbf{h}, \tau)$ of the space–time polynomial form (32) of Section 5. The parameters of the model (32) of Section 5 (orders ν and μ, as well as the coefficients $\alpha_{\rho\zeta}$, $\rho = 0, 1, \ldots, \nu$ and $\zeta = 0, 1, \ldots, \mu$) can be estimated on the basis of the available data by means of parameter estimation techniques such as least squares or maximum likelihood (see, e.g., Rao, 1973). Note that the estimated values of the coefficients $a_{\rho\zeta}$ should satisfy the conditions implied by Eq. (33), Section 5 above.

6

Space Transformations of Random Fields

"It isn't that they can't see the solution. It is that they can't see the problem."

G. K. Chesterton

1. Introduction

The analysis of some problems in the applied sciences is considerably simpler in one than in several dimensions. Such problems include the modeling of spatially distributed hydrogeologic data, groundwater contour mapping, and simulation of groundwater flow. In these circumstances, the mathematical operations of space transformations introduced in this chapter simplify the study of a physical process that takes place in several spatial dimensions by "conveying" its study to a suitable one-dimensional space.

This "conveyance" has both substance and depth, and is established in terms of suitable Radon operations (Radon, 1917) that act on the random field representing the physical process of interest. These operations, which will be termed space transformations (ST), apply in spatial and spatiotemporal random fields (SRF and S/TRF, respectively). ST have elegant and comprehensive representations both in the space (physical) and the frequency domains and preserve the second-order correlation structure of a random field (in the sense that a homogeneous SRF in R^n is transferred to an SRF in R^{n-k}, which is homogeneous too).

Random field representations show that a random field in space R^n can be represented by a linear combination of statistically uncorrelated random processes in space R^1. Expressions relating the corresponding spatial correlation characteristics of the two spaces are established.

The ST operators are very powerful tools in the study of multidimensional differential equation models governing natural processes, such as flow systems in spatially variable soils. Much that seemed problematic about the application of stochastic differential equation theory in the study of earth systems may be resolved by means of ST.

Furthermore, comprehensive and analytically tractable expressions of the most important criteria of permissibility (for ordinary as well as generalized correlation functions) can be developed with the help of ST. Finally, ST constitute a particularly attractive instrument in the context of random field (ordinary or generalized) simulation (Matheron, 1973; Journel, 1974; this matter deserves a detailed study, which is carried out in Chapter 8).

2. Space Transformations

Let us begin by defining the concept of space transformations in terms of functions $f_n(\mathbf{s})$, $\mathbf{s} \in R^n$, which belong to the Schwartz spaces K and S (see Chapter 3). These spaces possess a number of properties that make it easier to connect the general space transformation theory with concrete applications. Moreover, under certain circumstances the analysis applies to elements of the dual spaces K' and S' of distributions or generalized functions.

In particular, the space transformation of a function $f_n(\mathbf{s})$ is defined as follows.

Definition 1: Let $\{H_{n-1}; \boldsymbol{\theta}, \mathbf{s} \cdot \boldsymbol{\theta}\}$ be a set of hyperplanes passing through a given point in R^n, where the subscript $n-1$ denotes the dimension of the hyperplanes H_{n-1} in R^n (belonging to the space E^n of all hyperplanes in R^n). The unit vector $\boldsymbol{\theta} = (\theta_1, \ldots, \theta_n)$ defines the orientation of H_{n-1} and $\mathbf{s} \cdot \boldsymbol{\theta}$ is the inner product that defines its distance from the origin so that

$$\int_{R^n} \delta_t(\mathbf{s} \cdot \boldsymbol{\theta})\, d\mathbf{s} = 1; \qquad \delta_t(\mathbf{s} \cdot \boldsymbol{\theta}) = 0 \qquad \text{if} \quad t \neq \mathbf{s} \cdot \boldsymbol{\theta} \tag{1}$$

The *space transformation of first kind* (*ST-1*), T_n^1, maps a function $f_n(\mathbf{s})$, $\mathbf{s} \in R^n$ into the following set of functions

$$\hat{f}_{1,\boldsymbol{\theta}}(t) = T_n^1[f_n(\mathbf{s})] = \int_{R^n} f_n(\mathbf{s})\, \delta_t(\mathbf{s} \cdot \boldsymbol{\theta})\, d\mathbf{s} \tag{2}$$

on several hyperplanes $\{H_{n-1}; \boldsymbol{\theta}, \mathbf{s} \cdot \boldsymbol{\theta}\}$.

The ST-1 will be considered as completely defined if Eq. (2) is known for all $\mathbf{s}$ and $\boldsymbol{\theta}$.

Definition 2: The *space transformation of the second kind* (*ST-2*), Ψ_1^n, assigns to the set of functions $f_{1,\boldsymbol{\theta}}(t)$, defined on a set of hyperplanes $\{H_{n-1}; \boldsymbol{\theta}, t = \mathbf{s} \cdot \boldsymbol{\theta}\}$ passing through a given point in R^n, the function

$$f_n(\mathbf{s}) = \Psi_1^n[f_{1,\boldsymbol{\theta}}(t)] = \int_{\Theta_n} u(\mathbf{s}, \boldsymbol{\theta}) f_{1,\boldsymbol{\theta}}(\mathbf{s} \cdot \boldsymbol{\theta})\, d\boldsymbol{\theta} \tag{3}$$

where $u(\mathbf{s}, \boldsymbol{\theta})$ is a weight function and the integration is carried out over the closed surface Θ_n in R^n.

ST-1 is a Radon-type transformation (see, e.g., Helgason, 1980; Deans, 1983; Christakos, 1984a; 1986a and b). An interesting special case of ST-1 emerges if we set $\boldsymbol{\theta} = (1, 0, \ldots, 0)$. Then Eq. (2) becomes

$$\hat{f}_{1,\boldsymbol{\theta}}(s_1) = T_n^1[f_n(\mathbf{s})] = \int_{R^n} f_n(\mathbf{s})\, \delta_{s_1}(\mathbf{s} \cdot \boldsymbol{\theta})\, d\mathbf{s}$$

$$= \int_U f_n(s_1, s_2, \ldots, s_n)\, ds_2 \ldots ds_n \tag{4}$$

where $U \subset R^n$ (for simplicity, symbols under the integral are sometimes omitted).

Regarding ST-2, it may be convenient to assume that Θ_n is the surface S_n of the n-dimensional unit sphere (S_n denotes both the surface of the n-dimensional unit sphere and its surface area). The surface area is given by

$$\Theta_n = S_n = \frac{2\pi^{n/2}}{G\left(\frac{n}{2}\right)}$$

where G is the gamma function. Moreover let $u(\mathbf{s}, \boldsymbol{\theta}) = 1/S_n$, that is, uniform weight. In this case we have

$$f_n(\mathbf{s}) = \Psi_1^n[f_{1,\boldsymbol{\theta}}(t)] = \frac{1}{S_n} \int_{S_n} f_{1,\boldsymbol{\theta}}(\mathbf{s} \cdot \boldsymbol{\theta})\, d\boldsymbol{\theta} \tag{5}$$

with weight function $1/S_n$.

ST-1 and 2 may be extended to the case of hyperplanes H_{n-k} of dimension $n-k$, in general. For example, Eq. (4) may be generalized as follows:

$$\hat{f}_{n-k}(s_1, s_2, \ldots, s_{n-k}) = T_n^{n-k}[f_n(\mathbf{s})]$$

$$= \int f_n(s_1, s_2, \ldots, s_n)\, ds_{n-k+1} \ldots ds_n \tag{6}$$

An illustration of ST-1 and 2 for $n = 2$ and $k = 1$ is given in Fig. 6.1.

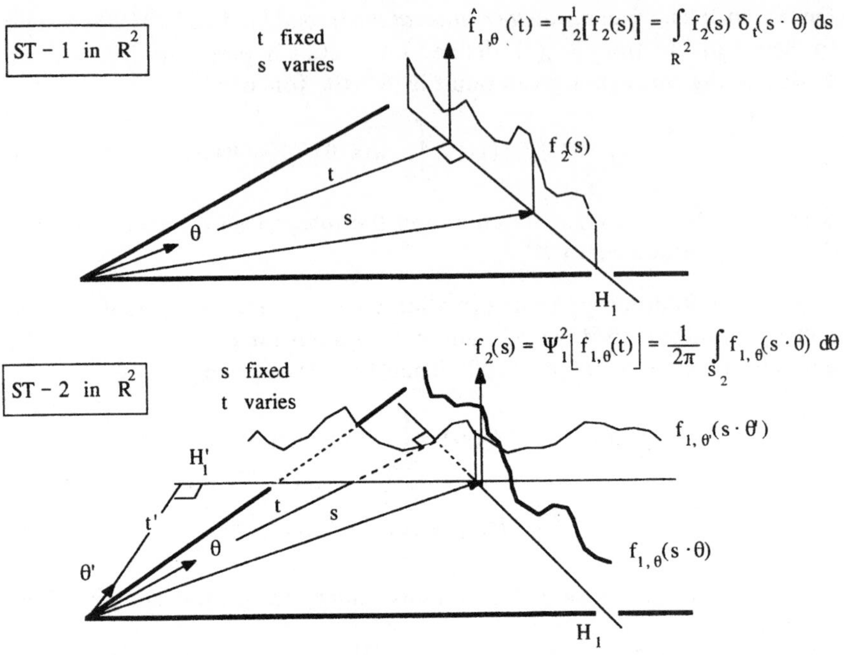

Figure 6.1 ST-1 and ST-2 in R^2

When $f(\mathbf{s})$ is an *isotropic* function, that is, $f_n(\mathbf{s}) = f_n(s = |\mathbf{s}|)$, the following expressions are obtained (Christakos, 1986a)

$$f_{n+1}(s) = T_n^{n+1}[f_n(s)] = -\frac{1}{\pi}\int_s^\infty \frac{1}{\sqrt{(u^2 - s^2)}} \frac{df_n(u)}{du}\, du \tag{7}$$

$$f_{n+2}(s) = T_n^{n+2}[f_n(s)] = -\frac{1}{2\pi s}\frac{df_n(s)}{ds} \tag{8}$$

where the T_n^{n+1} and T_n^{n+2} are the *inverse ST-1* (*IST-1*); and

$$f_{n-1}(s) = \Psi_n^{n-1}[f_n(s)] = \frac{2G\left(\dfrac{n+1}{2}\right)s^{2-n}}{(n-1)G\left(\dfrac{1}{2}\right)G\left(\dfrac{n}{2}\right)} \times \frac{d}{ds}\left[\int_0^s f_n(u)\frac{u^{n-1}}{(u^2 - s^2)}\, du\right] \tag{9}$$

$$f_{n-2}(s) = \Psi_n^{n-2}[f_n(s)] = \frac{s^{3-n}}{n-2}\frac{d}{ds}[s^{n-2}f_n(s)] \tag{10}$$

where the Ψ_n^{n-1} and Ψ_n^{n-2} are the *inverse ST-2* (*IST-2*).

Let us now consider ST in a stochastic framework, particularly in terms of covariance functions of homogeneous SRF, which are the fundamental sources of information regarding the spatial correlation structure. It will be shown later that the same concept applies in terms of generalized covariances of nonhomogeneous SRF, spectral density functions, and realizations $X_n(\mathbf{s})$ of SRF themselves.

We denote by $c_n(\mathbf{h})$, $\mathbf{h} \in R^n$ the covariance of the homogeneous SRF $X_n(\mathbf{s})$. Assume that $c_n(\mathbf{h}) \in K$ or S; it is also possible that $\in K'$ and S' [consider, for example, the case of a white-noise covariance $c_n(\mathbf{h}) = \delta(\mathbf{h})$, where $\delta(\cdot)$ is the delta function].

According to the preceding definitions, an ST-1 maps a covariance function $c_n(\mathbf{h})$, $\mathbf{h} \in R^n$ into the following set of covariances

$$\hat{c}_{1,\boldsymbol{\theta}}(t) = T_n^1[c_n(\mathbf{h})] = \int_{R^n} c_n(\mathbf{h})\, \delta_t(\mathbf{h} \cdot \boldsymbol{\theta})\, d\mathbf{h} \tag{11}$$

on the lines $\{H_{n-1}; \boldsymbol{\theta}, \mathbf{h} \cdot \boldsymbol{\theta}\}$.

Similarly, the ST-2 assigns to the set of covariances $c_{1,\boldsymbol{\theta}}(t)$, defined on a set of hyperplanes $\{H_{n-1}; \boldsymbol{\theta}, t = \mathbf{h} \cdot \boldsymbol{\theta}\}$ passing through a given point in R^n, the covariance

$$c_n(\mathbf{h}) = \Psi_1^n[c_{1,\boldsymbol{\theta}}(t)] = \frac{1}{S_n} \int_{S_n} c_{1,\boldsymbol{\theta}}(\mathbf{h} \cdot \boldsymbol{\theta})\, d\boldsymbol{\theta} \tag{12}$$

with weight function $1/S_n$.

To fix ideas, let us discuss the following examples.

Example 1: Illustrations of ST-1 and 2 for two-dimensional covariances are shown in Figs. 6.2 and 6.3. In Fig. 6.2, $\boldsymbol{\theta}$, t are fixed while $\mathbf{h}$ is varying on the line defined by $\boldsymbol{\theta}$, t. Then the ST-1 $\hat{c}_{1,\boldsymbol{\theta}}(t)$ is the integral of $c_2(\mathbf{h})$ along this line, viz.,

$$\hat{c}_{1,\boldsymbol{\theta}}(t) = \int_{R^2} c_2(\mathbf{h})\, \delta_t(\mathbf{h} \cdot \boldsymbol{\theta})\, d\mathbf{h} \tag{13}$$

In Fig. 6.3a, $\mathbf{h}$ is now fixed while the $\boldsymbol{\theta}$, t vary; that is, we consider all lines passing through the given point P. Then, ST-2, $c_2(\mathbf{h})$ is the integral of $c_{1,\boldsymbol{\theta}}(t)$ over the surface S_2 of the unit circle, that is,

$$c_2(\mathbf{h}) = \frac{1}{S_2} \int_{S_2} c_{1,\boldsymbol{\theta}}(\mathbf{h} \cdot \boldsymbol{\theta})\, d\boldsymbol{\theta} \tag{14}$$

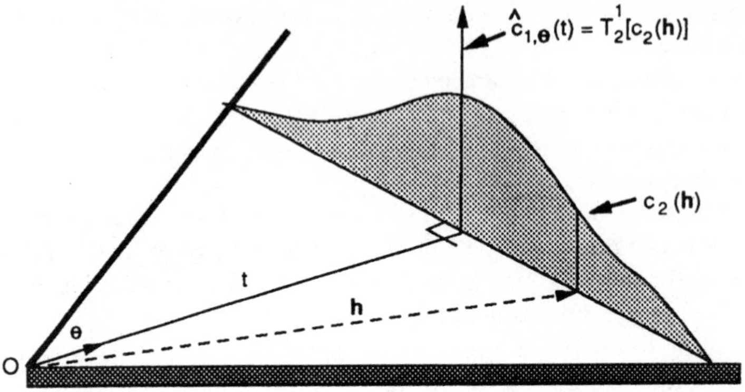

Figure 6.2 ST-1 representation of covariance functions in R^2

For computational convenience the equivalent representation of Fig. 6.3b may be used: Point P is now transferred to the origin, $\boldsymbol{\theta}$ is the unit vector along the line, and t is the projection of **h** onto the line.

Manipulations (11) through (14) motivate the following question. "How are covariances $\hat{c}_{1,\boldsymbol{\theta}}(t)$ and $c_{1,\boldsymbol{\theta}}(t)$ related?" The answer to this question is given by the proposition below.

Proposition 1: The inverse space transformation of the second kind (IST-2), Ψ_n^1, of the ST-2 Ψ_1^n is as follows:

$$c_{1,\boldsymbol{\theta}}(t) = \Psi_n^1[c_n(\mathbf{h})] = \Omega[\hat{c}_{1,\boldsymbol{\theta}}(t)] \tag{15}$$

in which

$$\Omega = \begin{cases} \dfrac{(-1)^m S_{2m+1}}{2(2\pi)^{2m}} \quad \dfrac{\partial^{2m}}{\partial t^{2m}}[\cdot] & \text{if} \quad n = 2m+1 \\[2ex] \dfrac{(-1)^{m-1} S_{2m}}{2(2\pi)^{2m-1}} \quad H\left\{\dfrac{\partial^{2m-1}}{\partial t^{2m-1}}[\cdot]\right\} & \text{if} \quad n = 2m \end{cases} \tag{16}$$

where H denotes Hilbert transform.

Proof: See Christakos (1987a, 1990b).

As an immediate consequence of Eq. (15), the inverse space transformation of the first kind (IST-1), T_1^n, of the ST-1 T_n^1, is written

$$c_n(\mathbf{h}) = T_1^n[\hat{c}_{1,\boldsymbol{\theta}}(t)] = \Psi_1^n\{\Omega[\hat{c}_{1,\boldsymbol{\theta}}(t)]\} \tag{17}$$

In other words, $T_1^n = \Psi_1^n \Omega$.

Some interesting results are obtained in the frequency domain (Christakos, 1987a, 1990b).

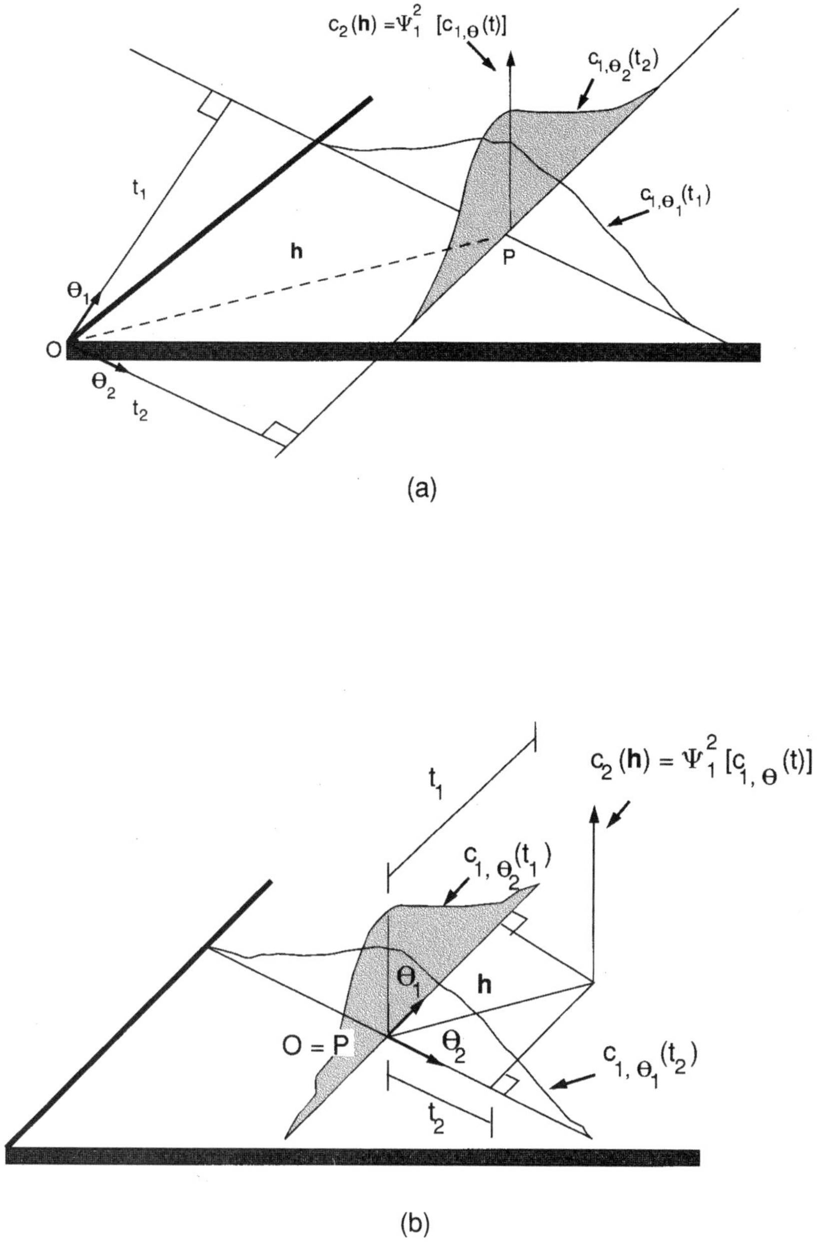

Figure 6.3 ST-2 representations of covariance functions in R^2 when (a) $O \neq P$ and (b) $O = P$

Proposition 2: Assumptions of Eqs. (11) and (12) above. In the frequency domain the space transformation operators are expressed by the following simple algebraic relationships.

$$\hat{C}_{1,\boldsymbol{\theta}}(\omega) = T_n^1[C_n(\mathbf{w})] = C_n(\mathbf{w}) \tag{18}$$

and

$$C_n(\mathbf{w}) = \Psi_1^n[C_{1,\boldsymbol{\theta}}(\omega)] = \frac{2}{S_n|\omega|^{n-1}} C_{1,\boldsymbol{\theta}}(\omega) \tag{19}$$

where $\mathbf{w} = \omega\boldsymbol{\theta}$ and $\hat{C}_{1,\boldsymbol{\theta}}(\omega)$ is the spectral density that corresponds to $\hat{c}_{1,\boldsymbol{\theta}}(t)$. (Note that depending on the expression of the n-fold Fourier transform used, the right-hand side of Eq. (18) is sometimes multiplied by a constant.)

Example 2: In the isotropic case, viz., $C_n(\mathbf{w}) = C_n(\omega = |\mathbf{w}|)$, we have

$$C_{n+k}(\omega) = T_n^{n+k}[C_n(\omega)] = C_n(\omega) \tag{20}$$

$$C_n(\omega) = \Psi_1^n[C_1(\omega)] = \frac{1}{S_n\omega^{n-1}} C_1(\omega) \tag{21}$$

Remark 1: In summary the space transformations are related as follows:
(a) In the space domain,

$$T_1^n[\cdot] = \Psi_1^n\{\Omega[\cdot]\} \qquad \text{and} \qquad \Psi_n^1[\cdot] = \Omega\{T_n^1[\cdot]\}$$

(b) In the frequency domain,

$$T_1^n[\cdot] = (T_n^1)^{-1}[\cdot] \qquad \text{and} \qquad \Psi_n^1[\cdot] = (\Psi_1^n)^{-1}[\cdot]$$

The next proposition is a straightforward application of the analysis above (Christakos and Panagopoulos, 1992).

Proposition 3: Assumptions of Eqs. (11) and (12) above. The covariance $c_n(\mathbf{h})$, $\mathbf{h} \in R^n$ is uniquely determined by means of its ST-1 or IST-2.

Example 3: In R^3 the $c_3(\mathbf{h})$ is uniquely defined if the values of $\hat{c}_{1,\boldsymbol{\theta}}(t)$ are known along all lines passing through a given point, or if the values of $\hat{c}_{2,\boldsymbol{\theta}}(t)$ are known on all planes passing through a given line. Similarly, $c_3(\mathbf{h})$ is uniquely determined if the $c_{1,\boldsymbol{\theta}}(t)$ or $c_{2,\boldsymbol{\theta}}(t)$ are known for all $\boldsymbol{\theta}$. (Note that uniqueness requires an infinite number of hyperplanes; that is, $c_3(\mathbf{h})$ is not uniquely determined by any finite set of $c_{1,\boldsymbol{\theta}}(t)$ or $c_{2,\boldsymbol{\theta}}(t)$.)

Assuming that $c_n(\mathbf{h})$ is everywhere continuous, and since $c_n(\mathbf{h})$ and $C_n(\mathbf{w})$ form a Fourier transform pair in R^n, there exists a random variable $\mathbf{w}$ in R^n that has $c_n(\mathbf{h})/c_n(\mathbf{0})$ as a characteristic function; that is,

$$\frac{c_n(\mathbf{h})}{c_n(\mathbf{0})} = E[\exp(i\mathbf{h} \cdot \mathbf{w})] \tag{22}$$

where the probability density of $\mathbf{w}$ is $f_{n,\mathbf{w}}(\boldsymbol{\omega}) = C_n(\mathbf{w})/c_n(\mathbf{0})$. In the isotropic case $f_{n,\mathbf{w}}(\boldsymbol{\omega})$ will be a function of $\omega = |\boldsymbol{\omega}|$ only, which is written as $f_{n,\mathbf{w}}(\omega)$. Let $f_\omega(\omega)$ be the probability density of ω, that is, one-dimensional. Then an interesting situation is described by the following proposition (Christakos and Panagopoulos, 1992).

Proposition 4: If $f_{1,\mathbf{w}}(\omega)$ is the one-dimensional probability density obtained by applying ST-2 on the density $f_{n,\mathbf{w}}(\boldsymbol{\omega})$, $f_{1,\mathbf{w}}(\omega)$ is related to $f_\omega(\omega)$ as follows:

$$f_{1,\mathbf{w}}(\omega) = \Psi_n^1[f_{n,\mathbf{w}}(\boldsymbol{\omega})] = \tfrac{1}{2} f_\omega(\omega) \tag{23}$$

where $\Psi_n^1[\cdot] = S_n \omega^{n-1}/2$ is the frequency domain expression of the IST-2.

3. Space Transformation Representations of Spatial Random Fields

3.1 Random Field Representations

On the basis of the above results it is evident that it is possible to apply the ST approach to find solutions to n-dimensional problems of SRF theory by transferring analysis to one-dimensional random processes (RP). The fundamental problem may be summarized as follows: "When can an SRF in R^n be represented as a linear combination of statistically uncorrelated RP in R^1?" Regarding this problem two results can be proven (Christakos and Panagopoulos, 1992).

Proposition 1 (*SRF representation; finite case*): The SRF $X_n^*(\mathbf{s})$, $\mathbf{s} \in R^n$ admits a linear representation

$$X_n^*(\mathbf{s}_k) = \sum_{i=1}^{N} a_{ki} X_{1,i}(\mathbf{s}_k \cdot \boldsymbol{\theta}_i), \qquad a_{ki} \in R^1 \tag{1}$$

of pairwise statistical uncorrelated one-dimensional RP $X_{1,i}$ along lines $\boldsymbol{\theta}_i$, $i = 1, 2, \ldots, N$ passing through a given point, if and only if the corresponding covariance admits the linear representation

$$c_n^*(\mathbf{h} = \mathbf{s}_k - \mathbf{s}_\lambda) = \sum_{i=1}^{N} a_{ki} a_{\lambda i} c_{1,i}(\mathbf{h} \cdot \boldsymbol{\theta}_i) \tag{2}$$

where $c_{1,i}(\mathbf{h} \cdot \boldsymbol{\theta}_i)$ is the one-dimensional covariance of the RP $X_{1,i}(\mathbf{s} \cdot \boldsymbol{\theta}_i)$, $i = 1, 2, \ldots, N$. (For simplicity, it is assumed that $X_{1,i}(\mathbf{s} \cdot \boldsymbol{\theta}_i)$ have zero mean and unit variance.)

In the general case, ST appear in the scene.

Proposition 2 (*SRF representation*; *general case*): The SRF $X_n(\mathbf{s})$, $\mathbf{s} \in R^n$ allows the representation

$$X_n(\mathbf{s}) = \Psi_1^n[X_{1,\boldsymbol{\theta}}(\tau)] \tag{3}$$

with weight function $a(\mathbf{s}, \boldsymbol{\theta})$, where $\tau = \mathbf{s} \cdot \boldsymbol{\theta}$ and $X_{1,\boldsymbol{\theta}}(\tau)$ are pairwise uncorrelated RP along lines $\boldsymbol{\theta}$ passing through a given point, if and only if the corresponding covariances are related by

$$c_n(\mathbf{h} = \mathbf{s}_k - \mathbf{s}_\lambda) = \Psi_1^n[c_{1,\boldsymbol{\theta}}(t)] \tag{4}$$

with weight function $a(\mathbf{s}_k, \boldsymbol{\theta}) \cdot a(\mathbf{s}_\lambda, \boldsymbol{\theta})$, where $t = \mathbf{h} \cdot \boldsymbol{\theta}$.

3.2 A Few Additional Remarks

A useful result—immediate consequence of the SRF representation results above and the central limit theorem—is introduced by the following corollary.

Corollary 1: Assumptions of Proposition 1. Moreover, in Eq. (1) let $a_{ki} = 1/\sqrt{N}$; that is,

$$X_n^*(\mathbf{s}) = \frac{1}{\sqrt{N}} \sum_{i=1}^{N} X_{1,i}(\mathbf{s} \cdot \boldsymbol{\theta}_i) \tag{5}$$

The corresponding covariance will be given by [see Eq. (2)]

$$c_n^*(\mathbf{h}) = \frac{1}{N} \sum_{i=1}^{N} c_1(\mathbf{h} \cdot \boldsymbol{\theta}_i) \tag{6}$$

Let

$$X_n(\mathbf{s}_k) = \lim_{N \to \infty} X_n^*(\mathbf{s}_k) \tag{7}$$

If the limit is finite, then $X_n(\mathbf{s}_k)$ is an SRF with Gaussian univariate density and covariance

$$c_n(\mathbf{h}) = \lim_{N \to \infty} c_n^*(\mathbf{h}) = \Psi_1^n[c_{1,\boldsymbol{\theta}}(t)] \tag{8}$$

where $t = \mathbf{h} \cdot \boldsymbol{\theta}$ and the weight function is $1/S_n$.

On the strength of Corollary 1 above, given an SRF $X_n(\mathbf{s})$ with covariance $c_n(\mathbf{h})$ the factorization of $X_n(\mathbf{s})$ as a weighted sum of N statistically uncorrelated random processes $X_{1,i}(\mathbf{s} \cdot \boldsymbol{\theta}_i)$, $i = 1, 2, \ldots, N$ is equivalent to a description of the factorization of $c_n(\mathbf{h})$ in the form of a weighted sum of the one-dimensional covariances $c_{1,\boldsymbol{\theta}}(t)$. The latter factorization tends, as $N \to \infty$, to the ST-2, $\Psi_1^n[c_{1,\boldsymbol{\theta}}(t)]$.

In some situations of practical importance, several deviations from the assumptions of Corollary 1 may apply, such as

(i) finite number of lines N, or
(ii) nonexistence of the second moments of the variables $X_{1,i}(\mathbf{s}\cdot\boldsymbol{\theta}_i)$, $i=1,2,\ldots,N$.

Regarding (i), it is important to mention that all the integral relations defined in the preceding sections hold exactly as an infinite set of projection angles $\boldsymbol{\theta}_i$ exists. Naturally, in practice only a finite number of $\boldsymbol{\theta}_i$ will be taken, so that certain of the results obtained are only approximations of the integral relations above. This is not a novel fact; all types of integral transforms, like the Fourier, the Abel, and the Radon transforms, experience similar problems. Nevertheless, it turns out that in most applications of practical importance these approximations are very satisfactory. In the case that the univariate densities of the variables $X_{1,i}(\mathbf{s}\cdot\boldsymbol{\theta}_i)$ are non-Gaussian and $N<\infty$, the SRF (5) may be only approximately Gaussian. Nevertheless, in most situations in the simulation practice of engineering processes such an approximation is satisfactory.

Regarding (ii), an interesting situation arises when the $X_{1,i}(\mathbf{s}\cdot\boldsymbol{\theta}_i)$, $i=1,2,\ldots,N$ all have Cauchy densities,

$$\frac{1}{\pi(1+X_{1,i}^2(\mathbf{s}\cdot\boldsymbol{\theta}_i))}$$

Second moments do not exist and the SRF $X_n(\mathbf{s})$ of Eq. (5) is distributed with density

$$\frac{1}{\sqrt{N}\,\pi\left(1+\dfrac{X^2}{N}\right)}$$

which under no circumstances tends to the Gaussian density. This aspect may be important in practice, for experimental results show that certain natural processes do not satisfy the Gaussian assumption.

In general, a certain amount of experimentation with various numerical approaches and different algorithms is essential for the efficient application of the ST approach in practice.

3.3 The Case of Nonhomogeneous Spatial Random Fields

In the case of *nonhomogeneous* SRF the variance might not exist, and then the semivariogram $\gamma_n(\mathbf{h})$ is used instead of the ordinary covariance $c_n(\mathbf{h})$ (Chapter 2). The ST results derived above still apply by simply replacing $c_n(\mathbf{h})$, $c_{1,\boldsymbol{\theta}}(t)$, and $\hat{c}_{1,\boldsymbol{\theta}}(t)$ with $\gamma_n(\mathbf{h})$, $\gamma_{1,\boldsymbol{\theta}}(t)=\Psi_n^1[\gamma_n(\mathbf{h})]$ and $\hat{\gamma}_{1,\boldsymbol{\theta}}(t)=T_n^1[\gamma_n(\mathbf{h})]$, respectively.

In more complex situations of spatial variation one can model the underlying nonhomogeneous SRF by means of ISRF-ν (Chapter 3). Again, the ST theory is seen to be valid simply by replacing $c_n(\mathbf{h})$, $c_{1,\boldsymbol{\theta}}(t)$, and $\hat{c}_{1,\boldsymbol{\theta}}(t)$ with the GSC-ν $k_n(\mathbf{h})$, $k_{1,\boldsymbol{\theta}}(t) = \Psi_n^1[k_n(\mathbf{h})]$ and $\hat{k}_{1,\boldsymbol{\theta}}(t) = T_n^1[k_n(\mathbf{h})]$, respectively.

4. Stochastic Differential Equation Models

4.1 The Space Transformation Approach

There are certain important problems in the application of the existing stochastic approaches in the study of hydrogeologic processes in space-time. Such problems include (i) The spatial multidimensionality of the stochastic differential equations. (ii) The existence of the solution of these differential equations, before any statistical moments or spectral density functions can be meaningfully derived. (iii) Physically inadequate approximations and unrealistic assumptions (e.g., small fluctuations, spatial homogeneity, infinite flow domains). (iv) The closure problem (when solutions in terms of statistical moments are derived).

Multidimensionality, in particular, creates serious mathematical and technical difficulties in the study of *stochastic partial differential equations* (*SPDE*) representing natural processes, such as pollutant transport in the atmosphere. It also makes the treatment of the corresponding equations computationally demanding (due to the large number of mesh points, etc.). Apart from the sheer size of the computation, the real deterrent to a computational attack on spatially multidimensional equations is the very serious convergence difficulties that this size implies. Computations in multiple dimensions may be most reluctant to converge, and can only be made to do so by using soundly judged initial conditions and sophisticated devices for accelerating convergence.

A better alternative to the purely computational approach is to try to simplify the complex SPDE analytically. This can be achieved most efficiently by means of the ST approach, which transforms the original multidimensional problem to a much simpler unidimensional space. Apart from its intrinsic interest, an ST analytical attack has two attractions. If there are difficulties in applying the stochastic differential equation methods to porous media hydrodynamics and solute transport, analysis will show their nature more clearly than a computational approach. Furthermore, even if the analysis is not wholly successful, it may give us enough information to start a convergent computation in the considerably simpler unidimensional setting. The stochastic ST approach uses spatial as well as spatiotemporal RF concepts and tools. Space-time is treated in a dynamical way, rather than being imposed on the analysis rigidly. This allows us to consider flow and transport under the general conditions of space nonhomogeneity and time nonstationarity. Also, the theory can be extended to more realistic situations of finite domains and boundary conditions.

Certain classes of solutions of SPDE may have the same structure as functions of the ST-families discussed above. (Herein, when we say that a function belongs to the T or the Ψ-family, we shall mean that the forms of the function in spaces of one, two, and three dimensions are related by the ST-1 or the ST-2 operator, respectively.) This observation is especially important in the context of transport-type models describing flow through porous media; first the multidimensional SPDE is transferred into a suitable unidimensional equation by means of ST, then the differential equation is solved in R^1, and finally, by applying IST, solutions of the corresponding multidimensional equation are constructed.

As already mentioned, the ST operators are valid in the case of spatiotemporal fields $X_n(\mathbf{s}, t)$. More precisely, it holds true that

$$\hat{X}_{1,\boldsymbol{\theta}}(s, t) = T_n^1[X_n(\mathbf{s}, t)] \tag{1}$$

$$X_n(\mathbf{s}, t) = T_1^n[\hat{X}_{1,\boldsymbol{\theta}}(s, t)] \tag{2}$$

$$\begin{aligned} X_n(s_1, s_2, \ldots, s_n, t) &= T_{n+k}^n[X_{n+k}(s_1, s_2, \ldots, s_{n+k}, t)] \\ &= \int X_{n+k}(s_1, s_2, \ldots, s_{n+k}, t)\, ds_{n+1} \ldots ds_{n+k} \end{aligned} \tag{3}$$

and

$$X_n(\mathbf{s}, t) = \Psi_1^n[X_{1,\boldsymbol{\theta}}(s, t)] \tag{4}$$

$$X_{1,\boldsymbol{\theta}}(s, t) = \Psi_n^1[X_n(\mathbf{s}, t)] \tag{5}$$

at $\mathbf{s} = s\boldsymbol{\theta}$. Additional expressions can be derived in the frequency domain (see preceding sections).

Consider the SPDE

$$L_{\nu/\mu}[X_n(\mathbf{s}, t)] = \left[\sum_{|\boldsymbol{\rho}|=\nu} \alpha_{\boldsymbol{\rho}} D^{(\nu)} - \beta_\mu \frac{\partial^\mu}{\partial t^\mu}\right] X_n(\mathbf{s}, t) = 0 \tag{6}$$

where $\alpha_{\boldsymbol{\rho}} = \alpha_{\rho_1 \ldots \rho_n}$, $|\boldsymbol{\rho}| = \sum_{i=1}^n \rho_i = \nu$, and β_μ are constant coefficients. Assume that the initial conditions consist of a set of equations

$$\frac{\partial^k}{\partial t^k} X_n(\mathbf{s}, t)\big|_{t=0} = f_k(\mathbf{s}) \tag{7}$$

where $k = 0, 1, \ldots, \mu - 1$.

Step 1: If we apply ST-1 to Eqs. (6) and (7) we obtain

$$L_{\nu/\mu}[\hat{X}_{1,\boldsymbol{\theta}}(s, t)] = \left[\sum_{|\boldsymbol{\rho}|=\nu} \alpha_{\boldsymbol{\rho}} \theta_{\boldsymbol{\rho}} \frac{\partial^\nu}{\partial s^\nu} - \beta_\mu \frac{\partial^\mu}{\partial t^\mu}\right] \hat{X}_{1,\boldsymbol{\theta}}(s, t) = 0 \tag{8}$$

where

$$\theta_{\mathbf{p}} = \theta_1^{\rho_1}\theta_2^{\rho_2}\ldots\theta_n^{\rho_n}, \qquad \sum_{i=1}^{n} \rho_i = \nu$$

and

$$\frac{\partial^k}{\partial t^k}\hat{X}_{1,\boldsymbol{\theta}}(s,t)\big|_{t=0} = \hat{f}_{k;1,\boldsymbol{\theta}}(s) \tag{9}$$

where $\hat{X}_{1,\boldsymbol{\theta}}(s,t) = T_n^1[X_n(\mathbf{s},t)]$ at $s = \mathbf{s}\cdot\boldsymbol{\theta}$.

Step 2: The one-dimensional differential equation (8) can be solved with respect to $\hat{X}_{1,\boldsymbol{\theta}}(s,t)$, assuming that the necessary conditions for the existence of the solution in the mean square sense are satisfied. A number of methods for solving unidimensional stochastic differential equations are reviewed in, for example, Gihman and Skorokhod (1972). Note that apart from solutions in terms of realizations of spatial and spatiotemporal RF representations, the SPDE can also be studied by means of

(a) their correlation functions, or
(b) their spectral density functions.

In case (a) we will be dealing with deterministic differential equations, and case (b) with deterministic algebraic equations.

Step 3: Finally, by using the IST-1, realizations of the multidimensional differential equation (6) are derived, viz.,

$$X_n(\mathbf{s},t) = T_1^n[\hat{X}_{1,\boldsymbol{\theta}}(s,t)] = \Psi_1^n\Omega[\hat{X}_{1,\boldsymbol{\theta}}(s,t)] \tag{10}$$

In addition to providing a useful way of solving important SPDE, the ST approach offers valuable insight into the interrelations between the one-dimensional and the multidimensional structure of the underlying natural processes.

Example 1: Consider the particular case of Eq. (6),

$$\begin{aligned} L_{3/0}[X_2(\mathbf{s})] &= \sum_{|\boldsymbol{\rho}|=3} \alpha_{\boldsymbol{\rho}}\, D^{(3)} X_2(\mathbf{s}) \\ &= \alpha_{30}\frac{\partial^3 X_2(\mathbf{s})}{\partial s_1^3} + \alpha_{21}\frac{\partial^3 X_2(\mathbf{s})}{\partial s_1^2\,\partial s_2} + \alpha_{12}\frac{\partial^3 X_2(\mathbf{s})}{\partial s_1\,\partial s_2^2} \\ &\quad + \alpha_{03}\frac{\partial^3 X_2(\mathbf{s})}{\partial s_2^3} = 0 \end{aligned}$$

In this case the application of ST-1 gives

$$\begin{aligned} L_{3/0}[\hat{X}_{1,\boldsymbol{\theta}}(s)] &= \sum_{|\boldsymbol{\rho}|=3} \alpha_{\boldsymbol{\rho}}\theta_{\boldsymbol{\rho}}\frac{\partial^3}{\partial s^3}\hat{X}_{1,\boldsymbol{\theta}}(s) \\ &= \alpha_{30}\theta_1^3\frac{\partial^3\hat{X}_{1,\boldsymbol{\theta}}(s)}{\partial s^3} + \alpha_{21}\theta_1^2\theta_2\frac{\partial^3\hat{X}_{1,\boldsymbol{\theta}}(s)}{\partial s^3} \\ &\quad + \alpha_{12}\theta_1\theta_2^2\frac{\partial^3\hat{X}_{1,\boldsymbol{\theta}}(s)}{\partial s^3} + \alpha_{03}\theta_2^3\frac{\partial^3\hat{X}_{1,\boldsymbol{\theta}}(s)}{\partial s^3} = 0 \end{aligned}$$

4.2 The Wave and Related Stochastic Partial Differential Equations—Some Examples from the Hydrosciences

Let us consider next the *wave equation.* The reason for this choice is that fundamental solutions of a unidimensional wave (hyperbolic) equation can give us solutions to a Laplace (elliptic) or diffusion (parabolic) equation. The following statement is a straightforward consequence of the ST approach above.

If $X_n(\mathbf{s}, t)$ is a solution of the wave equation

$$[\partial_t^2 - \nabla_{\mathbf{s},n}^2]X_n(\mathbf{s}, t) = \alpha(t)\,\delta(s_1)\,\delta(s_2)\ldots\delta(s_n) \tag{11}$$

where $\nabla_{\mathbf{s},n}^2 = \sum_{i=1}^{n} \partial^2/\partial s_i^2$ and $\alpha(t)$ is a given function of time t, then $X_n(\mathbf{s}, t)$ belongs to the T-family of ST-1 related by Eq. (3) above; that is, the realizations

$$X_{n+1}(\mathbf{s}, t) = T_n^{n+1}[X_n(\mathbf{s}, t)] \tag{12}$$

are fundamental solutions of the equation

$$[\partial_t^2 - \nabla_{\mathbf{s},n+1}^2]\,X_{n+1}(\mathbf{s}, t) = \alpha(t)\,\delta(s_1)\,\delta(s_2)\ldots\delta(s_n)\,\delta(s_{n+1}) \tag{13}$$

where now

$$\nabla_{\mathbf{s},n+1}^2 = \sum_{i=1}^{n+1} \frac{\partial^2}{\partial s_i^2}$$

To fix ideas we consider the following examples.

Example 2: Taking the ST-1 of

$$\left[\frac{\partial^2}{\partial t^2} - \nabla_{\mathbf{s},n}^2\right]X_n(\mathbf{s}, t) = 0$$

we find

$$\left[\frac{\partial^2}{\partial t^2} - \frac{\partial^2}{\partial s^2}\right]X_1(s, t) = 0, \qquad \text{where} \quad X_1(s, t) = T_n^1[X_n(\mathbf{s}, t)]$$

This can be factored as

$$\left[\frac{\partial}{\partial t} + \frac{\partial}{\partial s}\right]\left[\frac{\partial}{\partial t} - \frac{\partial}{\partial s}\right]X_1(s, t) = 0$$

which implies that

$$\frac{\partial X_1(s, t)}{\partial t} - \frac{\partial X_1(s, t)}{\partial s}$$

is a function of $s - t$ alone.

Example 3: In this example we will see how we can obtain solutions for the *Laplace* equation modeling the three-dimensional steady flow in soil when the condition of permeability is the same in all directions, viz.,

$$-\nabla^2_{s,3} h_3(s_1, s_2, s_3) = \delta(s_1)\,\delta(s_2)\,\delta(s_3) \tag{14}$$

where $h_3(\mathbf{s})$ is the hydraulic head and the delta functions represent a well at the space origin.

First note that in one dimension the solution of Eq. (11) is $X_1(s, t) = f(t-s)/2$, where $df(t)/dt = \alpha(t)$. Note that if $\alpha(t) = \delta(t)$, then $f(t) = 1$ when $t > 0$, $= 0$ otherwise. Hence $X_1(s, t) = \frac{1}{2}$ if $t > s$ and $= 0$ otherwise. We will derive isotropic solutions, that is, solutions satisfying $X_n(\mathbf{s}) = X_n(s = |\mathbf{s}|)$ for $n = 2$ and 3. More specifically,

$$X_2(s, t) = T_1^2[X_1(s, t)] = \frac{1}{2\pi}\int_s^\infty \frac{\alpha(t-u)}{\sqrt{u^2 - s^2}}\,du \quad \text{in} \quad R^2 \tag{15}$$

and

$$X_3(s, t) = T_1^3[X_1(s, t)] = \frac{\alpha(t-s)}{4\pi s} \quad \text{in} \quad R^3 \tag{16}$$

(Notice that, due to isotropy, the directional vector $\boldsymbol{\theta}$ is no longer present.) Alternatively, one may work in the frequency domain where, for example, if $\tilde{X}_1(\omega, t)$ and $\tilde{X}_3(\omega, t)$ are the spectral functions of the realizations $X_1(s, t)$ and $X_3(s, t)$, respectively, it is valid that

$$\tilde{X}_3(\omega, t) = \tilde{X}_1(\omega, t) = \frac{F(t-\omega)}{2}$$

where $F(t-\omega)$ is the Fourier transform of $f(t-s)$. Taking now the inverse Fourier transform in R^3 we find $X_3(s, t) = \alpha(t-s)/4\pi s$, which is the same result as in Eq. (16) above. Next, from Eqs. (11) and (16) and assuming $\alpha(t) = 1$ if $t > 0$, $= 0$ otherwise one gets the solution

$$h_3(s) = \lim_{t\to\infty} X_3(s, t) = \frac{1}{4\pi s} \quad \text{in} \quad R^3 \tag{17}$$

In R^2, following a similar procedure one finds

$$\frac{\partial}{\partial s} h_2(s) = \lim_{t\to\infty} \frac{\partial}{\partial s} X_2(s, t)$$

where

$$X_2(s, t) = \frac{1}{2\pi}\int_s^t \frac{du}{\sqrt{u^2 - s^2}}$$

And, by differentiating we get the solution

$$h_2(s) = -\frac{1}{2\pi}\log(s) \tag{18}$$

Example 4: Let us reconsider the three-dimensional flow model studied by Bakr *et al.* (1978), under the light of the ST approach. We assume, for the moment, that the "small-perturbation" assumptions of the power series expansions used in these studies are valid. Then, the perturbation approximation of the stochastic flow equation will be given by

$$\nabla^2_{\mathbf{s},3} h_3(\mathbf{s}) = J \frac{\partial f_3(\mathbf{s})}{\partial s_1} \tag{19}$$

where $h_3(\mathbf{s})$, $\mathbf{s} \in R^3$ is the hydraulic head, J is the mean hydraulic gradient in the s_1 direction, and $f_3(\mathbf{s}) = \log[K(\mathbf{s})]$, where $K(\mathbf{s})$ is the hydraulic conductivity. Model (19) allows perturbations to both the flow and the hydraulic conductivity in three dimensions.

By applying IST-2, Eq. (19) reduces to the one-dimensional differential equation

$$\frac{\partial^2}{\partial s^2} h_{1,\boldsymbol{\theta}}(s) = J\theta_1 \frac{\partial}{\partial s} f_{1,\boldsymbol{\theta}}(s) \tag{20}$$

where $h_{1,\boldsymbol{\theta}}(s) = \Psi_3^1[h_3(\mathbf{s})]$ and $f_{1,\boldsymbol{\theta}}(s) = \Psi_3^1[f_3(\mathbf{s})]$. Note that we could also apply ST-1 to obtain

$$\frac{\partial^2}{\partial s^2} \hat{h}_{1,\boldsymbol{\theta}}(s) = J\theta_1 \frac{\partial}{\partial s} \hat{f}_{1,\boldsymbol{\theta}}(s)$$

where

$$\hat{h}_{1,\boldsymbol{\theta}}(s) = T_3^1[h_3(\mathbf{s})] \qquad \text{and} \quad \hat{f}_{1,\boldsymbol{\theta}}(s) = T_3^1[f_3(\mathbf{s})]$$

Now Eq. (20) is easily solved, viz.,

$$h_{1,\boldsymbol{\theta}}(s) = J\theta_1 \int_0^s f_{1,\boldsymbol{\theta}}(u)\, du + a_1 s + a_0$$

(where, again, the coefficients a_1 and a_0 are to be determined on the basis of the initial conditions).

Realizations of $h_3(\mathbf{s})$ satisfying Eq. (19) can be constructed in terms of the ST-2; that is,

$$h_3(\mathbf{s}) = \Psi_1^3[h_{1,\boldsymbol{\theta}}(s)] \tag{21}$$

at $\mathbf{s} = s\boldsymbol{\theta}$.

In the context of stochastic correlation analysis, the ST-based approach can be applied to study the partial differential equations governing the corresponding covariances (ordinary or generalized), or the algebraic expressions relating the associated spectral density functions. More specifically, in the spectral domain the differential representation (19) leads to the following algebraic representation.

$$C_{h,3}(\mathbf{w}) = \frac{J^2 w_1^2}{\mathbf{w}^4} C_{f,3}(\mathbf{w}) \tag{22}$$

where $C_{h,3}(\mathbf{w})$ and $C_{f,3}(\mathbf{w})$ are the three-dimensional spectral density functions of $h_3(\mathbf{s})$ and $f_3(\mathbf{s})$, respectively. By applying IST-2, Eq. (22) yields

$$C_{h,1}(w_1) = \frac{J^2}{w_1^2} C_{f,1}(w_1) \tag{23}$$

which is the spectral equation governing unidimensional, steady-state, saturated groundwater flow in the s_1 direction. Note that the unidimensional spectral density of the hydraulic head fluctuations is related to $C_{h,3}(\mathbf{w})$ by

$$C_{h,1}(w_1) = \Psi_3^1[C_{h,3}(\mathbf{w})] = 2\pi w_1^2 C_{h,3}(\mathbf{w})$$

Also, the hydraulic conductivity spectrum satisfies

$$C_{f,1}(w_1) = \Psi_3^1[C_{f,3}(\mathbf{w})] = 2\pi w_1^2 C_{f,3}(\mathbf{w})$$

(Here the spectral densities are as usual evaluated at $\mathbf{w} = w_1\boldsymbol{\theta}$, where $\boldsymbol{\theta}$ is the unit vector determining the s_1 direction in R^3.)

By means of Eq. (23), the study of a three-dimensional hydrogeologic process has been reduced to the study of a unidimensional process along the s_1 direction. The corresponding unidimensional spectral and covariance functions have a definite physical meaning and can be calculated experimentally much easier than their three-dimensional counterparts. We can proceed further by assuming that the unidimensional covariance of the hydraulic conductivity is

$$c_{f,1}(r) = \sigma_f^2\left(1 - \frac{r}{a}\right) \exp\left[-\frac{r}{a}\right] \tag{24}$$

where a represents correlation distance. ST-2 yields the corresponding three-dimensional isotropic covariance, viz.,

$$c_{f,3}(r) = \Psi_1^3[c_{f,1}(r)] = \sigma_f^2 \exp\left[-\frac{r}{a}\right] \tag{25}$$

(Note that this result coincides with the covariance model assumed by Bakr *et al.*, 1978). Taking the Fourier transform of (24) one gets

$$C_{f,1}(w_1) = \frac{2\sigma_f^2 a^3 w_1^2}{\pi(1 + a^2 w_1^2)^2}$$

and the model (23) yields

$$C_{h,1}(w_1) = \frac{2\sigma_f^2 a^3 J^2}{\pi(1 + a^2 w_1^2)^2}$$

Since $C_{h,3}(\mathbf{w})$ and $C_{h,1}(w)$ are related by means of the ST-2 operator ($\omega = |\mathbf{w}|$),

$$C_{h,3}(\mathbf{w}) = \frac{\sigma_f^2 a^3 J^2 w_1^2}{\pi^2(1 + a^2\omega^2)^2\omega^4} \tag{26}$$

which is the required three-dimensional hydraulic head fluctuation spectrum (22). Taking the Fourier transform of (26) we obtain the corresponding covariance function. (Note that the covariance functions belong to the Ψ-family, as well.)

To verify the result (26) one can proceed in several ways: For example, we can take the Fourier transform of the covariance (25) and insert it into the three-dimensional model (22) to get for $C_{h,3}(\mathbf{w})$ the same expression as Eq. (26). Similar results can be obtained for other stochastic flow models.

Remark 1: As has been mentioned in previous chapters, when the statistical moments or the spectral functions approach is used, one should first deal with the question whether any solution of the corresponding SPDE in terms of SRF representations does actually exist. For if such a solution does not exist, the application of the above methods makes no sense. Another important aspect of the statistical moments approach is the closure problem; that is, we have a hierarchy of N equations with $N+1$ statistical moments. Then an appropriate approximation technique must be used that will convert the infinite hierarchy of equations into a closed set. In stochastic hydrogeology, the most widely used method of closure is the power series method (e.g., Ababou and Gelhar, 1990). However, in many practical circumstances the series method may be incapable of representing the fundamental characteristics of the flow and transport, and it may lead to quite unreasonable results (Christakos, 1990c).

4.3 Stochastic Partial Differential Equations of Intrinsic Spatial Random Fields of Order ν

Consider the SPDE in R^3 (*Poisson* differential equation)

$$\nabla^2_{\mathbf{s},3} X_3(\mathbf{s}) = Y_3(\mathbf{s}) \tag{27}$$

where $X_3(\mathbf{s})$ is an ISRF-1 and $Y_3(\mathbf{s})$ is a homogeneous SRF. (Poisson equation is widely used, for example, as a perturbation approximation of the steady-state flow model in a fully saturated spatially random medium.)

The ST approach yields solutions of the general form

$$\begin{aligned} X_3(\mathbf{s}) &= -\frac{1}{2\pi}\Psi_1^3 T_3^1[Y_3(\mathbf{s})] = -\frac{1}{2\pi}\Psi_1^3[\hat{Y}_{1,\boldsymbol{\theta}}(\mathbf{s}\cdot\boldsymbol{\theta})] \\ &= -\frac{1}{8\pi^2}\int_{S_2} \hat{Y}_{1,\boldsymbol{\theta}}(\mathbf{s}\cdot\boldsymbol{\theta})\, d\boldsymbol{\theta} \end{aligned} \tag{28}$$

The correlation structure associated with Eq. (27) is given by

$$\nabla^4_{\mathbf{s},3} k_3(\mathbf{h}) = c_3(\mathbf{h}) \tag{29}$$

Assuming isotropy, Eq. (29) becomes

$$\frac{1}{r}\frac{d^4}{dr^4}[rk_3(r)] = c_3(r) \tag{30}$$

where $r = |\mathbf{h}|$. Now apply IST-2 to Eq. (30) to get

$$\frac{d^4}{dr^4} k_1(r) = c_1(r) \tag{31}$$

where

$$k_1(r) = \Psi_3^1[k_3(r)] = \frac{d}{dr}[rk_3(r)]$$

$$c_1(r) = \Psi_3^1[c_3(r)] = \frac{d}{dr}[rc_3(r)]$$

Equation (31) can be solved to obtain

$$k_1(r) = \int_0^r \frac{(r-u)^3}{3!} c_1(u)\, du + a_1 r + a_0$$

where the coefficients a_1 and a_0 are to be determined on the basis of the initial conditions of the particular problem. Next, by employing the ST-2 we find that

$$k_3(r) = \Psi_1^3 k_1(r) = \frac{1}{r}\int_0^r k_1(u)\, du$$

$$= \frac{1}{6r}\int_0^r u(r-u)^3 c_3(u)\, du + \sum_{j=-1}^{2} b_{j+1} r^j \tag{32}$$

For example, if $c_3(r) = \exp[-br]$, $k_3(r)$ is given by

$$k_3(r) = \frac{3}{b^4} - \frac{4}{b^5 r} - \frac{r}{b^3} + \frac{r^2}{6b^2} + \frac{4\exp[-br]}{b^5 r} + \frac{\exp[-br]}{b^4}$$

(compare with Eq. (13), Section 5 of Chapter 3).

4.4 Stochastic Partial Differential Equations with Variable Coefficients

In addition to the analysis proposed above, the analytical apparatus of the ST approach is capable of studying more complicated SPDE with variable coefficients that represent, for example, realistic groundwater flow and solute transport problems.

Investigations in the context of SPDE with variable coefficients require that the ST and the IST be defined for products of spatial and spatiotemporal natural processes, as well as for their derivatives. For illustration let us study the following example.

Example 5: Suppose one needs to determine the IST $T_1^3[\cdot]$ of the product $X_{1,\boldsymbol{\theta}}(s)\,Y_{1,\boldsymbol{\theta}}(s)$, where $X_{1,\boldsymbol{\theta}}(s)$ and $Y_{1,\boldsymbol{\theta}}(s)$ are the unidimensional representations of the SRF $X_3(\mathbf{s})$ and $Y_3(\mathbf{s})$, respectively. Using the theory of the previous sections, the IST $T_1^3[\cdot]$ of the product $X_{1,\boldsymbol{\theta}}(s)\,Y_{1,\boldsymbol{\theta}}(s)$, is defined as

$$X_3(\mathbf{s})\,Y_3(\mathbf{s}) = T_1^3[X_{1,\boldsymbol{\theta}}(s)\,Y_{1,\boldsymbol{\theta}}(s)]$$
$$= -\frac{1}{2(2\pi)^3}\nabla_{\mathbf{s}}^2 \int_{R^3}\left\{\int_{R^1} \exp[-i\omega\boldsymbol{\theta}\cdot\mathbf{s}] \times \tilde{X}_{1,\boldsymbol{\theta}}(\omega) * \tilde{Y}_{1,\boldsymbol{\theta}}(\omega)\,d\omega\right\} d\boldsymbol{\theta} \tag{33}$$

where $*$ denotes convolution. Moreover, by using well-known convolution and Fourier transform properties, Eq. (33) can also be written as

$$X_3(\mathbf{s})\,Y_3(\mathbf{s}) = -\frac{1}{2(2\pi)^3}\nabla_{\mathbf{s}}^2 \int_{R^3}\int_{R^3} g(\mathbf{s},\mathbf{s}',\mathbf{s}'')X_3(\mathbf{s}')\,Y_3(\mathbf{s}'')\,d\mathbf{s}'\,d\mathbf{s}'' \tag{34}$$

where

$$g(\mathbf{s},\mathbf{s}',\mathbf{s}'') = \int_{R^3}\left\{\int_{R^1} d\omega \int_{R^1} \exp[i\omega\boldsymbol{\theta}\cdot(\mathbf{s}''-\mathbf{s})] \times \exp[i\omega'\boldsymbol{\theta}\ \ (\mathbf{s}'-\mathbf{s}'')]\,d\omega'\right\} d\boldsymbol{\theta} \tag{35}$$

Such ST formulas are important in the context, for example, of groundwater flow analysis represented by the steady-state SPDE

$$\nabla\cdot[K_3(\mathbf{s})\cdot\nabla H_3(\mathbf{s})] = 0 \tag{36}$$

where $H_3(\mathbf{s})$ is the hydraulic head in a three-dimensional saturated porous medium with spatially variable conductivity $K_3(\mathbf{s})$. [Note that if power series expansions together with small-perturbation assumptions are used to solve the closure problem (as, e.g., in Bakr *et al.*, 1978; Ababou and Gelhar, 1990), Eq. (36) reduces to the perturbation approximation expressed by Eq. (19) above.]

In the case of Eq. (36), the ST analysis above will provide the necessary expressions for the quantities $T_3^1[K_3(\mathbf{s})\ \ \nabla H_3(\mathbf{s})]$, $T_1^3[K_{1,\boldsymbol{\theta}}(s)\,\partial/\partial s\,H_{1,\boldsymbol{\theta}}(s)]$, etc. Here, $K_{1,\boldsymbol{\theta}}(s) = T_3^1[K_3(\mathbf{s})]$ is the unidimensional hydraulic conductivity and $H_{1,\boldsymbol{\theta}}(s) = T_3^1[H_3(\mathbf{s})]$ is the unidimensional hydraulic head along the $\boldsymbol{\theta}$ direction. If the flow domain is assumed to be finite, the ST of the original multidimensional finite flow domain will lead to a unidimensional domain, which is also finite. The ST should also be applied on the original boundary conditions. Then, the ST version of Eq. (36), together with the ST finite domain and boundary conditions, completely define the unidimensional flow problem. The solution of the latter is considerably simpler than the original multidimensional flow model. Certainly, a significant amount of work remains to be done along these lines of thought, for a number of properly chosen finite domains and boundary conditions. This will naturally include the development of a physically adequate closure method regarding the hierarchy of equations of higher order correlations, which will arise when the statistical moments of the ST-reduced unidimensional version of Eq. (36) are derived (Christakos, 1990c).

In summary, the space transformation approach to SPDE is as follows:

(a) Transform the study of multidimensional SPDE from the n-dimensional space ($n=2$ or 3) to the one-dimensional space, by means of the ST operators.

(b) Examine the existence and uniqueness of the solutions of the unidimensional stochastic differential equations. Conditions for the existence and uniqueness in n dimensions may be related to those in one dimension through the ST operators.

(c1) Either derive solutions of the unidimensional stochastic differential equations, or

(c2) when step (c1) is not achievable, one may work in terms of the equations relating the corresponding statistical moments or spectral density functions. In this case, the associated multidimensional closure problem will be reduced to a unidimensional problem.

(d) Finally, obtain solutions to the original multidimensional equations by means of the inverse ST.

5. Criteria of Permissibility

Within the context of the criteria of permissibility (COP, see previous chapters), the following result establishes a means of testing if a function that is a permissible ordinary covariance in R^1 is also a permissible covariance in R^n (Christakos and Panagopoulos, 1992).

Proposition 1: A function $c_n(\mathbf{h})$ is a covariance in R^n if and only if the corresponding ST functions, $\hat{c}_{1,\boldsymbol{\theta}}(t)$, $c_{1,\boldsymbol{\theta}}(t)$ are such in R^1.

Example 1: Let the candidate function in R^3 be

$$c_3(t) = \exp[-t^2] \tag{1}$$

where $t = |\mathbf{h}|$ (isotropic function). The corresponding ST-1 function in R^1 is

$$\hat{c}_1(t) = \pi \exp[-t^2] \tag{2}$$

(Note that, due to isotropy, the subscript $\boldsymbol{\theta}$ is no longer present.) The function $\hat{c}_1(t)$ is a covariance in R^1 (see Example 8, Section 7 of Chapter

2) and, therefore, so is $c_3(t)$ in R^3. Alternatively, one can apply IST-2 to find

$$c_1(t) = (1-2t^2)\exp[-t^2] \tag{3}$$

Again, since $c_1(t)$ is a covariance in R^1, the $c_3(t)$ is a covariance in R^3.

Proposition 1 is also valid in terms of semivariograms and generalized covariances. In the case of isotropic covariance, semivariogram, or generalized covariance models an alternative, and sometimes practically more comprehensive ST-based approach, is as follows: (i) Find the spectral function of the candidate model in R^1, and check its permissibility by using the appropriate COP. (ii) If it is permissible in R^1 apply the IST-1 to obtain the spectral function in the space of interest R^n, and apply COP to check its permissibility. To clarify things consider the following example (see also Chapter 7).

Example 2: Assume we are given the model

$$\gamma(r) = 1 - \exp\left[-\frac{r}{a}\right] \tag{4}$$

where $a > 0$. Its spectral function in R^1 is

$$\Gamma_1(\omega) = -\frac{a}{\pi(1+a^2\omega^2)}$$

Clearly the appropriate COP-4 (Chapter 2) is satisfied; that is, $-\Gamma_1(\omega) > 0$ for all ω. By applying ST-1, we obtain the spectral semivariogram function in R^2, which, again, satisfies the corresponding COP, namely,

$$-\Gamma_2(\omega) = -\frac{1}{\pi}\int_\omega^\infty \left\{\frac{d}{du}\left[\frac{1}{u}\frac{d\Gamma_1(u)}{du}\right]\right\}\sqrt{u^2-\omega^2}\,du$$

$$= \frac{a^2}{2\pi[\sqrt{1+a^2\omega^2}]^3} > 0 \qquad \text{for all} \quad \omega$$

In R^3,

$$-\Gamma_3(\omega) = \frac{1}{2\pi\omega}\frac{d\Gamma_1(\omega)}{d\omega}$$

$$= \frac{a^3}{\pi^2(1+a^2\omega^2)^2} > 0 \qquad \text{for all} \quad \omega$$

Proposition 1 can also be used to derive new covariance models in R^n starting from one-dimensional covariances. For example, let the covariance in R^1 be $c_1(t) = \exp[-t^2]$; ST-2 applied in R^2 becomes $c_2(t) = I_0(t) - L_0(t)$, where $I_0(\cdot)$ and $L_0(\cdot)$ are Bessel and Struve functions, respectively, both of order zero. Several other examples are discussed in Christakos (1984b). Finally, the application of the ST theory in the simulation of random fields will be discussed in detail in Chapter 8.

7 Random Field Modeling of Natural Processes

"The sciences do not try to explain, they hardly even try to interpret, they mainly make models. By a model is meant a mathematical construct which, with the addition of certain verbal interpretations, describes observed phenomena. The justification of such a mathematical construct is solely and precisely that it is expected to work."

J. von Neumann

1. Introduction

Hydrogeologic parameters such as permeability, porosity, hydraulic head, transmissivity, storage coefficient, and rainfall are all functions whose properties are coordinated with the algebraic structure of the space and/or time. The same happens with the vast majority of pertinent natural processes in environmental engineering, mining, meteorology, etc.

The study of the spatial and temporal behavior of such natural processes by applying concepts and methods from the theory of random fields (RF) presented in previous chapters constitutes an important part of the stochastic research program (see also Chapter 1). More specifically, description of the dominant features of a natural process is achieved by means of

(a) certain fundamental working hypotheses, that belong to the hard core of the stochastic data processing research program;

(b) a set of auxiliary hypotheses together with some duality relations, which relate the natural processes to the mathematics of the RF model;

(c) a heuristic—a set of methodological rules—for determining spatial correlation models by means of experimental procedures (in the case of ordinary covariance and semivariogram functions), as well as automatic procedures (when generalized models are involved).

The implication of this approach is that the model parameters and auxiliary hypotheses are assigned real counterparts that are operationally determined. As a consequence, the conclusions obtained about the spatial structure of the natural processes represented in terms of spatial RF (SRF) have an objective meaning. Simultaneous references to experimental data support the theoretical claims.

Certain ways to incorporate qualitative (soft) information into the analysis are explored. Experience with the problem's specific discipline is very important here. Since the subject includes serious pitfalls, one should proceed with great caution. Particularly interesting is the situation where the natural process of interest is undersampled or unobservable, but is well correlated to processes about which a significant amount of information is available. Relevant is the case of data interrupted by measurement errors.

In practice, another useful approach of incorporating soft data into analysis is provided by the so-called indicator formalism. Within the same framework, a technique that determines spatial correlation models by an approach that maximizes entropy subject to all the information available about the spatial variability of the specific hydrologic process, is presented. The approach attributes great significance to procedures translating qualitative knowledge into appropriate quantitative conditions. In relation to this, it is argued that the real power of subjective analysis is in fulfilling the need for normative rules, according to which these translations will be carried out.

The use of space transformations largely simplifies things through the reduction of multidimensional processes to spaces of lower dimensionality. In connection with this, we outline certain important issues concerning the effective application of the theoretical permissibility criteria of spatial correlation models developed in previous chapters.

Finally, the various stages of the identification of the spatial and temporal variability are combined into a comprehensive step-by-step procedure whose outcomes can be clearly interpreted and appreciated. Moreover, it is possible to make some tentative observations about which of the various auxiliary hypotheses and duality rules will function better in complex real situations. Also, we will find that the distinction between testable and nontestable auxiliary hypotheses is not always clear-cut. For example, while under certain circumstances homogeneity can be taken as a nontestable hypothesis, it will appear to be testable in many other cases; this suggests that the distinction between testable and nontestable hypotheses is to some extent a matter of how hypotheses apply rather than their inherent content. These observations emphasize the fact that the effective application of the spatial and temporal structure identification procedure requires a deep understanding of the physics of the natural process of interest, intuition, and a broad knowledge of the specific scientific discipline.

2. Descriptive Features of Natural Processes and the Basic Working Hypotheses

Natural processes distributed in space are physically unique and rather descriptive. However, as was pointed out in Chapter 1, this spatial distribution displays two distinct features, namely,

(a) The *overall causal spatial structure*, which includes geological trends, transition zones, as well as location, size, and orientation of the individual samples. For example, a soil parameter takes its values within a geometrically well-defined region such as an earth dam, or the foundation medium under a construction site. Some locations are especially important for engineering performance: when predicting the soil settlement under a building, a sample located beneath the centerline may be more important than one outside the stress bulb. One tests samples that are determined precisely with their location, geometrical shape, dimensions, and orientation within the soil deposit. The latter may affect the sort of test to be performed in the laboratory, say active, passive, or direct shear strength tests. Sample size is known to affect the measured strength of stiff fissured clays, where small samples may overpredict the average *in-situ* value. Also, when estimating the *in-situ* coefficient of consolidation or the permeability of soils like varved clays, one may need large laboratory samples to obtain reasonable values.

(b) The *local noncausal randomness*, in the sense that the values measured cannot be determined in advance in terms of, say, an analytical expression. For example, the soil settlement pattern shown in Fig. 7.1 exhibits an overall structure obeying some specific laws or mechanisms of nature but, at the same time, the local variation of soil settlement is irregular enough to be considered as random. Consider, also, Fig. 7.2 which describes the Zn concentrations (%) from a mine at Lavrion (Greece). Both features of structure and randomness are apparent in this case, too. That is, there are areas rich in Zn and areas with low concentration in Zn; on the other hand, the Zn concentration presents locally erratic fluctuations.

In spatial data analysis such natural processes are called *topographic* (see, e.g., Whittle, 1954; Gandin, 1963); in geostatistics they are called *regionalized* (Matheron, 1965; Journel and Huijbregts, 1978). As we discussed in Chapter 1, any scientific approach to the problem of modeling and analyzing such

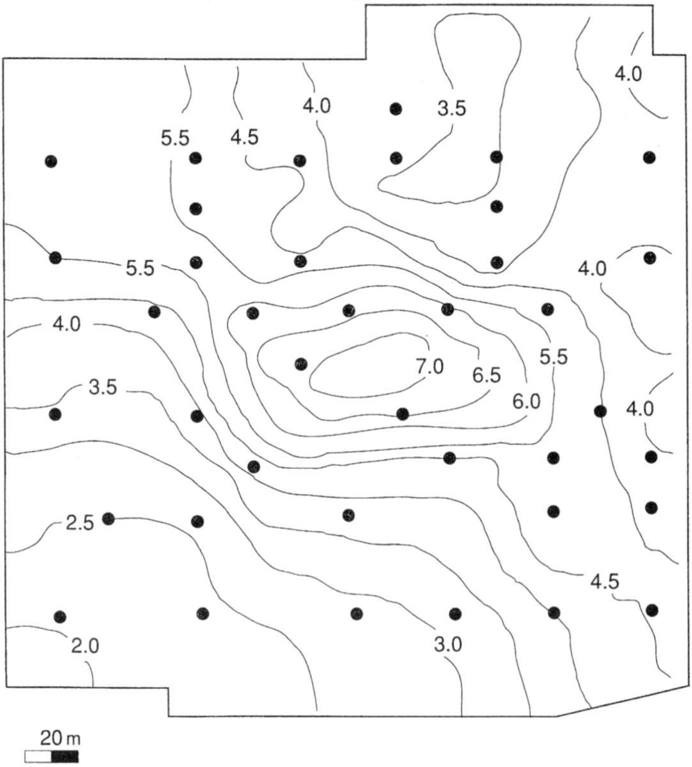

Figure 7.1 The pattern of expected soil settlements (centimeters)

natural processes should take both issues of randomness and spatial structure into account. In this context, the issue of randomness immediately excluded any deterministic model. Then, a representation of the spatial variation in statistical terms would not be appropriate, due to the assumption of spatial independence between observations. (Clearly, this assumption fails to incorporate in the analysis the most important issue of spatial structure mentioned above.) In view of this discussion, the conclusion of Chapter 1 was that the approach to be employed should be based on the interpretation of a natural process by means of the SRF model. The implications of this interpretation were fully understood by noticing that as a result of the functional nature of the SRF model, we were able to attach the appropriate mathematical content to features of both macroscopic structure and microscopic randomness.

To proceed further with the application of the SRF model developed in Chapters 2 through 6, it is necessary to make certain physically well-motivated *working hypotheses*. These are methodological hypotheses that belong to the *hard core* of the stochastic research program (see also Section

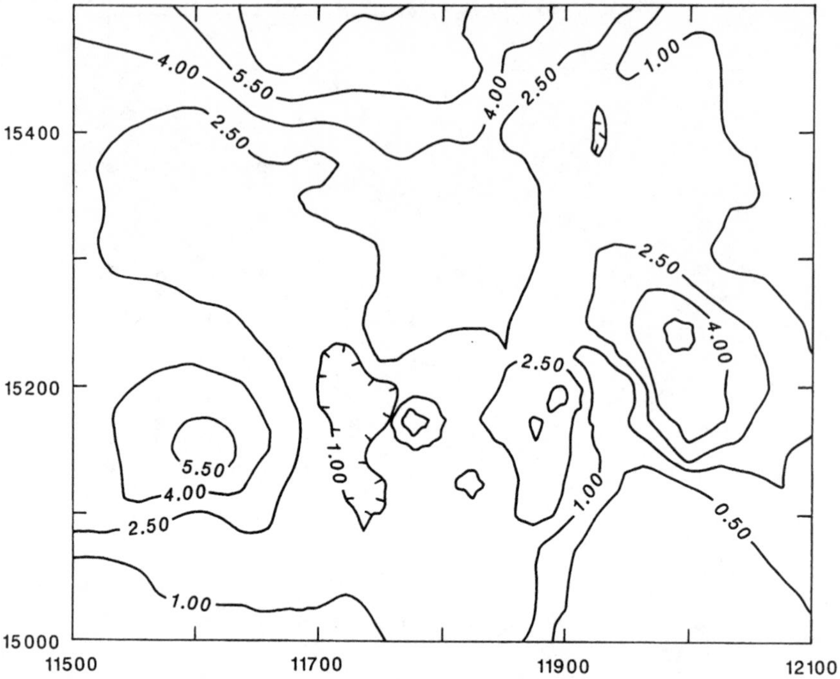

Figure 7.2 Zn concentrations (percent) from the mine of Lavrion

4, Chapter 1) and should be considered a modeling of reality rather than as the physical reality itself. The philosophy behind this view is that one cannot prove anything about the real world by means of a mathematical model. One develops such a model, studies its behavior, and uses it to make predictions about a natural process. Then one may evaluate the usefulness of the model in terms of the degree to which its behavior and predictions agree with reality.

Hypothesis H1: *The set of all available measurements in space is considered one of several realizations (sample functions) of an SRF.*

In other words, while locally the natural process is viewed as a random variable, globally it is characterized by the spatial dependence of the adjacent random variables that constitute the SRF (see, e.g., Fig. 2.1 of Chapter 2). Hypothesis H1 is a nontestable (after the event) hypothesis and should be considered a fundamental methodological choice that opens the way for applying the SRF model to our problem. In this connection, the next step is to choose, among the various parameters of an SRF model, those that will make the SRF representation of a natural process practically feasible.

Hypothesis H2: *All calculations pertinent to the stochastic structure of an SRF will be carried out on the basis of the statistical moments of order up to two, in the ordinary or in the generalized sense.*

By applying hypothesis H2 we essentially restrict ourselves to that part of the general theory of SRF—called *correlation theory*—which studies only those aspects of SRF determined by their statistical moments. To carry on, however, with practical aspects an additional hypothesis is indispensable.

Hypothesis H3: *All the information related to the statistical structure of the natural process is included in the single realization available.*

The methodological hypothesis H3 is heuristically most important: Since in reality the natural process is unique and, hence, only one realization of the SRF is available, H3 allows the determination of those model parameters having real counterparts that can be observed and measured, and, therefore, they are *testable* (after the event). More specifically, the theoretical statistical moments (mean value, covariance, and semivariogram functions) should now be expressed in terms of their *real counterparts*, namely,

$$m_x^*(\mathbf{h}) = \frac{1}{V} \int_U X(\mathbf{s}+\mathbf{h}) \, d\mathbf{s} \tag{1}$$

$$c_x^*(\mathbf{h}, \mathbf{h}') = \frac{1}{V} \int_U [X(\mathbf{s}+\mathbf{h}) - m_x^*(\mathbf{h})] \times [X(\mathbf{s}+\mathbf{h}') - m_x^*(\mathbf{h}')] \, d\mathbf{s} \tag{2}$$

$$\gamma_x^*(\mathbf{h}, \mathbf{h}') = \frac{1}{2V} \int_U [X(\mathbf{s}+\mathbf{h}) - X(\mathbf{s}+\mathbf{h}')]^2 \, d\mathbf{s} \tag{3}$$

where U is an appropriate domain of integration in R^n and V is the volume of the domain. The V is chosen so that its transverse dimension L is significantly larger than the correlation radius r_c (defined in Chapter 2). For example, in the three-dimensional domain, $L \approx \sqrt[3]{V} \gg r_c$.

From a heuristic viewpoint the important element here is that while the theoretical statistical moments are purely mathematical notions, their real counterparts (1) through (3) can be evaluated in practice (we will see how in Section 5 below) and, therefore, they are testable model parameters. According to the philosophical theses of the stochastic data processing program (Section 4 of Chapter 1), this implies that the parameters (1) through (3) represent a physical reality and that they have an objective meaning.

The *auxiliary hypotheses* of the heuristic part of the stochastic data processing program include SRF homogeneity, isotropy, and intrinsity of some order ν (see also Section 6 of Chapter 2). These hypotheses, which play a very significant role in the SRF-based representation of the natural process, are not theoretically testable on the basis of a single SRF realization; however, they now obtain real counterparts that can be operationally reconstructed and tested for the domain of interest U, in terms of the expressions (1) through (3). As a consequence, all the predictions and model outcomes derived in terms of these testable parameters and hypotheses (e.g., the estimates of a natural process at unknown locations in space and the associated estimation variance) can also be operationally determined and, therefore, they possess an objective meaning. Furthermore, as we shall see in the following section, the study of the experimental statistical moments reveals a significant amount of valuable information in the physical sense.

Note that in the specific situation where the SRF is homogeneous, H3, essentially implies that the corresponding SRF is *ergodic*; then the Eqs. (1) and (2) become (see also Section 12 of Chapter 2)

$$m_x^* = \frac{1}{V}\int_U X(\mathbf{s})\, d\mathbf{s} \tag{4}$$

$$c_x^*(\mathbf{h}-\mathbf{h}') = \frac{1}{V}\int_U [X(\mathbf{s}+\mathbf{h}) - m_x^*][X(\mathbf{s}+\mathbf{h}') - m_x^*]\, d\mathbf{s} \tag{5}$$

If the process is an ISRF-0 (Chapters 2 and 3), Eq. (3) becomes

$$\gamma_x^*(\mathbf{h}) = \frac{1}{2V(\mathbf{h})}\int_{U(\mathbf{h})} [X(\mathbf{s}+\mathbf{h}) - X(\mathbf{s})]^2\, d\mathbf{s} \tag{6}$$

where the domain of integrations now is $U(\mathbf{h}) = U \cap U_{-\mathbf{h}}$($U_{-\mathbf{h}}$ is the translation of U by $\mathbf{h}$), and $V(\mathbf{h})$ is the n-dimensional volume of $U(\mathbf{h})$.

In the case of spatiotemporal natural processes, the following real counterparts may be defined: If $X(\mathbf{s}, t)$ is a space-homogeneous/time-stationary RF, one can write

$$c^*(\mathbf{h}, \tau) = \frac{1}{2VT}\int_U \int_{-T}^{T} [X(\mathbf{s}, t) - m_x^*] \times [X(\mathbf{s}+\mathbf{h}, t+\tau) - m_x^*]\, d\mathbf{s}\, dt \tag{7}$$

where

$$m_x^* = \frac{1}{2VT}\int_U \int_{-T}^{T} X(\mathbf{s}, t)\, d\mathbf{s}\, dt \tag{8}$$

In Eq. (8) it is assumed that the transverse dimension L of V and the time period $2T$ are significantly larger than the correlation radius r_c and the correlation time τ_c, respectively; viz., $L \gg r_c$ and $2T \gg \tau_c$. If the random field $X(\mathbf{s}, t)$ is only space homogeneous, it is valid that

$$c^*(\mathbf{h}, t, t') = \frac{1}{V}\int_U [X(\mathbf{s}, t) - m_x^*(t)] \times [X(\mathbf{s}+\mathbf{h}, t') - m_x^*(t')]\, d\mathbf{s} \tag{9}$$

where

$$m_x^*(t) = \frac{1}{V} \int_U X(\mathbf{s}, t)\, d\mathbf{s} \tag{10}$$

with $L \gg r_c$. Finally, if $X(\mathbf{s}, t)$ is only time stationary,

$$c^*(\mathbf{s}, \mathbf{s}', \tau) = \frac{1}{2T} \int_{-T}^{T} [X(\mathbf{s}, t) - m_x^*(\mathbf{s})] \times [X(\mathbf{s}', t+\tau) - m_x^*(\mathbf{s}')]\, dt \tag{11}$$

where

$$m_x^*(\mathbf{s}) = \frac{1}{2T} \int_{-T}^{T} X(\mathbf{s}, t)\, dt \tag{12}$$

with $2T \gg \tau_c$.

Besides spatial variability, another important source of spatial uncertainty is attributed to *measurement errors.* Some factors contributing to measurement errors are random, while others are related to sample disturbance, simulation in the laboratory of *in-situ* conditions, etc. (Davis and Poulos, 1967; Milovic, 1970; Bishop *et al.*, 1973). The laboratory procedures for determining the relative density of sands offer a good example of serious errors and variability between laboratories (Tavenas *et al.*, 1972). Also, significant differences may exist between measurements taken in the laboratory and *in-situ* (Yucement *et al.*, 1973). When measurement errors are present, one does not observe the actual SRF $X(\mathbf{s}_i)$ at each location in space, but instead the measured SRF $Y(\mathbf{s}_i)$, which includes errors $V(\mathbf{s}_i)$. Usually, linear models taking into account such measurement errors are of the form

$$Y(\mathbf{s}_i) = aX(\mathbf{s}_i) + V(\mathbf{s}_i) \tag{13}$$

where a is a deterministic coefficient derived from testing experience and $V(\mathbf{s}_i)$ is a white-noise SRF uncorrelated with $X(\mathbf{s}_i)$ and $Y(\mathbf{s}_i)$; the $V(\mathbf{s}_i)$ is assumed to have zero mean and variance $\sigma_v^2(\mathbf{s}_i)$.

3. Duality Relations between the Natural Process and the Spatial Random Field Model—Examples from the Geosciences

The next step in developing our model is to examine how one can derive conclusions about the behavior of the natural process, starting from the study of the mathematical properties of the corresponding SRF model. In other words one seeks to establish a set of auxiliary hypotheses, which we will call *duality relations*, between the *in-situ* behavior of the natural process and the mathematical features of the SRF model. Hence, the fundamental role is played by the parameters associated with the spatial correlation functions (in the ordinary or the generalized sense).

The duality relations below stem from the theoretical results of Chapter 2 and are illustrated in Fig. 7.3 in the case of an ordinary semivariogram function $\gamma_x(\mathbf{h})$, but they are also valid for the corresponding covariance function.

Duality relation DR1: The variation of the semivariogram or the covariance function along *several directions in space* displays information about non-isotropic properties of the natural process.

That is, if the covariance varies significantly along different directions in space the SRF in *anisotropic*, and the same holds for the spatial variability of the natural process. For example, Fig. 7.4 shows the covariances of the soil settlement pattern of Fig. 7.1, computed along the four principal directions in space (the average covariance is also shown). Also, the directional semivariograms of marine clay strata thickness beneath a sand-fill embankment are plotted in Fig. 7.5. The implication of these plots is that

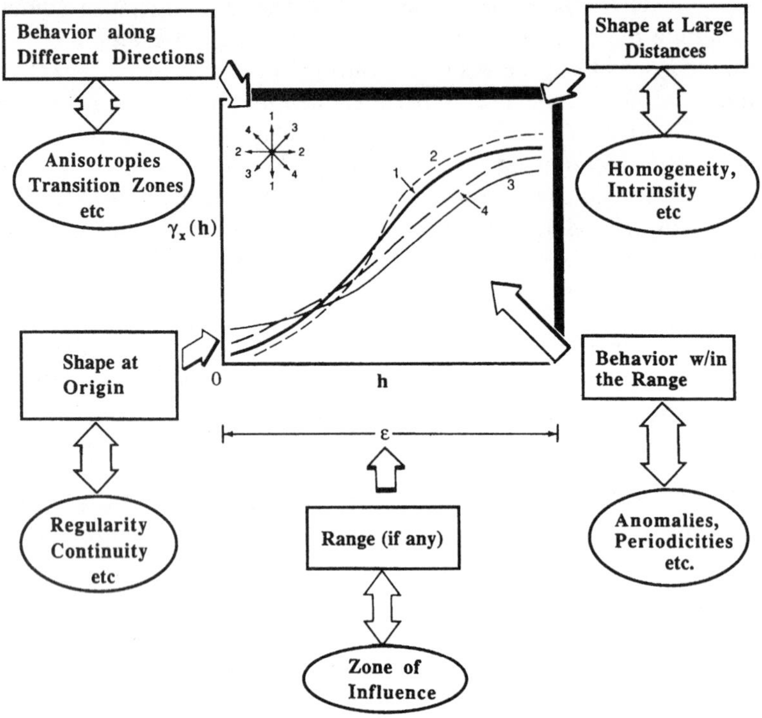

Figure 7.3 An illustration of the duality relations for the case of the semivariogram function

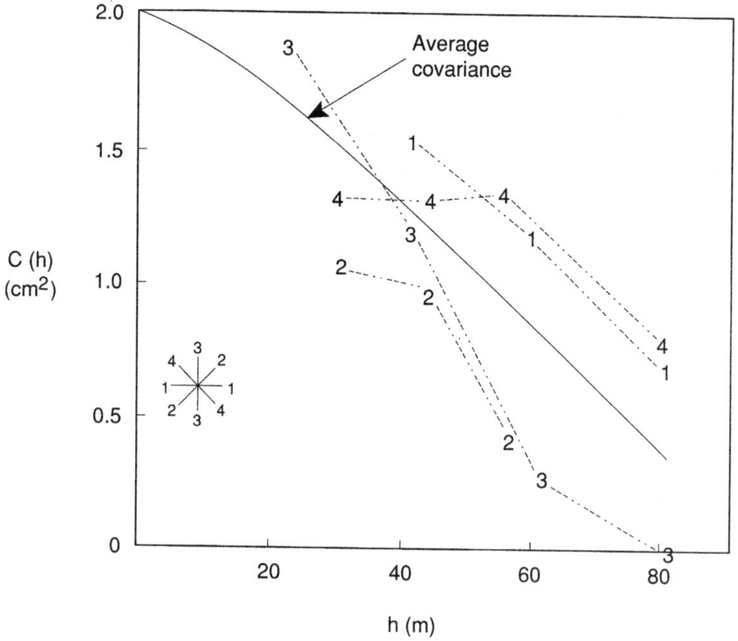

Figure 7.4 Spatial covariances of soil settlements computed along the four principal directions. The average spatial covariance is also shown

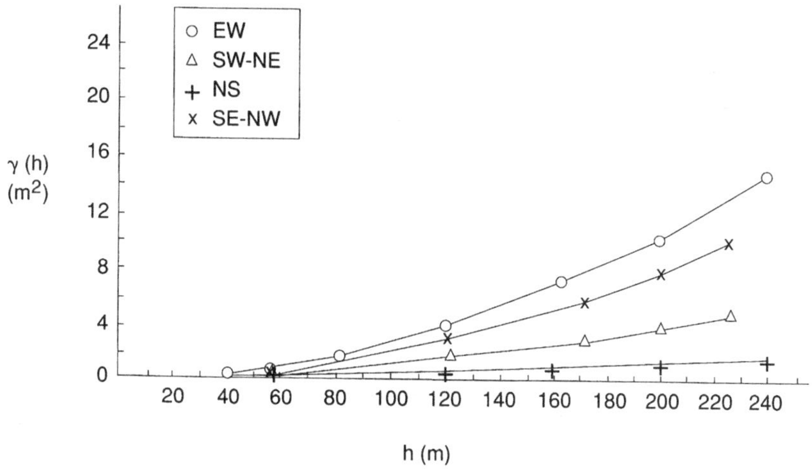

Figure 7.5 Semivariogram of clay thicknesses along the four principal directions

both the settlement pattern and the map of clay thickness are anisotropic. Anisotropy is due, in general, to the geological structure of the soil. For example, the undrained strength anisotropy of clays consists of two major components: inherent and stress-induced anisotropies, related to the sample orientation *in-situ* (Ladd *et al.*, 1977).

Duality relation DR2: The *range* ε of the semivariogram or the covariance function determines *the zone of influence* of a local value of the natural process.

The range is directly related to the rate of decrease of the covariance down from its initial value (variance) and confirms what earth scientists know by experience: The samples of a natural process that are close to each other are subject to stronger interactions than those that are far from each other. Naturally, samples whose distance from each other exceeds the range should be considered independent (and hence they will be omitted, for example, in estimation problems; see Chapter 9 later). In the case of an anisotropic process, the range may take different values along different directions; this will offer an additional parameter for determining the degree of the process lack of isotropy. The notion of range makes sense only in the case of homogeneous spatial distributions, such as the covariance of the standard penetration (SPT) blow counts data shown in Fig. 7.6. In the case of nonhomogeneous spatial variability the range of the semivariogram is not defined (see, for example, Fig. 7.5).

Duality relation DR3: The behavior of the covariance or the semivariogram at *large distances* determines the degree of homogeneity of the process.

In this way, an *asymptotic* behavior such as that of Fig. 7.6 predicts the homogeneous variation of the SPT blow counts. A *linear* covariance at large distance and a slow convergence toward zero, or the fast growth (usually linear) of the semivariogram at large distances, imply nonhomogeneous spatial processes, as in the soil settlement covariance and the clay thickness semivariogram in Fig. 7.4 and 7.5, respectively. To obtain some insight into how the presence of a trend may be evidenced by the behavior of the semivariogram at large lags $\mathbf{h}$, consider the case of a linear trend $m_x(\mathbf{s}) = \mathbf{a} \cdot \mathbf{s}$. The sample semivariogram $\hat{\gamma}(\mathbf{h})$ will be calculated by Eq. (23), Section 4 below and will be related to the actual semivariogram $\gamma(\mathbf{h})$ by Eq. (42) of Section 4. Note that it is valid that

$$\begin{aligned}\tfrac{1}{2}E[X(\mathbf{s}+\mathbf{h})-X(\mathbf{s})]^2 &= \gamma_x(\mathbf{h})+\tfrac{1}{2}[m_x(\mathbf{s}+\mathbf{h})-m_x(\mathbf{s})]^2\\ &= \gamma_x(\mathbf{h})+\tfrac{1}{2}[\mathbf{a}\cdot(\mathbf{s}+\mathbf{h})-\mathbf{a}\cdot\mathbf{s}]^2\\ &= \gamma_x(\mathbf{h})+\tfrac{1}{2}\mathbf{a}^2\mathbf{h}^2\end{aligned}$$

with

$$\gamma_x(\mathbf{h}) = \tfrac{1}{2}\,\mathrm{Var}[X(\mathbf{s}+\mathbf{h})-X(\mathbf{s})]$$

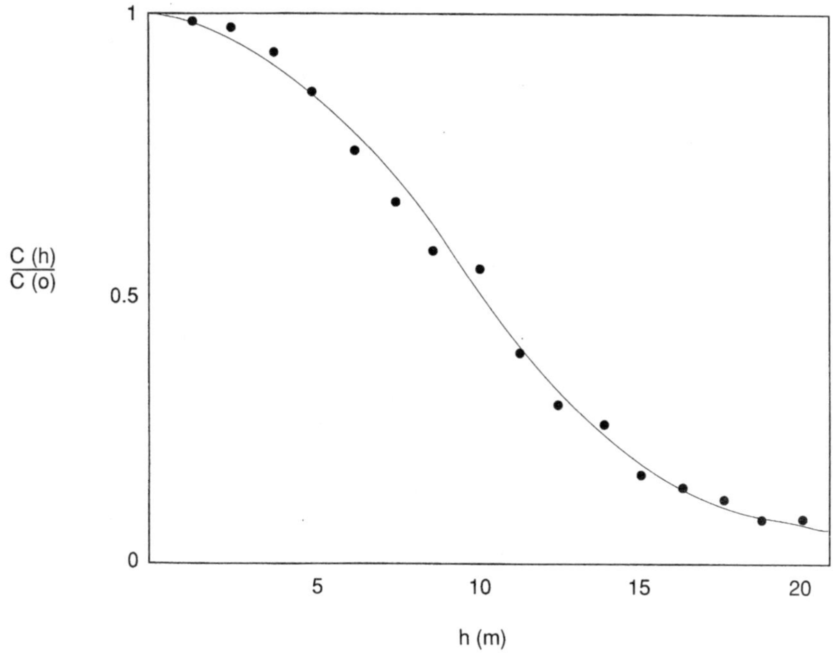

Figure 7.6 Covariance of SPT blow counts

At large distances, $\frac{1}{2}\mathbf{a}^2\mathbf{h}^2$ is the dominant (parabolic) term that causes the observed fast growth of the semivariogram.

Duality relation DR4: The behavior of the semivariogram or the covariance function at the *origin* determines the degree of regularity in the spatial variation of the process.

DR4 is a consequence of the theoretical result according to which the behavior of the semivariogram or the covariance at origin determines the m.s. continuity and differentiability of the SRF (Chapter 2). So, a *parabolic* (at origin) covariance represents a continuous and differentiable SRF, and a process modeled by means of such an SRF will have a very regular spatial variation. This is the case of the covariance of Fig. 7.6 and the semivariograms of Fig. 7.5 (the corresponding map of clay thickness verifying the regular spatial variability is shown in Fig. 7.7). A *linear* behavior characterizes a continuous but not differentiable SRF and, consequently, a nonsmooth process; in this case, the degree of irregularity will be directly dependent on the slope of the covariance or the semivariogram at the origin.

The semivariogram of the water content data plotted in Fig. 7.8 corresponds to a process with such an irregular spatial variability. A possible

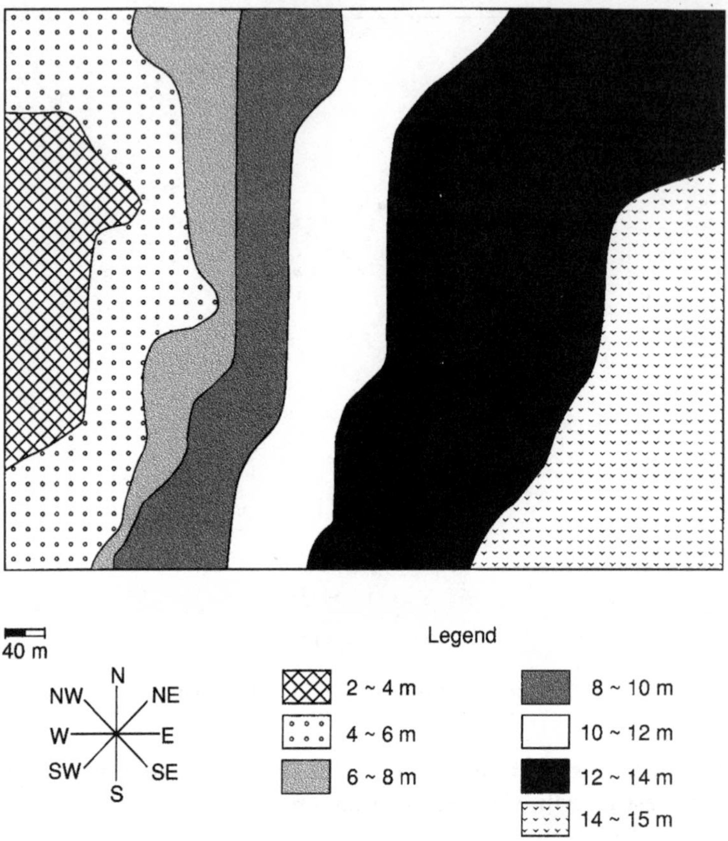

Figure 7.7 Spatial distribution of the thickness of marine clay stratum beneath a sand-fill embankment

discontinuity of the covariance at origin emphasizes the irregular features of the process. As we saw in previous chapters, this discontinuity is called the nugget effect in the geostatistical jargon (a discussion of the physical interpretation of the nugget-effect phenomenon and the RF models used to describe it mathematically will be given shortly); the process is then expected to be not continuously varying in space and highly irregular (see the covariance of cone penetration resistance in Fig. 7.9). The nugget effect may be due to microstructures, random variances, or measurement errors. An interesting case is that with zero covariance for distances greater than zero. This is a *white noise* or *pure nugget effect* and the variation of the process is purely random; see both the covariance and semivariogram functions of Fig. 7.10, corresponding to the time history of shear stresses during an earthquake (Fig. 7.11).

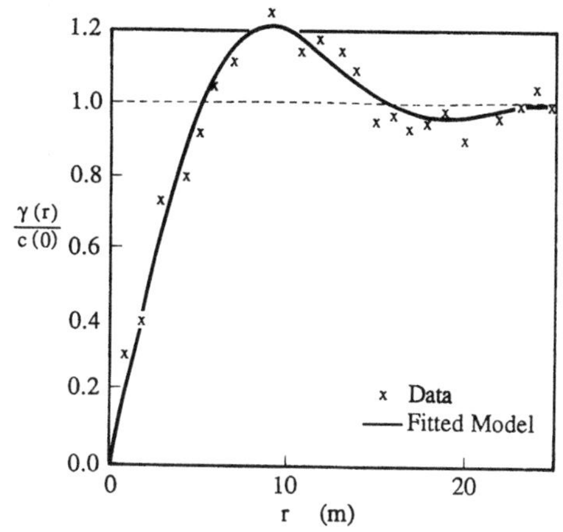

Figure 7.8 The semivariogram of water content data

At this point, it may be instructive to consider the following simple example: Let $\gamma_X(s)$ be the semivariogram of the unidimensional process $X(s)$ plotted in Fig. 7.12. The length of the line $P_1P_2\ldots P_{n-1}P_n$ can be considered a measure of the "smoothness" of $X(s)$ (that is, the smaller the length, the smoother the process), and we can write

$$L=\sum_{i=1}^{n-1}(P_iP_{i+1})^2=\sum_{i=1}^{n-1}[X((i+1)\,\Delta s)-X(i\,\Delta s)]^2+(n-1)\,\Delta s^2 \qquad (1)$$

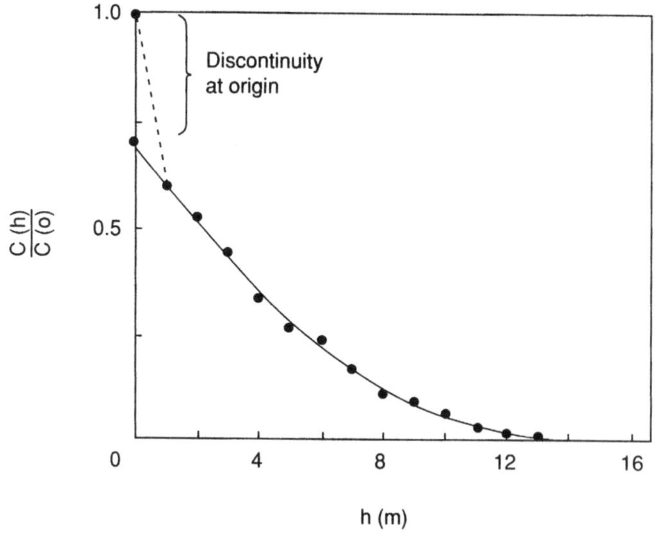

Figure 7.9 Covariance of cone penetration resistance with a discontinuity at origin

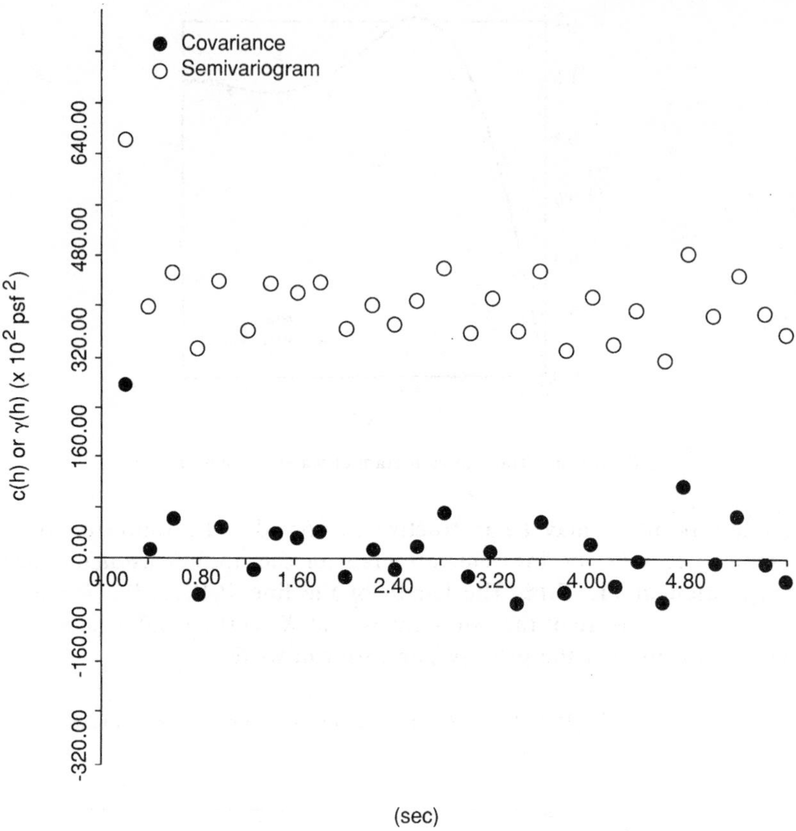

Figure 7.10 Semivariogram and covariance functions for the time history of shear stresses during an earthquake (shown in Fig. 7.11)

where Δs is the distance between the P_i's and

$$E[L] = 2\gamma_x(\Delta s) + (n-1)\,\Delta s^2 \tag{2}$$

Clearly, the smaller the value of the semivariogram at small distances Δs (e.g., a parabolic shape implies a smaller value than a linear one), the smaller is the mean value of L and, thus, the smoother is the $X(s)$.

Finally, notice that the behavior of the covariance or the semivariogram at very small distances affects substantially the smoothness of the estimated process (see Chapter 9). The estimation error may be significantly smaller for a covariance or a semivariogram that is parabolic at origin than for a linear one. Table 7.1 summarizes the duality relations in the space-time context.

Most of the preceding duality relations are associated with the mean square stochastic properties of the SRF $X(\mathbf{s})$. The following relation

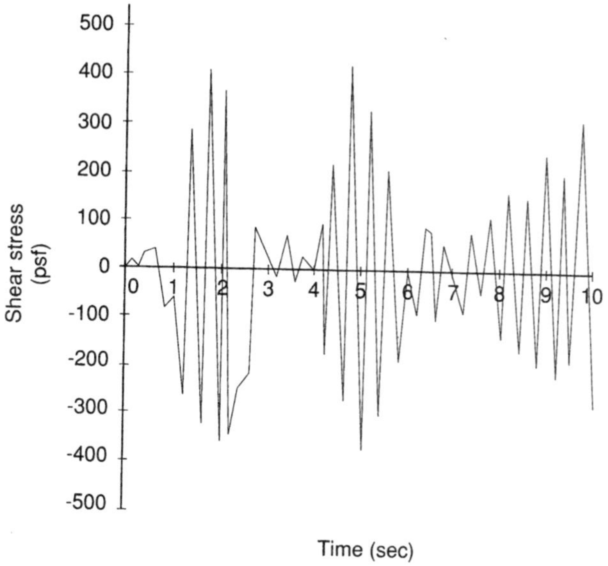

Figure 7.11 Time history of shear stresses during an earthquake

provides valuable information regarding the spatial behavior of a natural process by means of the sample function stochastic continuity and differentiability properties discussed in Chapter 2.

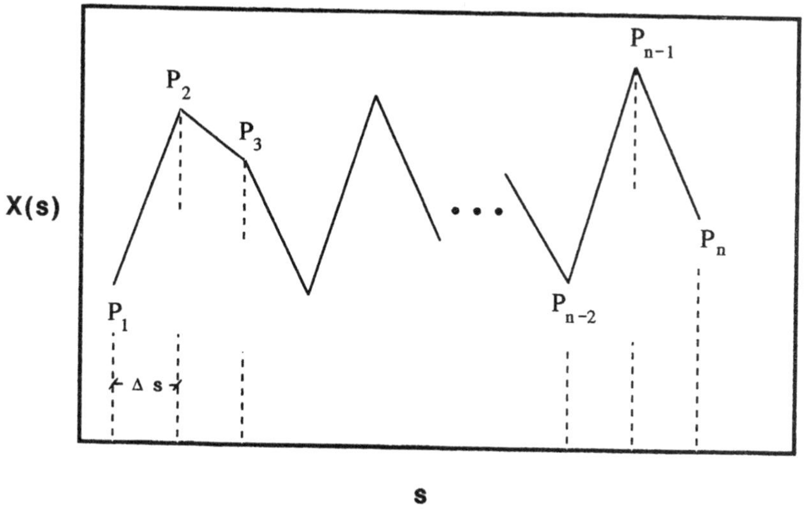

Figure 7.12 The unidimensional process $X(s)$

Table 7.1 Duality Relations

Covariance Model		In situ Behavior
Shape at the space-time origin	→	Regularity, Continuity in the spatiotemporal variation
Shape at large space-time distances	→	Spatial homogeneity, Time stationarity
Ranges in space-time (if any)	→	Zones of influence in space-time
Behavior within the range	→	Periodicities, anomalies, etc.
Behavior along different directions in space	→	Anisotropies in space, transition zones, etc.

Duality relation DR5: If it can be shown that

$$\gamma_x(\mathbf{h}) \leq \frac{c|\mathbf{h}|^{2n}}{\left|\log|\mathbf{h}|\right|^{1+\beta}}$$

where c is a positive constant and $\beta > 2$, the SRF $X(\mathbf{s})$, $\mathbf{s} \in R^n$ is a.s. continuous and, consequently, the underlying natural process is spatially continuous. If, in addition, the above inequality holds true in terms of the semivariograms of the m.s. partial derivatives of $X(\mathbf{s})$, the sample functions of these derivatives are a.s. continuous and we have a very smooth process.

Remark 1: From a physical point of view, a reasonable interpretation of sample function continuity is often associated with the notion of *scale.* For example, it is possible that while the sample functions of the (theoretical) SRF are changing discontinuously by discrete jumps, the real process is observable in a sufficiently large scale where these jumps are infinitesimal and the sample function can be approximated by a continuous surface.

For more complex spatial distributions, similar duality relations and comments are valid in terms of the generalized spatial covariances of order ν (GSC-ν; Chapter 3).

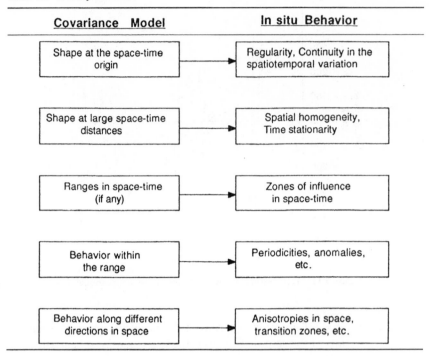

(i) A natural process characterized by a linear GSC, $k_x(\mathbf{h}) = -c_0|\mathbf{h}|$, is continuous but not very regular; the larger the slope c_0, the more irregular is the spatial variation. The corresponding SRF is m.s. continuous but not differentiable.

(ii) A GSC of the form $k_x(\mathbf{h}) = c_1|\mathbf{h}|^3$ characterizes an SRF that is m.s. differentiable once. The natural process is locally more regular than (i), but fluctuates significantly at the macroscopic scale.

(iii) The GSC $k_x(\mathbf{h}) = -c_3|\mathbf{h}|^5$ corresponds to a twice differentiable SRF and represents a natural process that, while very smooth at short distances, exhibits wide fluctuations at large distances.

Clearly, the higher the order of spatial variability ν, the more non-homogeneous is the spatial process. When the GSC-ν is a combination of the above powers [e.g., $k_x(\mathbf{h}) = -c_0|\mathbf{h}| + c_1|\mathbf{h}|^3 - c_3|\mathbf{h}|^5$], the spatial variability is characterized primarily by the component with the lowest power (here, $|\mathbf{h}|$).

All these features of the spatial correlation structure (ordinary covariance and semivariograms, generalized covariances, etc.) have several other important implications in practical applications. In general, knowledge of sources of spatial variability within a site allows improvement in sampling designs. This is why an expert in a field can sample more efficiently than other persons. Practically, the sampling network accuracy depends on the specific spatial variability characteristics of the natural process of interest. In many cases, the more irregular and discontinuous the spatial variability, the lower the estimation accuracy and the larger the required number of observations. Finally, it must be remarked that most of the results on purely spatial RF discussed above can be easily extended to spatiotemporal RF.

4. Certain Practical Aspects of Spatial and Temporal Variability Characterization

4.1 Application of the Permissibility Criteria

After the theoretical background provided by the various criteria of permissibility (COP) developed in Chapters 2, 3, and 6, we now pass to some practical aspects concerning their effective application. The following procedure is recommended:

Step 1: (This is a preliminary step.) Check whether the necessary conditions are satisfied. Particularly,

(i) In the case of *homogeneous* spatial processes the following must hold.

$$\lim_{|\mathbf{h}|\to\infty} \frac{c(\mathbf{h})}{|\mathbf{h}|^{(1-n)/2}} = 0 \tag{1}$$

in R^n. Also,

$$c(h) \leq c(0) \text{ in } R^1 \tag{2}$$

and if the SRF is isotropic, the covariance must satisfy

$$c(r) \geq -0.403c(0) \quad \text{in} \quad R^2 \tag{3}$$

$$c(r) \geq -0.218c(0) \quad \text{in} \quad R^3 \tag{4}$$

where $r = |\mathbf{h}|$. In terms of semivariograms it must hold true that

$$\gamma(h) \leq 2c(0) \tag{5}$$

$$|\gamma(h) - \gamma(h+h')| \leq \sqrt{2\gamma(h')} \tag{6}$$

$$\gamma(2^m h) \leq 4^m \gamma(h) \tag{7}$$

in R^1, where m is a nonnegative integer; also

$$\gamma(r) \leq 1.403c(0) \quad \text{in} \quad R^2; \tag{8}$$

$$\gamma(r) \leq 1.218c(0) \quad \text{in} \quad R^3 \tag{9}$$

(ii) For *nonhomogeneous* SRF it must hold true that

$$\lim_{|\mathbf{h}| \to \infty} \frac{\gamma(\mathbf{h})}{|\mathbf{h}|^2} = 0 \quad \text{in} \quad R^n \tag{10}$$

(ISRF-0), and

$$\lim_{|\mathbf{h}| \to \infty} \frac{k(\mathbf{h})}{|\mathbf{h}|^{2\nu+2}} = 0 \quad \text{in} \quad R^n \tag{11}$$

(ISRF-$\nu \geq 0$).

Step 2: Here we can distinguish among five possible cases.

(i) If the underlying process is *isotropic*, check whether the candidate covariance or semivariogram models satisfy the following *sufficient* conditions: At the origin it must hold true that

$$c'(r)|_{r=0} = \left.\frac{dc(r)}{dr}\right|_{r=0} < 0 \tag{12}$$

where $r = |\mathbf{h}| \in R^n$. At infinity Eq. (1) must be valid, and

$$c''(r) = \frac{d^2c(r)}{dr^2} \geq 0 \quad \text{in} \quad R^1 \tag{13}$$

$$\int_r^\infty \frac{u}{\sqrt{u^2 - r^2}}\, dc''(r) \geq 0 \quad \text{in} \quad R^2 \tag{14}$$

$$c''(r) - rc'''(r) \geq 0 \quad \text{in} \quad R^3 \tag{15}$$

Similar conditions are obtained in terms of the semivariogram $\gamma(r)$, by simply replacing $c(r)$ by $c(0)-\gamma(r)$. For example, the equivalent of Eq. (12) by means of $\gamma(r)$ is

$$-\gamma'(r)\big|_{r=0} = -\frac{d\gamma(r)}{dr}\bigg|_{r=0} < 0$$

(ii) If the candidate spatial correlation model is of a convenient, closed form (i.e., one can find its FT in R^1 by using the existing Fourier transform tables), calculate the particular spectral function in R^1 and check its permissibility by using the appropriate criterion of permissibility. If the answer is affirmative, calculate the spectral function in the field of interest R^2 or R^3 by applying the space transformation formulas of Chapter 6; then apply the appropriate criterion of permissibility again.

(iii) If the candidate semivariogram model is of a transitive, polygonal form (see Fig. 7.13) use the formula

$$\Gamma_1(\omega) = -\frac{1}{\pi\omega^2}\sum_{i=0}^{m}[\bar{\omega}(r_i)-\bar{\omega}(r_{i-1})]\cos(\omega r_i) \tag{16}$$

where $\bar{\omega}(r_i)$ is the slope at lag r_i, to find the unidimensional spectral semivariogram; then apply space transformations to find the spectral semivariogram in the field of interest R^2 or R^3; finally, apply the

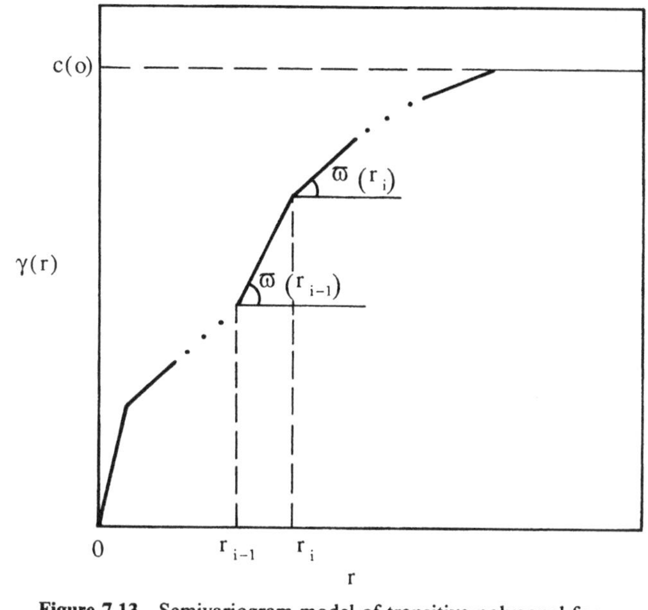

Figure 7.13 Semivariogram model of transitive polygonal form

corresponding criterion of permissibility. Notice that the polygonal models can be fit arbitrarily closely to the data, and this makes them popular among practitioners.

(iv) If the candidate model is of some complex, arbitrary form, we might have to approximate the spectral functions by using formulas of the form (Christakos, 1984b)

$$\Gamma_1(\omega)=\frac{1}{\omega_c}\sum_{k=0}^{m}\gamma\left(\frac{k\pi}{\omega_c}\right)\cos\left(\frac{k\pi\omega}{\omega_c}\right) \quad \text{in} \quad R^1 \tag{17}$$

$$\Gamma_2(\omega)=\frac{1}{\pi\omega_c^2}\sum_{k=1}^{m}\gamma\left(\frac{i_k}{\omega_c}\right)\frac{J_0\left(\frac{i_k\omega}{\omega_c}\right)}{[J_1^2(i_k)]} \quad \text{in} \quad R^2 \tag{18}$$

$$\Gamma_3(\omega)=\frac{1}{2\omega_c^2\omega}\sum_{k=1}^{m}k\gamma\left(\frac{k\pi}{\omega_c}\right)\sin\left(\frac{k\pi\omega}{\omega_c}\right) \quad \text{in} \quad R^3 \tag{19}$$

where J_u is the Bessel function of uth order ($u=0$ and 1) and i_k are the zeros of J_0. Similar formulas can be derived in terms of the ordinary spatial covariances and the generalized spatial covariances.

(v) In the case of generalized spatial covariances of order ν (GSC-ν) that are of the isotropic polynomial form (Chapter 3)

$$k_x(r)=a_0\,\delta(r)+\sum_{\rho=0}^{\nu}(-1)^{\rho+1}c_\rho r^{2\rho+1} \tag{20}$$

the application of the criteria of permissibility (in R^n) implies that the coefficients a_0 and c_ρ should satisfy the constraints summarized in Table 3.1 of Chapter 3.

(vi) If vector SRF is involved, one should apply the COP-2, -6, and -8 (see Sections 7.5 and 11.5 of Chapter 2 and Section 3.5 of Chapter 3).

(vii) In the case of spatiotemporal RF, one should consult COP-9 of Section 3.3, Chapter 5, and COP-10 of Section 5.2, Chapter 5.

4.2 Evaluation of Spatial Variability Characteristics

In practice, and to make inferences about a natural process that is unique but partially unknown in space, it is necessary to evaluate certain spatial variability characteristics like means, ordinary covariances, semivariograms and, in more complex situations, generalized covariances and semi-variograms, in terms of the fragmentary data available. In the context of stochastic data analysis and processing, this evaluation can be made by means of

(a) a heuristic—a set of empirical rules and tactics for choosing testable auxiliary hypotheses and model parameters; and

(b) physical models that describe the evolution of a natural process over space (algebraic operators, differential and difference equations, etc.).

4.2.1 Ordinary Spatial Covariance and Semivariogram

Although the ordinary covariance and semivariogram are similar in form and use, the latter is mean free and, therefore, its empirical calculation is subject to smaller errors. Moreover, the semivariogram may exist in cases where there is no finite *a priori* variance and the covariance does not exist (see, e.g., the semivariogram of the clay thickness values, Fig. 7.5).

Experimental Calculations

If data are available at regular intervals along transects, the sample covariance and semivariogram can be calculated by

$$\hat{c}_x(\mathbf{s}_i, \mathbf{s}_i + \mathbf{h}) = \frac{1}{N} \sum_{i=1}^{N} [\chi(\mathbf{s}_i) - \hat{m}_x(\mathbf{O})][\chi(\mathbf{s}_i + \mathbf{h}) - \hat{m}_x(\mathbf{h})] \tag{21}$$

where the sample mean value is

$$\hat{m}_x(\mathbf{h}) = \frac{1}{N} \sum_{i=1}^{N} \chi(\mathbf{s}_i + \mathbf{h}) \tag{22}$$

and

$$\hat{\gamma}_x(\mathbf{s}_i, \mathbf{s}_i + \mathbf{h}) = \frac{1}{2N(\mathbf{h})} \sum_{i=1}^{N(\mathbf{h})} [\chi(\mathbf{s}_i + \mathbf{h}) - \chi(\mathbf{s}_i)]^2 \tag{23}$$

respectively. There are N data available, and $N(\mathbf{h})$ is the number of data pairs sampled at an interval $\mathbf{h}$.

A particular situation arises when we have data at regular intervals on a two-dimensional grid; in such a case we can write (assuming constant mean)

$$\hat{c}_x(h_1, h_2) = \frac{1}{KL} \sum_{i=1}^{K} \sum_{j=1}^{L} [\chi(s_i, s_j) - \hat{m}_x][\chi(s_i + h_1, s_j + h_2) - \hat{m}_x] \tag{24}$$

and

$$\hat{\gamma}_x(h_1, h_2) = \frac{1}{2N(h_1)N(h_2)} \sum_{i=1}^{N(h_1)} \times \sum_{j=1}^{N(h_2)} [\chi(s_i + h_1, s_j + h_2) - \chi(s_i, s_j)]^2 \tag{25}$$

where $\mathbf{s} = (s_i, s_j)$, $\mathbf{h} = (h_1, h_2)$, K is the number of rows and L is the number of columns of the grid; $N(h_1)$ and $N(h_2)$ are the number of data pairs separated by h_1 and h_2, respectively.

When the data are irregularly spaced, it may be convenient to group them by distances and directions. The method is illustrated in Fig. 7.14: The

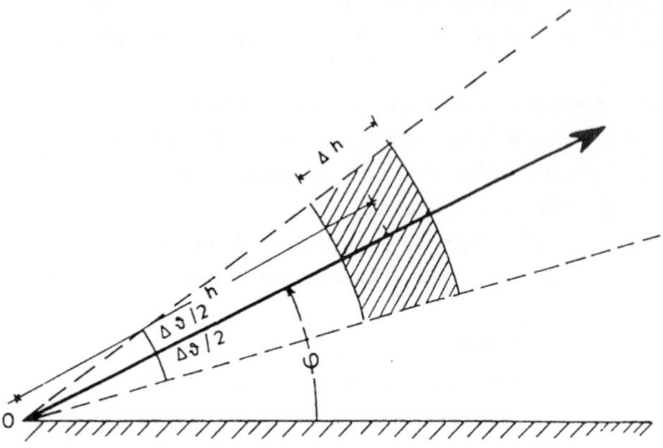

Figure 7.14 Experimental calculation of the semivariogram or covariance functions of irregularly spaced data

direction chosen forms an angle φ with the horizontal, and for each $\mathbf{h}$ all data within the element defined by the range $\Delta\mathbf{h}$ and the angle $\Delta\vartheta$ are used in the calculation of the covariance or the semivariogram. In relation to this method, common problems encountered in getting "sensible" experimental covariances or semivariograms include limited data, unsuitable choice of distance classes or working scale, and outliers (e.g., Gajem, 1981; Armstrong, 1984; Webster, 1985; Cressie, 1985). To overcome such problems it may be necessary to quantify geological information, rearrange data, or apply data cleaning and robust statistical techniques. Useful robust semivariogram estimators have been suggested by Cressie and Hawkins (1980). In practice, one fits to the data simply formed mathematical functions. Naturally, the final choice is affected by what is known about the underlying natural process, as well as by the graphic display of sampled values and the experience of the analyst. The fitting of the model chosen to the data can be achieved in a number of ways. Traditional methods include least squares, maximum likelihood, and weighted area (e.g., David, 1977; Marshal and Mardia, 1985; Stein, 1987). On the other hand, practitioners may find very convenient the "by eye" technique, which allows them to incorporate intuition and experience into the fit. However, not every function that appears to fit the data will serve as a covariance or a semivariogram model. It is necessary to satisfy the corresponding criteria of permissibility.

In conclusion, the heuristic here is that Eqs. (21) through (25) and the setting of Fig. 7.14 provide an approximate calculation of the spatial integrals of Eqs. (1) through (3), Section 2 above, on the basis of the fragmentary data available. The important element of such an approximation is that the mean value, covariance, and semivariogram are now measurable and, hence, testable (after the event) model parameters. In addition, the auxiliary hypotheses and duality relations established for the domain of interest V on the basis of these parameters (such as homogeneity, incremental homogeneity of order ν, behavior at origin, and range of influence) are, also testable physical realities. For example, if the calculations (21) through (23) yield $\hat{m}_x = \text{constant}$, and $\hat{c}_x(\mathbf{s}_i, \mathbf{s}_i + \mathbf{h}) = \hat{c}_x(\mathbf{h})$, one may conclude that the natural process is homogeneous within V. Several other practical aspects are covered in the geostatistical literature [see, e.g., Sophocleous (1983), Hohn (1988), and Isaaks and Srivastava (1989)]. An excellent source on the subject of semivariogram estimation is Cressie (1991).

Examples

Let us now discuss a few examples.

Example 1: Consider the water content data of Fig. 7.8. To the experimental semivariogram calculated by means of Eq. (23), we fitted the model

$$\gamma(r) = 1 - \exp\left[-\frac{r}{6.3}\right]\cos(0.3r) \tag{26}$$

The unidimensional spectral semivariogram function associated to model (26) is

$$-\Gamma_1(\omega) = \frac{39.78 + \omega^2}{[39.69 + (\omega - 0.3)^2][39.69 + (\omega + 0.3)^2]} > 0 \tag{27}$$

for all $\omega > 0$. Hence model (26) is a permissible semivariogram, and it can be used to describe unidimensional correlations of the water content data.

Example 2: The polygonal model shown in Fig. 7.15 is not a permissible

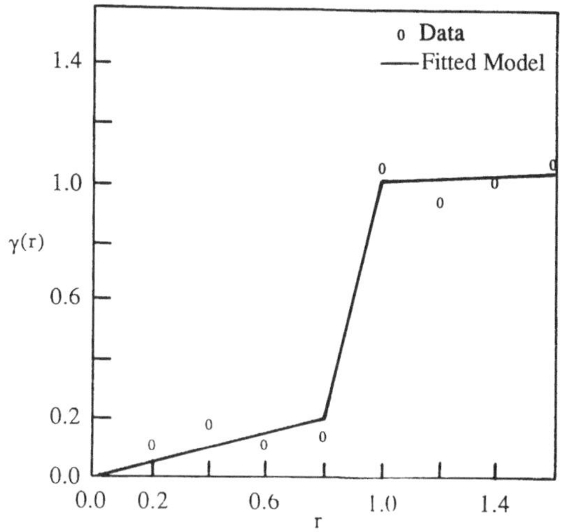

Figure 7.15 A nonpermissible semivariogram model of polygonal form

one, since the application of Eq. (16) above yields

$$-\Gamma_1(\omega) = \frac{\text{const.}}{\omega^2}[1 + 15\cos(4\omega) - 16\cos(5\omega)] \tag{28}$$

which takes negative values. The model of Fig. 7.15 is immediately rejected by application of the necessary condition (7) [for $m=1$, $h=0.5$ we get $\gamma(2\times 0.5)=\gamma(1)>4\gamma(0.5)$]. In connection with this example, the application of Eq. (16) can offer valuable hints concerning the permissibility of a variety of commonly used polygonal transitive correlation models. For example, a large initial slope contributes importantly to the permissibility of the candidate model (in Fig. 7.15 the initial slope is only 0.25). The same applies to a nonincreasing slope of the semivariogram [in Fig. 7.15 the slope increases from $r=0.8(-)$ to $0.8(+)$].

Example 3: Consider in R^3 the model

$$\gamma(r) = \frac{r^2}{1+r^2} \tag{29}$$

which is fitted to the experimentally calculated semivariogram shown in Fig. 7.16. The spectral function of the experimental semivariogram is calculated by means of Eq. (19) above, and is plotted in Fig. 7.17; the theoretical spectral function is also plotted, for comparison. Obviously, the spectral function is always nonnegative and hence the model (29) is permissible in R^3. (And, of course, also in R^1 and R^2.)

The Maximum-Entropy Formalism

The maximum entropy formalism can be used to assign any positive function, subject to the information we have about it (a detailed discussion of the maximum entropy formalism is given in Chapter 9). Here we will see how one can determine a spatial correlation function such as an ordinary covariance or an ordinary semivariogram by maximizing the entropy, subject to the information we have about the spatial variability.

Let $\gamma_x(\mathbf{s}_i, \mathbf{s}_j) = \gamma_x(r)$ ($r = |\mathbf{s}_i - \mathbf{s}_j|$) be an *isotropic* semivariogram function and let $C_x(\omega)$ be the corresponding spectral density function. Assume that the information we have regarding spatial variability is expressed by the constraints

$$\int_{\omega_1}^{\omega_2} C_x(\omega)[1 - e^{i\omega r_q}]\, d\omega = \gamma_x(r_q) \tag{30}$$

$q = 1, 2, \ldots, Q$, where ω_1 and ω_2 are suitable frequencies, given *a priori* or determined in some other way. These constraints express mathematically a rather common situation in practice; namely, when due to the spatial distribution of the available data the semivariogram can be calculated only

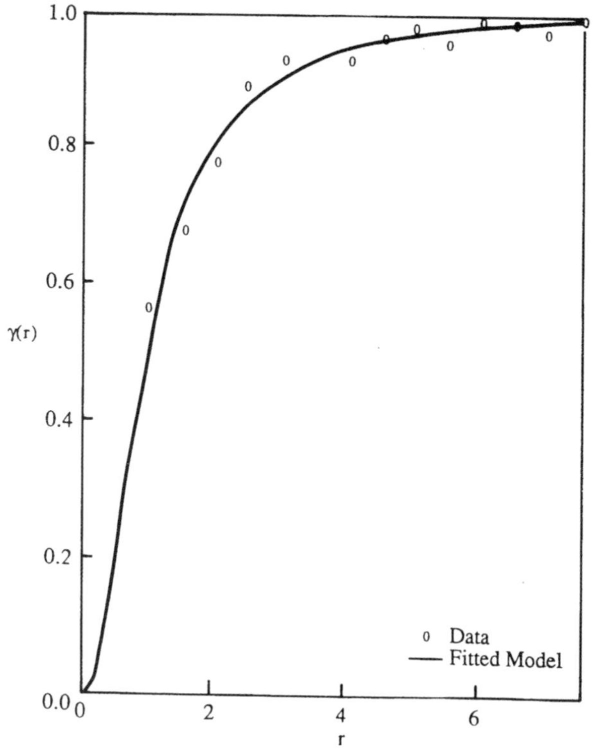

Figure 7.16 Calculated and fitted isotropic semivariogram models in R^3

for a particular number of intervals $r_q (q = 1, 2, \ldots, Q)$. An additional constraint is the normalization equation

$$\int_{\omega_1}^{\omega_2} C_x(\omega)\, d\omega = \gamma_x(\varepsilon) \tag{31}$$

where $C_x(\omega) \geq 0$ for all ω and ε is the range of influence of the semivariogram so that $\gamma_x(\varepsilon) = \sigma^2$; σ^2 is the variance.

Given the constraints (30) and (31), the problem of determining $\gamma_x(r)$ has been converted into the equivalent one of determining $C_x(\omega)$. In this case it is sufficient to maximize the so-called *Burg entropy function* (Burg, 1972)

$$\varepsilon(C_x) = \int_{\omega_1}^{\omega_2} \log[C_x(\omega)]\, d\omega \tag{32}$$

with respect to $C_x(\omega)$, subject to the constraints (30) and (31).

The general solution to this type of constrained maximization problem is well known (see, e.g., Ewing, 1969). By expressing the general solution

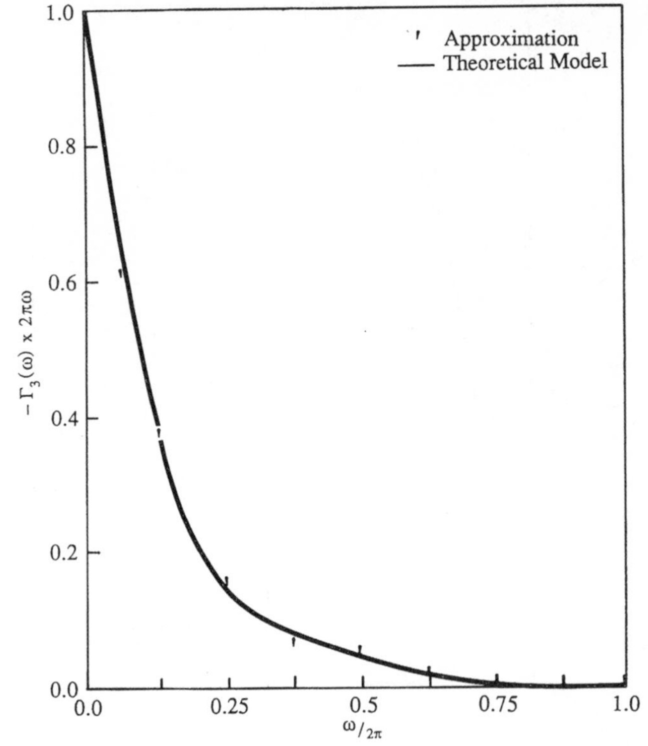

Figure 7.17 Calculated and theoretical spectral semivariogram functions for the semivariogram model of Fig. 7.16

in terms of the particular setting above, one finds that the spectral density must be of the form

$$C_x(\omega) = -\frac{1}{\mu_0 + \sum_{q=1}^{Q} \mu_q \, e^{i\omega r_q}} \tag{33}$$

where μ_q, $q = 1, 2, \ldots, Q$ are Lagrange multipliers. Then, the semivariogram function $\gamma_x(r)$ can be determined by applying inverse Fourier transform methods.

Example 4: In the unconstrained case where the only information available is the variance σ^2, the solution is a pure nugget-effect semivariogram, $\gamma_x(r) = \sigma^2 \, \delta(r)$. It is important to realize that the pure nugget-effect model is not necessarily the actual semivariogram. It is, however, the model that most honestly represents the given state of incomplete knowledge regarding spatial variability, without assuming anything else. Any other solution will necessarily take into consideration information not really available. Intuitively this implies that one needs more than just the variance to be able to derive sound conclusions about spatial correlation. Let us assume that in addition to the variance we obtain information about the frequencies $\omega_1 = -\omega_0$ and $\omega_2 = \omega_0$. Analysis based on Eqs. (30) through (33) will take into consideration the new information; the spectral density will now maximize (32) subject to the constraint (31), namely,

$$C_x(\omega) = \frac{\sigma^2}{2\omega_0} \quad (\text{if} \quad -\omega_0 \leq \omega \leq \omega_0), \qquad = 0 \quad \text{otherwise} \tag{34}$$

The corresponding semivariogram is

$$\gamma_x(r) = \sigma^2\left[1 - \frac{\sin(\omega_0 r)}{\omega_0 r}\right] \tag{35}$$

which is the so-called *hole-effect* model.

Model Cross-Validation

It is always useful to test the hypotheses made, and the parameter estimates obtained, by predicting the value of the natural process at known locations in space. More specifically, this concept consists of the following steps. First, consider the data points $\{\mathbf{s}_1, \mathbf{s}_2, \ldots, \mathbf{s}_n\}$ and remove one of them, say $\mathbf{s}_i$. Then, by using the form of the semivariogram or covariance model supposed, obtain an estimate $\hat{X}(\mathbf{s}_i)$ of the removed data value $X(\mathbf{s}_i)$ and the associated estimation error $\sigma_x^2(\mathbf{s}_i)$ on the basis of the remaining data, by applying the estimation method of Chapter 9. Moreover, since the actual value $X(\mathbf{s}_i)$ is known, one can calculate the actual estimation error

$$\sigma_x^*(\mathbf{s}_i) = \hat{X}(\mathbf{s}_i) - X(\mathbf{s}_i)$$

Do the same thing for all data points, one at a time, and calculate the quantities

$$\varepsilon = \sum_{i=1}^{n} \frac{\sigma_x^*(\mathbf{s}_i)}{n}$$

and

$$\eta = \frac{1}{n} \sum_{i=1}^{n} \left[\frac{\sigma_x^*(\mathbf{s}_i)}{\sigma_x(\mathbf{s}_i)}\right]^2$$

The more appropriate the hypotheses made and the semivariogram or the covariance chosen, the closer to 0 is the ε-value and the closer to 1 is the η-value.

4.2.2 Generalized Spatial Covariance and Semivariogram

Things are quite different with the statistical inference of the generalized spatial covariances of order ν (GSC-ν). As the theory (Chapter 3) indicates, there is not an analytical expression for the GSC-ν in terms of the data values and, also, there is not a unique GSC-ν for a particular ISRF-ν. Therefore, unlike the means, covariances, and semivariograms, the GSC-ν do not have a real counterpart that can be directly measured and tested (after the event).

The heuristic part of the stochastic data processing program is carried out in terms of an *automatic procedure* that determines GSC-ν by testing their compatibility with reality, solely on the basis of the successful estimates of known data points they lead to. Automatic fitting includes identification of both the order of spatial intrinsity ν—which corresponds to the degree of the spatial polynomial trend surface—and the GSC-ν, $k(\mathbf{h})$. Identification may require the use of some parameter estimation technique like ranking, or least squares, or maximum likelihood, and at this point empirical research is of great help in providing guidance. For practical reasons, analysis is usually restricted to moving neighborhoods instead of a unique one. That is, we partition the domain of the natural process into a number of subdomains and we consider local models of the SRF that take into account only data within a neighborhood. In such a setting notions like local homogeneity, local intrinsity of order ν, local behavior, etc., will arise naturally.

More specifically, a possible step-by-step automatic structure identification procedure may run as follows (a variety of procedures may be found in Delfiner, 1976; Kafritsas and Bras, 1980; Kitanidis, 1983):

Step 1: Data Input

(1.1) Set the data in matrix form: first the spatial coordinates, then the measured value of the natural process. Sometimes erroneous individual data must be rejected at this point, to ensure the objectivity of the analysis.

(1.2) Check for double points (i.e., points that are very close to one another in space and, hence, might cause numerical problems in subsequent calculations). Rearrange the data, if necessary, and let $D = \{\mathbf{s}_1, \mathbf{s}_2, \ldots, \mathbf{s}_n\}$ be the set of data points after the double points have been eliminated.

Step 2: Identification of the Order ν of Spatial Nonhomogeneity

(2.1) As an initial choice of a GSC-ν use the $k(r) = -r$. Such a choice is perfectly legitimate, since the linear model is a valid GSC-ν for all possible values of ν.

(2.2) For each data point $\mathbf{s}_i \in D$, $i = 1, 2, \ldots, n$, develop a local neighborhood N_i of surrounding points. This can be done (a) either by including in the neighborhood N_i the $n(i)$ data points closest to point $\mathbf{s}_i$, or (b) by including all neighboring points that lie within a circle having its center at the point $\mathbf{s}_i$ and radius r_i (see Fig. 7.18). The choice of the one or the other method depends on the spatial distribution of the data. If the data are evenly distributed in space, method (b) may be applied; in the case of unevenly distributed data, method (a) may be more appropriate.

(2.3) Remove the data points $\mathbf{s}_i \in D$ one at a time, and estimate them from the surrounding points $\mathbf{s}_j \in N_i$, $j \neq i$, assuming in turn $\nu = 0$, 1, and 2. Let the estimates and the associated error variances be denoted by

$$\hat{X}_\nu(\mathbf{s}_i) = \sum_j \lambda_j X_\nu(\mathbf{s}_j)$$

and

$$\sigma^2_{x,\nu}(\mathbf{s}_i) = E[\hat{X}_\nu(\mathbf{s}_i) - X_\nu(\mathbf{s}_i)]^2 = \sigma^2_{i,\nu}$$

respectively, where λ_j are coefficients to be determined by minimizing $\sigma^2_{i,\nu}$ (see Chapter 9).

(2.4) At each point $\mathbf{s}_i \in D$ compare the error variances $\sigma^2_{i,0}$, $\sigma^2_{i,1}$, and $\sigma^2_{i,2}$, and assign ranks 1, 2, and 3. [For example, if $\sigma^2_{i,0} > \sigma^2_{i,2} > \sigma^2_{i,1}$, then rank$(i, 0) = 3 >$ rank$(i, 2) = 2 >$ rank$(i, 1) = 1$.]

(2.5) Take average ranks over all points $\mathbf{s}_i \in D$, viz.,

$$\text{Avrank}(D, \nu) = \frac{\sum_{i \in D} \text{rank}(i, \nu)}{n}, \qquad \text{for each} \quad \nu = 0, 1, \quad \text{and} \quad 2$$

The ν that gives the smallest Avrank(D, ν) is the order of intrinsity.

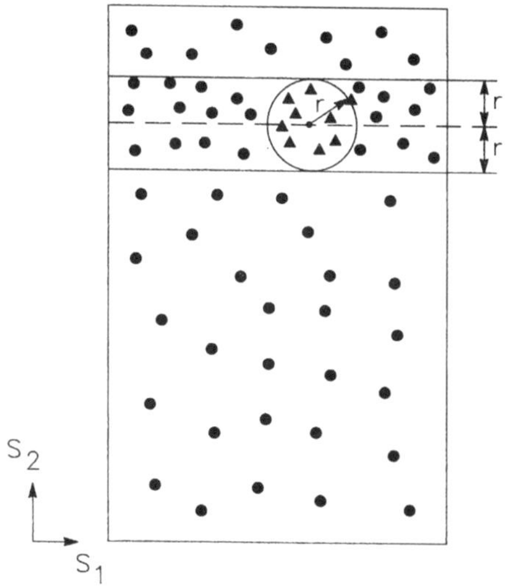

Figure 7.18 An illustration of a circular moving neighborhood

Step 3: Determination of the GSC-ν

Given the order ν, there is a set of suitable GSC-ν $\{k_p(r), p = 1, 2, \ldots, \mathcal{P}\}$ of the polynomial form of Eq. (20) above. The problem is to determine the corresponding coefficients $a_{p,0}$ and $c_{p,\rho}$, $\rho = 0, 1, \ldots, \nu$ so that the polynomial GSC-ν (20) above fits the data available, while the permissibility conditions of Table 3.1 of Chapter 3 are satisfied. This can be done as follows. For each $p = 1, 2, \ldots, \mathcal{P}$.

(3.1) Produce spatial increments of order ν (SI-ν)

$$Y(\mathbf{s}_i) = \sum_{j \in N_i \cup \{i\}} q_{ji} X(\mathbf{s}_j) \tag{36}$$

by letting

$$q_{ii} = -1 \quad \text{and} \quad q_{ji} = \lambda_j \, (i \neq j) \tag{37}$$

where λ_j are the estimation coefficients obtained when each point $\mathbf{s}_i$ is estimated from its neighborhood N_i, assuming that $k_p^{(0)}(r) = -r$ (see step 2 above).

(3.2) Write

$$A_i = E[Y(\mathbf{s}_i)]^2 = \left[\sum_{j_a \in N_i \cup \{i\}} \sum_{j_b \in N_i \cup \{i\}} q_{j_a i} q_{j_b i} k_p(\mathbf{s}_{j_a} - \mathbf{s}_{j_b}) \right] \tag{38}$$

for several $\mathbf{s}_i \in D$, where the weights $q_{j_a i}$ and $q_{j_b i}$ are given by means of the estimation coefficients λ_j, step (3.1) above. Then let

$$F = \sum_{i \in D} [Y(\mathbf{s}_i)^2 - A_i]^2 \tag{39}$$

(3.3) Insert the GSC-ν $k_p(r)$ into Eq. (38) and then calculate the covariance coefficients $a_{p,0}$ and $c_{p,\rho}$ ($\rho = 0, 1, \ldots, \nu$), by minimizing Eq. (39) with respect to these coefficients; that is, set

$$\frac{\partial F}{\partial a_{p,0}} = \frac{\partial F}{\partial c_{p,\rho}} = 0 \tag{40}$$

for all $\rho = 0, 1, \ldots, \nu$, and solve to find $a_{p,0}^{(1)}$ and $c_{p,\rho}^{(1)}$. The latter form a first approximation of the covariance coefficients, and the corresponding approximation of the GSC-ν is $k_p^{(1)}(r)$.

(3.4) Repeat the procedure (3.1)–(3.3), but this time use the GSC-ν $k_p^{(1)}(r)$. This should be done m times until convergence is achieved, that is,

$$a_{p,0}^{(m-1)} \approx a_{p,0}^{(m)}, \qquad c_{p,\rho}^{(m-1)} \approx c_{p,\rho}^{(m)} \qquad (\rho = 0, 1, \ldots, \nu)$$

and, thus, $k_p^{(m-1)}(r) \approx k_p^{(m)}(r)$.

Procedure (3.1)–(3.4) yields $\mathcal{P}' \leq \mathcal{P}$ permissible GSC-ν (that is, covariances that satisfy the permissibility conditions).

(3.5) Choose among the $\mathscr{P}'$ permissible GSC-ν determined above the one that best fits the data. This can be done if we let

$$\eta_p = \frac{\sum_{i \in D} Y(\mathbf{s}_i)^2}{\sum_{i \in D} A_i} \tag{41}$$

and calculate the η_p-values corresponding to each one of the $\mathscr{P}'$ GSC-ν. (Instead of η_p, one may use a jackknife estimator of η_p.) Then keep the three GSC-ν that yield η_p-values closest to one. Finally, using in turn these three covariances repeat step 2, where now the order ν is known. The GSC-ν that gives the lowest Avrank(D, ν) is selected.

Example 5: We now apply the above algorithm to a specific case study (Christakos and Olea, 1992). The spatial hydrologic process of interest here is the water table elevation (in feet) in an area that includes most of the Equus Beds, a major aquifer in Kansas (see Fig. 7.19). The Equus beds are stream-laid deposits of the Pliocene Blanco and the Pleistocene Meade and Sanborn formations. These consist of poorly sorted sediments ranging in size from silt to gravel with abundant clay lenses. Thickness varies from 0

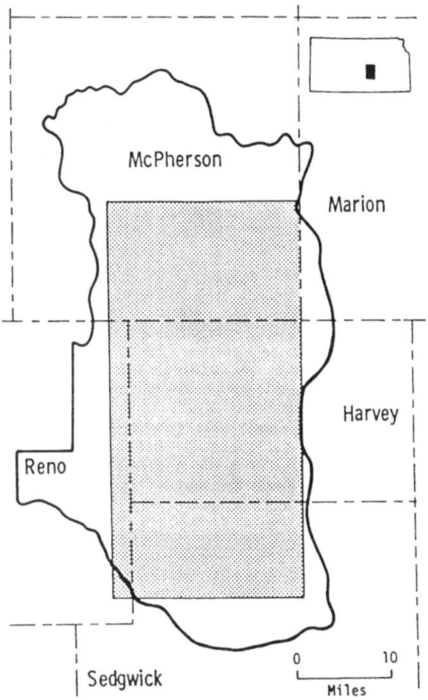

Figure 7.19 The Equus Beds in Kansas

to about 280 ft. The aquifer is the main water supply in the area extending for about 1400 square miles. The largest single user is the city of Wichita, pumping close to 50,000 gal/day. Other agricultural, industrial, municipal, and domestic users account for an additional 100,000 gal/day. The aquifer faces lowering of the water table due to pumpage in excess to the recharge as well as threats to the chemical quality of the groundwater caused mainly by past brine disposal practices by the oil industry, seepage of poor quality water from the Arkansas river, and leaks from the underlying Wellington aquifer high in mineral content (Spinazola et al, 1985). The water table elevation is of great importance for the agriculture of Kansas; lowering of water table levels can result in a significant reduction of the agricultural production. In this context, the identification of the spatial correlation structure of water levels is essential for the prediction of future systematic local or global changes in the water table.

In relation to this, the identification of the spatial correlation structure of the water table elevations is a necessary prerequisite for the mapping of water table levels over the entire region of interest and the design of a sampling network of future wells, as we will see in the continuation of this example in Chapters 9 and 10.

Data used in this study were chosen by analyzing measurements taken in the context of a statewide survey. To avoid measurement errors due to the excessive pumping in the summer months, only measurements taken during the winter have been selected. Also, a number of wells were not taken into consideration due to their uneven distribution in space, or to avoid "double points." As a result, the January 1981 data, which were available at 226 locations in an area of 800 square miles, were selected for analysis.

More specifically the numerical outputs of the step-by-step algorithm above are as follows:

Step 1: Here, $D = \{\mathbf{s}_1, \mathbf{s}_2, \ldots, \mathbf{s}_{226}\}$ is the set of data points to be used (there are no double points in this case).

Step 2: The local neighborhoods N_i developed include $n(i) = 12$ to 16 wells each. The choice of the wells in each local neighborhood was made on the basis of two requirements: to be as close as possible to the point to be estimated and to surround the point on all sides. Initially it is assumed that $k(r) = -r$, and the average ranks over all points $\mathbf{s}_i \in D$, $i = 1, 2, \ldots, 226$ are found to be

$$\text{Avrank}(D, \nu) = \frac{\sum_{i \in D} \text{rank}(i, \nu)}{226} = 1.9634$$

for $\nu = 0$; $= 1.9589$ for $\nu = 1$; $= 2.0731$ for $\nu = 2$. The $\nu = 1$ gives the smallest Avrank$(D, 0) = 1.9589$ and, hence, it is chosen as the order of spatial nonhomogeneity in this case.

Step 3: By solving the system of the corresponding equations we find $c_{1,0}^{(1)} = 39.66$, $c_{2,1}^{(1)} = -33.15$, and $c_{3,3}^{(1)} = 4.91$. These are a first approximation of the covariance coefficients, and the corresponding approximations of the GSC-1 are $k_1^{(1)}(r) = 39.66\ \delta(r)$, $k_2^{(1)}(r) = -33.15r$ and $k_3^{(1)}(r) = 4.91r^3$. Other possible covariance models were eliminated since they failed to satisfy the GSC permissibility conditions. By using these new GSC-1 we repeat this procedure until the values obtained converge; this is achieved after $m = 6$ repetitions, $k_1^{(6)}(r) = 28.75\ \delta(r)$, $k_2^{(6)}(r) = -24.22r$ and $k_3^{(6)}(r) = 2.33r^3$. For these three permissible GC-1 the corresponding η_p-values are $n_1 = 0.72$, $n_2 = 0.83$, and $n_3 = 1.81$. The GSC-1 with an η_p-value closest to one is the $k_2(r) = -24.22r$. This covariance also gives the lowest Avrank$(D, 1) = 1.80$ (step 2 above) from all three GSC-1. Therefore, the GSC-1 model that best fits the data is $k(r) = -24.22r$. As a cross-validation test for the order ν we repeat step 2, this time using $k(r) = -24.22r$ instead of $k(r) = -r$. The corresponding results are Avrank$(D, 0) = 2.11$, Avrank$(D, 1) = 1.80$, and Avrank$(D, 2) = 2.09$. The lowest value is obtained for $\nu = 1$, which coincides with the result of step 2.

The conclusion of the spatial variability identification part is that the spatial variation of the water table elevations in the Equus Beds is nonhomogeneous but continuous in space and rather regular, characterized by a locally linear trend. The results of the analysis above will serve as inputs to subsequent stages of stochastic data processing, such as estimation (see Chapter 9).

Remark 1: In some situations in practice, the automatic procedure may be more reliable than the experimental calculation of the ordinary covariance or semivariogram models, especially when a limited number of data are available. In this case, while the number of the experimental data pairs required for each point on a two-dimensional semivariogram may be greater than 30, inference in terms of GSC requires moving neighborhoods of about 8–16 points. Furthermore, when a nonconstant trend is present the experimental semivariogram may be biased. Indeed, it holds true that

$$E[\hat{\gamma}_x(\mathbf{h})] = \gamma_x(\mathbf{h}) + \frac{1}{2N(\mathbf{h})} \sum_{i=1}^{N(\mathbf{h})} \{E[X(\mathbf{s}_i + \mathbf{h})] - E[X(\mathbf{s}_i)]\}^2 \tag{42}$$

In other words, the sample semivariogram $\hat{\gamma}_x(\mathbf{h})$ is a biased estimator of the true one, $\gamma_x(\mathbf{h})$. Equation (42) shows how $\hat{\gamma}_x(\mathbf{h})$ may be distorted by the presence of a nonconstant trend leading, therefore, to misinterpretations regarding the nature of the spatial structure (see also Section 3 above). The situation is even worse with the experimental covariance. As Cressie (1991) shows, even a small amount of trend contamination can have disastrous effects in the estimation of the covariance. This disturbing situation does not exist with the automatic fitting procedure, which relies on homogeneity of the increments and eliminates question of trend. The latter does not need to be estimated and subtracted from the data; instead, it is simply identified by its degree ν.

As we saw above, the GSC-ν cannot be calculated experimentally from the data in terms of analytical expressions, such as those of Eqs. (21)-(25). However, when the data are along lines and sampled at regular intervals, the so-called *generalized spatial semivariogram of order* ν may be computed experimentally by means of the formula (e.g., Chiles, 1979)

$$\gamma_x(r) = \frac{1}{C_{2\nu+2}^{\nu+1}} \operatorname{Var}\left[\sum_{t=0}^{\nu+1} (-1)^t C_{\nu+1}^t X[\mathbf{s}+(\nu+1-t)r]\right] \tag{43}$$

where $r = |\mathbf{h}|$, and $C_k^m = k!/(k-m)!m!$. Definition (43) is essentially a generalization of the ordinary semivariogram. The generalized semivariogram is well determined as a linear function of the GSC-ν; that is, it is true that

$$\gamma_x(r) = \frac{1}{C_{2\nu+2}^{\nu+1}} \left[\sum_{t=-\nu-1}^{\nu+1} (-1)^t C_{2\nu+2}^{t+\nu+1} k_x(tr)\right] \tag{44}$$

The reverse is generally not true, except when $\nu = 0$; then $\gamma_x(r) = -k_x(r)$. Hence, while one can determine the corresponding permissible generalized spatial semivariogram of order ν from a permissible GSC-ν, one cannot, in general, do the reverse. In this case, to determine a permissible generalized semivariogram one may occasionally operate indirectly. For example, if the fitted generalized semivariogram is of polynomial form, one can start by determining its order ν. Then find the form of the corresponding GSC-ν from Eq. (20); but the coefficients a_0 and c_ρ will still be unknown. To estimate them, Eq. (44) is used; if the so-estimated a_0 and c_ρ satisfy the conditions of Table 3.1, Chapter 3, the sample semivariogram is considered permissible.

4.3 The Nugget-Effect Phenomenon and Its Modeling

When making reference to the term "nugget effect," it is important to distinguish between two things: the nugget-effect phenomenon and the modeling of the nugget effect. In geostatistics nugget effect is a term used to describe a phenomenon, not a random field model. Random fields are the discrete noise and the white noise, which are used to model nugget effects in the discrete parameter and the continuous parameter case, respectively. Let us be more specific.

The nugget-effect phenomenon is defined as a discontinuity of the semivariogram or the covariance function at the origin. The physical interpretation of the nugget-effect is that if measurements of a natural process represented by an SRF $X(\mathbf{s})$ are taken at locations $\mathbf{s}$ and $\mathbf{s}+\mathbf{h}$, which are very close to each other, the increment $Y_{\mathbf{h}} = X(\mathbf{s}+\mathbf{h}) - X(\mathbf{s})$ does not tend to zero (in the mean square sense); instead, it continues to fluctuate with an irreducible dispersion (see, e.g., Delfiner, 1979).

The modeling of the nugget effect is another, considerably more complicated issue. In general, two types of models can generate a nugget effect, a discrete parameter model called discrete noise, and a continuous parameter model called white noise.

(i) A discrete noise SRF consists of discretely valued, uncorrelated random variables $e(\mathbf{s}_i)$ $(i = 1, 2, \ldots, m)$ with zero mean and common variance, say c. In this case the nugget effect is generated by adding $e(\mathbf{s}_i)$ to the data, as measurement errors (not to the natural process itself); that is, assuming that $e(\mathbf{s}_i)$ are uncorrelated with $X(\mathbf{s}_i)$,

$$X^*(\mathbf{s}_i) = X(\mathbf{s}_i) + e(\mathbf{s}_i) \tag{45}$$

Then, if the $X(\mathbf{s}_i)$ has a continuous semivariogram $\gamma(\mathbf{h})$, we can write

$$\gamma^*(\mathbf{h}) = \begin{cases} 0 & \text{if } \mathbf{h} = \mathbf{O} \\ \gamma(\mathbf{h}) + c & \text{if } \mathbf{h} \neq \mathbf{O} \end{cases} \tag{46}$$

In other words, the nugget effect is modeled here in terms of a semivariogram with zero range. The special case of a pure nugget effect is modeled in the discrete domain by means of the semivariogram

$$\gamma_e(\mathbf{h}) = c[1 - \delta(\mathbf{h})] \tag{47}$$

where $\delta(\mathbf{h})$ denotes the Kronecker-delta [$\delta(\mathbf{h}) = 0$ if $\mathbf{h} \neq \mathbf{O}$; $= 1$ if $\mathbf{h} = \mathbf{O}$]. There should be no confusion with the so-called delta function or Dirac function of measure theory, for here the analysis is in the discrete domain. The delta or Dirac function, also denoted by $\delta(\mathbf{h})$, is used in the continuous case (see also discussion in Chapter 2).

(ii) A white-noise SRF is a generalization of the discrete noise in the continuous parameter case. It is defined as a zero-mean random field with a constant spectral density over all frequencies. White noise does not exist in the sense of ordinary functions. It is, however, well defined in terms of distributions or generalized functions (Schwartz, 1950–51). A generalized RF does not necessarily have point values, but if we take its convolution with a well-behaved function $q(\mathbf{h}) \in Q$ (for the definition of the space Q see Chapter 3) one can obtain an ordinary RF. The covariance of a white-noise SRF $e(\mathbf{h})$ is

$$c_e(\mathbf{h}) = c\,\delta(\mathbf{h}) \tag{48}$$

where $\delta(\mathbf{h})$ denotes now the delta or Dirac function. The latter is defined by the continuous linear functional

$$\delta(q) = \langle \delta, q \rangle = \int \delta(\mathbf{h})\, q(\mathbf{h})\, d\mathbf{h} = q(\mathbf{O})$$

where $q(\mathbf{h}) \in Q$. In practice, the data may have a finite support v; that is,

$$X_v(\mathbf{s}) = \frac{1}{v} \int_v X(\mathbf{s}+\mathbf{u})\, d\mathbf{u}$$

Then, the nugget effect is imposed on the semivariogram of $X_v(\mathbf{s})$ by means of a semivariogram reaching a sill $a = c/v$ after a transition zone reflecting the geometry of the support v. Finally, the white noise by itself can serve as a continuous parameter model of a pure nugget-effect phenomenon.

4.4 Stochastic Inferences by Means of Physical Models

In some cases, stochastic inference may also use a body of information that comes from physical principles and models, such as stochastic partial differential equations governing the spatial or the spatiotemporal evolution of a natural process.

Let us consider a few examples (some of these have been presented in detail in previous chapters). As we saw in Chapter 6, the isotropic generalized covariance associated to the stochastic Poisson differential equation modeling, for instance, a perturbation approximation of steady-state flow in a fully saturated spatially random medium in R^3, is given by

$$k_x(r) = \frac{1}{6r} \int_0^r u(r-u)^3 c_Y(u)\, du + b_3 r^2 + b_2 r + b_1 + b_0 r^{-1}$$

where the input covariance $c_Y(u)$ is known and the coefficients can be determined from the initial conditions of the particular problem.

In Chapter 5 we found that the covariances of the space–time model

$$\nabla_{\mathbf{s}}^2 \frac{\partial^2}{\partial t^2} X(\mathbf{s}, t) = Y(\mathbf{s}, t)$$

are related as

$$\nabla_{\mathbf{s}}^2 \nabla_{\mathbf{s}'}^2 \frac{\partial^4}{\partial t^2 \partial t'^2} c_x(\mathbf{s}, t; \mathbf{s}', t') = c_Y(\mathbf{h}, \tau)$$

In Example 1, Section 3 of Chapter 5 we discussed the stochastic $K\,dV$ equation modeling dispersive waves in a nonlinear one-dimensional medium, arising in hydrodynamical, meteorological, and geophysical applications. The mean value of $X(s, t)$ is approximated by

$$m_x(s, t) \approx -\frac{4k}{\sqrt{2\pi\sigma_w^2(t)}} \exp\left[-\frac{(ks-4k^3t)^2}{2k^2\sigma_w^2(t)}\right]$$

The stochastic flow equation studied in Example 4, Section 4 of Chapter 6, gave rise to the following expression of spectral density functions (in R^3)

$$C_h(\mathbf{w}) = \frac{J^2 w_1^2}{\mathbf{w}^4} C_f(\mathbf{w})$$

The two-dimensional hyperbolic stochastic differential equation,

$$\frac{\partial^2 X(s_1, s_2)}{\partial s_1 \, \partial s_2} + a_1 \frac{\partial X(s_1, s_2)}{\partial s_1} + a_2 \frac{\partial X(s_1, s_2)}{\partial s_2} + a_1 a_2 X(s_1, s_2) = W(s_1, s_2)$$

in which $W(s_1, s_2)$ is a white-noise SRF, represents natural processes with separable covariances of the form $c_x(h_1, h_2) = \exp[-a_2|h_1|] \exp[-a_1|h_2|]$ where $E[W^2(s_1, s_2)] = 4/a_1 a_2$.

A useful means for performing stochastic inferences is the simulation approach (Chapter 8). This approach allows one to use valuable information contained in physical laws. Consider an example from the study of flow and transport in porous media: First, simulated alternative realizations of log-transmissivities $z(\mathbf{s})$ with known mean $m_z(\mathbf{s})$ and covariance $c_z(\mathbf{h})$ or semivariogram $\gamma_z(\mathbf{h})$ provide the input to groundwater flow models; then, the model is solved to yield an ensemble of realizations of the piezometric head $h(\mathbf{s})$; Finally, the mean $m_h(\mathbf{s})$ and the covariance $c_h(\mathbf{h})$ or the semivariogram $\gamma_h(\mathbf{h})$ of the piezometric head can be calculated in terms of these realizations, using the experimental formulas presented in a previous section.

Other interesting applications may be found in the literature. Beran (1968), for example, applies the theory of functionals to derive the statistical moments associated with the Navier–Stokes equations and the continuity equation for an incompressible fluid. Locaiciga (1989), uses the advection-dispersion equation governing mass transport to obtain the mean and covariance of chloride concentration over space–time.

It must be remarked that this section is closely related to the discussion in Section 5 that follows. Also, some related results may be found in Chapter 8 on simulation.

5. Qualitative (Soft) Information

During data processing, it may be advisable that consideration be given not only to quantitative information in the form of a limited number of field observations or laboratory tests (also called hard data), but also to a significant amount of qualitative information (also called soft data), such as geologic mapping, soil mechanics, intuition, and experience with similar site conditions. The combination of these two types of information may be an important part of spatial correlation analysis which, if properly handled, might contribute, among others, to the reduction of statistical uncertainty (Christakos, 1987b and d).

When working with SRF it is necessary to express the qualitative information in a consistent, quantitative manner. Experience and intuition will improve one's ability to translate qualitative knowledge into explicit mathematical constraints under a suitable format. In connection with this, it seems that the real power of subjective analysis is in fulfilling the need for normative rules according to which such translations will be carried out. Nevertheless, one should generally avoid using qualitative knowledge that is too vague to be expressed in quantitative form. Modern approaches to the subject include the maximum-entropy approach (Section 8 of Chapter 9) and the indicator approach (Section 5.2 below). But first we restrict discussion to empirical, occasionally useful, approaches to utilizing qualitative information.

5.1 Empirical Approaches

One such approach is to develop criteria of favorable soil performance for the particular site and then to assign to each criterion a weight that reflects its importance relative to the other criteria as a soil performance indicator. Differences in opinion among engineers as to weights naturally result. The preparation of a list of criteria is closely related to the information level of the site. If investigation is initiated in a completely unknown site, the favorable criteria may be stratigraphy, slope geometry, external loading, etc. In a partially explored area, however, the criteria may include engineering properties, stress history, potential failure mechanisms, etc.

A practically more efficient approach is to utilize empirical relationships between groups of soil properties or different testing techniques to create subjective data for an unobserved or undersampled property. For instance, index tests are correlated with engineering properties; undrained strength may be measured through field vane tests or through controlled laboratory tests related to the former via empirical charts. In the context of groundwater contamination, the cost of analysis for an inorganic contaminant is usually much lower than that for an organic contaminant. Let the empirical model be of the linear form

$$X(\mathbf{s}) = \sum_i a_i Y_i(\mathbf{s}) \qquad (1)$$

where $X(\mathbf{s})$ is the soil property about which we do not have sufficient data, and $Y_i(\mathbf{s})$ are properties about which a significant amount of easily collected information is available. The spatial correlation structure of $X(\mathbf{s})$ can be evaluated in terms of the correlation structure of the $Y_i(\mathbf{s})$. For example the corresponding semivariogram and cross-semivariogram functions are related by

$$\gamma_X(\mathbf{h}) = \sum_i \sum_j a_i a_j \gamma_{Y_i Y_j}(\mathbf{h}) \tag{2}$$

where the cross-semivariogram is defined by

$$\gamma_{Y_i Y_j}(\mathbf{h}) = \tfrac{1}{2} E\{[Y_i(\mathbf{s}+\mathbf{h}) - Y_i(\mathbf{s})] \times [Y_j(\mathbf{s}+\mathbf{h}) - Y_j(\mathbf{s})]\} \tag{3}$$

The semivariogram (2) may then be used to generate $X(\mathbf{s})$-values, by applying some SRF simulation method (see Chapter 8). Similar relations may be derived in terms of spatial cross-covariances. The latter are related to the cross-semivariograms by

$$2\gamma_{Y_i Y_j}(\mathbf{h}) = 2c_{Y_i Y_j}(\mathbf{O}) - c_{Y_i Y_j}(\mathbf{h}) - c_{Y_j Y_i}(\mathbf{h}) \tag{4}$$

Note that the cross-covariance is in general nonsymmetric, and the $\gamma_{Y_i Y_j}(\mathbf{h})$ and $c_{Y_i Y_j}(\mathbf{h})$ are no longer equivalent spatial correlation functions.

Example 1: Figure 7.20 shows experimental data relating standard penetration resistance $X(s)$ and vertical stresses $Y(s)$, and also the estimated

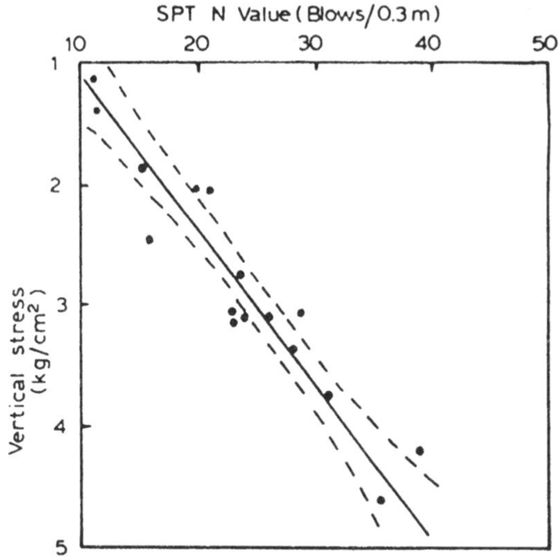

Figure 7.20 Empirical relationship between standard penetration resistance and vertical stresses for a cohesionless soil

empirical relationship $X(s) = 1.35 + 7.9Y(s)$. The loci of the confidence limits for a 5% significance level are curved with a minimum separation occurring at about the mean $m_Y = 2.9$ [some data fall outside the 95% confidence limits, but these limits are for the mean of a sample of $X(s)$ at a given value of $Y(s)$, and we do not have replicate measurements]. Assuming the parameter of vertical stresses to be statistically well established, the aforementioned empirical relationship leads to the mean and covariance of the standard penetration resistance, respectively, $m_x = 1.35 + 7.9 \times m_Y$ and $c_x(h) = 62.4 \times c_Y(h)$. From sensitivity analysis it was found that 10% change in $c_Y(h)$ will cause a 0.6% change in the variance $E[X(s) - m_x]^2$ and about 0.75% change in the coefficient 7.9; this indicates that the empirical model used was, in fact, quite good.

In many applications, especially in the geostatistical context, the analysis above may be used to obtain optimal linear estimates of one natural process in terms of the measured values of other processes. This matter will be discussed in Chapter 9 (see, also, Journel and Huijbregts, 1978; Myers, 1982; Ahmed and De Marsily, 1987; Yates and Warrick, 1987).

When the empirical model is of some nonlinear form

$$X(\mathbf{s}) = \sum_i f_i[Y_i(\mathbf{s})] \tag{5}$$

the nonlinearities may be approximated by means of orthogonal polynomial expansions. Orthogonal polynomials have properties that are theoretically and computationally attractive, and may be compared favorably to Taylor series or conventional statistical linearization expansions (see, e.g., Christakos, 1988a and b; 1989). For example, if Hermite polynomials are used and by assuming Gaussian distributions we can write

$$X(\mathbf{s}) = \sum_i \sum_k f_{ik} H_k[Y_i(\mathbf{s})] \tag{6}$$

where $H_k[y]$ are Hermite polynomials of degree k and f_{ik} are Hermite coefficients (Gradshteyn and Ryzhik, 1965). The above setting leads to the following

$$\gamma_x(\mathbf{h}) = \sum_i \sum_j \sum_k f_{ik} f_{jk} [c^k_{Y_i Y_j}(\mathbf{O}) - c^k_{Y_i Y_j}(\mathbf{h})] \tag{7}$$

Finally, a popular method among practitioners utilizes the fact that experimental findings and empirical values are available for statistical parameters such as coefficient of variation

$$\mathrm{COV} = \frac{\sqrt{\mathrm{Var}[X(\mathbf{s})]}}{E[X(\mathbf{s})]} \tag{8}$$

for various soils (see Singh and Lee, 1970; Kraft and Mukhopadhyay, 1977;

Haldar and Tang, 1979). For example, a typical value for the overall COV of the dynamic cyclic stress required to cause liquefaction is about 0.35. The COV of the shear strengths along a failure surface typically ranges between 0.1 and 0.5, depending on whether the slope analysis is made using total or effective stresses. With the COV known and the average of the soil field calculated for the particular site, the variance can be obtained from Eq. (8). Then, random soil data may be generated by the Monte Carlo simulation method. An application of this empirical method is given below. However, the method may best be appreciated within the context of spatial estimation (see Chapter 9).

Example 2: The purpose of this project was to provide assessments of the spatial variability of undrained shear strength (S_u) data and obtain reliable estimates for geotechnical analysis (Christakos, 1987b). Figure 7.21 and 7.22a show the site with the sampling locations and the foundation conditions, respectively. The marine clay stratum had the Atterberg limits and natural water content values also depicted in Fig. 7.22a. An essential task of the project was the determination of the strength of the marine clay over the site. This was achieved through a testing program consisting of 49 field vane tests at locations labeled VH-1 to VH-49, over a four-year period. A typical strength profile obtained along the VH-23 is shown in Fig. 7.22b. To describe horizontal variabilities, it was decided to assign at each vane

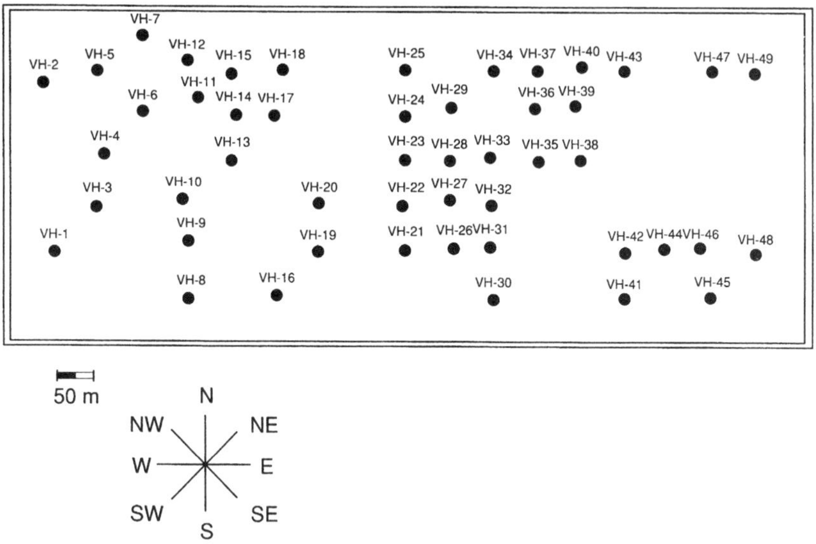

Figure 7.21 Site with the sampling locations of the field vane tests

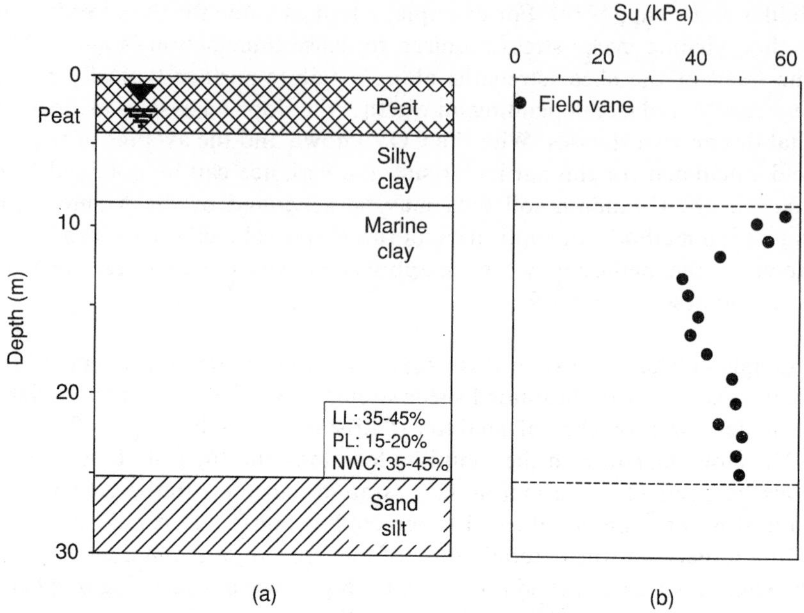

Figure 7.22 (a) Foundation conditions of the site. (b) Field vane profile at VH-23

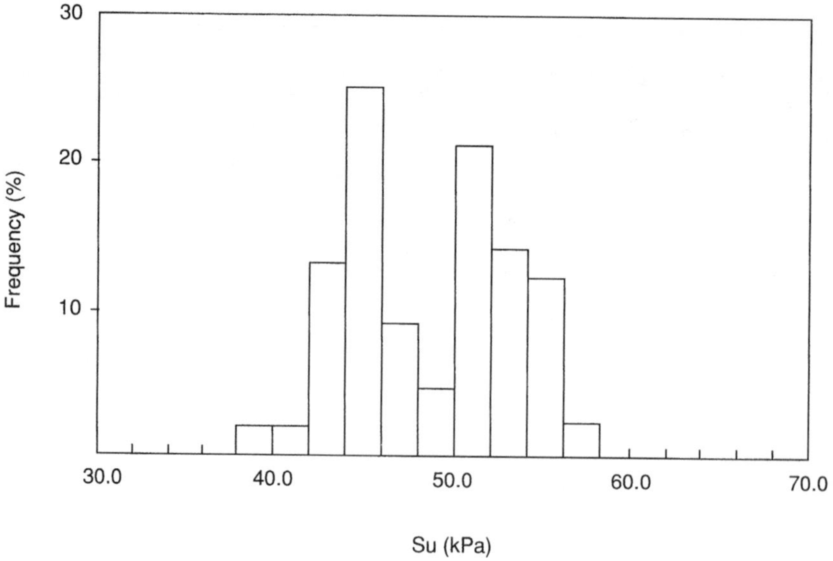

Figure 7.23 Histogram of the 49 field vane tests showing a bimodal shape

values come from two statistically different populations. These two populations may belong to geologically distinct zones, or they may arise by mixing field vane data from the two different exploration campaigns carried out a few years apart. The experimental semivariograms along four principal directions in space are plotted in Fig. 7.24. The graphic appearance of the semivariograms implies that the spatial variation of the surface strength is nonhomogeneous and experiences a moderate regularity. It also reveals anisotropies, since the plots at different directions display different variations. All semivariograms increase linearly at large lag intervals except the one along the NW–SE direction, which seems to exhibit a transitive character; that is, it oscillates around a sill of about $32\,\mathrm{kPa}^2$.

Due to nonhomogeneities, further study of the spatial structure requires the use of generalized SRF. Several of the field vane measurements, say $Y(\mathbf{s}_i)$, are interrupted by measurement errors modeled by a linear model

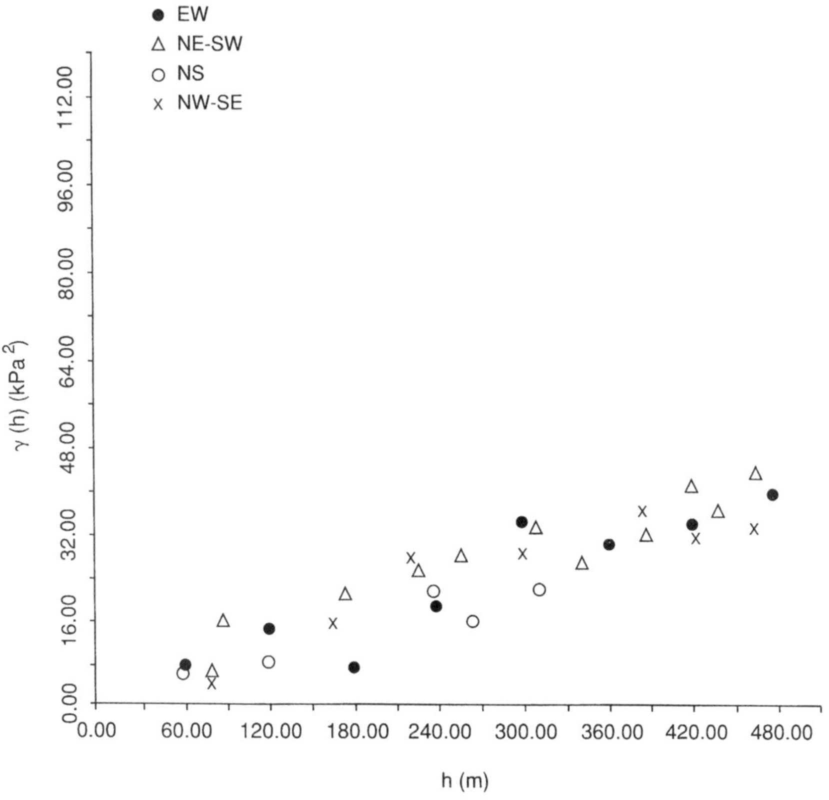

Figure 7.24 Experimental semivariogram of shear strengths based on the 49 field vane tests along four principal directions

of the form of Eq. (1) of Section 2 above, where $X(\mathbf{s}_i)$ are the actual soil strength values, $a=1$ and $V(\mathbf{s}_i)$ is a white-noise measurement error uncorrelated with $X(\mathbf{s}_i)$ and $Y(\mathbf{s}_i)$; the $V(\mathbf{s}_i)$ is assumed to have a zero mean and variance $\sigma_v^2(\mathbf{s}_i) = 12.25\ \mathrm{kPa}^2$ (this value has been chosen on the basis of empirical charts of the field vane correction factor versus plasticity index values (see Ladd *et al.*, 1977). In this case the generalized covariances of interest are $k_x(\mathbf{s}_i - \mathbf{s}_j) + \sigma_v^2(\mathbf{s}_i)\ \delta_{ij}$, where δ_{ij} is the Kronecker delta ($=1$ if $i=j$, $=0$ if $i \neq j$).

Following the steps of the algorithm discussed in previous sections, the order ν of nonhomogeneity and the associated GSC-ν must be determined. We decided to work with moving neighborhoods instead of a unique one, because due to the spatial nature of the field vane values the degree of nonhomogeneity of the local structure is smaller than the overall model of undrained strengths may suggest, thus corresponding to a lower degree trend at short distances. The limited number and the configuration of the vane holes imply that spatial variability is best known at short lags where the sample semivariograms are less uncertain. Working with moving averages of about 12 data points, we find that for $\nu = 0$, 1, and 2, the average error rank in estimating the 49 data points is 2.06, 1.88, and 2.17, respectively. Since the value $\nu = 1$ gives the lowest error rank it will be chosen as the order of field vane nonhomogeneity (that is, the underlying field vane surface is locally linear).

The corresponding permissible GSC-1 are $k_1(\mathbf{h}) = -3.5542 \times 10^{-2}|\mathbf{h}|$ and $k_2(\mathbf{h}) = 0.1352 \times 10^{-5}|\mathbf{h}|^3$. To decide which one of these two covariances best represent the available field vane data, we perform the tests discussed in Section 4.2.2. The covariance $k_1(\mathbf{h})$ leads to a jackknife estimator $j_1 = 0.9875$, which is closer to unity than the $j_2 = 1.7473$ derived by $k_2(\mathbf{h})$. In addition, the average error rank produced by the first covariance is lower than that of the second (1.4082 against 1.5918). Consequently, we conclude that $k_1(\mathbf{h})$ is the most proper GSC-1.

Since the cost of extensive sampling is usually prohibitive even in extensive geotechnical investigation programs, the application of a method that produces "subjective" strength values based on experimental findings and understanding of the geology of the site becomes very useful. In the present project the reliability of such a method could be tested, since the results already obtained are very satisfactory due to the sufficient number of observations available as well as the adequate geometrical configuration of the installed vane holes. Since the source of strength data is field vane tests, results of extensive studies on the subject (see, e.g., Bjerrum, 1973; Ladd *et al.*, 1977) were manipulated with the data of the particular site to yield a value for the coefficient of strength variation, COV $= 0.25$. With this value and since the mean strength is known, the variance can be derived from

Eq. (8) above. Then the Monte Carlo simulation procedure (e.g., Johnson, 1987) is employed to generate random strength values at 119 points on a square grid. The spatial structure identification of the 168 data (49 "objective" field measurements and 119 "subjective" data) leads to the semivariograms of Fig. 7.25.

Comparing Fig. 7.24 and 7.25, we see that the latter also exhibits the characteristics of a nonhomogeneous spatial structure of moderate continuity. Notice that, as happened in Fig. 7.24, the semivariogram along the NW–SE direction oscillates around a sill of about 32 kPa^2. In this case the spatial structure identification is determined by the same order $\nu = 1$ and the GSC-1, $k(\mathbf{h}) = -2.9752 \times 10^{-3}|\mathbf{h}|$. The slope of the GSC-1 is now smaller than before, thus implying a more regular spatial variability; this should also be expected, due to the significantly larger number of data used.

5.2 The Indicator Approach

This approach has been developed in geostatistics by Journel (1984, 1986,

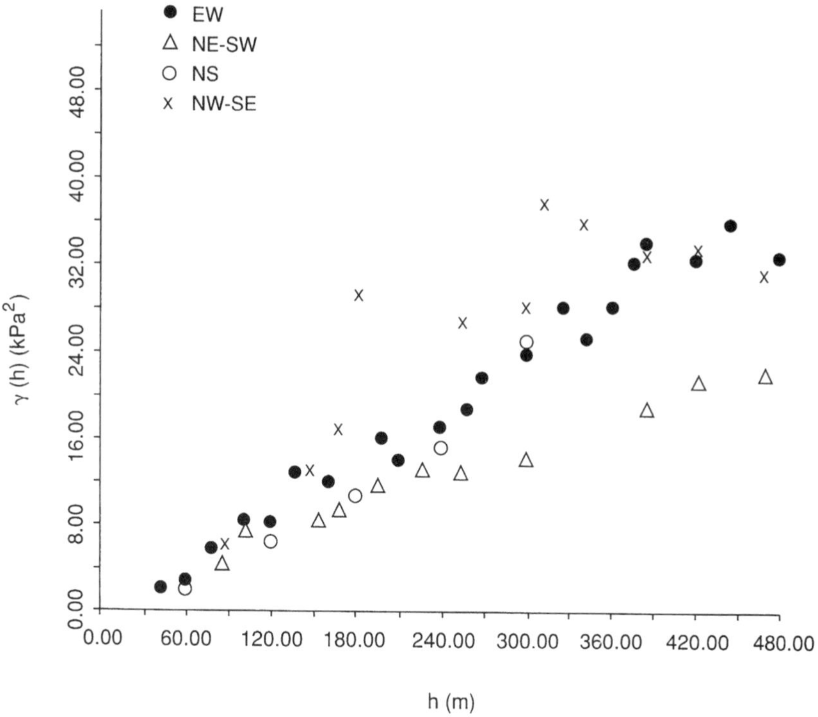

Figure 7.25 Experimental semivariograms of the shear strengths after the addition of the 119 simulated values

1989), Journel and Alabert (1988), and others. The basic idea here is to transform the original spatial RF into a derived RF that can still throw light on the spatial variability characteristics of the data, but whose analysis is simpler and faster.

5.2.1 Indicator Random Fields

Let $X(\mathbf{s})$ be a homogeneous spatial RF, with univariate and bivariate probability distribution functions $F_x(\chi) = P[X(\mathbf{s}) \leq \chi]$, and $F_x(\chi, \chi') = P[X(\mathbf{s}) \leq \chi, X(\mathbf{s}') \leq \chi']$, respectively.

Definition 1: The *indicator* RF is defined by

$$I_x(\mathbf{s}, \zeta) = \begin{cases} 1 & \text{if} \quad X(\mathbf{s}) \leq \zeta \\ 0 & \text{otherwise} \end{cases} \tag{9}$$

which is a homogeneous RF, as well.

Note that while $X(\mathbf{s})$ is a continuously valued function, the $I_x(\mathbf{s}, \zeta)$ is a discretely valued function. Moreover, the indicator RF (9) has certain attractive features:

(i) It allows nonparametric assessments of local uncertainty. The uncertainty model takes the form of a law concerning the probability that the unknown value at any unmeasurable location is greater than a given threshold value. This probability law, which does not need to be of the Gaussian type, can be expressed by means of the indicator function (9).

(ii) The data can be statistically characterized by means of a few simple moments that can be computed easily and fast.

(iii) The indicator formalism reduces significantly any "noise" contaminating the original data.

(iv) The problem of outliers is essentially eliminated.

(v) A variety of sources of soft information can be taken into consideration.

The last feature is the concern of this section. However, before we proceed with this subject, let us present a brief summary of the correlation theory of an indicator RF.

The mean value of the indicator RF (9) for each threshold value ζ is given by

$$m_I(\mathbf{s}, \zeta) = E[I_x(\mathbf{s}, \zeta)] = P[X(\mathbf{s}) \leq \zeta] = F_x(\mathbf{s}, \zeta) \tag{10}$$

When the SRF is homogeneous, $F_x(\mathbf{s}, \zeta)$ is independent of the position s; that is, $F_x(\mathbf{s}, \zeta) = F_x(\zeta)$ for all $\mathbf{s}$. Then, the noncentered and the centered covariances can be written as

$$\begin{aligned} \sigma_I^2(\mathbf{h}; \zeta) &= E[I_x(\mathbf{s}, \zeta) I_x(\mathbf{s}+\mathbf{h}, \zeta)] \\ &= P[X(\mathbf{s}) \leq \zeta, X(\mathbf{s}+\mathbf{h}) \leq \zeta] \\ &= F_x(\mathbf{h}; \zeta, \zeta' = \zeta) \end{aligned} \tag{11}$$

and

$$c_{\mathrm{I}}(\mathbf{h}; \zeta) = \sigma_{\mathrm{I}}^2(\mathbf{h}; \zeta) - m_{\mathrm{I}}^2 \tag{12}$$

respectively. Clearly,

$$\mathrm{Var}[I_x(\mathbf{s}, \zeta)] = F_x(\zeta)[1 - F_x(\zeta)]$$

and the indicator correlogram is given by

$$\rho_{\mathrm{I}}(\mathbf{h}; \zeta) = \frac{F_x(\mathbf{h}; \zeta) - F_x^2(\zeta)}{F_x(\zeta)[1 - F_x(\zeta)]} \in [-1, 1] \tag{13}$$

Finally, an important second-order spatial moment is the indicator semivariogram

$$\gamma_{\mathrm{I}}(\mathbf{h}; \zeta) = \tfrac{1}{2} E[I_x(\mathbf{s}+\mathbf{h}, \zeta) - I_x(\mathbf{s}, \zeta)]^2 \tag{14}$$

Just as for ordinary RF, it is not difficult to show that in the case of indicator RF the semivariogram and the covariance are related by

$$\gamma_{\mathrm{I}}(\mathbf{h}; \zeta) = c_{\mathrm{I}}(\mathbf{O}; \zeta) - c_{\mathrm{I}}(\mathbf{h}; \zeta) = F_x(\zeta) - \sigma_{\mathrm{I}}^2(\mathbf{h}; \zeta) \tag{15}$$

One of the most interesting outcomes of the analysis above is that the univariate and the bivariate probability distributions can be calculated in terms of the first and second-order moments of the indicator RF. In addition, the indicator statistical moments above can be expressed in terms of the corresponding moments of the original SRF $X(\mathbf{s})$. For example, the mean values and the centered covariances are related by

$$E[X(\mathbf{s})] = \int_0^{\infty} \{1 - E[I(\mathbf{s}, \zeta)]\}\, d\zeta \tag{16}$$

and

$$c_x(\mathbf{h}) = \int_0^{\infty} \int_0^{\infty} c_{\mathrm{I}}(\mathbf{h}; \zeta, \zeta')\, d\zeta\, d\zeta' \tag{17}$$

respectively.

Assume that $X(\mathbf{s})$ is a bivariate Gaussian RF, and let $\zeta = 0$. The indicator correlogram of Eq. (13) can be expressed in terms of the correlogram of $X(\mathbf{s})$ as follows:

$$\rho_{\mathrm{I}}(\mathbf{h}; 0) = \frac{2}{\pi} \arcsin \rho_x(\mathbf{h}) \tag{18}$$

The last equation implies that

$$|\rho_{\mathrm{I}}(\mathbf{h}; 0)| \leq |\rho_x(\mathbf{h})| \tag{19}$$

In addition, it leads to the following interesting corollary.

Corollary 1: A bivariate Gaussian RF $X(\mathbf{s})$ is an independent RF if and only if the corresponding indicator RF $I_x(\mathbf{s}, \zeta)$ is independent.

5.2.2 Bayesian Coding of a Priori Information

Typically, the measurement space of a natural process consists of *hard data* of the form

$$X(\mathbf{s}_i) = \chi_i, \qquad i \in A_1 \tag{20}$$

The hard data of Eq. (20) can be readily coded in terms of indicator RF, namely,

$$I_x(\mathbf{s}_i, \zeta) = \begin{cases} 1 & \text{if} \quad \chi_i \leq \zeta \\ 0 & \text{otherwise} \end{cases} \qquad (i \in A_1)$$

Apart from hard data, several other sources of high-quality knowledge are usually available. The first operation applied to the *a priori* qualitative knowledge should convert it to the appropriate quantitative *relationship space* $\mathfrak{R}$. In general, elements of the space $\mathfrak{R}$ are provided by scientists and engineers as a result of their expertise about a scientific field. The information thus conveyed could be, in many cases, equivalent to a large number of measurements. In geosciences, a particularly useful relationship space $\mathfrak{R}$ includes the following types of *soft data*:

(a) Ranges for values of $X(\mathbf{s})$; typical examples are the constrained interval data

$$X(\mathbf{s}_j) \in (a_j, b_j], \qquad j \in A_2 \tag{21}$$

(b) Truncated or contaminated values $X(\mathbf{s}_q) = \chi_q$, $q \in A_3$ (e.g., indirect estimation of porosity by means of seismic inversion techniques). In this case, some sort of uncertainty measure, usually highly subjective, must be associated with the truncated values.

(c) Subjective prior probability distributions expressing degress of expert beliefs, geological interpretations, experience, etc. For example, given information about the depth of the peat underlying the levee, an expert may assign a probability of slope failure.

(d) Correlations between the values of $X(\mathbf{s})$ at various locations in space. Correlations between $X(\mathbf{s})$ and other natural processes may be available through calibration scattergrams. For example, seismic travel data supply indirect information about local porosity values through prior calibration statistics represented by a scattergram. Nonlinear relationships will be discussed shortly.

Remark 1: At this point, some scientists may argue that techniques for constructing relationships from qualitative knowledge is not good theoretical research. One should not forget, however, that in practice there is strong need for such relationships, because important qualitative knowledge can be communicated only by means of these relationships.

The formation of the relationship space $\mathfrak{R}$ leads naturally to the following extension of the indicator RF.

Definition 2: The *generalized indicator RF* is defined by

$$Y_x(\mathbf{s}, \zeta) = P[X(\mathbf{s}) \leq \zeta | \mathfrak{R}] = F_x(\zeta | \mathfrak{R}) \tag{22}$$

The concept of the probability of a probability is intrinsic in the formalism of Eq. (22). That is, to the probability distribution $F_x(\zeta|\mathfrak{R})$ we are assigning a probability density $f_Y(\psi|\mathfrak{R})$. As we shall see shortly the probability density $f_Y(\psi|\mathfrak{R})$ is a useful instrument in the estimation of some calibration parameters. Note that while $F_x(\zeta)$ is independent of the spatial location $\mathbf{s}$, the $F_x(\zeta|\mathfrak{R})$ depends on $\mathbf{s}$; on the other hand, the $Y_x(\mathbf{s}, \zeta)$ is assumed to be a homogeneous RF. Moreover, the indicator RF (9) is related to the generalized indicator RF as follows:

$$Y_x(\mathbf{s}, \zeta) = E[I_x(\mathbf{s}, \zeta) | \mathfrak{R}] \tag{23}$$

Clearly, if $\mathfrak{R}$ consists of a hard datum at location $\mathbf{s}$, then Eq. (23) reduces to $Y_x(\mathbf{s}, \chi) = I_x(\mathbf{s}, \chi)$.

The indicator formalism [Eqs. (9) and (22)] can provide particularly useful means for incorporating soft data into spatial variability assessments. Prior knowledge offers, in turn, valuable information about the preposterior probability distribution $Y_x(\mathbf{s}, \zeta)$, before any measurements are taken into consideration. To obtain some insight, let us consider a few examples.

Example 3: Constrained interval data [Eq. (21)] can be expressed as follows

$$Y_x(\mathbf{s}_j, \zeta) = \begin{cases} 0 & \text{if } \zeta \leq a_j \\ \text{undefined} & \text{if } \zeta \in (a_j, b_j] \qquad (j \in A_2) \\ 1 & \text{if } \zeta > b_j \end{cases}$$

Other types of soft data can also be expressed in terms of Eq. (22). For example, a prior probability distribution concerning the attribute $X(\mathbf{s})$ of slope failure at point $\mathbf{s}$ given the depth $z(\mathbf{s}) = h$ of peat underlying the levee is equal to $Y_x(\mathbf{s}, \zeta)] = F_x(\zeta|h) \in [0, 1]$.

Let us see now how Eq. (22) can be calculated in practice. Assume that $\mathfrak{R}$ consists of a number of soft data at several locations in space. Then, the $Y_x(\mathbf{s}, \zeta)$ can be approximated on the basis of these data by means of a minimum mean square error estimation technique, such as those discussed in Chapter 9 (before proceeding, the reader is advised to review the relevant material in Chapter 9). More specifically, the present problem can be considered an estimation problem where one seeks the estimation of $I_x(\mathbf{s}, \chi)$ at a location $\mathbf{s}$, in terms of data at neighboring locations $\mathbf{s}_i$ $(i = 1, \ldots, m)$. The best minimum mean square error estimator is the nonlinear estimator

$$\hat{I}_x(\mathbf{s}, \zeta) = F_x(\zeta | \mathfrak{R}) = E[I_x(\mathbf{s}, \zeta) | \mathfrak{R}] \tag{24}$$

Equation (24) is an unbiased estimator of $I_x(\mathbf{s}, \chi)$. Indeed,

$$E_{\mathrm{I}}[\hat{I}_x(\mathbf{s}, \zeta) - I_x(\mathbf{s}, \zeta] = E_{\mathfrak{R}}\{E_{\mathrm{I}}[I_x(\mathbf{s}, \zeta)|\mathfrak{R}]\} - E_{\mathrm{I}}[I_x(\mathbf{s}, \zeta)]$$
$$= E_{\mathrm{I}}[I_x(\mathbf{s}, \zeta)] - E_{\mathrm{I}}[I_x(\mathbf{s}, \zeta)] = 0$$

It is, also, a conditionally unbiased estimator, for

$$E_{\mathrm{I}}\{[\hat{I}_x(\mathbf{s}, \zeta) - I_x(\mathbf{s}, \zeta)]|\mathfrak{R}\} = E_{\mathfrak{R}}\{E_{\mathrm{I}}[I_x(\mathbf{s}, \zeta)|\mathfrak{R}]|\mathfrak{R}\}$$
$$- E_{\mathrm{I}}[I_x(\mathbf{s}, \zeta)|\mathfrak{R}] = E_{\mathrm{I}}[I_x(\mathbf{s}, \zeta)|\mathfrak{R}] - E_{\mathrm{I}}[I_x(\mathbf{s}, \zeta)|\mathfrak{R}] = 0$$

In most applications, however, we usually restrict ourselves to linear estimators of the form

$$\hat{I}_x(\mathbf{s}, \zeta) = \sum_{i=1}^{m} \lambda_i(\mathbf{s}, \zeta)\, Y_x(\mathbf{s}_i, \zeta) \tag{25}$$

with weighted coefficients $\lambda_i(\mathbf{s}, \zeta) \geq 0$ $(i = 1, 2, \ldots, m)$, and such that

$$\sum_{i=1}^{m} \lambda_i(\mathbf{s}, \zeta) = 1$$

A more complete linear estimator is provided by

$$\hat{I}_x(\mathbf{s}, \zeta) = \sum_{j=1}^{k} \sum_{i=1}^{m} \lambda_i(\mathbf{s}, \zeta_j)\, Y_x(\mathbf{s}_i, \zeta_j) \tag{26}$$

where indicator data at various threshold values $\zeta_j \neq \zeta$ $(j = 1, \ldots, k)$ are included, as well.

Remark 2: Apart from minimum mean square error estimation, another way to calculate the probability distribution of Eq. (22) is suggested by the maximum entropy approach already presented in Section 4 above. The idea is to maximize the entropy of the corresponding probability density function subject to all data available, hard and soft. This is a very promising approach, regarding which significant research effort is currently underway (see also Chapter 9).

5.2.3 The Markovian Approximation

The use of Eqs. (25) and (26) in practice involves complicated inference and joint modeling of covariances and cross-covariances. Thus, the efficient implementation of the estimation process above may require some additional assumptions, such as the following *Markovian approximation* (Journel and Zhu, 1990): "Hard information always prevails over any soft collocated information." In quantitative terms this approximation implies that

$$E[I_x(\mathbf{s}', \zeta)|I_x(\mathbf{s}, \zeta) = 1, Y_x(\mathbf{s}, \zeta) = y] = E[I_x(\mathbf{s}', \zeta)|I_x(\mathbf{s}, \zeta) = 1] \tag{27}$$

and

$$E[I_x(\mathbf{s}', \zeta)|I_x(\mathbf{s}, \zeta) = 0, Y_x(\mathbf{s}, \zeta) = y] = E[I_x(\mathbf{s}', \zeta)|I_x(\mathbf{s}, \zeta) = 0] \tag{28}$$

for all $y \in [0, 1]$ and all $\mathbf{s}, \mathbf{s}' \in R^n$. In this way, the covariances required for the application of Eqs. (25) and (26) reduce to the covariances $c_I(\mathbf{h}; \zeta)$. In particular, the following proposition can be proven (Journel and Zhu, 1990).

Proposition 1: Under the Markovian approximation, the spatial correlation structure of the indicator RF and the generalized indicator RF are given by

$$c_{IY}(\mathbf{h}; \zeta) = \text{Cov}[I_x(\mathbf{s}+\mathbf{h}; \zeta), Y_x(\mathbf{s}; \zeta)] = B(\zeta) c_I(\mathbf{h}; \zeta) \quad (29)$$

for all $\mathbf{h}$;

$$c_Y(\mathbf{h}; \zeta) = \text{Cov}[Y_x(\mathbf{s}+\mathbf{h}; \zeta), Y_x(\mathbf{s}; \zeta)] = B^2(\zeta) c_I(\mathbf{h}; \zeta) \quad (30)$$

for all $\mathbf{h} > \mathbf{O}$; and

$$E[Y_x(\mathbf{s}, \zeta)] = F_x(\zeta) m^{(1)}(\zeta) + [1 - F_x(\zeta)] m^{(0)}(\zeta) \quad (31)$$

The $B(\zeta)$ and $m^{(i)}(\zeta)$ $(i = 0, 1)$ are calibration parameters defined by

$$B(\zeta) = m^{(1)}(\zeta) - m^{(0)}(\zeta) \in [-1, 1] \quad (32)$$

and

$$m^{(i)}(\zeta) = E[Y_x(\mathbf{s}; \zeta) | I_x(\mathbf{s}; \zeta) = i] \in [0, 1], \qquad i = 0 \quad \text{and} \quad 1 \quad (33)$$

On the basis of Proposition 1 the proof of the following corollary is straightforward.

Corollary 2: It is valid that

$$c_Y(\mathbf{O}; \zeta) = \text{Var}[Y_x(\mathbf{s}; \zeta)] = V_c^2(\zeta) + V_f^2(\zeta) \quad (34)$$

where

$$V_c^2(\zeta) = F_x(\zeta)[1 - F_x(\zeta)] B^2(\zeta) \quad (35)$$

and

$$V_f^2(\zeta) = F_x(\zeta) \sigma_{(1)}^2(\zeta) + [1 - F_x(\zeta)] \sigma_{(0)}^2(\zeta) \quad (36)$$

where

$$\sigma_{(i)}^2(\zeta) = \text{Var}[Y_x(\mathbf{s}; \zeta) | I_x(\mathbf{s}; \zeta) = i], \qquad i = 0 \quad \text{and} \quad 1 \quad (37)$$

The parameters $B(\zeta)$, $m^{(i)}(\zeta)$ and $\sigma_{(i)}^2(\zeta)(i = 0 \text{ and } 1)$, $V_c^2(\zeta)$, and $V_f^2(\zeta)$ are independent of $\mathbf{h}$. Furthermore, they provide valuable measures of accuracy and precision regarding the soft data. More specifically,

(a) The parameters $m^{(1)}(\zeta)$ and $\sigma_{(1)}^2(\zeta)$ are, respectively, the mean and the variance of those soft data $Y_x(\mathbf{s}; \zeta) = y_x(\mathbf{s}; \zeta)$ for which the actual value is known to be indeed no greater than the threshold value ζ. Hence, $m^{(1)}(\zeta)$ provides a measure of the accuracy of the soft data $y_x(\mathbf{s}; \zeta)$ in predicting that $X(\mathbf{s}) \leq \zeta$, and $\sigma_{(1)}^2(\zeta)$ offers a measure of the precision of the prediction. For perfectly accurate and precise soft information, these parameters become $m^{(1)}(\zeta) = 1$, and $\sigma_{(1)}^2(\zeta) = 0$; that is, the soft data are equivalent to hard indicator information.

(b) The parameters $m^{(0)}(\zeta)$ and $\sigma^2_{(0)}(\zeta)$ measure the accuracy and precision of the soft data $y_x(\mathbf{s}; \zeta)$ in predicting that $X(\mathbf{s}) > \zeta$. The best case, again, occurs when $m^{(0)}(\zeta) = 1$ and $\sigma^2_{(0)}(\zeta) = 0$.

(c) The parameter $B(\zeta) = m^{(1)}(\zeta) - m^{(0)}(\zeta)$ is an accuracy index for the soft data $y_x(\mathbf{s}; \zeta)$; the most informative case corresponds to $B(\zeta) = 1$ or -1, and the less informative to $B(\zeta) = 0$.

Note that the $m^{(i)}(\zeta)$ and $\sigma^2_{(i)}(\zeta)$ ($i = 0$ and 1) are the means and variances of $Y_x(\mathbf{s}; \zeta)$ with respect to the conditional probability densities $f_Y(\psi|\mathfrak{R}_i)$. The $\mathfrak{R}_i$ correspond to the relationships (prior information) $I_x(\mathbf{s}; \zeta) = i$, $i = 0$ and 1. In practice, one needs first to estimate the probability densities $f_Y(\psi|\mathfrak{R}_i)$, usually through the appropriate scattergrams, and then to calculate the calibration parameters.

Remark 3: From Eq. (30) one obtains

$$\lim_{|\mathbf{h}| \to 0^+} c_Y(\mathbf{h}; \zeta) = V_c^2(\zeta)$$

By comparing this result with Eq. (34) one finds that the covariance of $Y_x(\mathbf{s}; \zeta)$ can be decomposed as

$$c_Y(\mathbf{h}; \zeta) = V_f^2(\zeta)\, \delta(\mathbf{h}) + c_Y^*(\mathbf{h}; \zeta) \tag{38}$$

where $c_Y^*(\mathbf{h}; \zeta)$ is proportional to $c_I(\mathbf{h}; \zeta)$, viz., $c_Y^*(\mathbf{h}; \zeta) = B^2(\zeta) c_I(\mathbf{h}; \zeta)$. Note that even if $c_I(\mathbf{h}; \zeta)$ does not have a nugget effect and, hence, the $c_Y^*(\mathbf{h}; \zeta)$ is everywhere continuous, the $c_Y(\mathbf{h}; \zeta)$ will always have a discontinuity at origin.

5.2.4 Nonlinear a Priori Relationships

In certain situations in practice it may be appropriate to express prior knowledge by forming relationship functions of an arbitrary form $\Phi(\aleph)$; $\aleph$ is a vector consisting of the natural process of interest, $X(\mathbf{s})$, as well as other natural processes $Z_i(\mathbf{s})$ ($i = 1, \ldots, m$) related to $X(\mathbf{s})$. These functions may express prior knowledge about $X(\mathbf{s})$, such as information provided by a variety of, in general, nonlinear scattergrams between $X(\mathbf{s})$ and the processes $Z_i(\mathbf{s})$.

Example 4: The functions $\Phi(\aleph)$ may include (see Fig. 7.26)

(a) Ratio,

$$\frac{Z_i(\mathbf{s})}{X(\mathbf{s})} \quad \text{and} \quad \frac{Z_i(\mathbf{s}) - b}{X(\mathbf{s}) - a}$$

(b) Arc,

$$\sqrt{X^2(\mathbf{s}) + Z_i^2(\mathbf{s})}$$

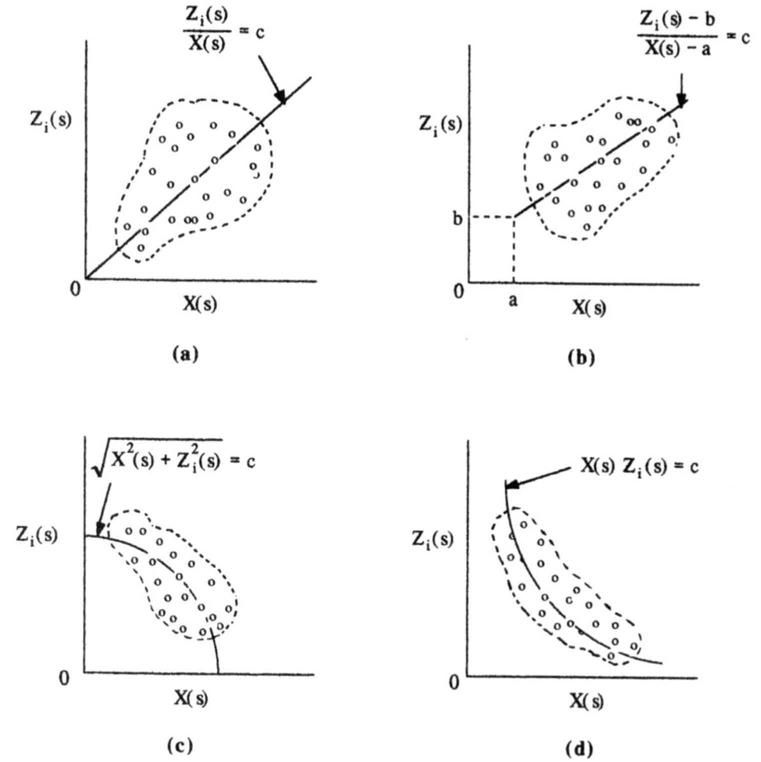

Figure 7.26 Examples of functions Φ

(c) Product,

$$X(\mathbf{s}) \times Z_i(\mathbf{s})$$

In cases such as the above, to determine the function $Y_x(\mathbf{s}, \zeta)$ one can proceed as follows:

(i) First, the probability density $f_\Phi(\varphi|\mathfrak{R})$ is estimated using prior knowledge.
(ii) Next, one can determine the density

$$f_\aleph(\boldsymbol{\nu}|\mathfrak{R}) = \frac{f_\Phi(\varphi|\mathfrak{R})}{|J(\aleph)|} \tag{39}$$

where $|J(\aleph)|$ is the appropriate Jacobian.
(iii) Integration of Eq. (39) with respect to $Z_i(\mathbf{s})$ $(i = 1, \ldots, m)$ gives the probability density $f_x(\chi|\mathfrak{R})$.

(iv) Finally, integration from $-\infty$ to ζ yields the required probability distribution $F_x(\zeta|\Re) = Y_x(\mathbf{s}, \zeta)$.

6. Some Final Comments about the Stochastic Research Program

When real life is regarded as an inseparable guide of theoretical developments, one is inevitably forced to consider what appear to be desirable goals of a scientific application; we may call them *desiderata*. Such desiderata are as follows:

(a) We know from experience that the primary emphasis on the *logic* of the problem, rather than the mathematics, is necessary in the early stages of any scientific study. Before any formal mathematical technique is applied, the relevance of conceptual matters and cause-and-effect relationships on the conclusions that we should draw from a given set of facts must be clarified and understood. For example, the inductive basis of site exploration rests on developing hypotheses and assigning degrees of plausibility to them, while the conclusions drawn from exploration contain more than the samples themselves. These are tasks of logic and judgment, not of mathematical formalisms.

(b) In many practical situations we deal with a limited amount of hard data and, hence, it is important that other sources of *prior information* (in the form of qualitative models, physics, geological interpretations, intuition, and experience, etc.) should be taken into account. For instance, environmental hazards cannot be judged rationally if we look only at the set of "hard" data available and ignore the prior information about the mechanism at work. Also, it is only by fully appreciating the implications of incomplete prior knowledge that hydrologists can successfully manage their groundwater resources.

(c) The *problem-solving power* of a quantitative model comes from the knowledge it processes and the physically meaningful requirements it accounts for, not only from the mathematical formalisms and inference schemes it uses. It is, therefore, important that information processing approaches incorporate into analysis the knowledge one may have about the natural process to be studied and the specific physical or mathematical properties one wishes the stochastic data processing to account for.

The implication of desiderata (a) through (c) is that, apart from the physical properties of the hydrogeology of a region, stochastic analysis should also account for an important characteristic of scientific development, namely *intelligence*. The latter is neither an inherent property of a theory or a model in itself, but rather an ability exhibited by the scientist as a person who puts things together, based on knowledge—understanding of an area of expertise. Therefore, to examine the various sources of knowledge in hydrogeology, we may be bound to consider the hydrogeologist as the person developing and using (i) a theory or a model, as well as (ii) his knowledge in the domain of application. These considerations show another interesting dimension of the stochastic approach in the context of *expert knowledge formalization and propagation.*

To account for these desiderata, a powerful heuristic is adopted by the stochastic data processing program leading to a *Bayesian maximum-entropy* formalism. According to this heuristic, stochastic data processing may be considered an approach

(a) for conducting scientific inferences in real problems;
(b) for providing the technical means that will take into account incomplete prior information and the consistent rules for improving the existing state of knowledge as soon as new data become available;
(c) for quantifying and conveying physical models all the way to earth systems design and engineering.

One example of the application of this approach was discussed in Section 4. Some additional results will be discussed in Chapter 9.

8 Simulation of Natural Processes

"If one wishes to learn what are the methods theoretical physicists use do not listen to their words, fix your attention on their deeds."

A. Einstein

1. Introduction

Engineers, geologists, and other applied scientists deal with complex and changeable natural phenomena such as multiphase flow in porous media, sea waves, oil reservoir characteristics, and ore deposits. These phenomena have several aspects. Among the most important and at the same time most complex aspects are spatial and spatiotemporal variabilities. The latter are immensely important for interpreting spatially and spatiotemporally distributed observations, respectively, and for predictive performance. For example, in the context of hazardous wastewater site management and decision-making, some of the most difficult problems are related to model uncertainties caused by the spatial variability of model parameters. The adequate assessment of the spatial and temporal variability of geological reservoir processes is a requisite for an efficient reservoir characterization and oil production planning. It is, therefore, necessary that practicing scientists and engineers have a working understanding of spatial variability and its implications.

This chapter presents spatial and spatiotemporal variability in the light of the stochastic simulation concept, and should be considered as a continuation of the modeling issues discussed in the previous chapter. As we saw in preceding chapters, spatial variability is characterized in practice through the spatial covariance, semivariogram, and generalized covariance functions or by means of their frequency domain equivalents, spectral density, spectral semivariogram, and generalized spectral density functions. In this circumstance, spatial random field (SRF) simulation can be a very powerful and indispensable tool for constructing realizations that take into account the structural dependence in space of the natural process. Particularly, both the actual (but unknown) natural process and the simulated process are considered as realizations of the same SRF. They share the same mean, covariance, or semivariogram (ordinary, or generalized in the case of complex nonhomogeneous spatial variability) and univariate probability distribution, and under certain circumstances (e.g., no measurement errors) they will honor the measured values at the data points. Certain simulation methods may take into consideration other important sources of prior information ("soft" data, see Chapter 7). By studying the morphology of the simulated process one may derive valuable information about the variation in space of the actual process.

The space transformation (ST) approach of Chapter 6 enables one to establish a theory of random field simulation where a group of useful spatial simulation methods, such as the turning bands method and the spectral method, can be derived in a unified and powerful general setting.

Apart from ST simulation, other random field simulation techniques, which have proved useful in the study of earth systems, are discussed. These include frequency-domain methods; the lower–upper triangular matrix method, the Karhunen–Loeve expansion method, stochastic partial differential equation methods; and sequential indicator simulation. The simulation of vector random fields is also discussed.

In most cases the underlying probability distributions are assumed to be Gaussian, but non-Gaussian distributions can also be considered. Spatiotemporal natural processes are generated in terms of ST and other simulation techniques. Theoretically and technically interesting results are established in the physical as well as in the frequency domains. These include the simulation of integrated natural processes, the study of stochastic transport systems, and the effect of measurement errors.

The advantages and disadvantages of each one of the above methods can best be judged by means of specific cases. Finally, it must be remarked that there is a considerable literature devoted to simulation techniques, and the intent in this chapter is certainly not to review them all; however, several references are provided for the interested reader.

2. The Physical Significance of Simulation

Before proceeding with technical details, and to give a more concrete feeling of the simulation approach, we discuss a few examples: In ocean engineering, maps of sea waves' surface elevations like that of Fig. 8.1, which is the output of computer simulation using random fields concepts (for the underlying theory see following sections), closely resemble real sea wave topography as given by aerial stereophotogrammetry methods. In applying simulation techniques, the necessary spectral functions (directional wave spectrum, frequency spectrum, etc.) are obtained either on the basis of field measurements or they are of a standard functional form (e.g., Borgman, 1969; Goda, 1980). Simulation of Fig. 8.1 contributes to a detailed understanding of sea waves transformation toward the shore, after they have been developed by the wind in the offshore region. This understanding is a prerequisite for the reliable estimation of the action of sea waves on maritime structures.

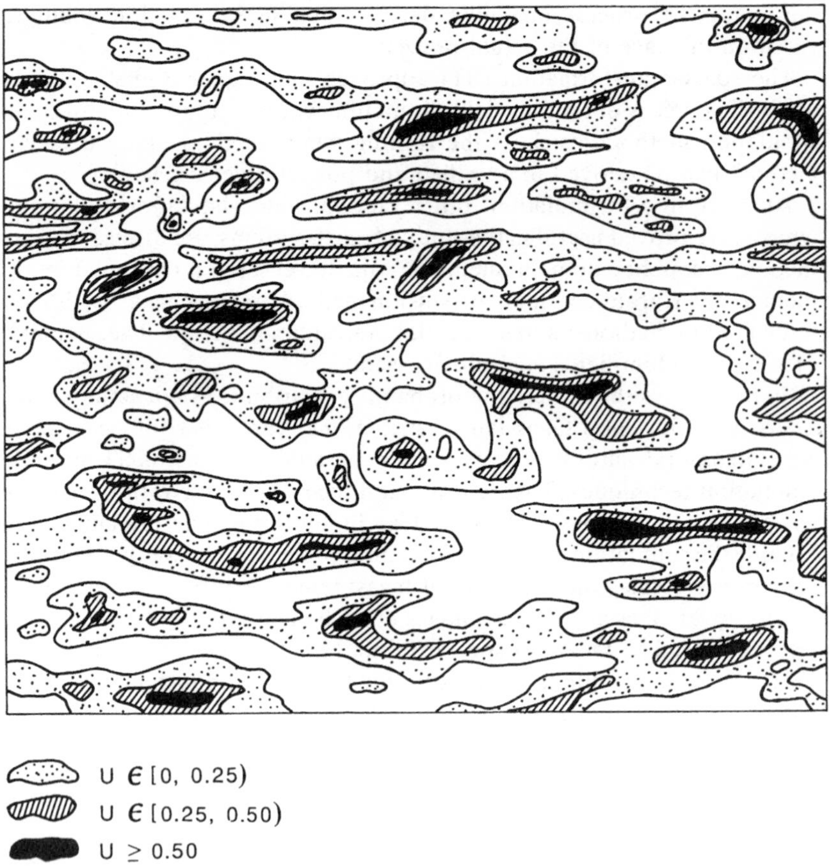

Figure 8.1 Simulated map of sea waves surface; $u = h/h_s$, where h is the water surface elevation (in meters) and h_s is the significant wave height (in meters)

In mining engineering, spatial variability of ore characteristics exercises a significant influence on predictions concerning ore favorability as well as on the choice of mining and haulage technology. Fig. 8.2 depicts a simulated contour map of Zn concentrations (%) from the mine of Lavrion (Greece). This is one possible realization of the SRF modeling the real deposit. The realization exhibits the same spatial characteristics as the real deposit (its covariance is calculated on the basis of the data available and geologic information from similar deposits worked out) and is forced to meet the measured Zn values at the data points, by applying a conditioning method. The usefulness of such maps is significant: They enable richer areas (in Zn) to be identified and ranked. Production planning is optimized taking into account technical constraints and monetary factors. By essence of SRF simulation, there can be several realizations like those of Fig. 8.2 that will honor the data points but which nevertheless will differ from each other. By studying their differences one may derive a measure of the uncertainties in the spatial distribution of the Zn concentrations. Furthermore, based on expertise about the particular natural phenomenon, one will be able to select the realization that best represents knowledge, past experience, and observation.

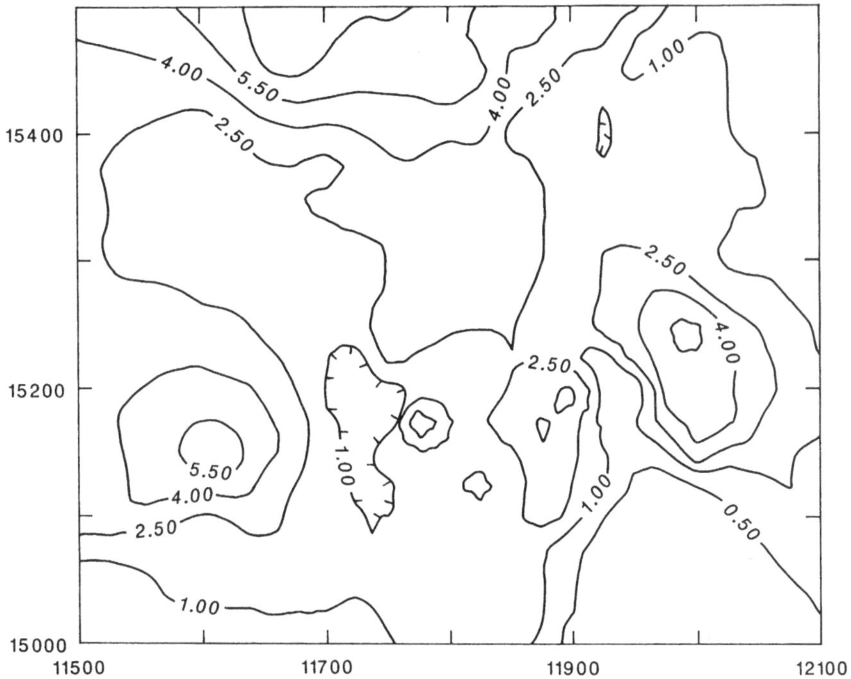

Figure 8.2 Simulated map of Zn concentrations (%) of the mine of Lavrion

In geology the simulation approach is used to visualize fluctuations in major geologic patterns, to investigate the morphology of fossils, and to map stratigraphical and structural surfaces. Recent research in the Arctic, motivated by the existence of rich natural resources (minerals, hydrocarbons, etc.), involves extensive use of numerical simulation to study the deformation and progressive failure of ice during ice-structure interaction and ice penetration.

Simulation may not be an end in itself. In many cases, the simulated values, once obtained, are used to initiate a course of action or generate values of another natural process. In this larger context, simulation may be considered a model-dependent process in the sense that the simulated surface may be controlled by laws that govern the underlying physical mechanisms, the morphology of the deposit or the operational program of research and production. In hydrology, for example, simulated alternative realizations of a spatial process (e.g., permeability), which share in common the statistical information available regarding the process, can provide the inputs to flow models (Fig. 8.3). The latter, which are usually stochastic partial differential equations, will then be solved numerically with respect to the output variables (e.g., hydraulic head), and the ensemble of output realizations will be studied statistically to yield the mean value, the covariance or semivariogram functions, as well as the probability distribution of the output variable at each point in space. This sort of application of the

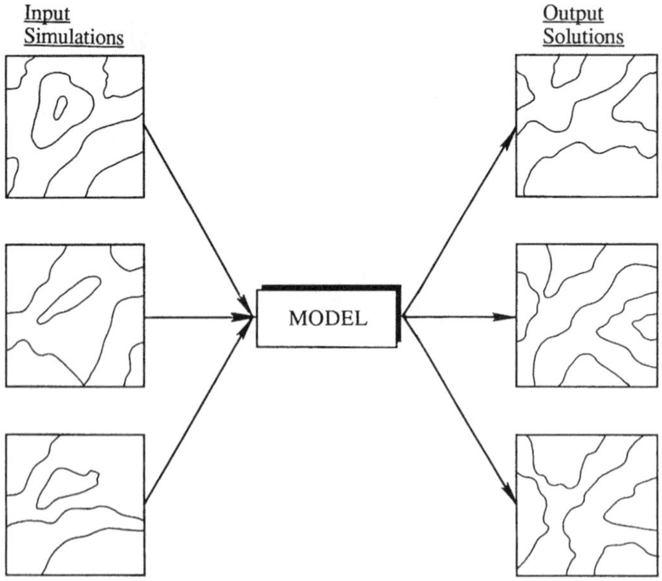

Figure 8.3 The input–output simulation process

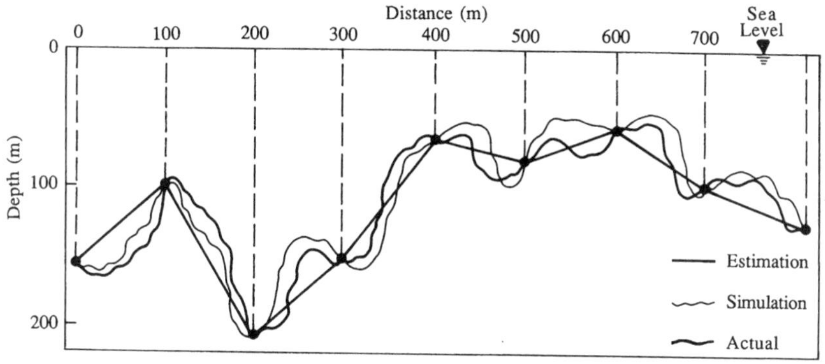

Figure 8.4 The sea-depth cable example

stochastic simulation concept constitutes one of its most important advantages over spatial estimation. Another situation in practice where SRF simulation is a more realistic approach than SRF estimation is illustrated in Fig. 8.4. Estimation techniques are designed to minimize local uncertainties and not to reproduce global spatial patterns. Simulation yields a better reproduction of the fluctuations of the sea depth and, therefore, a more accurate evaluation of the necessary total length of the cable.

3. Simulation of Random Fields

3.1 Unconstrained Simulation

The random field simulation methods to be discussed in this section are called *unconstrained*, to be distinguished from simulations where the realizations derived are constrained by the available values of the natural process.

Let us first consider a general-purpose method of random field simulation. Assume that $X(\mathbf{s})$ is an SRF with probability density $f_x(\chi_1, \ldots, \chi_m)$. As we saw in Section 2 of Chapter 2, the following general expression is valid

$$f_x(\chi_1, \ldots, \chi_m) = f_x(\chi_1) f_x(\chi_2 | \chi_1) \ldots f_x(\chi_m | \chi_1, \ldots, \chi_{m-1}) \tag{1}$$

The $X(\mathbf{s})$ can be simulated through Eq. (1) as follows:

Step 1: Given the $f_x(\chi_1, \ldots, \chi_m)$, one can calculate the probability densities $f_x(\chi_1)$, $f_x(\chi_2|\chi_1), \ldots,$ and $f_x(\chi_m|\chi_1, \ldots, \chi_{m-1})$. The corresponding probability distributions can also be found; namely,

$$F_x(\chi_1) = \int_0^{\chi_1} f_x(u)\, du$$

$$F_x(\chi_2|\chi_1) = \int_0^{\chi_2} f_x(u|\chi_1)\, du$$

and so on.

Step 2: Next, the following system of equations of marginal and conditional distributions is developed.

$$
\begin{aligned}
F_x(\chi_1) &= V_1 \\
F_x(\chi_2 | \chi_1) &= V_2 \\
&\vdots \\
F_x(\chi_m | \chi_1, \ldots, \chi_{m-1}) &= V_m
\end{aligned}
\tag{2}
$$

where V_i $(i = 1, 2, \ldots, m)$ are independent random numbers.

Step 3: Finally, realizations of the SRF $X(\mathbf{s})$ can be generated by solving the system (2) with respect to $\chi_1, \ldots,$ and χ_m.

Example 1: An SRF $X(\mathbf{s})$ with a multivariate Cauchy probability distribution can be simulated by means of the above method. More specifically, the marginal distributions in Eq. (2) are univariate Cauchy, and the conditional distributions are univariate student's t with m degrees of freedom. A variety of procedures generating realizations of random variables with these univariate distributions (as well as several others) are reviewed in Bratley *et al.* (1983) and in Johnson (1987).

When the conditional distributions in Eq. (2) are hard to determine, a simulation method based on a convenient *transformation* of the $X(\mathbf{s})$ in terms of suitable univariate random variables (usually independent) may be used. "Suitable" means that procedures generating these variables are already available in the literature.

Example 2: An SRF with a multivariate Gaussian distribution can be simulated through a linear transformation of independent Gaussian random variables. A multivariate Cauchy $X(\mathbf{s})$ is simulated by means of the transformation $\chi_i = \zeta_i / \gamma$, where ζ_i and γ are independent Gaussian variables and the square root of a gamma $G(\frac{1}{2}, 2)$ variable, respectively.

The formulas of Table 8.1 generate some important univariate probability densities and are useful tools in SRF simulation studies.

Remark 1: Formula (1) above leads to the sequential indicator simulation (see Section 13 below), which is a very useful method, especially in situations where the underlying probability distribution is non-Gaussian. The latter can be estimated with the help of the indicator coding of "soft" data discussed in Chapter 7.

Table 8.1 Formulas for the Generation of Random Variables with a Specific Probability Density Function[a]

Probability density	Simulation formula
Standard Gaussian:	
$f_x(\chi)=\frac{1}{\sqrt{2\pi}}\exp\left[-\frac{\chi^2}{2}\right]$	$\chi=\sum_{i=1}^{n} u_i$
Gaussian:	
$f_x(\chi)=\frac{1}{\sqrt{2\pi}\sigma_x}\exp\left[-\frac{(\chi-m_x)^2}{2\sigma_x^2}\right]$	$\chi=\sqrt{\frac{3\sigma_x^2}{n}}\sum_{i=1}^{n}(2u_i-1)+m_x$
Exponential:	
$f_x(\chi)=\lambda\exp[-\lambda\chi]$, $(\chi\geq 0, \lambda>0)$	$\chi=\zeta_1^2+\zeta_2^2$, $\left(\lambda=\frac{\sigma_x^2}{2}\right)$
Standard Rayleigh:	
$f_x(\chi)=\chi\exp\left[-\frac{\chi^2}{2}\right]$, $(\chi\geq 0)$	$\chi=\sqrt{-2\log u}$
Rayleigh:	
$f_x(\chi)=\frac{\chi}{\alpha^2}\exp\left[-\frac{1}{2}\frac{\chi^2}{\alpha^2}\right]$, $(\chi\geq 0)$	$\chi=\sqrt{\zeta_1^2+\zeta_2^2}$, $(\alpha=\sigma_x)$
Chi-squared with 2*n* degrees of freedom:	
$f_x(\chi)=\frac{(\chi^2)^{(n/2)-1}}{2^{n/2}G(n/2)}\exp\left[-\frac{\chi^2}{2}\right]$	$\chi=\sum_{i=1}^{n}\zeta_i^2$, or $\chi=2\sum_{i=1}^{n}\eta_i$

[a] The values of u_i, ζ_i, and η_i are (0, 1) uniformly distributed, $(0, \sigma_x^2)$ Gaussian, and exponentially distributed random numbers, respectively.

3.2 Constrained Simulation

From a physical intuition point of view, in addition to the above characteristics, it is desirable that the SRF simulations honor the measured values at the data locations (assuming, of course, that there is no measurement error). The latter requires the implementation of some sort of a conditioning approach, so that the generated SRF realizations are constrained by the available data. In other words, assuming that values of the SRF $X(\mathbf{s})$ are available at locations $\mathbf{s}_i$ $(i=1,\ldots,m)$, one generates realizations of the constrained SRF

$$X^*(\mathbf{s})=X(\mathbf{s})\,|\,X(\mathbf{s}_i)=\chi_i \qquad (i=1,\ldots,m) \tag{3}$$

The conditioning can be achieved by means of one of the spatial estimation approaches discussed in Chapter 9. Constrained simulation is also called conditional simulation; however, this terminology is not used here, to avoid confusion with the conditional distribution methods of the previous subsection.

Of particular importance is the case where the SRF $X(\mathbf{s})$ is Gaussian. Realizations of the conditional SRF $X^*(\mathbf{s})$ can be obtained by means of the expression

$$\chi^*(\mathbf{s}) = \hat{\chi}(\mathbf{s}) + \varepsilon(\mathbf{s}) \tag{4}$$

where $\hat{\chi}(\mathbf{s})$ is an estimate of the actual process (e.g., Kriging; see Chapter 9), and $\varepsilon(\mathbf{s}) = \chi'(\mathbf{s}) - \hat{\chi}'(\mathbf{s})$ is the error between a realization of the unconditional $X(\mathbf{s})$ and its estimator. It is easily shown that the SRF (4) shares the same second-order statistics with the original SRF and passes through the data points.

In the majority of practical applications, however, very rarely the multivariate probability density is known or can be estimated on the basis of the data available. Hence, one is often restricted to a more limited objective, namely, the generation of random field simulations solely on the basis of its *second-order statistics.* In particular, most of the SRF simulation techniques used in earth sciences produce realizations of Gaussian SRF. These realizations preserve the mean, covariance, and semivariogram (ordinary or generalized), and the univariate probability distribution of the actual natural process.

The purpose of the following sections is to present some of the most powerful of the second-order statistics simulation techniques.

4. Simulation of Spatial Random Field by Space Transformations—Examples

SRF simulation in more than one dimension is more difficult, in general. The difficulties are of both mathematical and technical nature. As we shall see next, the space transformation theory of Chapter 6 provides a general and powerful means of transforming a simulation technique from one to several dimensions.

4.1 The Turning Bands Method

The first example is the *turning bands method* of simulation introduced by Matheron (1973) and Journel (1974) for isotropic Gaussian random fields in the space domain, and by Mantoglou and Wilson (1982) in the spectral domain. The method has been generalized within the framework of space transformations by Christakos (1984a; 1987a and c).

The basic steps of the turning bands method of simulation are as follows:

Step 1: Determine the covariance model $c_n(\mathbf{h})$ of the SRF $X_n(\mathbf{s})$ [or the spectral density $C_n(\mathbf{w})$] that characterizes the spatial variability of the natural process of interest. In practice the covariance model will be the result of a (sometimes tedious) procedure of fitting spatial functions to experimental data. The fitted function must be nonnegative-definite and there are permissibility criteria to ensure this (Chapter 7). The spectral density model is determined as the n-dimensional Fourier transform of the covariance model. Generally, previous experience with the particular natural process and formal cross-validation methods contribute to the choice of the proper covariance or spectral density function model.

Step 2: Given the covariance $c_n(\mathbf{h})$ [or the spectral density $C_n(\mathbf{w})$] of the SRF $X_n(\mathbf{s})$, find the one-dimensional covariance $c_{1,\boldsymbol{\theta}_i}(t)$ [or the spectral density $C_{1,\boldsymbol{\theta}_i}(\omega)$] on each line $\boldsymbol{\theta}_i$ [the necessary formulas to do this are Eqs. (15) and (19), respectively, Section 2 in Chapter 6].

Step 3: Generate several on-line realizations $X_{1,i}(\mathbf{s}\cdot\boldsymbol{\theta}_i)$, $i=1, 2, \ldots, N$ lines, on the basis of $c_{1,\boldsymbol{\theta}_i}(t)$ [or $C_{1,\boldsymbol{\theta}_i}(\omega)$]. For this purpose, certain of the most important unidimensional simulation techniques are presented in Section 5 below. Regarding the number N of on-line simulations necessary in practice, it has been found that for two and three-dimensional applications, the generation of a maximum of $N=16$ and $N=15$ on-line realizations, respectively, produces excellent results.

Step 4: The idea now is to simulate an SRF in R^n by summing contributions from random processes in R^1. In doing so, use Eq. (5) of Section 3 of Chapter 6 to produce realizations of $X_n(\mathbf{s})$. The corresponding simulated covariance $c_n^*(\mathbf{h})$ will be given by Eq. (6), Section 3 of Chapter 6.

Step 5: Carry out sensitivity analysis of the results obtained, by comparing the initial model $c_n(\mathbf{h})$ and the simulated covariance $c_n^*(\mathbf{h})$, or by investigating the influence on the produced simulations $X_n(\mathbf{s})$ of the approximations involved in steps 1 through 4 like, for example, the number N of lines in summations (5) and (6) of Section 3, Chapter 6.

For computational purposes, instead of vectors it is usually convenient to work with spherical coordinates, namely, $\mathbf{w}=(w_1, \ldots, w_k, \ldots, w_n)$ with

$$
\begin{aligned}
w_1 &= \omega \cos \phi_1 \\
w_k &= \omega \cos \phi_k \prod_{i=1}^{k-1} \sin \phi_i \qquad (1)\\
w_n &= \omega \prod_{i=1}^{n-1} \sin \phi_i
\end{aligned}
$$

where $k = 2, 3, \ldots, n-1$, and the angles ϕ_i, $i = 1, 2, \ldots, n$ define the direction of the corresponding line $\boldsymbol{\theta}_i$ in R^n. More details concerning the computations and assumptions of the procedure above will be given in the applications to be discussed next.

Example 1: An illustration of the simulation method is shown in Figs. 8.5 through 8.9 for the case $n = 2$. Figure 8.5 depicts the simulation geometry; the line at angle ϕ is the domain of the one-dimensional realization $X_{1,i}(\mathbf{s} \cdot \boldsymbol{\theta}_i)$.

According to the simulation algorithm discussed above and working in the frequency domain, we proceed as follows. First, the SRF $X_2(\mathbf{s})$, $\mathbf{s} \in R^2$ to be simulated is assumed to have a mean equal to six and an anisotropic covariance of the form

$$c_2(h_1, h_2) = \exp[-\sqrt{0.7225h_1^2 + 0.3025h_2^2}] \tag{2}$$

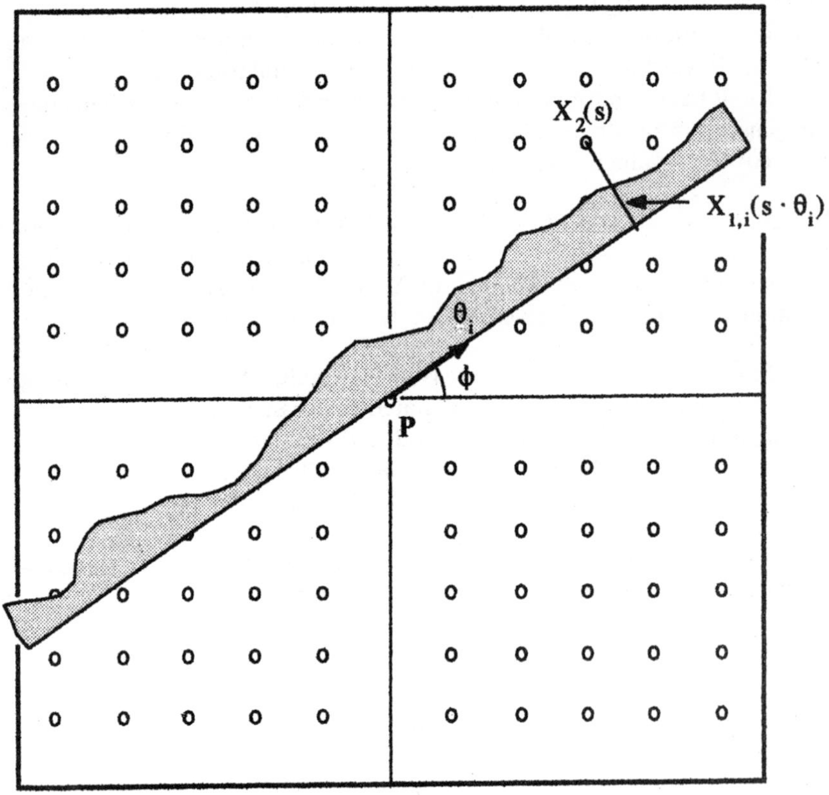

Figure 8.5 Simulation geometry on the plane. Line at angle ϕ is the domain of the one-dimensional realization $X_{1,\boldsymbol{\theta}_i}(\mathbf{s} \cdot \boldsymbol{\theta}_i)$; $X_2(\mathbf{s})$ is the simulated two-dimensional SRF

(Fig. 8.6). This theoretical example has been chosen to illustrate sensitivity analysis of the quality of the simulation results obtained. In practice, however, the model (2) will be the outcome of a process that combines experience with the specific physical variates and sometimes tedious experimental calculations.

Next, the two-dimensional Fourier transform of Eq. (2) is calculated, inserted into Eq. (19), Section 2 of Chapter 6, and spherical coordinates applied to yield the one-dimensional spectral density along the lines at angle ϕ,

$$C_{1,\phi}(\omega)=\frac{0.1093\omega}{[0.2186+(0.3025\cos^2\phi+0.7225\sin^2\phi)\omega^2]^{3/2}} \tag{3}$$

Equation (3) is plotted in Fig. 8.7 for $\omega \geq 0$ and $0 \leq \phi \leq 2\pi$.

From this one-dimensional simulation $X_{1,\boldsymbol{\theta}_i}(\mathbf{s}\cdot\boldsymbol{\theta}_i)$ are generated along several lines using the spectral method (for details, see Section 5 below), viz.,

$$X_{1,\boldsymbol{\theta}_i}(s_i)=\sqrt{2}\sum_{j=1}^{M} A_j\cos(\omega'_j s_i+\tilde{\phi}_j) \tag{4}$$

where $s_i=\mathbf{s}\cdot\boldsymbol{\theta}_i$, $M=400$ terms and $A_j=\sqrt{C_{1,\phi}(\omega_j)\,\Delta\omega}$ (the $C_{1,\phi}(\omega_j)$ has negligible value outside the interval $[-\omega_0, \omega_0]$, $\omega_0=100$); $\omega'_j=\omega_j+\delta\omega_j$, where $\omega_j=-\omega_0+(j-\frac{1}{2})\,\Delta\omega$ and $\Delta\omega=2\omega_0/M=0.5$ is the discretization frequency; $\tilde{\phi}_j$ are random angles distributed uniformly and independently inside the interval $[0, 2\pi]$. To avoid periodicities, a small random frequency

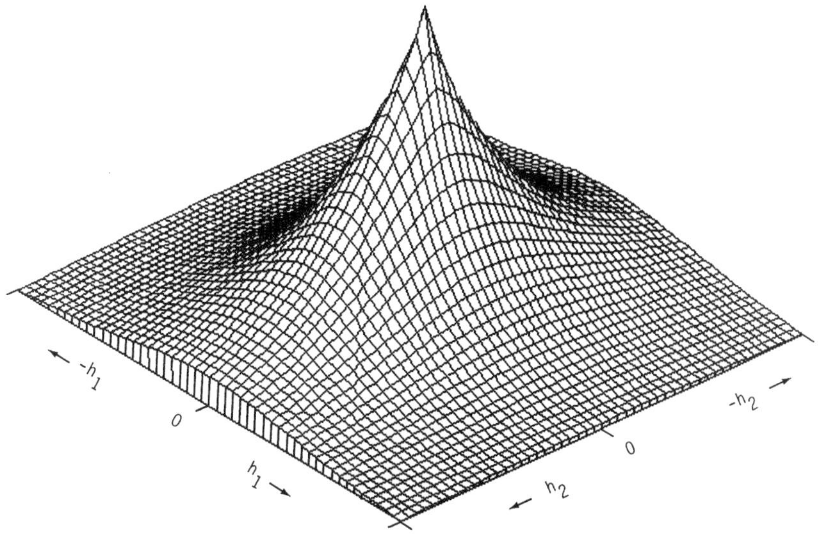

Figure 8.6 The two-dimensional covariance of Eq. (2)

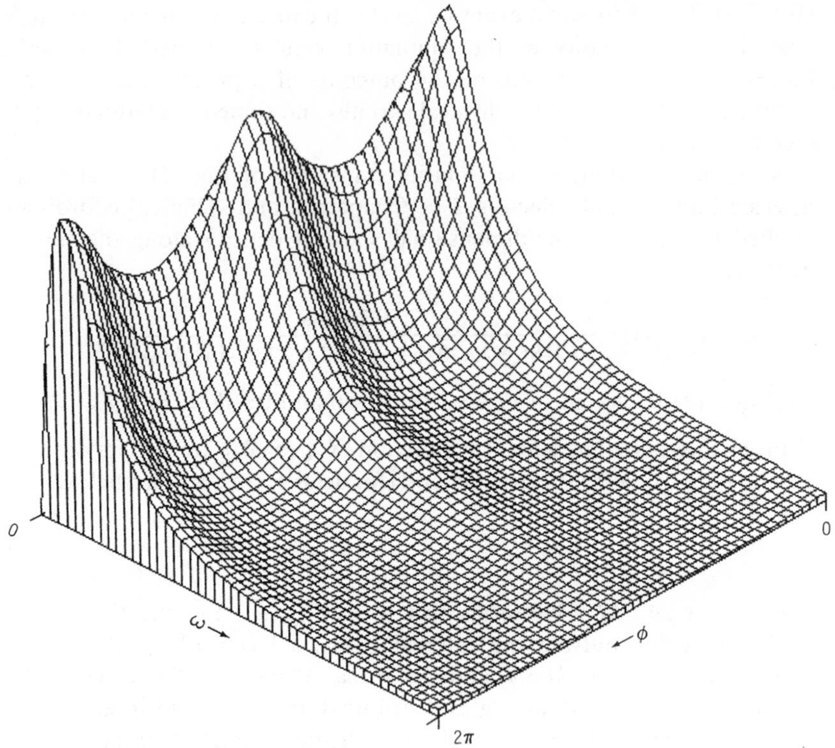

Figure 8.7 The one-dimensional spectral density function of Eq. (3)

increment $\delta\omega_j$ is uniformly distributed in the interval $(-\Delta\omega_j'/2, \Delta\omega_j'/2)$, where $\Delta\omega_j' \ll \Delta\omega_j$. In this application $N = 16$ equally spaced lines have been used.

We can now apply summation (5), Section 3 of Chapter 6 for $n = 2$ and $N = 16$, viz.,

$$X_2(\mathbf{s}) = \frac{1}{\sqrt{16}} \sum_{i=1}^{16} X_{1,\boldsymbol{\theta}_i}(s_i) \qquad (5)$$

where $s_i = \mathbf{s} \cdot \boldsymbol{\theta}_i$, to produce simulated values on a grid of 4800 points (80×60), leading to the contour map of Fig. 8.8. The corresponding perspective plot is shown in Fig. 8.9.

The sampled mean of the simulation map of Fig. 8.8 is about 6.015, which is close to the model (6.0). The simulated covariance, that is, the covariance that corresponds to the SRF (5), is given by [Eq. (6), Section 3 of Chapter 6 for $n = 2$]

$$c_2^*(\mathbf{h}) = \frac{1}{N} \sum_{i=1}^{N} c_1(\mathbf{h} \cdot \boldsymbol{\theta}_i) \qquad (6)$$

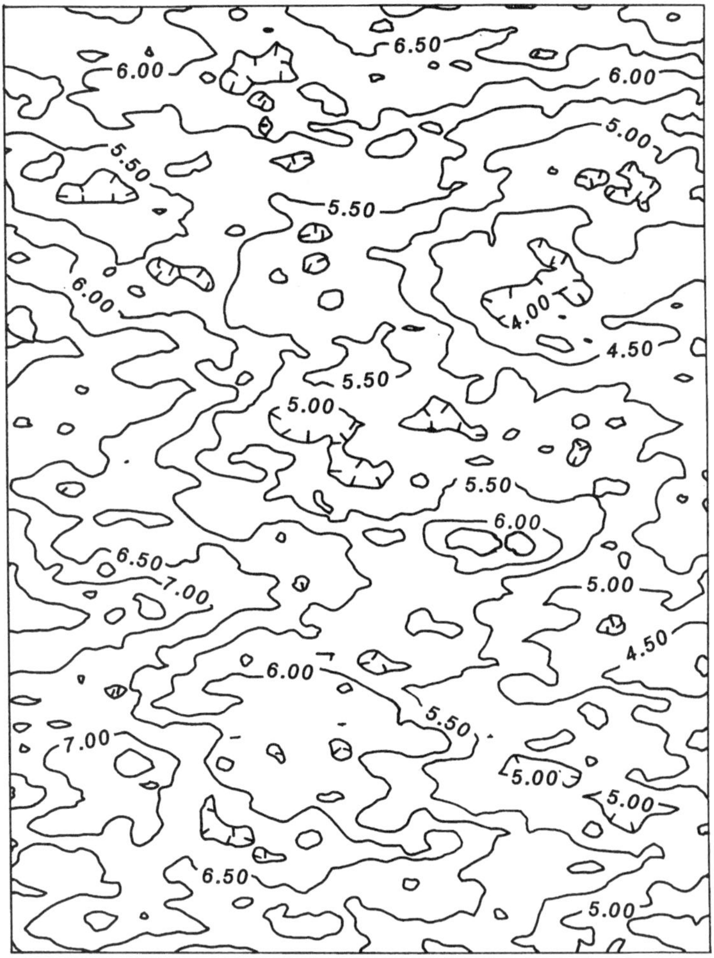

Figure 8.8 Simulated contour map

Simulated covariances calculated for various values of the number N of lines used in summations (5) and (6), converge to the theoretical model of Eq. (2) very fast as the number N increases. The simulated covariances corresponding to $N = 16$ lines show excellent agreement with the theoretical ones. For this number of lines, the simulated marginal distribution of the realizations $X_2(\mathbf{s}), \mathbf{s} \in R^2$ is found to be practically Gaussian. Several simulated maps like that of Fig. 8.8 were produced and then used to calculate the ensemble statistics of the underlying process. As the number of the simulated maps taken increases, the ensemble statistics converge to the theoretical ones, thus proving the ergodicity of the method (Christakos, 1987a). An interesting result is suggested by the following proposition.

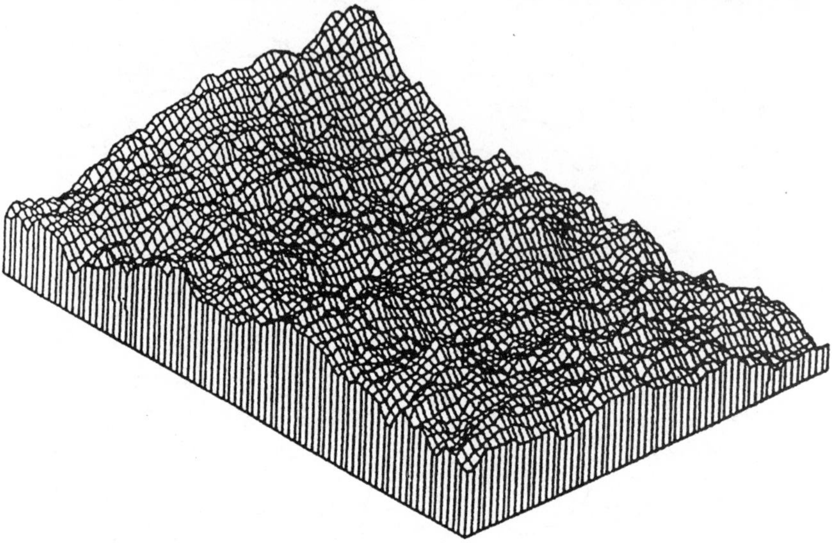

Figure 8.9 Perspective plot for the contour map of Fig. 8.8

Proposition 1: By means of the space transformation concept, a white-noise process in R^n can be simulated by summing contributions from random processes in R^1 that are no longer white noises.

Proof: A white-noise SRF $V(\mathbf{s})$, $\mathbf{s} \in R^n$, is generated by a covariance function of the form

$$c_n(\mathbf{h}) = \sigma_v^2 \, \delta(\mathbf{h}) \tag{7}$$

where $\mathbf{h} \in R^n$ and σ_v^2 is the variance of the white noise. The corresponding spectral function is

$$C_n(\mathbf{w}) = \frac{\sigma_v^2}{(2\pi)^n} \tag{8}$$

By applying Eq. (19) of Section 2, Chapter 6 the one-dimensional spectral density for simulations along lines $\boldsymbol{\theta}$ can be written

$$C_{1,\boldsymbol{\theta}}(\omega) = \sigma_v^2 \frac{S_n |\omega|^{n-1}}{2(2\pi)^n} \tag{9}$$

which clearly is the spectral density of a non-white noise process. □

For completeness, the one-dimensional covariance corresponding to Eq. (9) is given below (Christakos, 1987a; Christakos and Panagopoulos, 1992).

Corollary 1: Let Eq. (7) be the covariance function of a white-noise SRF in R^n. The corresponding one-dimensional covariance defined by means of the ST formalism of Chapter 6 is as follows:

$$c_{1,\boldsymbol{\theta}}(t) = \sigma_v^2 \frac{(-1)^m S_{2m+1}}{2(2\pi)^{2m}} \frac{\partial^{2m}}{\partial t^{2m}} \delta(t) \qquad \text{if} \quad n = 2m+1 \tag{10}$$

and

$$c_{1,\boldsymbol{\theta}}(t) = \sigma_v^2 \frac{(-1)^{3m-1}(2m-2)!}{(2\pi)^{2m}} S_{2m} \frac{1}{t^{2m}}, \qquad \text{if} \quad n = 2m \tag{11}$$

The five-step simulation approach above also applies in the case of nonhomogeneous physical processes, by simply replacing the covariance $c_n(\mathbf{h})$ with the semivariogram $\gamma_n(\mathbf{h})$ or the generalized covariance model $k_n(\mathbf{h})$ (see Fig. 8.10). In this case, if

$$k_n(r) = \sum_{\rho=0}^{\nu} (-1)^{\rho+1} c_\rho r^{2\rho+1} \tag{12}$$

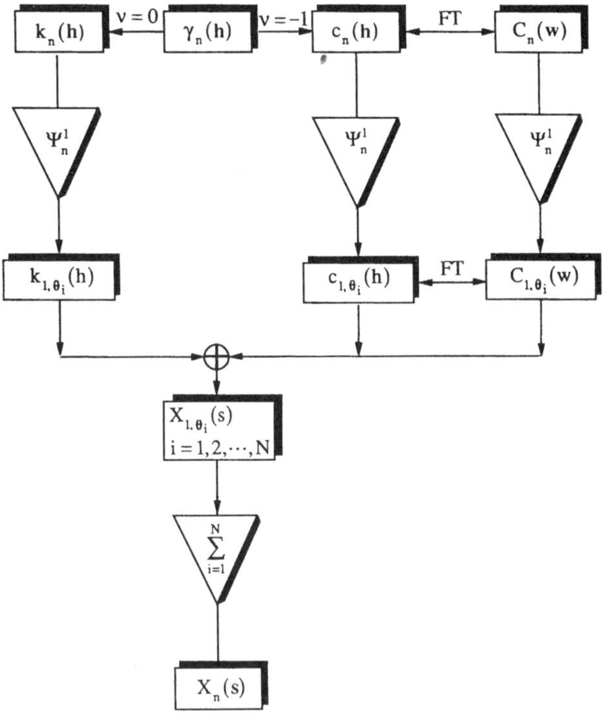

Figure 8.10 Flowchart of the turning bands simulation procedure

($r=|\mathbf{h}|$) is the GSC-ν in R^n, the corresponding GSC-ν in R^1 is given by

$$k_1(r)=\sum_{\rho=0}^{\nu}(-1)^{\rho+1}\frac{c_\rho}{B_{n,\rho}}r^{2\rho+1} \tag{13}$$

where

$$B_{n,\rho}=\frac{\rho!\,G\left(\frac{n}{2}\right)}{\sqrt{\pi}\,G\left(\frac{2\rho+n+1}{2}\right)}$$

Example 2: The measured quantity in the mine of Lavrion (Greece) is the Zn concentration (%). In the area of interest (Fig. 8.11) there are seventy five (75) measurements available, arbitrarily distributed in space and not interrupted by considerable measurement errors. The spatial variation of the Zn values is nonhomogeneous and, thus, it must be represented in terms of ISRF-ν (Chapter 3). The identification of the spatial variability of Zn concentrations is performed by means of the procedure detailed in Section 4.2.2 of Chapter 7. The results are order of intrinsity $\nu=0$ and a two-dimensional generalized spatial covariance (GSC-0)

$$k_2(r)=-0.1375r \tag{14}$$

where $r=|\mathbf{h}|$. The corresponding unidimensional GSC-0 can be found by applying IST-2, viz.,

$$k_{x,1D}(t)=\Psi_2^1[k_x(r)]=-\frac{\pi}{2}0.1375t \tag{15}$$

Next, simulations of the one-dimensional ISRF-0 $X_{1,\boldsymbol{\theta}_i}(\mathbf{s}\cdot\boldsymbol{\theta}_i)$, having the GSC-0 of Eq. (15), are produced along 16 lines. As it is shown in Section 5 below, the one-dimensional ISRF-0 on line i, $X_{1,\boldsymbol{\theta}_i}(\tau)$, $\tau=\mathbf{s}\cdot\boldsymbol{\theta}_i$, can be generated by

$$X_{1,\boldsymbol{\theta}_i}(\tau)=\sqrt{c_0}\,W_0(\tau)=\sqrt{0.1375\frac{\pi}{2}}\,W_0(\tau) \tag{16}$$

where $W_0(\tau)$ is a Wiener process. Realizations of $W_0(\tau)$ can be constructed by means of $W_0(\tau+\Delta\tau)=W_0(\tau)+\sqrt{24\,\Delta\tau}\,V(\tau)$, where $V(\tau)$ is a random variable uniformly distributed inside the interval $[-\frac{1}{2},\frac{1}{2}]$.

The summation formula, Eq. (5) above, is applied to generate Zn values on a grid of 3200 points (80×40). Among the several possible Zn realizations we keep one that honors the measurements at the 75 data locations (assuming, of course, that there is no measurement error). The constraining of the simulation process to the data is achieved by means of the method described in Section 3.2 above. The resulting simulated contour map is shown in Fig. 8.2.

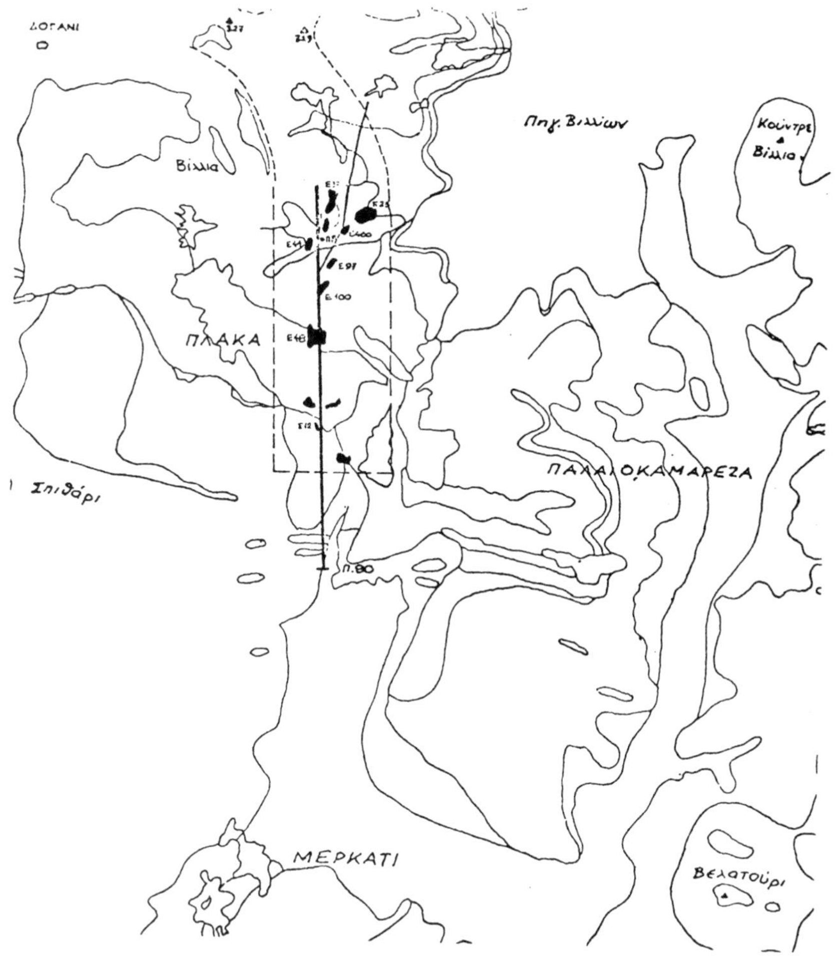

Figure 8.11 The area of the mine of Lavrion, Greece

To examine the sensitivity of the simulation results, the following tests were performed:

(a) We repeat the procedure above, this time using the 3200 simulated Zn values: We again obtain the same order of intrinsity $\nu = 0$, while the two-dimensional GSC-0 is now $k_2^*(r) = -0.1419r$. This is very close to the original GSC-0, $k_2(r) = -0.1375r$, which was obtained before using the 75 Zn values. Hence, the produced simulation maps restore the spatial variability characteristics of the Zn data available.

(b) Simulation maps of Zn concentration were generated by means of the procedure above, using first $N = 32$ and then $N = 64$ lines. Then, the spatial structure identification procedure was applied to the two new sets of simulated Zn values. The spatial variability characteristics (order of intrinsity and GSC) obtained were, for the first set of simulated Zn values, $\nu = 0$ and $k_2^*(r) = -0.1406r$; and for the second set of values, $\nu = 0$ and $k_2^*(r) = -0.1396r$. Both GSC-0 are very close to the original one, implying that the 16 lines used are in fact sufficient for the particular case study.

Example 3: In Fig. 8.12 various cases of spatial simulation are shown, including isotropic, anisotropic, and nonhomogeneous SRF.

Remark 1: All methods of SRF simulation involve approximations. The turning bands method, in particular, produces SRF that are approximately Gaussian; other approximations are due to the finite number of simulation lines, the discretization along the lines $\Delta\tau$, the discretization frequency $\Delta\omega$, and the number M of the terms in the series (4). A detailed sensitivity analysis of the influence of the above parameters on the simulation accuracy can be found in Mantoglou and Wilson (1982). From a practical point of view, perhaps the most useful way to demonstrate whether a simulation method works is to investigate its statistical behavior on synthetic data generated from models with known parameters.

4.2 The Spectral Method

Another useful simulation technique can be developed as follows. Bartlett (1955) and Matern (1960) have shown that, if $\mathbf{w}$ is a random variable with isotropic probability density $f_{n,\mathbf{w}}(\omega)$, $(\omega = |\boldsymbol{\omega}|)$ and u is a one-dimensional random variable independent of $\mathbf{w}$ and uniformly distributed between 0 and 2π, the SRF

$$X_n(\mathbf{s}) = \exp[i(\mathbf{s} \cdot \mathbf{w} + u)] \tag{17}$$

has

$$c_n(t) = c_n(0) E\{\exp[i(\mathbf{h} \cdot \mathbf{w})]\} \tag{18}$$

where $t = |\mathbf{h}|$, as covariance.

By applying the ST concept, the analysis above leads to the following simulation approach.

Step 1: Given $c_n(t)$, find $f_{n,\mathbf{w}}(\omega)$ [Section 2 of Chapter 6].

Step 2: Apply Eq. (23), Section 2 of Chapter 6 to get $f_{1,\mathbf{w}}(\omega)$.

Step 3: Then generate simulations

$$X_n(\mathbf{s}) = \exp[i(\omega \mathbf{s} \cdot \boldsymbol{\theta} + u)] \tag{19}$$

$$m_x = 6.0, \quad c_x(r) = \exp[-r], \quad r = |\mathbf{h}|$$

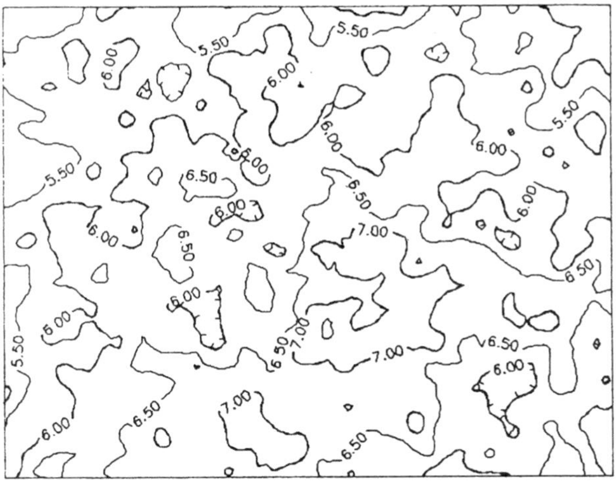

$$m_x = 0.0, \quad c_x(r) = 0.8r\, K_1(0.8r), \quad r = |\mathbf{h}|$$

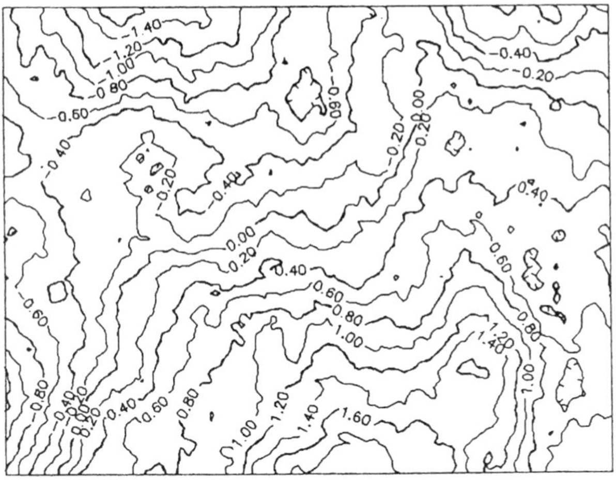

(a)

Figure 8.12 Examples of spatial simulation: (a) Isotropic SRF; (b) anisotropic SRF; (c) nonhomogeneous SRF, ISRF-0; (d) nonhomogeneous SRF, ISRF-1; and (e) nonhomogeneous SRF, ISRF-2. (*Figure continues*)

$$m_x = 6.0, \quad c_x(h_1,h_2) = \exp\left[-\sqrt{0.85^2 \times h_1^2 + 0.20^2 \times h_2^2}\right], \quad \mathbf{h} = (h_1,h_2)$$

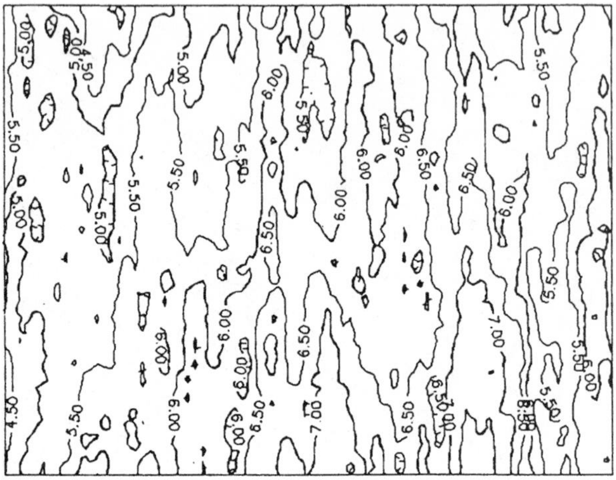

(b)

$$c_x(s,s') = k_x(r) + p_0(s,s'), \quad \nu = 0 \quad \text{and} \quad k_x(r) = -r, \quad r = |s - s'| = |\mathbf{h}|$$

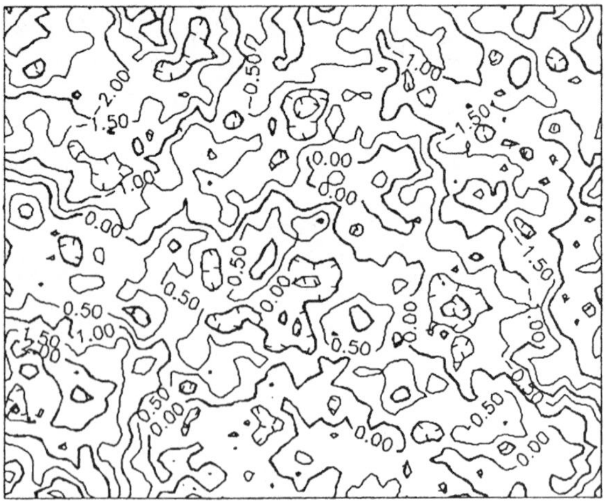

(c)

Figure 8.12 (*Continued*)

$$c_x(s,s') = k_x(r) + p_1(s,s'), \quad \nu = 1 \text{ and } k_x(r) = r^3, \; r = |s - s'| = |\mathbf{h}|$$

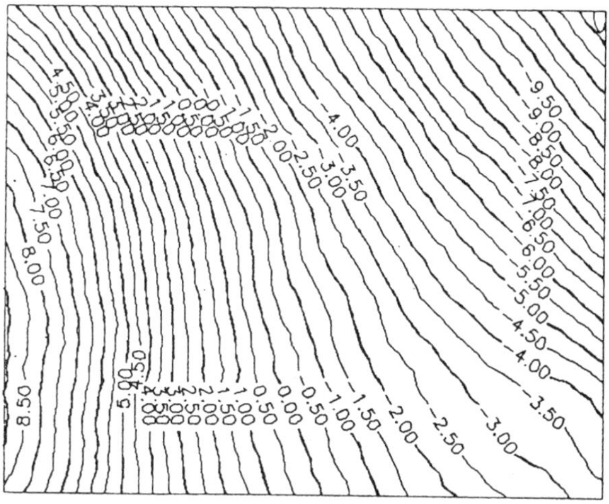

(d)

$$c_x(s,s') = k_x(r) + p_2(s,s'), \quad \nu = 2 \text{ and } k_x(r) = -r^5, \; r = |s - s'| = |\mathbf{h}|$$

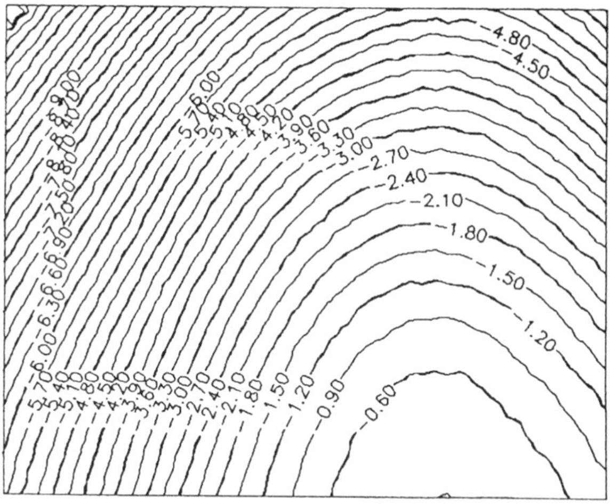

(e)

Figure 8.12 (*Continued*)

where ω has one-dimensional probability density $f_{1,w}(\omega)$ and $\boldsymbol{\theta}$ has a uniform density $1/S_n$.

Remark 2: This method has been used extensively in hydrologic sciences for the generation of spatial processes (e.g., Mejia and Rodriguez-Iturbe, 1974).

5. Techniques of One-Dimensional Simulation

In the preceding section, a multidimensional simulation was shown to be technically reducible to a much simpler one-dimensional simulation. The problem of generating one-dimensional realizations remains. Recall that the terms "random process," "stationary," etc., are the one-dimensional counterparts of the terms "random field," "homogeneous," etc., which are used when more than one spatial dimension is involved.

The most important of the various existing one-dimensional simulation methods will be reviewed briefly. These methods can be classified into more than one way. For example, one classification considers two major groups of simulation techniques: space-domain and frequency-domain (spectral) techniques. Another classification distinguishes between techniques for simulating stationary random processes, and techniques for simulating nonstationary random processes.

5.1 Convolution Techniques

An extensively used technique for simulating stationary random processes is one where the unidimensional covariance is expressed by the convolution of a properly chosen function $p(v)$, viz.,

$$c_1(r) = \int_{R^1} p(v)\, p(v+r)\, dv \tag{1}$$

Line realizations are obtained as the *weighted summation*

$$X_1(s) = \sum_{j=-M}^{M} p(j\, \Delta s)\, U(s+j) \tag{2}$$

where $U(s)$ are random variables. The latter are independent of each other and have the same distribution with zero mean and known variance (e.g., Journel, 1974).

5.2 Moving-Average–Autoregressive Techniques

Unidimensional stationary simulations also may be produced by a *moving-average* process similar to Eq. (2), namely,

$$X_1(s) = \sum_{j=-M}^{M} p_j U(s-j) \tag{3}$$

where $U(s)$ are Gaussian white-noise processes with zero mean and unit variance; M is chosen so that

$$\left| 1 - \frac{1}{\sigma_x^2} \sum_{j=-M}^{M} p_j^2 \right| < \varepsilon$$

where ε is a small positive number and the coefficients p_j are calculated by

$$p_j = \frac{\Delta s}{\pi} \int_0^{\pi/\Delta s} \sqrt{\frac{C_1(\omega)}{\Delta s}} \cos(j\omega\, \Delta s)\, d\omega$$

where $C_1(\omega)$ is the spectral density of the process $X_1(s)$ and Δs is the discretization step along the simulation line (note that $p_j = p_{-j}$). Since the $U(s)$ are Gaussian, Eq. (3) generates Gaussian processes too.

Stationary simulations can be generated by the *autoregressive process*

$$X_1(s) = \sum_{j=1}^{M} a_j X_1(s-j) + W(s) \tag{4}$$

where a_j are coefficients and $W(s)$ is a zero-mean Gaussian process with specified variance, whereas the state $\{X_1(s-j); j = s-1, \ldots, s-M\}$ must be determined before start of recursion.

Furthermore, combining Eqs. (3) and (4), on-line simulations can be generated by an autoregressive–moving-average process. The same role may be played by random differential equations whose solutions may serve as realizations of the process (e.g., Ripley, 1981).

5.3 Spectral Techniques

Spectral techniques concentrate on the spectral function of the process. Rice (1954) created a popular method in which the simulation is expressed as a series of cosine functions

$$X_1(s) = \sum_{j=1}^{M} \sqrt{2C_1(\omega_j)\, \Delta\omega}\, \cos(\omega_j s + \phi_j) \tag{5}$$

where $C_1(\omega_j)$ is the unidimensional spectral density that vanishes outside the interval $[-\omega_0, \omega_0]$, $\Delta\omega = 2\omega_0/M$, $\omega_j = -\omega_0 + (j - \frac{1}{2})\Delta\omega$, and ϕ_j are random phase angles uniformly and independently distributed in the interval $(0, 2\pi)$. Due to the central limit theorem, $X_1(s)$ is a stationary Gaussian random process; it is, also, periodic. The covariance and spectrum of the process can be shown to approach their true values at the rate $1/M^2$, where M is the number of terms in the series.

To avoid periodicities, Shinozuka and Jan (1972) recommend another version of Eq. (5) where in place of ω_j they use $\omega_j' = \omega_j + \delta\omega_j$, viz.,

$$X_1(s) = \sum_{j=1}^{M} \sqrt{2C_1(\omega_j)\,\Delta\omega}\,\cos(\omega_j' s + \phi_j) \tag{6}$$

where $\delta\omega_j$ is a small random frequency increment, uniformly distributed in the interval $(-\Delta\omega_j'/2, \Delta\omega_j'/2)$, where $\Delta\omega_j' \ll \Delta\omega_j$. Again, the generated random process $X_1(s)$ is stationary and Gaussian; it is, also, ergodic irrespective of N. This technique was used in Example 1 of Section 4 above.

5.4 Fast Fourier Transform Techniques

Some spectral techniques are based on *fast Fourier transform* algorithms. In fact, both convolution and cosine series methods can be written in forms suitable for the fast Fourier transform.

For example, Eq. (5) may be written as

$$X_1(s) = \mathrm{Re}\left\{\sum_{j=1}^{M} [\sqrt{2C_1(\omega_j)\,\Delta\omega}\,\exp[i\phi_j]]\exp[i\omega_j s]\right\} \tag{7}$$

where Re means "real part of." Fast Fourier transform techniques may be applied with efficiency to convolution (2) too. The Fourier transform $Z_1(\omega)$ of $X_1(s)$ is first specified from the Fourier transforms of $p(v)$ and $U(s)$. Then, $X_1(s)$ is obtained by an inverse Fourier transform

$$X_1(s) = \mathrm{FT}^{-1}[Z_1(\omega)] \tag{8}$$

The choice of one or the other method is based on several considerations: the ease in calculating the one-dimensional covariance or spectral functions, the desired accuracy, the relative rates of convergence, and the cost of the numerical simulator for the particular case. Furthermore, in practice, evaluating the function $p(v)$ of the convolution form Eq. (1) in the space domain is not always possible. This may not be a problem in the frequency domain were Eq. (1) becomes $C_1(\omega) = |P(\omega)|^2$ and $P(\omega)$ is the Fourier transform of $p(v)$. [Obtaining $P(\omega)$ from $C_1(\omega)$ may be easy if $C_1(\omega)$ is rational.] If the spectrum $C_1(\omega)$ is known, an inverse Fourier transform will yield $p(v)$, which may be used for on-line simulations in the space domain. The cost of a multidimensional simulation is directly proportional to the cost of the on-line simulations and, of course, the number of lines used.

5.5 Nonstationary Random Processes

For the purposes of this book the most important case of nonstationary simulation is that of unidimensional ISRF-ν (or, equivalently, intrinsic random process of order ν, IRP-ν).

To simulate an IRP-ν $X_1(s)$ with a generalized covariance of order ν $k_1(h)$, one needs to apply the representation (73), Section 3.4 of Chapter 3, viz.,

$$X_1(s) = b_0 W_0(s) + b_1 \int_0^s W_0(v)\, dv + \cdots + b_\nu \times \int_0^s \frac{(s-v)^{\nu-1}}{(\nu-1)!} W_0(v)\, dv \tag{9}$$

where

(a) The $W_0(s)$ is an IRP-0 with $k_{w_0}(h) = -|h|$. Usually, $W_0(s)$ is taken to be a Wiener random process. Realizations of $W_0(\tau)$ are constructed through the simple recursive formula

$$W_0(s+h) = W_0(s) + \sqrt{24h}\; V(s) \tag{10}$$

where $V(\tau)$ is a random variable uniformly distributed inside the interval $[-\frac{1}{2}, \frac{1}{2}]$ (Orfeuil, 1972; Dimitrakopoulos, 1990).

(b) The b_ρ ($\rho = 0, 1, \ldots, \nu$) are coefficients that can be calculated in terms of the coefficients of $k_1(h)$. Suppose that the $k_1(h)$ is of the form

$$k_1(h) = \sum_{\rho=0}^{\nu} (-1)^{\rho+1} c_\rho |h|^{2\rho+1}/(2\rho+1)!$$

Then the coefficients b_ρ are related to the covariance coefficients c_ρ as follows:

$$\sum_{\rho=0}^{\nu} c_\rho u^{2(\nu-\rho)} = \left| \sum_{\rho=0}^{\nu} b_\rho (iu)^{\nu-\rho} \right|^2 \tag{11}$$

where $i = \sqrt{-1}$.

For illustration, Table 8.2 provides simulation formulas for the important cases of $\nu = 0$, 1, and 2. This technique was used in Example 2 of Section 4 above.

A method of constructing nonstationary simulations, not necessarily IRP-ν, is provided by extending Eq. (5), where the random phase ϕ_j is now taken to be nonuniformly distributed. Then, by assuming that a relationship between the nonhomogeneous characteristics of $X_1(s)$ and the probability density of ϕ_j can be established, Eq. (5) generates nonstationary processes. Some authors (e.g., Veneziano, 1980) suggest the replacement of the spectral

Table 8.2 Simulation Formulas for IRP-ν

$\nu=0$:

$$X_1(s)=\sqrt{c_0}\,W_0(s)$$

$\nu=1$:

$$X_1(s)=\sqrt{c_0}\,W_0(s)+\sqrt{c_1}\int_0^s W_0(v)\,dv$$

$\nu=2$:

$$X_1(s)=\sqrt{c_0}\,W_0(s)+\sqrt{c_1+2\sqrt{c_0c_2}}\times\int_0^s W_0(v)\,dv+\sqrt{c_2}\times\int_0^s (s-v)\,W_0(v)\,dv$$

density function $C_1(\omega_j)$ in Eq. (5) by the evolutionary mean power spectral density function $C_1(s, \omega_j)$ (for definitions see Section 5.3 of Chapter 2), so that the formula of nonstationary simulation becomes

$$X_1(s)=\sum_{j=1}^{M}\sqrt{2C_1(s,\omega_j)\,\Delta\omega}\,\cos(\omega_j s+\phi_j) \tag{12}$$

Another approach, also based on an evolutionary spectral density function, is suggested by the expression

$$X_1^2(s)=\frac{\partial}{\partial s}\int_{R^1}\lim_{L\to\infty}[LC_1(s,\omega)]\,d\omega \tag{13}$$

where the evolutionary spectral density function is defined as

$$C_1(s,\omega)=\int_{R^1}c_1(s,t)\exp[-i\omega t]\,dt$$

and

$$c_1(s,t)=\frac{1}{L}\int_{s-L}^{s}X_1(u)X_1(u+t)\,du$$

However, the practical application of this approach needs some additional information regarding the correlation structure of the process $X_1(s)$.

Finally, a powerful nonstationary simulation method in terms of the *impulse response function* $h(u)$ is as follows. Assume that one seeks the simulation of the random process $X_1(s)$ with a given spectral density function $C_x(s, \omega)$. Choose a process $Z_1(s)$ that can be simulated easily, say a nonstationary white-noise process with a spectral density $C_z(s, \omega)$. The Fourier transform of $C_z(s, \omega)$ determines the covariance $c_z(s, u)$ of $Z_1(s)$. If

$$\beta(s, \omega) = \frac{c_z(s-\omega, u)}{c_z(s, u)}$$

the following system of equations must be solved with respect to $h(u)$, $H(\omega)$, $\theta(s, u)$, and $\Theta(s, \omega)$ (see, e.g., Cacko *et al.*, 1988),

$$\begin{aligned} H(\omega) &= \int_0^\infty h(u) \exp[-i\omega u]\, du \\ \theta(s, u) &= \beta(s, u) h(u) \\ \Theta(s, \omega) &= \int_0^\infty \theta(s, u) \exp[i\omega u]\, du \\ C_x(s, \omega) &= H(\omega)\Theta(s, \omega) C_z(s, \omega) \end{aligned} \tag{14}$$

Assuming that the solution of the system (14) yields a function $h(u)$ such that $h(u) \xrightarrow[u\to\infty]{} 0$, the process $X_1(s)$ can be simulated by

$$X_1(s) = \sum_{j=0}^{M} p_j Z(s-j) \tag{15}$$

where the coefficients $p_j = h(j\,\Delta s)\,\Delta s$ are such that $p_j \to 0$ for all $j > M$.

6. Simulation of Integrated Natural Processes

The ST concept may introduce valuable insights into the study of spatially integrated patterns. Assume that the ST-1 projections of a geologic process $X_n(\mathbf{s})$ can be measured along lines or over planes, as, for example, the total or integrated bulk density along well logs. Then the ST formulas developed in Chapter 6 connect these projections with the original process in both the space and frequency domains. Consequently, the geologic process may be simulated by means of these projections (see also Christakos, 1987a).

The ST-1 integrations are defined over hyperplanes (i.e., over the entire range of the process). However, in practical applications, the integral or the average of a geologic process $X_n(\mathbf{s})$ may be taken over finite sets such as mining blocks or polluted regions. If this is the case, the factor

$$\frac{A(\mathbf{c}, \mathbf{w})}{\{\exp[i(\mathbf{c}\cdot\mathbf{w})]\}^2}$$

where $A(\mathbf{c}, \mathbf{w})$ depends on geometric characteristics of the sets, may be inserted in the calculation. To evaluate this factor, the finite ST-1 projection must be defined. Specifically, let

$$X_v(Q_\mathbf{y}) = \frac{1}{V(Q_\mathbf{y})} \int_{Q_\mathbf{y}} X_n(\mathbf{s})\, d\mathbf{s} \tag{1}$$

be the ST-1 projection of $X_n(\mathbf{s})$ over the finite set $Q_\mathbf{y}$, which belongs to a linear subset of R^n; vector $\mathbf{y}$ defines the position of $Q_\mathbf{y}$ and the symbol $V(Q_\mathbf{y})$ denotes its surface.

Matern (1960) has shown that if $\mathbf{s}$, $\mathbf{s}'$ are points chosen independently and at random in the set $Q_\mathbf{y}$ and $Q'_\mathbf{y}$, respectively, and $\mathbf{c} = \mathbf{y} - \mathbf{y}'$, then $A(\mathbf{c}, \mathbf{w})$ is the characteristic function of the difference $\mathbf{h} = \mathbf{s} - \mathbf{s}'$. Moreover, analytical expressions for certain shapes of $Q_\mathbf{y}$ were derived: for example, when $Q_\mathbf{y}$ is an n-dimensional rectangular block with sides d_i, $i = 1, \ldots, n$

$$A(\mathbf{c}, \mathbf{w}) = \left[2^n \exp[i(\mathbf{c} \cdot \mathbf{w})] \prod_{j=1}^{n} \frac{\sin \dfrac{d_j w_j}{2}}{d_j w_j} \right]^2 \tag{2}$$

Using this result,

$$\tilde{C}_Q(\mathbf{w}) = \left[2^n \prod_{j=1}^{n} \frac{\sin \dfrac{d_j w_j}{2}}{d_j w_j} \right]^2 C_n(\mathbf{w}) \tag{3}$$

$\tilde{C}_Q(\mathbf{w})$ is now the spectral function of the average process Eq. (1) to be simulated and $C_n(\mathbf{w})$ is the spectral function of the point process $X_n(\mathbf{s})$.

The quantity within brackets in Eq. (3) is the *operator transfer function* of the linear system defined in Eq. (1) for the particular geometry of the set $Q_\mathbf{y}$. Taking into account Eq. (19), Section 2 of Chapter 6, Eq. (3) gives

$$\tilde{C}_Q(\mathbf{w}) = \frac{2}{S_n \omega^{n-1}} \left[2^n \prod_{j=1}^{n} \frac{\sin \dfrac{d_j w_j}{2}}{d_j w_j} \right]^2 C_{1,\boldsymbol{\theta}}(\omega) \tag{4}$$

which implies that the $X_v(Q_\mathbf{y})$ may be simulated by applying the turning bands operator directly on the one-dimensional point process $X_{1,\boldsymbol{\theta}}$. Equations (3) and (4) illustrate certain important features of the frequency domain approach of simulating geologic processes: for example, the statistical properties of the output process $X_v(Q_\mathbf{y})$ may be evaluated in terms of the properties of the input process $X_n(\mathbf{s})$.

7. Simulation of Dynamic Stochastic Systems

As we saw in Chapter 6, space transformations may lead to an interesting method of studying transport-type equations such as those representing hydrologic processes. Examples included partial differential equations used to model flow in porous media. By decomposition into ST-projections, the multidimensional equation can be reduced formally to an equation of the same type in R^1; after this has been solved, solutions of the original equation can easily be constructed.

A similar approach can be applied within the framework of stochastic simulation, aiming at the study of dynamic and uncertain environments in several dimensions. Realizations of an output spatial process may be generated by summing contributions from one-dimensional simulations of the input process.

Example 1: Consider the two-dimensional equation governing perturbations in steady groundwater flow (Mizell *et al.*, 1982)

$$\frac{\partial^2 h(s_1, s_2)}{\partial s_1^2} + \frac{\partial^2 h(s_1, s_2)}{\partial s_2^2} - \frac{\partial \tau(s_1, s_2)}{\partial s_1} J_1 = 0 \tag{1}$$

where $h(s_1, s_2)$, $\tau(s_1, s_2)$ are fluctuations about the means $M_{\mathrm{H}}(s_1)$ and M_{T} of the hydraulic head $H(s_1, s_2)$ and the log transmissivity $\log T(s_1, s_2)$. The mean $M_{\mathrm{H}}(s_1)$ is assumed to be a function of only the s_1 direction, whereas the mean M_{T} is constant, $J_1 = -\partial M_{\mathrm{H}}(s_1)/\partial s_1$ is the mean gradient. Assuming homogeneity of the processes involved and working in the convenient frequency domain, Eq. (1) becomes

$$C_{h,2}(w_1, w_2) = \frac{J_1^2 w_1^2}{\pi \omega^5} C_{\tau(1,\boldsymbol{\theta})}(\omega) \tag{2}$$

where $C_{h,2}(w_1, w_2)$, and $C_{\tau(1,\boldsymbol{\theta})}(\omega)$ are the two-dimensional head spectrum and the one-dimensional log transmissivity spectrum. Equation (2) allows the generation of two-dimensional simulations of hydraulic head directly from on-line simulations of the corresponding unidimensional log transmissivity. The latter is simply the IST-2 of $\tau(s_1, s_2)$.

Example 2: Let $X(s_1, s_2)$ be an SRF representing the concentration of an aerosol substance transported with an air flow in the atmosphere. Assume that the phenomenon is two-dimensional and is governed by the equation (e.g., Marchuk, 1986)

$$V_1 \frac{\partial X(s_1, s_2)}{\partial s_1} + V_2 \frac{\partial X(s_1, s_2)}{\partial s_2} - \mu \nabla^2 X(s_1, s_2) = W \delta(\mathbf{s} - \mathbf{s}_0) \tag{3}$$

where V_1 and V_2 are the random velocity components, W is the capacity of the source exhausting the aerosol into the atmosphere, μ is the diffusion coefficient, and $\mathbf{s}_0$ determines the location of the exhaust point.

The SRF V_1 and V_2 can be expressed in terms of their means m_{v_1}, m_{v_2} and the perturbations V_1', V_2', namely, $V_i = m_{v_i} + V_i'$, $i = 1, 2$. For an infinite domain, the solution of Eq. (3) is given by

$$X(s_1, s_2) = \frac{W}{2\pi\mu} \exp\left[\frac{(m_{v_1} + V_1')(s_1 - s_{0,1}) + (m_{v_2} + V_2')(s_2 - s_{0,2})}{2\mu}\right]$$
$$\times K_0\left[\frac{\sqrt{(m_{v_1} + V_1')^2 + (m_{v_2} + V_2')^2}}{2\mu}|\mathbf{s} - \mathbf{s}_0|\right] \tag{4}$$

where

$$K_0(\chi) = \int_0^\infty \exp[-\chi\, ch\psi]\, d\psi, \qquad \chi > 0$$

is the McDonald function. Assuming that their covariances (or spectral density functions) are known, one may first generate realizations of the SRF V_1 and V_2, and then use Eq. (4) to produce simulations of the SRF $X(s_1, s_2)$.

The discussion above gives rise to other important features of the simulation approach. The quality of the simulated processes may be further improved by incorporating into the analysis laws known to govern the controlling mechanisms in the field. The simulations, in turn, may provide numerical solutions for the corresponding equations that model transport phenomena in porous media, turbulent motions in the atmosphere, the action of sea waves on structures, as well as in other applications. Finally, the simulation maps of aerosol transport can be used in the study of optimum location of industrial plans; similar maps of polluting hydrosol transport are valuable in the determination of the location of pollution sources in water bodies and coastal seas.

8. The Effect of Measurement Error

If the statistical information about the spatial process includes measurement errors, these may play an important role in the simulation. To simplify the analysis, the process is assumed statistically isotropic and the measurement model linear

$$Y_n(\mathbf{s}) = X_n(\mathbf{s}) + V_n(\mathbf{s}) \tag{1}$$

$Y_n(\mathbf{s})$ are the measured values of the actual process $X_n(\mathbf{s})$, and the measurement error $V_n(\mathbf{s})$ is a zero-mean white-noise with variance σ_V^2, uncorrelated to $X_n(\mathbf{s})$ and $Y_n(\mathbf{s})$. Covariances of the processes in Eq. (1) then satisfy

$$c_{Y_n}(r) = c_{X_n}(r) + \sigma_V^2\, \delta(r) \tag{2}$$

Expressions similar to Eq. (2) may arise, also, when simulating from covariances having an impulse at the origin (nugget effect). In the frequency domain, the n-dimensional Fourier transform pair $[\delta(\mathbf{s}), 1/(2\pi)^n]$ can be taken into account together with Proposition 1 and Corollary 1 of Section 4 above to yield

$$\Psi_n^1\left[\frac{1}{(2\pi)^n}\right]=\frac{S_n\omega^{n-1}}{2(2\pi)^n}$$

which relates the Fourier transform of the $\delta(\mathbf{s})$ and its Ψ_n^1 space transform. The Fourier transform of Eq. (2) can now be formed

$$C_{Y_n}(\omega)=C_{x_n}(\omega)+C_{V_n} \tag{3}$$

where $C_{V_n}=\sigma_V^2/(2\pi)^n$ is the spectrum of the white noise $V_n(\mathbf{s})$, which is constant over the interval $(0, 2\pi)$ assuming discrete sampling concepts. Applying the Ψ_n^1 operator in Eq. (3)

$$C_{Y_1}(\omega)=C_{x_1}(\omega)+C_{V_1}(\omega) \tag{4}$$

where

$$C_{V_1}(\omega)=\frac{S_n\omega^{n-1}}{2(2\pi)^n}\sigma_V^2$$

is the spectrum of the unidimensional error process $V_1(s)$. As we see above, a white error process in R^n is generated by summing contributions from on-line simulations of a process no longer white (see also Proposition 1 of Section 4). Similar conclusions may be derived for covariances that contain an impulse at the origin. For dealing with these situations, simulation in the frequency domain may be preferable, as the spatial covariances are not always easy to simulate (Christakos, 1987a).

9. Simulation of Spatial Random Fields by Means of Frequency Domain Techniques

The multidimensional frequency-domain simulation techniques are extensions of the unidimensional techniques discussed in Section 5 above. Multidimensional frequency-domain simulation techniques have been applied extensively in numerous applications (see, e.g., Borgman, 1969; Shinozuka, 1971; Shinozuka and Jan, 1972).

Let us consider first *homogeneous* SRF. The simulation method that follows generates Gaussian SRF. The fundamental simulation equation in R^n is

$$X(\mathbf{s})=\sum_{j_1=1}^{M_1}\cdots\sum_{j_n=1}^{M_n}A(j_1,\ldots,j_n)\cos\left[\sum_{i=1}^{n}w'_{ij_i}s_i+\phi_{j_1\ldots j_n}\right] \tag{1}$$

where

$$A(j_1, \ldots, j_n) = \sqrt{2C_x(w_{1j_1}, \ldots, w_{nj_n})\,\Delta w_1 \ldots \Delta w_n} \tag{2}$$

the spectral density function $C_x(w_{1j_1}, \ldots, w_{nj_n})$ is assumed to be negligible outside a given domain

$$W = \{\mathbf{w}: -\omega_i \leq w_i \leq \omega_i,\ \omega_i \in R_+^1;\ i = 1, \ldots, n\}$$

$$w_{ij_i} = -\omega_i + (j_i - \tfrac{1}{2})\,\Delta w_i$$

$$\Delta w_i = \frac{2\omega_i}{M_i}, \qquad w'_{ij_i} = w_{ij_i} + \delta w_i$$

where δw_i are random frequencies uniformly distributed between $-\Delta w'_i/2$ and $\Delta w'_i/2$ ($\Delta w'_i \ll \Delta w_i$); the $\phi_{j_1 \ldots j_n}$ is a random angle distributed uniformly between 0 and 2π.

For *nonhomogeneous* SRF the spectral density function $C_x(w_{1j_1}, \ldots, w_{nj_n})$ in Eq. (2) should be replaced by the corresponding evolutionary mean power spectral density function $C_x(\mathbf{s}, w_{1j_1}, \ldots, w_{nj_n})$ (Section 5.3 of Chapter 2), viz.,

$$A(\mathbf{s}, j_1, \ldots, j_n) = \sqrt{2C_x(\mathbf{s}, w_{1j_1}, \ldots, w_{nj_n})\,\Delta w_1 \ldots \Delta w_n} \tag{3}$$

Fast Fourier transform algorithms are useful tools in SRF simulation. In the case of Eq. (1), for example, the application of the fast Fourier transform algorithm leads to

$$X(\mathbf{s}) = \mathrm{Re}\left\{\sum_{j_1=1}^{M_1} \cdots \sum_{j_n=1}^{M_n} A(j_1, \ldots, j_n) \exp\left[i\left(\sum_{i=1}^{n} w'_{ij_i} s_i + \phi_{j_1, \ldots, j_n}\right)\right]\right\} \tag{4}$$

A frequency domain technique that has been used with considerable success in hydrology (e.g., Bras and Rodriguez-Iturbe, 1985) is based on the expression

$$X(\mathbf{s}) = \sqrt{\frac{2c_x(\mathbf{O})}{M}} \sum_{j=1}^{M} \cos(\mathbf{w}_j \cdot \mathbf{s} + \phi_j) \tag{5}$$

where $\mathbf{w}_j$ are independent random vectors with probability density function $C_x(\mathbf{w})/c_x(\mathbf{O})$, and ϕ_j are random angles distributed independently and uniformly between 0 and 2π, as before.

10. The Lower–Upper Triangular Matrix Technique

The lower-upper triangular matrix technique (LU; see, e.g., Elishakoff, 1983, Alabert, 1987) is a technique that produces Gaussian SRF, as well.

Let $c_x(\mathbf{s}_i, \mathbf{s}_j)$, be the covariance of a zero mean SRF $X(\mathbf{s})$. Assume that we seek simulations at m grid points. The application of the LU approach to this kind of problem consists of three steps:

Step 1: For the given covariance model $c_x(\mathbf{s}_i, \mathbf{s}_j)$ $(i, j = 1, 2, \ldots, m)$, develop the corresponding covariance matrix of size $m \times m$ as follows:

$$\mathbf{C}_x = \begin{bmatrix} c_x(\mathbf{s}_1, \mathbf{s}_1) \ldots c_x(\mathbf{s}_1, \mathbf{s}_m) \\ c_x(\mathbf{s}_2, \mathbf{s}_1) \ldots c_x(\mathbf{s}_2, \mathbf{s}_m) \\ \vdots \qquad\qquad \vdots \\ c_x(\mathbf{s}_m, \mathbf{s}_1) \ldots c_x(\mathbf{s}_m, \mathbf{s}_m) \end{bmatrix} \tag{1}$$

Step 2: Equation (1) is a nonnegative-definite symmetric matrix that can be decomposed into a product of a lower triangular matrix $\mathbf{L}$ and an upper triangular matrix $\mathbf{U}$ by means of the Cholesky algorithm (see also Example 1 below); that is,

$$\mathbf{C}_x = \mathbf{LU} \tag{2}$$

where $\mathbf{U} = \mathbf{L}^{\mathrm{T}}$.

Step 3: Suppose that $\mathbf{V}$ is a vector of m independent standard Gaussian random variables, and define the vector

$$\mathbf{X} = \mathbf{LV} \tag{3}$$

where $\mathbf{X}^{\mathrm{T}} = [X(\mathbf{s}_1), \ldots, X(\mathbf{s}_m)]$. The simulations generated by Eq. (3) have zero mean and covariance

$$E[(\mathbf{LV})(\mathbf{LV})^{\mathrm{T}}] = \mathbf{L}E[\mathbf{VV}^{\mathrm{T}}]\mathbf{U} = \mathbf{LIU} = \mathbf{C}_x$$

as required. Moreover, since (3) is a linear combination of independent identically distributed random variables, according to the central limit theorem it will produce simulations of Gaussian SRF.

Remark 1: Let $Z(\mathbf{s})$ be an SRF with a nonzero mean value $E[Z(\mathbf{s})]$. By setting $Z(\mathbf{s}) = X(\mathbf{s}) + E[Z(\mathbf{s})]$, where $X(\mathbf{s})$ is a zero mean SRF, one can generate realizations of $Z(\mathbf{s})$ by

$$\mathbf{Z} = \mathbf{LV} + \mathbf{M} \tag{4}$$

where $\mathbf{Z}^{\mathrm{T}} = [Z(\mathbf{s}_1), \ldots, Z(\mathbf{s}_m)]$, $\mathbf{M}^{\mathrm{T}} = [E[Z(\mathbf{s}_1)], \ldots, E[Z(\mathbf{s}_m)]]$ is the corresponding mean vector, and the $X(\mathbf{s})$ is simulated through the LU-matrix technique discussed above.

Example 1: Consider an SRF $X(\mathbf{s})$ with a covariance matrix $\mathbf{C}_x$ having elements c_{ij} $(i, j = 1, \ldots, m)$. The elements of the corresponding lower triangular matrix $\mathbf{L}$ will be given by (Cholesky decomposition):

$$\lambda_{i1} = c_{i1}(c_{11})^{-1/2} \qquad \text{and} \quad \lambda_{ii} = \left[c_{ii} - \sum_{k=1}^{i-1} \lambda_{ik}^2\right]^{1/2}$$

for $1 \leq i \leq m$; and

$$\lambda_{ij} = \begin{cases} \left[c_{ij} - \sum_{k=1}^{j-1} \lambda_{ik}\lambda_{jk} \right] \lambda_{jj}^{-1} & \text{for} \quad 1 < j < i \leq m \\ 0 \quad \text{for} \quad i < j \leq m \end{cases}$$

Then, realizations of the SRF are generated by means of Eq. (3) above.

11. The Karhunen–Loeve Expansion Technique

This method is based on the Karhunen–Loeve theorem (e.g., Loeve, 1953). According to this theorem, an SRF $X(\mathbf{s})$ defined on a domain U has the orthogonal expansion

$$X(\mathbf{s}) = \sum_{i=1}^{\infty} \lambda_i x_i \phi_i(\mathbf{s}) \tag{1}$$

where λ_i are constant coefficients, x_i are random variables to be determined such that

$$E[x_i x_j] = \delta_{ij} \tag{2}$$

and $\phi_i(\mathbf{s})$ are deterministic functions such that

$$\int_U \phi_i(\mathbf{s})\phi_j(\mathbf{s})\, d\mathbf{s} = \delta_{ij} \tag{3}$$

if and only if $|\lambda_i|^2$ and $\phi_i(\mathbf{s})$ are, respectively, the eigenvalues and eigenfunctions of the integral equation

$$\int_U c_x(\mathbf{s}, \mathbf{s}')\, \phi_i(\mathbf{s}')\, d\mathbf{s}' = |\lambda_i|^2 \phi_i(\mathbf{s}) \tag{4}$$

Then the series (1) converges in the mean square sense uniformly on U.

For practical purposes the series (1) is truncated at the mth term giving

$$X(\mathbf{s}) = \sum_{i=1}^{m} \lambda_i x_i \phi_i(\mathbf{s}) \tag{5}$$

where

$$x_i = \frac{1}{\lambda_i} \int_U X(\mathbf{s})\phi_i(\mathbf{s})\, d\mathbf{s} \tag{6}$$

The mean and covariance are

$$E[x_i] = \frac{1}{\lambda_i} \int_U E[X(\mathbf{s})]\phi_i(\mathbf{s})\, d\mathbf{s} \tag{7}$$

and

$$c_x(\mathbf{s}, \mathbf{s}') = \sum_{i=1}^{m} \sum_{j=1}^{m} \psi_{ij} \lambda_i \lambda_j \phi_i(\mathbf{s}) \phi_j(\mathbf{s}') \tag{8}$$

respectively, where

$$\begin{aligned} \psi_{ij} &= E[\{x_i - E[x_i]\}\{x_j - E[x_j]\}] \\ &= \frac{1}{\lambda_i \lambda_j} \int_U \int_U c_x(\mathbf{s}, \mathbf{s}') \phi_i(\mathbf{s}) \phi_j(\mathbf{s}') \, ds \, ds' \end{aligned} \tag{9}$$

$(i, j = 1, \ldots, m)$. Using the results above the simulation problem can be solved by means of the LU technique discussed in the previous section.

More specifically, the steps of the Karhunen–Loeve simulation technique are as follows:

Step 1: For the given covariance model $c_x(\mathbf{s}, \mathbf{s}')$, solve the integral equation (4) to obtain the eigenvalues λ_i and the eigenfunctions $\phi_i(\mathbf{s})$. In some cases a closed-form analytical solution is possible; in other cases a numerical approach may be necessary.

Step 2: Substitute the solutions into Eqs. (7) and (9) to obtain $E[x_i]$ and $\psi_{ij} (i, j = 1, \ldots, m)$.

Steps 3, 4, and 5: Apply steps 1, 2, and 3 of the LU technique, with (Remark 1)

$$\mathbf{Z}^{\mathrm{T}} = [x_1, \ldots, x_m], \qquad \mathbf{M}^{\mathrm{T}} = [E[x_1], \ldots, E[x_m]]$$

and

$$\mathbf{C}_x = \begin{bmatrix} \psi_{11} & \ldots & \psi_{1m} \\ \psi_{21} & \ldots & \psi_{2m} \\ \vdots & & \vdots \\ \psi_{m1} & \ldots & \psi_{mm} \end{bmatrix}$$

12. Simulation of Vector Spatial Random Fields

12.1 The Turning Bands Method

Mathematically, the extension of a scalar SRF simulation technique to a vector SRF simulation technique usually poses little difficulty. For the turning bands technique, such an extension involves merely notational changes.

Consider, for example, the case of homogeneous SRF. The scalar SRF $X_n(\mathbf{s})$ is replaced by the vector SRF

$$\mathbf{X}_n(\mathbf{s}) = [X_1(\mathbf{s}) X_2(\mathbf{s}) \ldots X_k(\mathbf{s})]^{\mathrm{T}} \tag{1}$$

the covariance $c_x(\mathbf{s}_i, \mathbf{s}_j)$ is replaced by the covariance matrix

$$\mathbf{C}_x(\mathbf{h}) = \begin{bmatrix} c_{x_1}(\mathbf{h}) & \ldots & c_{x_1x_k}(\mathbf{h}) \\ c_{x_2x_1}(\mathbf{h}) & \ldots & c_{x_2x_k}(\mathbf{h}) \\ \vdots & & \vdots \\ c_{x_kx_1}(\mathbf{h}) & \ldots & c_{x_k}(\mathbf{h}) \end{bmatrix} \tag{2}$$

and the summation (5), Section 3 of Chapter 6 becomes

$$\mathbf{X}_n(\mathbf{s}) = \frac{1}{\sqrt{N}} \sum_{i=1}^{N} \mathbf{X}_{1,\boldsymbol{\theta}_i}(\mathbf{s} \cdot \boldsymbol{\theta}_i) \tag{3}$$

where $\mathbf{X}_{1,\boldsymbol{\theta}_i}$ are vectors of unidimensional random processes associated to Eq. (1) through space transformation (e.g., Mantoglou, 1987).

Similarly, in the case of nonhomogeneous SRF the semivariogram $\gamma_n(\mathbf{h})$ and the generalized covariances $k_n(\mathbf{h})$ must be replaced by the corresponding semivariogram and generalized covariance matrices.

12.2 Stochastic Partial Difference Equation Techniques

Stochastic partial difference equation techniques (SPDE; Larimore, 1977) can be used to simulate *vector homogeneous Gaussian* SRF. The concept of this technique is based on two aspects:

(a) the spectral representation of a homogeneous SRF $X(\mathbf{s})$ is a well-defined RF $\tilde{X}(\mathbf{w})$ on the frequency domain that has independent increments; and

(b) several discrete SRF can be represented by stochastic partial difference equations excited by white-noise SRF.

Assume that the vector SRF

$$\mathbf{X}(\mathbf{s}) = [X_1(\mathbf{s}) X_2(\mathbf{s}) \ldots X_k(\mathbf{s})]^T$$

is expressed by

$$L_1(S_\mathbf{h})\mathbf{X}(\mathbf{s}+\mathbf{h}) = L_2(S_{\mathbf{h}'})\mathbf{U}(\mathbf{s}+\mathbf{h}') \tag{4}$$

where

$$L_1(S_\mathbf{h}) = \sum_{\mathbf{h}\in\Omega_1} \mathbf{B}(\mathbf{h}) S_\mathbf{h}$$

$$L_2(S_{\mathbf{h}'}) = \sum_{\mathbf{h}'\in\Omega_2} \mathbf{A}(\mathbf{h}') S_{\mathbf{h}'}$$

$S_\mathbf{h}$ is the familiar shift operator $[S_\mathbf{h}\mathbf{X}(\mathbf{s}) = \mathbf{X}(\mathbf{s}+\mathbf{h})]$; $\mathbf{A}(\mathbf{h}')$ and $\mathbf{B}(\mathbf{h})$ are coefficient matrices of size $k \times k$ such that $\mathbf{A}(\mathbf{O}) = \mathbf{B}(\mathbf{O}) = \mathbf{I}$, the identity matrix; the Ω_1 and Ω_2 are finite subdomains of the spatial simulation domain; and $\mathbf{U}(\mathbf{s}) = [U_1(s) U_1(\mathbf{s}) \ldots U_k(\mathbf{s})]^T$ is a white-noise vector SRF having a covariance $k \times k$ matrix $\mathbf{Q}$ with elements $\delta(\mathbf{s}-\mathbf{s}')E[U_i(\mathbf{s})U_j(\mathbf{s}')]$, $i, j = 1, 2, \ldots, k$.

The measured vector SRF is

$$\mathbf{Y}(\mathbf{s}) = \mathbf{X}(\mathbf{s}) + \mathbf{V}(\mathbf{s}) \tag{5}$$

where $\mathbf{Y}(\mathbf{s}) = [Y_1(\mathbf{s})\, Y_2(\mathbf{s}) \ldots Y_k(\mathbf{s})]^{\mathrm{T}}$, and $\mathbf{V}(\mathbf{s}) = [V_1(\mathbf{s})\, V_2(\mathbf{s}) \ldots V_k(\mathbf{s})]^{\mathrm{T}}$ is a vector of additive measurement noises independent of $\mathbf{X}(\mathbf{s})$. On the basis of Eq. (5), the spectral representation of the vector SRF $\mathbf{X}(\mathbf{s})$ is related to that of $\mathbf{U}(\mathbf{s})$ by

$$\tilde{\mathbf{X}}(d\mathbf{w}) = \mathbf{H}(\mathbf{w})\tilde{\mathbf{U}}(d\mathbf{w}) \tag{6}$$

where $\mathbf{H}(\mathbf{w})$ is the so-called matrix transfer function of size $k \times k$ (Chapter 2),

$$\mathbf{H}(\mathbf{w}) = L_1^{-1}\{\exp[i\mathbf{w} \cdot \mathbf{h}]\}L_2\{\exp[i\mathbf{w} \cdot \mathbf{h}]\} \tag{7}$$

Consequently, the cross-spectral density matrix of $\mathbf{Y}(\mathbf{s})$ is related to that of $\mathbf{X}(\mathbf{s})$ and $\mathbf{U}(\mathbf{s})$ by

$$\tilde{\mathbf{C}}_Y(\mathbf{w}) = \tilde{\mathbf{C}}_X(\mathbf{w}) + \tilde{\mathbf{C}}_V(\mathbf{w}) = \mathbf{H}(\mathbf{w})\tilde{\mathbf{C}}_U(\mathbf{w})\overline{\mathbf{H}(\mathbf{w})} + \tilde{\mathbf{C}}_V(\mathbf{w}) \tag{8}$$

where the bar denotes the complex conjugate transpose and $\tilde{\mathbf{C}}_V(\mathbf{w})$ is the cross-spectral density matrix of the noise vector $\mathbf{V}(\mathbf{s})$. It can be shown that the likelihood function of the measurements is approximated by

$$\Im[\tilde{\mathbf{Y}}(\mathbf{w}), \mathbf{w} \in \Omega_m, \boldsymbol{\beta}] = -\frac{m}{2}\log\{2\pi\}$$

$$-\frac{1}{2}\sum_{\mathbf{w}\in\Omega_m}[\log|\tilde{\mathbf{C}}_Y(\mathbf{w})|] + \overline{\mathbf{Y}(\mathbf{s})}\tilde{\mathbf{C}}_Y^{-1}(\mathbf{w})\mathbf{Y}(\mathbf{s})] \tag{9}$$

where the dependence of $\tilde{\mathbf{C}}_Y(\mathbf{w})$ on the vector of the unknown parameters $\boldsymbol{\beta}$ is provided by Eqs. (6) and (7), and the $\mathbf{A}(\mathbf{h}')$, $\mathbf{B}(\mathbf{h})$, and $\mathbf{Q}$ are assumed to be parameterized as functions of $\boldsymbol{\beta}$. Also, m is the number of measurements at the set of points $\mathbf{s} \in R^n$ in a rectangle of the simulation domain of interest with components $0 \le s_i < m_i$ and $m = m_1 m_2 \ldots m_n$; Ω_m is the set of discrete frequencies $\mathbf{w}$ with components

$$w_i = \frac{2\pi\lambda_i}{m_i} \qquad \text{for} \quad 0 \le \lambda_i < m_i$$

In the light of the analysis above, the SPDE algorithm consists of the following three steps:

Step 1: Use numerical optimization techniques to maximize the likelihood function in Eq. (9) with respect to the vector of the unknown parameters $\boldsymbol{\beta}$.

Step 2: Express the unknown elements of the matrices $\mathbf{A}(\mathbf{h}')$, $\mathbf{B}(\mathbf{h})$, and $\mathbf{Q}$ in terms of $\boldsymbol{\beta}$.

Step 3: Substitute matrices into Eq. (4) to produce simulations of the vector SRF $\mathbf{X}(\mathbf{s})$.

Finally, the frequency domain methods of Section 9 can be generalized with efficiency to simulate vector SRF (see, e.g., Shinozuka and Jan, 1972; Wittig and Sinha, 1975).

13. Simulation of Non-Gaussian Spatial Random Fields

13.1 The Nonlinear Transformation Approach

As a result of the central limit theorem, the turning bands method generates realizations of SRF with Gaussian probability distribution (or at least approximately Gaussian). As we saw above, the same holds true for most simulation methods. In some cases, however, the probability distribution of the simulated SRF cannot be approximated by a Gaussian distribution. Then, some sort of nonlinear transformation must be applied before the simulation method is used.

In particular, let $X(\mathbf{s})$ be an SRF with a non-Gaussian univariate probability distribution $f_x(\chi)$ and a covariance $c_x(\mathbf{h})$. For simplicity, assume that the $X(\mathbf{s})$ has zero mean and unit variance.

We can always define a transformation G so that

$$X(\mathbf{s}) = G[Z(\mathbf{s})] \tag{1}$$

where $Z(\mathbf{s})$ is a Gaussian SRF with zero mean and unit variance. Assuming that G is strictly monotonic, it is valid that

$$G(\zeta) = F_x^{-1}[F_z(\zeta)] \tag{2}$$

where $F_z(\zeta)$ is the standard Gaussian distribution. The covariance $c_x(\mathbf{h})$ is given by

$$c_x(\mathbf{h}) = \frac{1}{2\pi\sqrt{1-c_z^2(\mathbf{h})}} \times \int_{R^1}\int_{R^1} G(\zeta)G(\zeta') \times \exp\left[-\frac{\zeta^2 - 2c_z(\mathbf{h})\zeta\zeta' + \zeta'^2}{2[1-c_z^2(\mathbf{h})]}\right] d\zeta\, d\zeta' \tag{3}$$

where $f_z(\zeta, \zeta')$ and $c_z(\mathbf{h})$ are the bivariate probability density and the covariance of $Z(\mathbf{s})$.

The simulation of the SRF $X(\mathbf{s})$ could now be achieved by means of the following procedure:

Step 1: Solve Eq. (3) with respect to $c_z(\mathbf{h})$, viz.,

$$c_z(\mathbf{h}) = B[c_x(\mathbf{h})] \tag{4}$$

where $B[\cdot]$ is a suitable function.

Step 2: Apply the turning bands method to generate simulations of the Gaussian SRF $Z(\mathbf{s})$ with covariance given by Eq. (4).

Step 3: Simulations of the original non-Gaussian SRF $X(\mathbf{s})$ are provided by Eq. (1), where the transformation G is determined by Eq. (2).

The implementation of the simulation procedure is straightforward when analytical expressions are available for Eqs. (2) and (4).

Example 1: Assume that the SRF $X(\mathbf{s})$ has a log-normal distribution. Equations (2) and (4) give, respectively (see, e.g., Chapter 2),

$$\chi = G(\zeta) = \exp[\zeta] \tag{5}$$

and

$$c_z(\mathbf{h}) = \log\left[\frac{c_x(\mathbf{h})}{m_x^2} + 1\right] \tag{6}$$

Example 2: Let $X(\mathbf{s})$ be a uniformly distributed within $[-b, b]$. In this case Eqs. (2) and (4) yield

$$\chi = G(\zeta) = b[2\,\mathrm{erf}(\zeta) - 1], \tag{7}$$

where $\mathrm{erf}(\zeta)$ is the error function (e.g., Gradshteyn and Ryzhik, 1965), and

$$c_z(\mathbf{h}) = 2c_z(\mathbf{O})\sin\left[\frac{\pi c_x(\mathbf{h})}{2b^2}\right] \tag{8}$$

Problems associated with the practical implementation of the above simulation approach arise from the fact that in many situations it is not possible to define the transformation G analytically, in terms of Eq. (2), or to obtain a solution of Eq. (3) in the closed form suggested by Eq. (4). Consequently, one may have to resort to some sort of numerical approximation.

As regards transformation G, a possible numerical approximation follows: First the standard Gaussian distribution of $Z(\mathbf{s})$ is confirmed to a properly chosen interval $[-b, b]$; this leads to a so-called *truncated standard Gaussian* distribution. Then, several values of ζ are selected using the experimental cumulative frequency distribution together with equation

$$F_x(\chi_i) = F_z(\zeta_i) \tag{9}$$

Note that a common approximation is $F_x(\chi_i) \approx i/(m+1)$, where m is the number of the data values sorted as $\chi_1 \leq \chi_2 \leq \cdots \leq \chi_m$.

For the calculation of $c_z(\mathbf{h})$, the expansion of the bivariate Gaussian distribution in terms of Hermite polynomials can be applied. More specifically, by inserting Eq. (10), Section 4 of Chapter 4 into Eq. (3) one finds

$$c_x(\mathbf{h}) = \sum_{k=0}^{M} a_k^2 c_z^k(\mathbf{h}) \tag{10}$$

where

$$a_k = \int_{R^1} G(\zeta) H_k(\zeta) \frac{\exp[-\zeta^2/2]}{\sqrt{2\pi}} \, d\zeta \tag{11}$$

and $H_k(\zeta)$ are Hermite polynomials. As soon as the transformation G has been calculated as above, the coefficients a_k can be found from Eq. (11). The last step is to obtain a numerical solution of Eq. (10) with respect to $c_z(\mathbf{h})$.

13.2 The Sequential Indicator Approach

This approach, which is useful when a significant amount of quantifiable "soft" data is available, is based on the fundamental multivariate distribution formula (1) of Section 3 above and the indicator coding of "soft" data discussed in Chapter 7.

In particular, the *sequential indicator simulation* steps are as follows (Journel, 1989):

Step 1: Assuming that one seeks simulations at m grid points, start at any point $\mathbf{s}_1$ of the grid and with the aid of Section 5.2 of Chapter 7 derive the indicator function

$$Y_x(\mathbf{s}_1, \zeta) = P[X(\mathbf{s}_1) \leq \zeta \,|\, \mathfrak{R}] \tag{12}$$

where $\mathfrak{R}$ is the relationship space generated from the information available.

Step 2: From $Y_x(\mathbf{s}_1, \zeta)$ generate a value of $X(\mathbf{s}_1)$, say $\chi_1^{(1)}$ [the superscript (1) denotes that the value belongs to the first realization of the SRF $X(\mathbf{s})$].

Step 3: Consider another point $\mathbf{s}_2$, and derive the indicator function

$$Y_x(\mathbf{s}_2, \zeta) = P[X(\mathbf{s}_2) \leq \zeta \,|\, \mathfrak{R}(\mathbf{s}_1)] \tag{13}$$

where $\Re(\mathbf{s}_1)$ denotes that the information available now includes the value $\chi_1^{(1)}$.

Step 4: From $Y_x(\mathbf{s}_2, \zeta)$ derive a value of $X(\mathbf{s}_2)$, say $\chi_2^{(1)}$.

Step 5: Repeat steps 2 to 4 using all remaining grid points $\mathbf{s}_i$, $i = 3, 4, \ldots, m$.

The set of values $\chi_i^{(1)}$ $(i = 1, 2, \ldots, m)$ constitutes one possible realization of the SRF $X(\mathbf{s})$.

14. Simulation in Space–Time

Several of the simulation approaches considered above can be extended to produce realizations of spatiotemporal random fields (S/TRF).

By means of the ST simulation concept, for example, the space n-dimensional × time random field $X_n(\mathbf{s}, t)$, where $(\mathbf{s}, t) \in R^n \times T$, can be simulated by summing contributions from several random processes $X_{1,\boldsymbol{\theta}_i}(s_i, t)$, where $s_i = \mathbf{s} \cdot \boldsymbol{\theta}_i$, $(s_i, t) \in R^1 \times T$; viz.,

$$X_n(\mathbf{s}, t) = \frac{1}{\sqrt{N}} \sum_{i=1}^{N} X_{1,\boldsymbol{\theta}_i}(s_i, t) \tag{1}$$

in which N is the number of simulation lines. On-line realizations of the S/TRF $X_{1,\boldsymbol{\theta}_i}(s_i, t)$ can be generated in terms of its spectral density function

$$C_{1,\boldsymbol{\theta}}(\omega, \lambda) = \Psi_n^1[C_n(\mathbf{w}, \lambda)] \tag{2}$$

where $\mathbf{w} = \omega\boldsymbol{\theta}$, by using the simulation formula

$$X_1(s, t) = \sum_{j=1}^{M} \sum_{k=1}^{K} \sqrt{2C_{1,\boldsymbol{\theta}}(\omega_j, \lambda_k)\, \Delta\omega_j\, \Delta\lambda_k}\, \cos(\omega_j s - 2\pi\lambda_k t + \phi_{j,k}) \tag{3}$$

where the phase angles $\phi_{j,k}$ are distributed randomly but uniformly within $[0, 2\pi]$. Equation (3) is the space–time generalization of the spectral techniques of Section 5. Of course, if values of the S/TRF are available at certain points in space–time, Eq. (1) should be constrained to honor these values by means of a technique similar to that described in Section 3.2 above.

Another useful space–time simulation approach can be developed on the basis of the frequency-domain techniques of Section 9. For example, an $R^2 \times T$ extension of Eq. (1) of Section 9 is

$$X(s_1, s_2, t) = \sum_{i=1}^{M_1} \sum_{j=1}^{M_2} A(i, j) \cos(w_i s_1 \cos\theta_j + w_i s_2 \sin\theta_j - \lambda_i t + \phi_{i,j}) \tag{4}$$

where

$$A(i, j) = \sqrt{2C_x(\lambda_i, \theta_j)\, \Delta\lambda_i\, \Delta\theta_j} \tag{5}$$

$C_x(\lambda_i, \theta_j)$ is the so-called directional wave spectral density function, the θ_j denote angles between the s_1-axis and the propagation direction of component waves, λ_i denote wave frequencies, and $\phi_{i,j}$ are random angles distributed independently but uniformly between 0 and 2π (Goda, 1980).

Spatiotemporal simulation is a valuable tool in the context of random moving surfaces studies, such as sea waves and their action on structures, atmospheric pollutants, and meteorological elements. Also, the simulation method may be used to develop a spatiotemporal model for rainfall generation. Space–time rainfall simulations can be used in evaluating strategies for satellite remote sensing of rainfall and for studying storm runoff problems.

We will conclude this chapter by noticing that other useful simulation techniques include Boolean methods, simulated annealing, and Markovian procedures. For more discussion, the reader is referred to publications like, for example, Ripley (1981), Johnson (1987), and Deutsch and Journel (1991).

9

Estimation in Space and Time

"Never let yourself be goaded into taking seriously problems about words and their meanings. What must be taken seriously are questions of fact, and assertions about facts; theories and hypotheses; the problems they solve; and the problems they raise."

K. R. Popper

1. Introduction

In environmental engineering, the design of any remedial measure regarding groundwater pollution caused by industries and municipalities requires knowledge of the extent of subsurface contamination. It is thus important to create predictive subsurface contamination maps that will cover the whole area of interest and will provide information regarding the shape, the size, and the existing trends in the spatial variability of the contaminant plume. In geotechnical engineering, procedures for predicting soil performance consists of three parts: (1) constitutive models (stress strain, rheologic, etc.); (2) methods for the estimation of soil parameters used in the above models; and (3) numerical approaches to apply the models in practice (finite elements, finite differences etc.). The accuracy in estimating the soil parameters obviously affects the reliability of any prediction made and is, therefore, of practical importance in earth sciences. These are two typical estimation problems that can be handled by means of spatial random field.

The first part of this chapter deals with the spatial estimation of natural processes. The general spatial estimation problem is defined and its various forms are examined. Several approaches to the problem are reviewed.

Spatial estimation can be studied in the context of the stochastic research program, by using results from the previous chapters. At this point, the hard core of the stochastic research program discussed in Chapter 1 includes a few additional methodological assumptions. In particular, the estimators considered should meet certain fundamental stochastic optimality criteria. Both linear and nonlinear estimators are discussed. These include the traditional Wiener–Kolmogorov type estimators, kriging estimators, and nonlinear factorable estimators. Their applicability covers a wide range of homogeneous as well as nonhomogeneous spatial random fields. Important properties of the estimation approaches are examined in detail, and their mathematical and physical interpretations are investigated. The outcomes of the estimation approach are maps representing values of the natural processes over the entire region of interest, together with the maps of the accuracy (or estimation error variance) with which the processes are represented. These maps can be valuable inputs to decision-making processes, risk evaluation methods, control and investment policies.

A factorization scheme of the estimation error variance is discussed. This scheme possesses certain attractive properties that allow significant savings in the computations. An important application of this factorization scheme is developed within the context of sampling design, to be discussed in Chapter 10.

The spatiotemporal estimation problem is defined and a stochastic solution is discussed, by means of the theory of spatiotemporal random fields developed in Chapter 5. An interesting feature of the space–time correlation decomposition discussed in that chapter is that stochastic inferences can be made and optimal linear estimators of discrete-valued spatiotemporal random fields can be derived solely in terms of the generalized spatiotemporal covariances. Estimation takes into account important time-related information and, therefore, under certain circumstances it provides improved results compared to those obtained by purely spatial estimation techniques.

The last part of this chapter presents a different heuristic adopted by the stochastic data processing program regarding the spatial estimation problem. This is a Bayesian/maximum-entropy approach according to which the estimation equations emerge through a process that balances two requirements: high prior information about the spatial variability and high posterior probability about the reconstructed map. The first requirement enables one to use a variety of sources of prior information and involves the maximization of a so-called Bayes function. This approach yields, in general, nonlinear estimates and does not call for any Gaussian-type hypothesis.

Spatiotemporal problems can also be considered in the light of the aforementioned Bayesian/maximum-entropy approach. The latter probably does not suffice, in its present form, to account for all sorts of prior information, but it does significantly restrict the range of arbitrariness, and it has several other important properties that contribute to the progressiveness of the stochastic research program.

2. A Brief Review of Nonstochastic Estimators and the Emergence of Stochastic Estimation

We do not give a detailed review of the various nonstochastic estimation techniques here, but we consider briefly some of the most important of them; for the interested reader, several references are provided. Common deterministic linear estimators of spatial processes are those of least squares, of Lagrange, and of weighted coefficients.

The *least squares* approach assumes that the estimator of the value of a natural process at an unobservable location in space $\mathbf{s} \in R^n$ is the linear combination of the base functions $p_k(\mathbf{s})$ (usually polynomials); that is,

$$\lambda_i = \sum_k \lambda_{k_i} p_k(\mathbf{s})$$

at each $\mathbf{s}$. The coefficients λ_i are determined by means of the least squares difference between the estimators above and the true values (see, e.g., Daniel *et al.*, 1971; Davis, 1973). A crucial disadvantage of the method is that it leads to estimates that do not coincide with the known values at the data points (this happens because least squares actually yield estimates of the mean value of the spatial process). This is a purely formal feature of the method without any physical motivation behind it. For instance, in estimating a porosity field hydrologists will not give credit to an estimation method that is not consistent with the known values of the field. Moreover, the method does not take into consideration the structure of the natural processes and does not provide information regarding the accuracy of the obtained estimates. This is a serious shortcoming when it comes to applying optimal exploration strategies, decision analysis, etc.

Lagrange's method makes similar assumptions concerning the nature of the estimator, but the coefficients λ_i are calculated so that the estimated surface passes through the data points (Davis, 1975). The results are satisfactory only when the spatial variation is very regular. Also, the use of polynomials $p_k(\mathbf{s})$ of higher degrees often lead to nonrealistic estimates, which are in no way related to the properties of the natural processes under study. This method does not provide any information about the accuracy of its estimates, either.

The method of *weighted coefficients* is based on the prior selection of λ_i, which usually are distance functions between the point $\mathbf{s}$ where an estimate of the natural process is needed, and the data points $\mathbf{s}_i$, $i = 1, 2, \ldots, m$ (e.g.,

$\lambda_i = |\mathbf{s}_i - \mathbf{s}|^{-a}$, $= \exp(-a|\mathbf{s}_i - \mathbf{s}|)$, $a \in R^1$). Weak points of the method are the subjective choice of the coefficients λ_i, and the fact that it does not account for the relative positions of the data points and spatial variabilities of the natural processes. Just as for the previous ones, the weighted coefficients method lacks any physical content and has a rather conventional and purely instrumental character.

Certain of the above methods can also be formulated in *statistical* terms. Even then, however, they fail to incorporate inportant features of the natural process. Consider, for example, the statistical version of the least squares method (Davis, 1973). Each sample χ_i is viewed as a realization of the corresponding random variable x_i, the distribution of which depends on the nonrandom coordinates $\mathbf{s}_i$. Fundamental is the assumption that there is no correlation among the random variables. This, however, is a rather

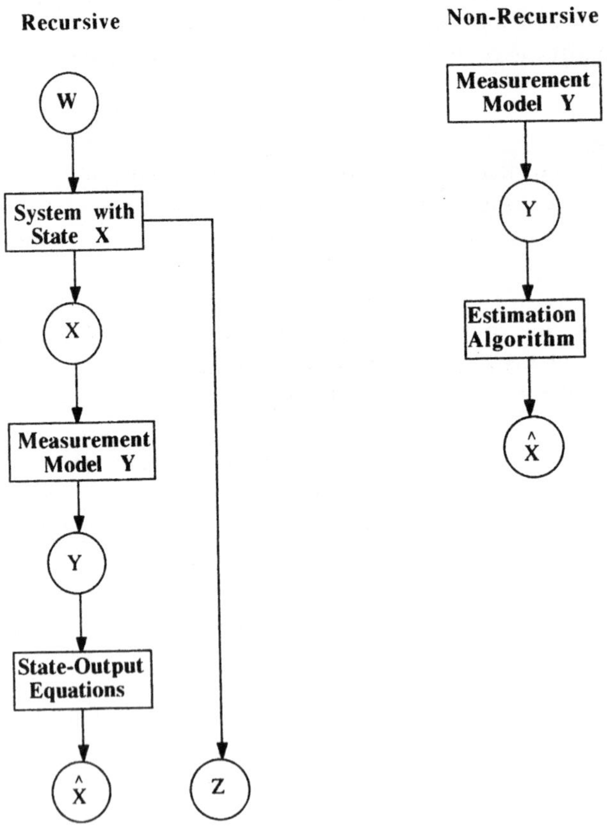

Figure 9.1 Recursive versus nonrecursive estimation procedures. W is state fluctuations; X is system state; V is measurement disturbance; Y is measurements; X is system state estimator (output); and Z is system output

nonrealistic modeling of the spatial process since it fails to account for interactions and dependencies of neighboring samples. Furthermore, the method still cannot, by its nature, reproduce the known values at the data points and yields mean values rather than local estimates, especially when the natural process exhibits irregularities in its spatial variation.

In view of the foregoing considerations, it becomes apparent that the traditional solutions to the estimation problem evade central physical issues rather than face them. On the other hand, by taking into account the modeling considerations discussed in Chapter 1, physical intuition guides one to use estimation methods where the natural processes of interest are modeled as spatial random fields. Random field estimation methods are generally known as *optimum stochastic estimation methods.* These methods are classified in various ways, such as

(i) By means of the form of the estimator assumed (i.e., linear versus nonlinear estimators).
(ii) By means of the optimality criterion used (i.e., minimum mean square error, maximum likelihood, Bayesian, etc., estimators).
(iii) By means of constraints imposed on the estimation process (i.e., unbiased versus biased estimators).
(iv) By means of the operational format of the estimation process (for instance, in Fig. 9.1 the flowcharts of recursive versus nonrecursive estimation procedures are compared).

The concept underlying optimal stochastic estimation was first introduced by Kolmogorov (1941) and Wiener (1949), and subsequently applied in various fields like physics (e.g., Davis, 1952), electrical engineering (e.g., Wainstein and Zubakov, 1962), meteorology (e.g., optimum interpolation, Gandin, 1963), mining and petroleum engineering (e.g., kriging, Matheron, 1965; David, 1977; Journel and Huijbregts, 1978; Hohn, 1988), and civil engineering (e.g., Veneziano, 1980).

3. Optimum Stochastic Spatial Estimation

3.1 General Considerations

The general setting of the stochastic spatial estimation problem to be considered here is as follows:

Problem 1: Let $X(\mathbf{s})$ and $Y(\mathbf{s})$ be two jointly distributed SRF. We would like to estimate $X(\mathbf{s})$ at a location $\mathbf{s}_k \in R^n$ from measurements of $Y(\mathbf{s})$, say $\Psi_{1,m} = \{Y(\mathbf{s}_i) = \psi_i;\ i = 1, \ldots, m\}$ at the locations $\mathbf{s}_i \in R^n$, $i = 1, 2, \ldots, m$. The estimator $\hat{X}(\mathbf{s}_k)$ of $X(\mathbf{s}_k)$ is assumed to be an arbitrary function of $Y(\mathbf{s}_i)$, $i = 1, 2, \ldots, m$, namely,

$$\hat{X}(\mathbf{s}_k) = F[Y(\mathbf{s}_i), Y(\mathbf{s}_2), \ldots, Y(\mathbf{s}_m)] = F(Y_{1,m}) \tag{1}$$

Let $L[\hat{X}(\mathbf{s}_k), X(\mathbf{s}_k)]$ be the *loss-function* associated to the estimator (1); the $L[\hat{X}(\mathbf{s}_k), X(\mathbf{s}_k)]$ is a suitable function that expresses the "loss" of guessing the $\hat{X}(\mathbf{s}_k)$ when in fact the real field is $X(\mathbf{s}_k)$. We seek an estimator of the form (1), such that the following *optimality criterion* is satisfied: the expected value of the loss-function

$$E[L] = \underbrace{\int \cdots \int}_{m+1 \text{ times}} L[F(\psi_1, \ldots, \psi_m), \chi_k] f_{X,Y}(\chi_k, \psi_1, \ldots, \psi_m)\, d\chi_k\, d\psi_1 \ldots d\psi_m$$

$$= \underbrace{\int \cdots \int}_{m+1 \text{ times}} L[F(\Psi_{1,m}), \chi_k] f_{X,Y}(\chi_k, \Psi_{1,m})\, d\chi_k\, d\Psi_{1,m} \tag{2}$$

is minimum with respect to F (unless stated otherwise, the integration ranges will be assumed to vary from $-\infty$ to ∞).

Note that since $Y(\mathbf{s})$ is a random quantity we cannot say which value ψ it will take in any specific realization; hence, the estimator $\hat{X}(\mathbf{s}_k)$ of Eq. (1) is a random field as well.

We now proceed to the solution of Problem 1. By minimizing $E[L]$ with respect to F one obtains the following fundamental integral equation

$$\int \frac{\partial L[F(\Psi_{1,m}), \chi_k]}{\partial F} f_{X,Y}(\chi_k, \Psi_{1,m})\, d\chi_k = 0 \tag{3}$$

The general solution to Problem 1 is obtained by solving (3) with respect to F and simultaneously using Eq. (1). Naturally, a sufficient condition for a minimum is given by

$$\frac{\partial^2 L[F(\Psi_{1,m}), \chi_k]}{\partial F^2} \geq 0$$

To proceed further, let us consider some interesting special cases of the loss function. First, assume that the loss function is of the quadratic form

$$L[\hat{X}(\mathbf{s}_k), X(\mathbf{s}_k)] = [\hat{X}(\mathbf{s}_k) - X(\mathbf{s}_k)]^2 \tag{4}$$

The optimality criterion (2) is then called the *minimum mean square error criterion*, and Eq. (3) yields the *conditional mean estimator*

$$\hat{X}(\mathbf{s}_k) = E_X[X(\mathbf{s}_k) | \Psi_{1,m}] \tag{5}$$

If we choose the loss function to be the absolute error, that is,

$$L[\hat{X}(\mathbf{s}_k), X(\mathbf{s}_k)] = |\hat{X}(\mathbf{s}_k) - X(\mathbf{s}_k)| \tag{6}$$

then Eq. (3) becomes

$$\int_{-\infty}^{\hat{\chi}_k} f_{X,Y}(\chi_k, \Psi_{1,m})\, d\chi_k = \int_{\hat{\chi}_k}^{\infty} f_{X,Y}(\chi_k, \Psi_{1,m})\, d\chi_k$$

which implies that the $\hat{X}(\mathbf{s}_k) = \hat{\chi}_k$ is the *median estimator.*

Finally, if we let

$$L[\hat{X}(\mathbf{s}_k), X(\mathbf{s}_k)] = \begin{cases} 0 & \text{if} \quad \hat{X}(\mathbf{s}_k) = X(\mathbf{s}_k) \\ 1 & \text{otherwise} \end{cases} \tag{7}$$

the integral equation (3) leads to the *mode estimator.*

Remark 1: In the following we will use mainly the minimum mean square error criterion. There are several reasons for doing so, theoretical as well as practical. Some of these reasons are as follows:

(i) In the case of Gaussian SRF, which are very popular models for a variety of applications, the minimum mean square error estimator turns out to be a linear combination of the data. This result has many convenient properties from both the analytical and the computational points of view.

(ii) In the general case the minimum mean square error estimator is determined explicitly in terms of a conditional mean. This fact establishes illuminating connections with several aspects of the theory of SRF.

(iii) When analysis is restricted to the classes of linear estimators (e.g., linear kriging), the results obtained are probability distribution-free, and depend solely on the means, the covariances, and the semivariograms of the natural processes involved.

(iv) Linear minimum mean square error estimators have several other attractive features, as well. For example, the estimation system of equations and the estimation error variances do not depend on the specific values of the data, but only on the spatial locations of the data and on information in terms of means, covariances, and semivariograms.

(v) Linear minimum mean square error concepts are closely related to important topics from the areas of linear differential and difference equations, as well as matrix and integral equations.

3.2 More on Conditional Mean Spatial Estimation

It is instructive to consider another way of obtaining the solution of Problem 1, in the special case that of the loss-function (4). Then Eq. (2) can also be written in the familiar form

$$\sigma_x^2(\mathbf{s}_k) = E[e(\mathbf{s}_k)]^2 = E[\hat{X}(\mathbf{s}_k) - X(\mathbf{s}_k)]^2 \tag{8}$$

where $e(\mathbf{s}_k) = \hat{X}(\mathbf{s}_k) - X(\mathbf{s}_k)$ is the estimation error; or as

$$\begin{aligned}\sigma_x^2(\mathbf{s}_k) &= E_{XY}[F(Y_{1,m}) - X(\mathbf{s}_k)]^2 \\ &= E_Y\{E_X[\{F(Y_{1,m}) - X(\mathbf{s}_k)\}^2 | \Psi_{1,m}]\}\end{aligned} \tag{9}$$

In the light of Eq. (9), the minimization of $\sigma_x^2(\mathbf{s}_k)$ is equivalent to the minimization of

$$p_x^2(\mathbf{s}_k) = E_X\{[\hat{X}(\mathbf{s}_k) - X(\mathbf{s}_k)]^2 | \Psi_{1,m}\} \tag{10}$$

with respect to $\hat{X}(\mathbf{s}_k)$. This requires that

$$\frac{dp_x^2(\mathbf{s}_k)}{d\hat{\chi}_k} = 0$$

or

$$\begin{aligned}&2E_X\{[\hat{X}(\mathbf{s}_k) - X(\mathbf{s}_k)] | \Psi_{1,m}\} \\ &\quad = 2E_X[\hat{X}(\mathbf{s}_k) | \Psi_{1,m}] - 2E_X[X(\mathbf{s}_k) | \Psi_{1,m}] = 0\end{aligned}$$

which leads to the conditional mean estimator (5). Furthermore,

$$\frac{d^2 p_x^2(\mathbf{s}_k)}{d\hat{\chi}_k^2} = 2 > 0$$

which assures that the estimator (5) is a global minimum over all possible functions $F[\cdot]$.

The above estimator is *unbiased* (i.e., it assures absence of systematic over-estimation or under-estimation), for

$$\begin{aligned}&E_X[\hat{X}(\mathbf{s}_k) - X(\mathbf{s}_k)] \\ &\quad = E_Y\{E_X[X(\mathbf{s}_k) | \Psi_{1,m}]\} - E_X[X(\mathbf{s}_k)] \\ &\quad = E_X[X(\mathbf{s}_k)] - E_X[X(\mathbf{s}_k)] = 0\end{aligned}$$

which leads to the equation

$$E[\hat{X}(\mathbf{s}_k)] = E[X(\mathbf{s}_k)] \tag{11}$$

Remark 2: The $\hat{X}(\mathbf{s}_k)$ is a *conditionally unbiased* estimator, too; indeed,

$$\begin{aligned}&E_X\{[\hat{X}(\mathbf{s}_k) - X(\mathbf{s}_k)] | \Psi_{1,m}\} \\ &\quad = E_Y\{E_X[X(\mathbf{s}_k) | \Psi_{1,m}] | \Psi_{1,m}\} - E_X[X(\mathbf{s}_k) | \Psi_{1,m}] \\ &\quad = E_X[X(\mathbf{s}_k) | \Psi_{1,m}] - E_X[X(\mathbf{s}_k) | \Psi_{1,m}] = 0\end{aligned}$$

Other important properties of the estimation process above are as follows: First, on the basis of the preceding analysis we can conclude that

$$\begin{aligned} E_X[e(\mathbf{s}_k)\hat{X}(\mathbf{s}_k)|\Psi_{1,m}] &= E_X\{[\hat{X}(\mathbf{s}_k) - X(\mathbf{s}_k)]F(Y_{1,m})|\Psi_{1,m}\} \\ &= F(\Psi_{1,m})E_X\{[\hat{X}(\mathbf{s}_k) - X(\mathbf{s}_k)]|\Psi_{1,m}\} \\ &= F(\Psi_{1,m})\{E_X[X(\mathbf{s}_k)|\Psi_{1,m}] - \hat{X}(\mathbf{s}_k)\} = 0 \end{aligned}$$

In other words,

$$E_X[e(\mathbf{s}_k)\hat{X}(\mathbf{s}_k)|\Psi_{1,m}] = E_X[e(\mathbf{s}_k)F(Y_{1,m})|\Psi_{1,m}] = 0 \tag{12}$$

In addition, by expecting the last equation with respect to $Y(\mathbf{s})$ we find that

$$E_Y\{E_X[e(\mathbf{s}_k)F(Y_{1,m})|\Psi_{1,m}]\} = 0$$

or

$$E_{XY}[e(\mathbf{s}_k)\hat{X}(\mathbf{s}_k)] = E_{XY}[e(\mathbf{s}_k)F(Y_{1,m})] = 0 \tag{13}$$

Equations (12) and (13) are called the *orthogonality conditions.*

Remark 3: Notice that the estimator provided by Eq. (5) constitutes the most general case of minimum mean square error spatial estimation, where no restrictions are imposed regarding the functional form of the estimator, the underlying probability laws, or the regularity (homogeneity, etc.) characteristics of the SRF $X(\mathbf{s})$ and $Y(\mathbf{s})$. Its application, on the other hand, requires information regarding the $m+1$ variate probability distributions of the SRF involved. In practice, this sort of information is usually inaccessible, with the notable exception of the multivariate Gaussian and homogeneous SRF (in the Gaussian case, the conditional mean estimator reduces to a linear estimator that can be constructed on the basis of second-order statistical moments). In addition, most practical applications will require the establishment of a model relating the field of interest $X(\mathbf{s})$ with the measured field $Y(\mathbf{s})$.

3.3 Functional Estimation

In more complicated applications one may seek the estimation of a linear functional $\mathfrak{F}[X(\mathbf{s})]$ of $X(\mathbf{s})$ over a specific region V_k, as well as an assessment of the associated statistical error of the estimation. The data points $\mathbf{s}_i$ are usually, but not always, located within V_k.

Generally, the functional $\mathfrak{F}[.]$ may take one of the following forms:

$$\mathfrak{F}[X(\mathbf{s}_k)] = X(\mathbf{s}_k) \tag{14}$$

which is the case of *point estimation* considered above;

$$\mathfrak{F}[X(\mathbf{s}_k)] = \frac{1}{V_k}\int_{V_k} X(\mathbf{s})\,d\mathbf{s} \tag{15}$$

that is, the *mean* process value within a volume $V_k \subset R^n$, where V_k is the domain of the natural process;

$$\mathfrak{F}[X(\mathbf{s}_k)] = \int_V X(\mathbf{s}_k + \mathbf{s}) g(\mathbf{s}) \, d\mathbf{s} \tag{16}$$

that is, the *moving mean* process value weighted by $g(\mathbf{s})$; and

$$\mathfrak{F}[X(\mathbf{s}_k)] = \text{grad } X(\mathbf{s})|_{\mathbf{s}=\mathbf{s}_k} \tag{17}$$

that is, the *slope* of the process at point $\mathbf{s}_k$. Other forms of the estimation problem are possible, too. In this book we will deal in detail mainly with the case of point estimation. Nonetheless, wherever it is necessary, we will point the way for extending the results to other cases.

Estimation problems of the forms considered above can be found in almost any area of applied physical science, such as hydrology (e.g., estimation of the transmissivities of an aquifer, prediction of subsurface contamination), geology (e.g., mapping the depth to bedrock), geotechnical engineering (e.g., prediction of the soil settlement pattern), environmental engineering (e.g., forecasting of atmospheric pollutants), and meteorology (e.g., construction of prognostic charts).

4. Certain Classes of Linear Spatial Estimators

In this section we will consider the most important classes of *linear* minimum mean square error spatial estimators. These include the Wiener–Kolmogorov estimator and its various derivatives, such as the linear kriging estimators of geostatistics. An excellent presentation of the various types of kriging estimators can be found in Journel (1989).

Despite certain problems related to linearity (see, e.g., Wilde and Beightler, 1967; Beveridge and Schechter, 1970), linear estimators continue to be popular because they have certain important advantages over nonlinear ones. More precisely, they require substantially fewer assumptions and simpler mathematics; they are much faster and computationally more efficient; they usually lead to unique solutions; and last but not least, based on significant experience we know that they function properly in the majority of applications (Whittle, 1963; Aoki, 1967; Kailath, 1974; Christakos and Paraskevopoulos, 1986).

4.1 Unconstrained Wiener–Kolmogorov Estimator

Consider the linear estimator of the form (see Fig. 9.2 for a two-dimensional illustration)

$$\hat{X}(\mathbf{s}_k) = \mathbf{\Lambda}^T \mathbf{Y} \tag{1}$$

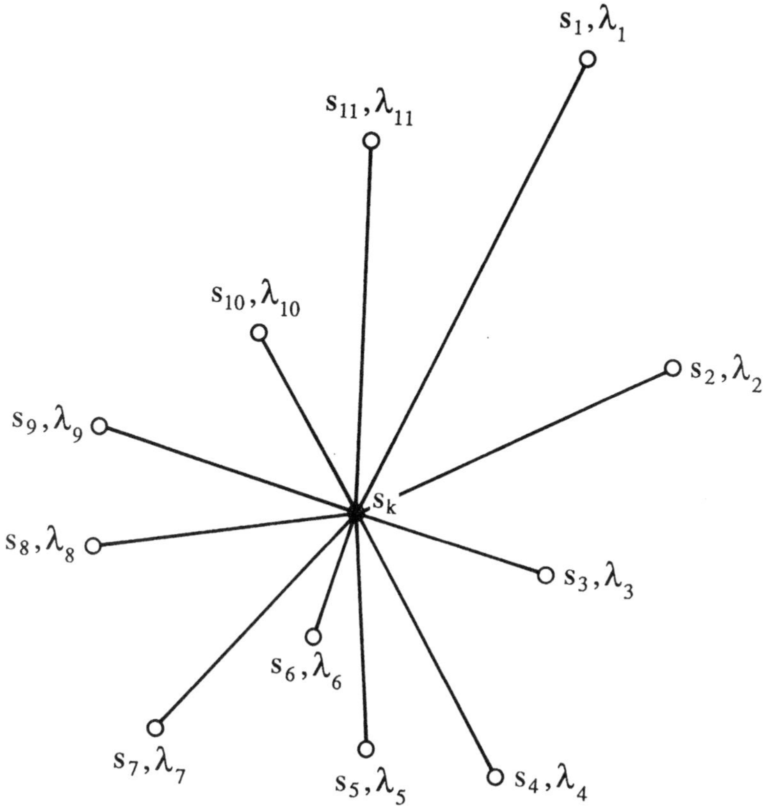

Figure 9.2 Linear estimation of the natural process $X(\mathbf{s})$ at point $\mathbf{s}_k$ by means of data values of the process $Y(\mathbf{s})$ at points $\mathbf{s}_i$, $i=1,\ldots,11$; λ_i = weights assigned by the estimator to the data points

where

$$\mathbf{Y}^{\mathrm{T}}=[Y(\mathbf{s}_1)\,Y(\mathbf{s}_2)\ldots Y(\mathbf{s}_m)]$$

and

$$\boldsymbol{\Lambda}^{\mathrm{T}}=[\lambda_1\lambda_2\ldots\lambda_m]$$

is a vector of coefficients to be determined through the estimation process. This setting leads us to our second estimation problem.

Problem 1: Let $X(\mathbf{s})$ and $Y(\mathbf{s})$ be two jointly distributed SRF. Find estimates $\hat{X}(\mathbf{s}_k)$ of $X(\mathbf{s}_k)$ at points $\mathbf{s}_k$ given data $\Psi=\{Y(\mathbf{s}_i)=\psi_i;\ i=1,2,\ldots,m\}$ at the points $\mathbf{s}_i\in R^n$, $i=1,2,\ldots,m$, such that the estimates (i) are linear combinations of the data [see Eq. (1)] and (ii) minimize the estimation error variance given by Eq. (8) of Section 3 above.

This is an unconstrained optimization problem, which can be solved with the help of the following proposition.

Proposition 1: The solution of Problem 1 is given by those coefficients λ_i $(i=1,2,\ldots,m)$ that satisfy the set of equations

$$\mathbf{C}\boldsymbol{\Lambda}=\mathbf{H} \tag{2}$$

where

$$\mathbf{C}=\begin{bmatrix} E[Y(\mathbf{s}_1)]^2 & \ldots & E[Y(\mathbf{s}_1)Y(\mathbf{s}_m)] \\ \vdots & & \vdots \\ E[Y(\mathbf{s}_m)Y(\mathbf{s}_1)] & \ldots & E[Y(\mathbf{s}_m)]^2 \end{bmatrix} \qquad \boldsymbol{\Lambda}^T=[\lambda_1\lambda_2\ldots\lambda_m]$$

and

$$\mathbf{H}^{\mathrm{T}}=[E[Y(\mathbf{s}_1)X(\mathbf{s}_k)]\ldots E[Y(\mathbf{s}_m)X(\mathbf{s}_k)]]$$

The corresponding estimation error variance is given by

$$\sigma_x^2(\mathbf{s}_k)=E[X(\mathbf{s}_k)]^2-\boldsymbol{\Lambda}^{\mathrm{T}}\mathbf{H}=E[X(\mathbf{s}_k)]^2-\mathbf{H}^{\mathrm{T}}\mathbf{C}^{-1}\mathbf{H} \tag{3}$$

Proof: The proof is straightforward. In particular, by inserting the linear estimator (1) into Eq. (8) of Section 3 above we get

$$\sigma_x^2(\mathbf{s}_k)=E[X(\mathbf{s}_k)]^2-2\boldsymbol{\Lambda}^{\mathrm{T}}\mathbf{H}+\boldsymbol{\Lambda}^{\mathrm{T}}\mathbf{C}\boldsymbol{\Lambda} \tag{4}$$

By minimizing Eq. (4) with respect to the coefficients λ_i $(i=1,\ldots,m)$, we find Eq. (2). The solution of system (2) yields $\boldsymbol{\Lambda}=\mathbf{C}^{-1}\mathbf{H}$. Substituting this solution into Eq. (4) we obtain Eq. (3) above. □

Important theoretical features of the *unconstrained Wiener–Kolmogorov* estimator above include the following: (i) the estimator is mean value-free; (ii) it is not an unbiased estimator, in general (this means that when *a priori* statistical information is available, it can be exploited to produce more accurate results than an unbiased estimator); and (iii) in theory, there is no need to make any assumption regarding the regularity characteristics (homogeneity, etc.) of the SRF involved in the estimation process. Nevertheless, when the Wiener–Kolmogorov approach is applied in practice certain modifications may be necessary, such as (i) the development of a model relating the SRF of interest $X(\mathbf{s})$ with the observed SRF $Y(\mathbf{s})$ and (ii) the restriction of estimation to homogeneous (in the wide sense) SRF.

Remark 1: In the special case where the measurable process and the natural process under estimation coincide, the symbol Y in the right-hand side of Eqs. (1), (2), and (3) must be replaced by the symbol X. In this case, of course, we must have $k\neq i$ $(i=1,2,\ldots,m)$. The stochastic inference part of the analysis is considerably simplified, for the latter is now restricted to only one SRF.

Example 1: Suppose that the SRF of interest $X(\mathbf{s})$ and the observed SRF $Y(\mathbf{s})$ are related by means of the model $Y(\mathbf{s}) = X + U(\mathbf{s})$, where $U(\mathbf{s})$ is a zero-mean white noise uncorrelated to X and $Y(\mathbf{s})$; notice that in this case X is independent of $\mathbf{s}$ and $E[Y(\mathbf{s})] = E[X]$. The system of equations (2) becomes

$$\sum_{i=1}^{m} \lambda_i E[Y(\mathbf{s}_i)Y(\mathbf{s}_j)] = E[XY(\mathbf{s}_j)] \qquad (j = 1, 2, \ldots, m)$$

where

$$E[Y(\mathbf{s}_i)Y(\mathbf{s}_j)] = E[X^2] + E[U^2]\,\delta_{ij}$$

and

$$E[XY(\mathbf{s}_j)] = E[X^2]$$

The solution of this system yields

$$\lambda_i = \frac{E[X^2]}{mE[X^2] + E[U^2]}, \qquad i = 1, 2, \ldots, m$$

By substituting these values into Eqs. (1) and (3) we obtain the optimal estimate and the associated estimation error variance

$$\hat{X} = \frac{E[X^2]}{mE[X^2] + E[U^2]} \sum_{i=1}^{m} Y(\mathbf{s}_i)$$

and

$$\sigma_x^2 = \frac{E[X^2]E[U^2]}{mE[X^2] + E[U^2]}$$

respectively.

4.2 Constrained Wiener–Kolmogorov and Ordinary Kriging Estimators

The estimator assumed here is, again, of the form of Eq. (1), but now some additional constraints are imposed.

Problem 2: Find estimates $\hat{X}(\mathbf{s}_k)$ of $X(\mathbf{s}_k)$ at points $\mathbf{s}_k$ given data $Y(\mathbf{s}_i) = \psi_i$ $(i = 1, 2, \ldots, m)$, such that the estimates (i) are linear combinations of the data, as in Eq. (1) above; (ii) satisfy the unbiasedness condition (11) of Section 3; and (iii) minimize the estimation error variance given by Eq. (8) of Section 3.

Proposition 2: The solution of Problem 3 is given by those coefficients λ_i $(i = 1, 2, \ldots, m)$ that satisfy the set of equations

$$E[X(\mathbf{s}_k)] = \mathbf{\Lambda}^T \mathbf{M}_Y \tag{5}$$

where $\mathbf{M}_Y^T = [E[Y(\mathbf{s}_1)] \ldots E[Y(\mathbf{s}_m)]]$, and

$$\mathbf{C\Lambda} - \mu \mathbf{M}_Y = \mathbf{H} \tag{6}$$

where μ is a Lagrange multiplier. The estimation error variance is

$$\begin{aligned} \sigma_x^2(\mathbf{s}_k) &= E[X(\mathbf{s}_k)]^2 - \mathbf{\Lambda}^T[\mathbf{H} - \mu \mathbf{M}_Y] \\ &= E[X(\mathbf{s}_k)]^2 - [\mathbf{H} + \mu \mathbf{M}_Y]^T \mathbf{C}^{-1} [\mathbf{H} - \mu \mathbf{M}_Y] \end{aligned} \tag{7}$$

Proof: This is a constrained minimization problem, where we seek the minimization of the quantity

$$\begin{aligned} \mathfrak{F} &= \sigma_x^2(\mathbf{s}_k) + 2\mu \left\{ E[X(\mathbf{s}_k)] - \sum_{i=1}^m \lambda_i E[Y(\mathbf{s}_i)] \right\} \\ &= E\left[X(\mathbf{s}_k) - \sum_{i=1}^m \lambda_i Y(\mathbf{s}_i) \right]^2 + 2\mu \left\{ E[X(\mathbf{s}_k)] - \sum_{i=1}^m \lambda_i E[Y(\mathbf{s}_i)] \right\} \end{aligned}$$

with respect to λ_i and μ. This yields the system of Eqs. (5) and (6). Finally, taking into account Eq. (6), Eq. (4) yields Eq. (7). □

The *constrained Wiener–Kolmogorov* estimator above is an unbiased estimator. Just as for the unconstrained estimator, in practical applications one needs to develop a model relating $X(\mathbf{s})$ and $Y(\mathbf{s})$; moreover, estimation is usually performed for homogeneous spatial RF.

Example 2: Consider, again, Example 1. Since $E[Y(\mathbf{s})] = E[X]$, Eq. (5) gives $\sum_{i=1}^m \lambda_i = 1$. Also, the system of equations in (6) becomes

$$\sum_{i=1}^m \lambda_i E[Y(\mathbf{s}_i) Y(\mathbf{s}_j)] - \mu E[X] = E[XY(\mathbf{s}_j)] \qquad (j = 1, 2, \ldots, m)$$

The solution of this system yields $\lambda_i = 1/m$ $(i = 1, 2, \ldots, m)$, and $\mu = E[U^2]/E[X]m$. Then, the optimal estimate and the associated estimation error variance are

$$\hat{X} = \frac{1}{m} \sum_{i=1}^m Y(\mathbf{s}_i)$$

and

$$\sigma_x^2 = \frac{E[U^2]}{m}$$

respectively.

Remark 2: The so-called *ordinary kriging* of geostatistics is a special case of the constrained Wiener–Kolmogorov estimator. More specifically, in its most popular form, ordinary kriging makes two additional assumptions:

(a) the measurable process is the SRF $X(\mathbf{s})$ itself (in this case, of course, it is supposed that the data are available at locations different than those under estimation); and (b) the SRF $X(\mathbf{s})$ is homogeneous, with mean value $E[X(\mathbf{s}_i)] = m_x$ for all $i = 1, 2, \ldots, m$. For more details see Remark 4 below.

4.3 Simple Kriging Estimator

This is a linear estimator of the form

$$\hat{X}(\mathbf{s}_k) = \mathbf{\Lambda}_0^{\mathrm{T}} \mathbf{Y}_0 \tag{8}$$

where $\mathbf{Y}_0^{\mathrm{T}} = [1\, Y(\mathbf{s}_1)\, Y(\mathbf{s}_2) \ldots Y(\mathbf{s}_m)]$ and $\mathbf{\Lambda}_0^{\mathrm{T}} = [\lambda_0 \lambda_1 \lambda_2 \ldots \lambda_m]$ are coefficients that remain to be determined. This leads to a slightly different form of the estimation Problem 1 above.

Problem 3: Find estimates $\hat{X}(\mathbf{s}_k)$ of the SRF $X(\mathbf{s}_k)$ at points $\mathbf{s}_k$ given data $\Psi = \{Y(\mathbf{s}_i) = \psi_i;\ i = 1, 2, \ldots, m\}$, such that the estimates (i) are linear combinations of the data of the form of Eq. (8) above, (ii) satisfy the unbiasedness condition (11) of Section 3 above, and (iii) minimize the estimation error variance (8) of Section 3.

This constrained optimization problem can be solved by means of the following proposition.

Proposition 3: The solution of Problem 3 is given by these coefficients λ_i $(i = 0, 1, \ldots, m)$, which satisfy the set of equations

$$E[X(\mathbf{s}_k)] = \mathbf{\Lambda}_0^{\mathrm{T}} \mathbf{M}_1 \tag{9}$$

where $\mathbf{M}_1^{\mathrm{T}} = [1\, E[Y(\mathbf{s}_1)] \ldots [Y(\mathbf{s}_m)]]$, and

$$\mathbf{C}^* \mathbf{\Lambda} = \mathbf{H}^* \tag{10}$$

in which

$$\mathbf{C}^* = \begin{bmatrix} c_Y(\mathbf{s}_1, \mathbf{s}_1) & \ldots & c_Y(\mathbf{s}_1, \mathbf{s}_m) \\ \vdots & & \vdots \\ c_Y(\mathbf{s}_m, \mathbf{s}_1) & \ldots & c_Y(\mathbf{s}_m, \mathbf{s}_m) \end{bmatrix}$$

and

$$\mathbf{H}^{*\mathrm{T}} = [c_{XY}(\mathbf{s}_1, \mathbf{s}_k) \ldots c_{XY}(\mathbf{s}_m, \mathbf{s}_k)]$$

where

$$c_Y(\mathbf{s}_i, \mathbf{s}_j) = E[Y(\mathbf{s}_i) Y(\mathbf{s}_j)] - E[Y(\mathbf{s}_i)] E[Y(\mathbf{s}_j)]$$

and

$$c_{XY}(\mathbf{s}_i, \mathbf{s}_k) = E[Y(\mathbf{s}_i) X(\mathbf{s}_k)] - E[Y(\mathbf{s}_i)] E[X(\mathbf{s}_k)]$$

The estimation variance is given by

$$\sigma_x^2(\mathbf{s}_k) = c_X(\mathbf{s}_k, \mathbf{s}_k) - \mathbf{H}^{*\mathrm{T}} \mathbf{C}^{*-1} \mathbf{H}^* \tag{11}$$

Proof: The proof here is similar to that of Proposition 1 and offers no conceptual difficulty. More precisely, by inserting the linear estimator (8) into Eq. (8) of Section 3 above, and by minimizing with respect to the coefficients λ_i, $i=0, 1, \ldots, m$, subject to the constraint (11) of Section 3, we find the system of Eqs. (9) and (10). The solution of this system gives $\mathbf{\Lambda}=\mathbf{C}^{*-1}\mathbf{H}^*$. Substituting this solution into Eq. (8) of Section 3, we obtain Eq. (11) above. □

The *simple kriging* estimator has the following properties: (i) the mean values of the spatial RF involved must be known, and (ii) it is an unbiased estimator. The practical application of simple kriging is restricted to homogeneous fields, and requires the construction of a model relating $X(\mathbf{s})$ with $Y(\mathbf{s})$.

Remark 3: Equation (8) can also written as $\hat{X}(\mathbf{s}_k)=E[X(\mathbf{s}_k)]+\mathbf{\Lambda}^T\mathbf{Y}^*$, where

$$\mathbf{Y}^*=[Y(\mathbf{s}_1)-E[Y(\mathbf{s}_1)]\ldots Y(\mathbf{s}_m)-E[Y(\mathbf{s}_m)]]^T$$

That is, the simple kriging is a linear regression-type estimator.

Example 3: Consider, once more, Example 1. In this case, Eqs. (10) become

$$\sum_{i=1}^{m} \lambda_i c_Y(\mathbf{s}_i, \mathbf{s}_j)=c_{XY}(\mathbf{s}_j) \qquad (j=1, 2, \ldots, m)$$

where

$$\begin{aligned} c_Y(\mathbf{s}_i, \mathbf{s}_j) &= E[Y(\mathbf{s}_i)Y(\mathbf{s}_j)]-E[Y(\mathbf{s}_i)]E[Y(\mathbf{s}_j)] \\ &= E[X^2]-\{E[X]\}^2+E[U^2]\,\delta_{ij} \end{aligned}$$

and

$$c_{XY}(\mathbf{s}_j)=E[XY(\mathbf{s}_j)]-E[X]E[Y(\mathbf{s}_j)]=E[X^2]-\{E[X]\}^2$$

The solution of this system yields the following values for the estimation coefficients

$$\lambda_i=\frac{E[X^2]-\{E[X]\}^2}{m[E[X^2]-\{E[X]\}^2]+E[U^2]}, \qquad i=1, 2, \ldots, m$$

Then, from Eq. (9) we get

$$\lambda_0=E[X]\left\{1-\frac{m[E[X^2]-\{E[X]\}^2]}{m[E[X^2]-\{E[X]\}^2]+E[U^2]}\right\}$$

By substituting these values into Eqs. (8) and (11) we obtain the optimal estimate and the associated estimation error variance

$$\hat{X} = \frac{E[X]E[U^2] + [E[X^2] - \{E[X]\}^2] \sum_{i=1}^{m} Y(\mathbf{s}_i)}{m[E[X^2] - \{E[X]\}^2] + E[U^2]}$$

and

$$\sigma_x^2 = \frac{E[U^2][E[X^2] - \{E[X]\}^2]}{m[E[X^2] - \{E[X]\}^2] + E[U^2]}$$

respectively.

If we compare the last error estimation variance with the ones obtained in Examples 1 and 2, we find that, in this case:
σ_x^2 (simple kriging) $< \sigma_x^2$ (unconstrained Wiener–Kolmogorov) $< \sigma_x^2$ (constrained Wiener–Kolmogorov or ordinary kriging).

Remark 4: It may be instructive to summarize the above estimators under the following conditions: (a) the measurable process is the RF $X(\mathbf{s})$ itself (in this case, of course, it is supposed that the data are available at locations different than those under estimation); (b) the RF $X(\mathbf{s})$ is homogeneous, with mean value $E[X(\mathbf{s}_i)] = m_x$ for all $i = 1, 2, \ldots, m$. To emphasize the fact that the estimation coefficients of the various estimators are not equal, different notations have been used (see Table 9.1).

4.4 Intrinsic Kriging Estimator

This, is perhaps, the most interesting of all kriging estimators, from a theoretical point of view. Assume that the spatially nonhomogeneous, in general, natural process is modeled as an ISRF-ν. This is the key assumption, and the solution of the estimation problem emerges naturally from the theory developed in previous chapters. Initially, we will assume that the measurable process is the SRF $X(\mathbf{s})$ itself. In the subsequent sections we shall also deal with the situation where the available measurements come from another process, $Y(\mathbf{s})$.

Problem 4: Let $X(\mathbf{s})$ be an ISRF-ν. Find estimates $\hat{X}(\mathbf{s}_k)$ of $X(\mathbf{s}_k)$ at points $\mathbf{s}_k$ given data $X(\mathbf{s}_i) = \chi_i$ $(i = 1, 2, \ldots, m;\ i \neq k)$, such that the estimates (i) are of the linear form

$$\hat{X}(\mathbf{s}_k) = \mathbf{\Lambda}^{\mathrm{T}}\mathbf{X} \tag{12}$$

where $\mathbf{X}^{\mathrm{T}} = [X(\mathbf{s}_1)X(\mathbf{s}_2)\ldots X(\mathbf{s}_m)]$; (ii) satisfy the unbiasedness condition (11) of Section 3 above; and (iii) minimize the estimation error variance given by Eq. (8) of Section 3.

Table 9.1 Summary of Estimators[a]

(i) Estimator form:

Unconstrained Wiener-Kolmogorov:

$$\hat{X}(\mathbf{s}_k) = \sum_{i=1}^{m} \lambda_i X(\mathbf{s}_i)$$

Ordinary kriging:

$$\hat{X}(\mathbf{s}_k) = \sum_{i=1}^{m} \zeta_i X(\mathbf{s}_i)$$

Simple kriging:

$$\hat{X}(\mathbf{s}_k) = E[X(\mathbf{s}_k)] + \sum_{i=1}^{m} \xi_i \{X(\mathbf{s}_i) - E[X(\mathbf{s}_i)]\}$$

(ii) Estimation system:

Unconstrained Wiener-Kolmogorov:

$$\sum_{i=1}^{m} \lambda_i E[X(\mathbf{s}_i)X(\mathbf{s}_j)] = E[X(\mathbf{s}_k)X(\mathbf{s}_j)] \qquad \text{or}$$

$$\sum_{i=1}^{m} \lambda_i c_x(\mathbf{s}_i, \mathbf{s}_j) = c_x(\mathbf{s}_k, \mathbf{s}_j) - m_x^2 \left[\sum_{i=1}^{m} \lambda_i - 1\right] \qquad j = 1, 2, \ldots, m$$

Ordinary kriging:

$$\sum_{i=1}^{m} \zeta_i c_x(\mathbf{s}_i, \mathbf{s}_j) - \mu = c_x(\mathbf{s}_k, \mathbf{s}_j), \qquad \text{or}$$

$$\sum_{i=1}^{m} \zeta_i E[X(\mathbf{s}_i)X(\mathbf{s}_j)] - \mu = E[X(\mathbf{s}_k)X(\mathbf{s}_j)], \qquad j = 1, 2, \ldots, m, \qquad \text{and}$$

$$\sum_{i=1}^{m} \zeta_i = 1$$

Simple kriging:

$$\sum_{i=1}^{m} \xi_i c_x(\mathbf{s}_i, \mathbf{s}_j) = c_x(\mathbf{s}_k, \mathbf{s}_j), \qquad \text{or}$$

$$\sum_{i=1}^{m} \xi_i E[X(\mathbf{s}_i)X(\mathbf{s}_j)] = E[X(\mathbf{s}_k)X(\mathbf{s}_j)] + m_x^2 \left[\sum_{i=1}^{m} \xi_i - 1\right], \qquad j = 1, 2, \ldots, m$$

[a] No measurement error is involved.

Proposition 4: If $X(\mathbf{s})$ is an ISRF-ν, the solution of Problem 4 above is provided by those coefficients λ_i $(i = 1, 2, \ldots, m)$ that satisfy the set of equations

$$\mathbf{K\Xi} = \mathbf{\Theta} \tag{13}$$

where

$$\mathbf{K} = \begin{bmatrix} k_x(0) & \dots & k_x(r_{1m}) & 1 & p_1(\mathbf{s}_1) & \dots & p_\rho(\mathbf{s}_1) \\ \vdots & & & & & & \\ k_x(r_{m1}) & \dots & k_x(0) & 1 & p_1(\mathbf{s}_m) & \dots & p_\rho(\mathbf{s}_m) \\ 1 & \dots & 1 & 0 & 0 & \dots & 0 \\ p_1(\mathbf{s}_1) & \dots & p_1(\mathbf{s}_m) & 0 & 0 & \dots & 0 \\ \vdots & & \vdots & \vdots & \vdots & & \vdots \\ p_\rho(\mathbf{s}_1) & \dots & p_\rho(\mathbf{s}_m) & 0 & 0 & \dots & 0 \end{bmatrix}$$

$$r_{ij} = |\mathbf{s}_i - \mathbf{s}_j| \qquad \text{and} \quad \rho \le \nu$$

$$\boldsymbol{\Xi}^{\mathrm{T}} = [\lambda_1 \lambda_2 \dots \lambda_m \mu_0 \mu_1 \dots \mu_\rho]$$

μ_ρ are Lagrange multipliers, and

$$\boldsymbol{\Theta}^{\mathrm{T}} = [k_x(r_{1k}) \dots k_x(r_{mk}) 1 p_1(\mathbf{s}_k) \dots p_\rho(\mathbf{s}_k)]$$

Proof: We know (Chapter 3) that the SRF $Z(\mathbf{s}_k) = \hat{X}(\mathbf{s}_k) - X(\mathbf{s}_k)$ is a spatial increment of order ν (SI-ν) and its variance is given by

$$\begin{aligned} E[Z(\mathbf{s}_k)]^2 = E[\hat{X}(\mathbf{s}_k) - X(\mathbf{s}_k)]^2 &= E\left[\sum_{i=1}^{m,k} \lambda_i X(\mathbf{s}_i)\right]^2 \\ &= \sum_{i=1}^{m} \sum_{j=1}^{m} \lambda_i \lambda_j k_x(\mathbf{s}_i - \mathbf{s}_j) \\ &\quad - 2 \sum_{i=1}^{m} \lambda_i k_x(\mathbf{s}_i - \mathbf{s}_k) + k_x(\mathbf{O}) \end{aligned} \tag{14}$$

($\lambda_k = -1$). Moreover, since $Z(\mathbf{s}_k)$ is a spatial increment of order ν it holds that (let $p_\rho(\mathbf{s}) = \mathbf{s}^{\boldsymbol{\rho}}$)

$$\sum_{i=1}^{m} \lambda_i \mathbf{s}_i^{\boldsymbol{\rho}} = \mathbf{s}_k^{\boldsymbol{\rho}} \tag{15}$$

where $0 \le |\boldsymbol{\rho}| \le \nu$. By minimizing Eq. (14) with respect to the λ_i subject to the constraint (15) we obtain the system of Eqs. (13). □

In view of the preceding analysis, the step by step application of the intrinsic kriging is as follows (see Fig. 9.3):

(a) Solve the system of Eqs. (13) for λ_i ($i = 1, 2, \dots, m$) and μ_j ($j = 0, 1, \dots, \rho$).
(b) Substitute the coefficients λ_i into Eq. (12), to find the optimal estimates $\hat{X}(\mathbf{s}_k)$, and
(c) then into Eq. (14) to find the estimation error variance $\sigma_x^2(\mathbf{s}_k)$.

Another useful expression for the latter is given by

$$\sigma_x^2(\mathbf{s}_k) = E[Z(\mathbf{s}_k)]^2 = k_x(\mathbf{O}) - \boldsymbol{\Xi}^{\mathrm{T}} \boldsymbol{\Theta} \tag{16}$$

Example 4: Let us reconsider the Equus Beds case study of Example 5, Section 4, Chapter 7. Using the information derived in that example, we

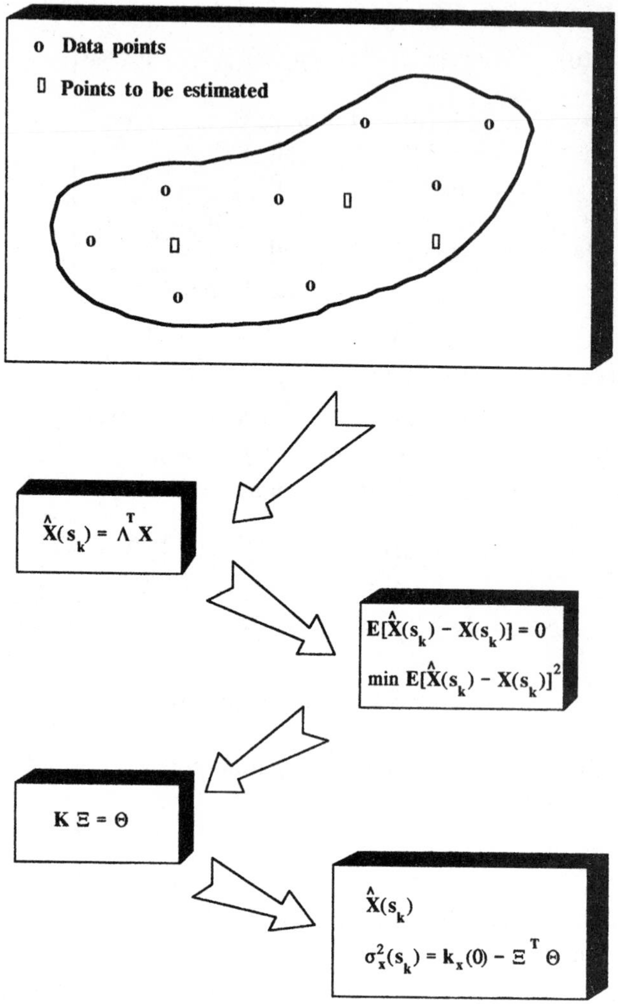

Figure 9.3 The intrinsic kriging procedure

solve the estimation system (13) and obtain values for the water table elevations and the associated error variances (16) at numerous locations on a properly specified grid. These values are then inserted into a plotting computer program to produce the contour map of water table elevations (in feet, Fig. 9.4) together with the contour map of error variances (in feet^2, Fig. 9.5).

When an interactive graphics terminal is available, the procedure can be controlled by the user, who can intervene to sort data, refine the estimates, or gauge the visual effects. In connection with this, when a specified level

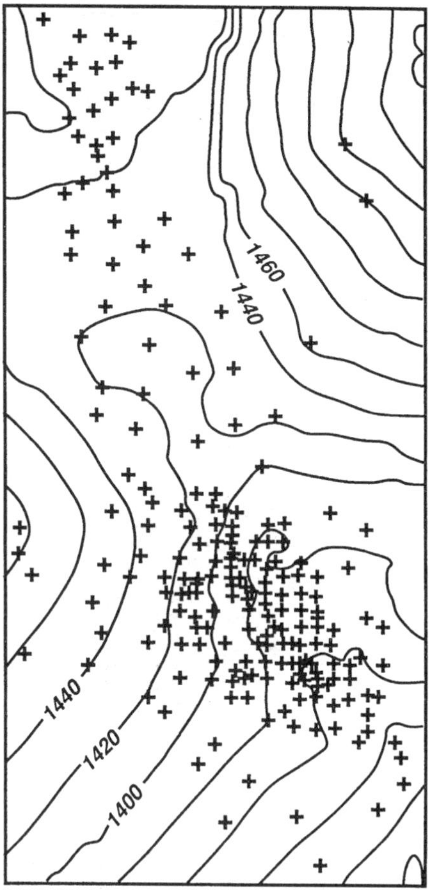

Figure 9.4 Contour map of water table elevation (in feet). The locations of the 226 observation wells are also shown

of accuracy of water table predictions is needed, one may increase the number of observations in areas showing high error variance and decrease the number of observations in areas showing low error variance.

5. Properties and Physical Interpretations of Linear Spatial Estimators

A careful examination of the estimation equations leads to interesting conclusions concerning both the mathematical and the physical components of the linear spatial estimation approaches.

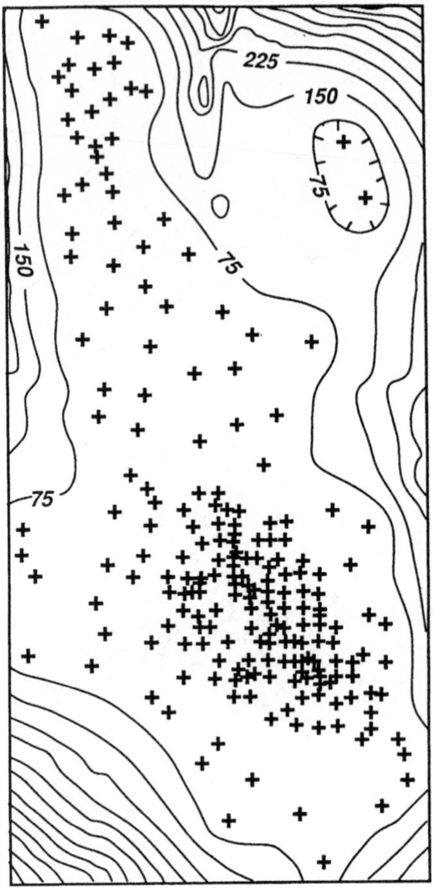

Figure 9.5 Contour map of the estimation error variance (in feet^2) for the water table elevation surface of Fig. 9.4

5.1 Certain Features of the Estimation System

As we have seen (Chapter 3) $c_x(\mathbf{s}_i, \mathbf{s}_j)$ is, in general, decomposed into $k_x(r = |\mathbf{s}_i - \mathbf{s}_j|)$ and a polynomial component $Q(\mathbf{s}_i, \mathbf{s}_j)$ that cannot be estimated from the data. One of the most appealing features of the estimation scheme above is that we need only the homogeneous GSC-ν $k_x(r)$ and not the nonhomogeneous ordinary covariance $c_x(\mathbf{s}_i, \mathbf{s}_j)$ for the optimal estimation of the SRF $X(\mathbf{s})$.

The solution estimation equations (13) of Section 4 are independent of the coordinate system. If two of the data points $\mathbf{s}_i$ coincide, the system of equations has no solution, since two of the estimation equations will be identical. Furthermore, the solution will not be unique for certain symmetric combinations of the points $\mathbf{s}_i$ that make at least two of the Eqs. (13) of Section 4 above identical.

Tips regarding the reduction of the number of equations in the estimation system include: (i) Use symmetry of the data points configuration. (ii) Ignore, if possible, a certain number of data points. (iii) Force the system to give the same weight to groups of points.

5.2 The Exact Interpolation Property

The estimation scheme leads to *exact interpolations*, if no measurement error is present. This simply follows from the fact that the minimum value of $E[\hat{X}(\mathbf{s}_k)-X(\mathbf{s}_k)]^2$, when $\mathbf{s}_k$ is a data point, is attained for $\hat{X}(\mathbf{s}_k)=X(\mathbf{s}_k)$, when we have $\min E[\hat{X}(\mathbf{s}_k)-X(\mathbf{s}_k)]^2=0$.

5.3 Confidence Intervals of the Estimates

We can often assume that the estimation error $\hat{X}(\mathbf{s}_k)-X(\mathbf{s}_k)$ is *normally* distributed with zero mean and variance $\sigma_x^2(\mathbf{s}_k)$. Then it is possible to define *confidence intervals* for the estimates $\hat{X}(\mathbf{s}_k)$. For example, the 95% confidence interval will be $[-2\sqrt{\sigma_x^2(\mathbf{s}_k)},\ 2\sqrt{\sigma_x^2(\mathbf{s}_k)}]$, and the estimate of $X(\mathbf{s}_k)$ with 95% confidence will be $\hat{X}(\mathbf{s}_k)\pm 2\sqrt{\sigma_x^2(\mathbf{s}_k)}$.

5.4 The Data-Independence Property

The estimation system and the estimation error variance [Eqs. (13) and (16), Section 4 above] do not depend on the specific values of the data. Therefore the system can be solved and the variance can be computed as long as the positions of the field observations are known. Herein we will refer to this as the *data-independence* property of the estimation variance. But the extrapolation to the limit might not be valid: No design is possible without knowledge of ν and $k_x(\cdot)$, for which a minimum sampling or extrapolation of structural identification from a nearby area is required.

5.5 Linear Spatial Estimation as a Filtering Process

As we saw in Chapter 3, in the case where $X(\mathbf{s})$ is an ISRF-ν, the

$$Z(\mathbf{s}_k)=\hat{X}(\mathbf{s}_k)-X(\mathbf{s}_k)=\sum_{i=1}^{m}\lambda_i X(\mathbf{s}_i)-X(\mathbf{s}_k)$$

$$=\sum_{i=1}^{m,k}\lambda_i X(\mathbf{s}_i)$$

is a SI-ν which, because $\sum_{i=1}^{m,k}\lambda_i=0$, can be considered as a *high-pass filter* with point spread function $\mathrm{PSF}_{\mathrm{H}}(\mathbf{s}_i)=\sum_{i=1}^{m,k}\lambda_i$.

High-pass filtering is a property associated with an estimation process that enhances detail of the spatial pattern; in other words, the derived maps contain high-frequency information. In light of these considerations the spatial estimator can be written as

$$\hat{X}(\mathbf{s}_k) = \sum_{i=1}^{m,k} \lambda_i X(\mathbf{s}_i) + X(\mathbf{s}_k) \tag{1}$$

where $\lambda_k = -1$. Furthermore, since the coefficient of $X(\mathbf{s}_k)$ is one, Eq. (1) yields (see also Fig. 9.6)

$$\hat{X}(\mathbf{s}_k) = \mathrm{PSF_H}(\mathbf{s}_k) * X(\mathbf{s}_k) + \mathrm{PSF_L}(\mathbf{s}_k) * X(\mathbf{s}_k) \tag{2}$$

where the $\mathrm{PSF_L}(\mathbf{s}_k) = 1$ represents a *low-pass filter* spread function and $*$ denotes the convolution operation. (A low-pass filter smooths the detail in the spatial pattern.)

In the light of Eq. (2), the intrinsic spatial estimator can be seen as the sum of a high-pass and a low-pass filter. If $\mathbf{s}_k$ is a data point, in view of the exact interpolation property of the spatial estimation concept $[\hat{X}(\mathbf{s}_k) = X(\mathbf{s}_k)]$, Eq. (2) gives

$$X(\mathbf{s}_k) = \mathrm{PSF_H}(\mathbf{s}_k) * X(\mathbf{s}_k) + \mathrm{PSF_L}(\mathbf{s}_k) * X(\mathbf{s}_k)$$

This is a fundamental identity of image processing-related filtering theory (e.g., Schowengerdt, 1983; Carr, 1990) and establishes an interesting link between the latter and the spatial estimation theory discussed above.

5.6 Possible Modifications of the Estimation Scheme

The intrinsic estimation scheme can be adjusted to other forms of the $\mathfrak{F}[\cdot]$ function [see Eqs. (15)-(17), Section 3]. For example, in the case of Eq. (15) of Section 3, we replace $k_x(\mathbf{s}_i - \mathbf{s}_k)$ by $k_x(\mathbf{s}_i, V_k)$, where the latter is the mean value of the GSC-ν between the point $\mathbf{s}_i$ and the volume V_k; also, $k_x(0)$ is replaced by $k_x(V_k, V_k)$, the mean value of the GSC-ν within the volume V_k.

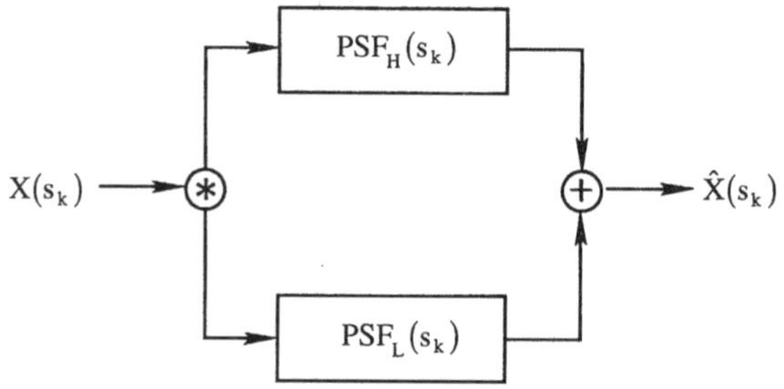

Figure 9.6 Spatial estimation as a filtering process

It is possible that the measured values, say $Y(\mathbf{s}_i)$, include some *measurement error* $V(\mathbf{s}_i)$ so that [see also Eq. (13), Section 2 of Chapter 7]

$$Y(\mathbf{s}_i) = aX(\mathbf{s}_i) + V(\mathbf{s}_i) \tag{3}$$

where a is a known deterministic coefficient and $V(\mathbf{s}_i)$ is a white-noise process with zero mean and variance $\sigma_v^2(\mathbf{s}_i)$, uncorrelated with $X(\mathbf{s}_i)$. In this case we should replace the terms $k_x(\mathbf{s}_i - \mathbf{s}_j)$ and $k_x(\mathbf{s}_i - \mathbf{s}_k)$ in the estimation system by $a^2 k_x(\mathbf{s}_i - \mathbf{s}_j) + \sigma_v^2(\mathbf{s}_j)\, \delta_{ij}$ and $ak_x(\mathbf{s}_i - \mathbf{s}_k)$, respectively. The associated error estimation variance is written as

$$\sigma_x^2(\mathbf{s}_k) = k_x(\mathbf{O}) - a \sum_{i=1}^{m} \lambda_i k_x(\mathbf{s}_i - \mathbf{s}_k) + \frac{1}{a} \sum_{0 \leq |\boldsymbol{\rho}| \leq \nu} \mu_\rho p_\rho(\mathbf{s}_k) \tag{4}$$

Obviously, if only a subset of the data points includes measurement errors, the $\sigma_v^2(\mathbf{s}_i)$ should be equal to zero at the points that are not interrupted by measurement errors.

Example 5: This example illustrates the quantitative analysis of the spatial variability of soil settlements, and also the prediction of settlement patterns coupled with the associated error maps. The insight thus obtained will be used in the design of the foundations, the optimal location of the buildings, and the site exploration. In the area of interest a heavy industry was to be founded. The subsoil structure is dominated by clay soils with sand layers in some parts of the site. Borings were spaced at locations where footings of the buildings were to be placed. Measurements were made of the settlements at each point, based on conventional stress–strain analysis and laboratory tests with simulated loading conditions (see Lambe and Whitman, 1969). Thus we implicitly included in the analysis dependencies of settlements on the nonuniform variation of stresses imposed by the foundations on the subsoil and on the spatial distribution of other soil parameters.

Starting with the spatial variability identification, we note that the settlement variability shows a complex trend. This is apparent from the five different concentrations in the histogram of the data (Fig. 9.7). The experimental covariances along different directions in space as well as the average covariance are shown in Fig. 7.4 of Chapter 7. The inspection of the behavior of the covariances at the origin and at large distances led to the conclusion that the spatial variability of the settlement surface is expected to be nonhomogeneous but regular. Also, the covariance presents the highest variability along directions 2-2 and 3-3, and the lowest variability in directions 1-1 and 4-4.

Several of the settlement measurements included measurement errors modeled by a linear model of the form of Eq. (3) above, where $Y(\mathbf{s}_i)$ are the measured settlements, $X(\mathbf{s}_i)$ are the actual settlements, $a = 1$ and $V(\mathbf{s}_i)$ are a white-noise measurement error with zero mean and variance $\sigma_v^2(\mathbf{s}_i) = 0.35\ \text{cm}^2$. In this case the generalized covariances of the settlement pattern

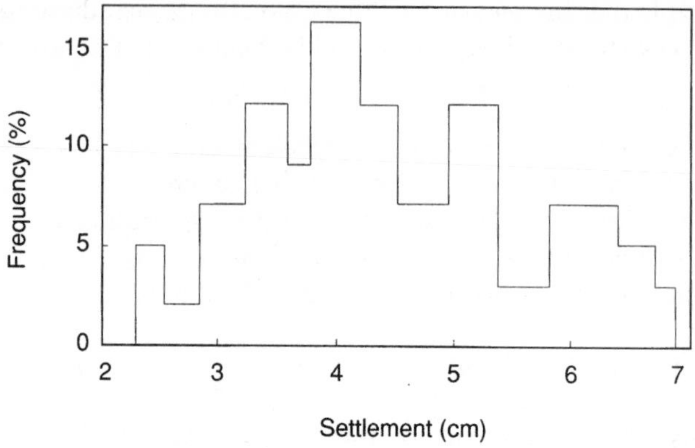

Figure 9.7 Histogram of soil settlements with five peaks evidencing the existence of local trends

are $k_x(\mathbf{s}_i - \mathbf{s}_j) + \sigma_v^2(\mathbf{s}_i)\ \delta_{ij}$. Areas containing 16 data points each were selected so that the settlement increments obtained are homogeneous. In this respect, the calculated order of settlement variability is $\nu = 1$, while the corresponding GSC-1 is

$$k_x(\mathbf{h}) = -0.4648|\mathbf{h}| + 3.835 \times 10^{-3}|\mathbf{h}|^3$$

For the specific stratification we observe linearly varying settlement trends, while the term in $|\mathbf{h}|^3$ evidences the regularity of the settlement surface, which is in agreement with the previous findings.

The next step in the soil settlement data processing deals with the production of predictive soil settlement patterns over the area of interest. Such a map was plotted in Fig. 7.1 of Chapter 7, where the boring locations were also shown. Clearly, samples located inside the stress bulb are especially important for predicting soil settlements in space. The estimation procedure was carried out using measurements within a bounded neighborhood with a radius of about 65 m. These neighborhoods were selected according to physically motivated criteria, such as data quality and smooth trends.

Clearly, the size of the memory required in this case is much smaller than if we had employed all the available data. The accuracy obtained is optimal, given the data we want to use. This is justified in practice where, after a certain number, the taking of further data may have a negligible effect on the estimation accuracy gained, and the use of a unique neighborhood would be useless. Moreover, the SRF-based hypotheses of stochastic inference (incremental homogeneity, isotropy, etc.) are more realistic in a local-scale settlement pattern than in a global one. Figure 7.1 shows that some areas are expected to settle more than others, due either to heavy buildings or to different soil properties. However, no excessive differential settlements are observed. The corresponding map of estimation error variances (Fig. 9.8) depicts relatively large values at the side parts of the site, which are clearly due to the small number of observations.

Once more, by making use of the "data value-independence" property of the estimation error variance, the reduction of the estimation error variance was computed for an additional number of five "fictitious" borings, before installing new borings with the associated costs. This would help determine whether such an installation is worthwhile, where the additional borings should be located, etc. In this case the locations of the five "fictitious" borings are shown in Fig. 9.9, together with the map of the new estimation error variances obtained. Comparing this map with the previous one (Fig. 9.8), we can clearly see the improvement in accuracy. This sort of sensitivity analysis may also guide the geotechnical engineer in choosing among different techniques of soil settlement prediction; certainly, the measurement errors introduced by each one of these techniques will play an important role here. Furthermore, the careful study of these maps will offer valuable information regarding the foundation design, the optimal location of the various buildings, and the site exploration.

If the spatial variation is *homogeneous* the estimation can be significantly simplified. The GSC-ν $k_x(\cdot)$ are reduced to the ordinary covariances $c_x(\cdot)$

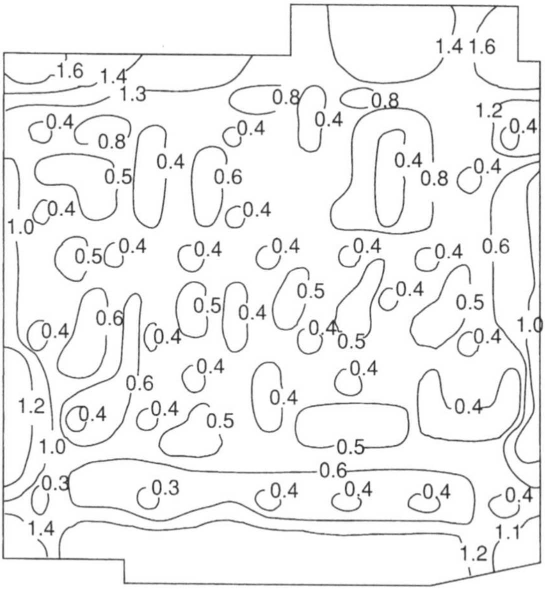

Figure 9.8 Contour map of spatial estimation error variances in predicting the settlement pattern of Fig. 7.1 of Chapter 7

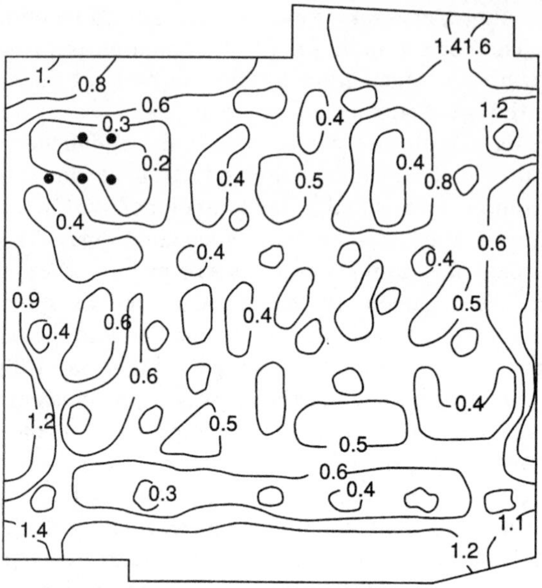

Figure 9.9 Contour map of spatial estimation error variances of the settlement pattern after the addition of five boreholes

and $p_\rho(\cdot)=1$, $\mu_0=-\mu$, and $\mu_\rho=0$, $\rho\geq 1$. Therefore, Eq. (13) of Section 4 yields

$$\mathbf{CM}=\mathbf{\Psi},$$

where

$$\mathbf{C}=\begin{bmatrix} c_x(0) & \dots & c_x(r_{1m}) & 1 \\ \vdots & & \vdots & \vdots \\ c_x(r_{m1}) & \dots & c_x(0) & 1 \\ 1 & \dots & 1 & 0 \end{bmatrix}$$

$$\mathbf{M}^{\mathrm{T}}=[\lambda_1\lambda_2\dots\lambda_m-\mu]$$

and

$$\mathbf{\Psi}^{\mathrm{T}}=[c_x(r_{1k})\dots c_x(r_{mk})1]$$

We have already seen in Chapter 7 that, in several practical situations in addition to the natural process of interest, say $X_0(\mathbf{s})$, there might exist a significant amount of data regarding other natural processes, say $X_i(\mathbf{s})$, $i=1, 2, \dots, \mu$, which are well correlated with $X_0(\mathbf{s})$. In these circumstances, it is natural to estimate $X_0(\mathbf{s})$ based on the measurements not only of $X_0(\mathbf{s})$, but also of $X_i(\mathbf{s})$, $i=1, 2, \dots, \mu$. For example, in petroleum engineering the hydrocarbon pore volume can be estimated by well log measurements and seismic data, where the latter are well correlated with the former. In hydrologic studies, the transmissivity in an aquifer can be estimated by means of pumping tests, as well as specific capacity measurements. The necessary modifications of the estimation scheme, which in the case of geostatistics is called *co-kriging* (e.g., Marechal, 1970; Francois-Bongarcon, 1981; Myers, 1982; Stein *et al.*, 1988), are relatively straightforward. One essentially replaces the scalar natural process $X(\mathbf{s})$ with the *vector* of the natural processes involved, viz.,

$$\mathbf{X}(\mathbf{s})^{\mathrm{T}} = [X_0(\mathbf{s}), X_1(\mathbf{s}), \ldots, X_\mu(\mathbf{s})] \tag{5}$$

The estimator $\hat{X}_0(\mathbf{s}_k)$ of $X_0(\mathbf{s}_k)$ at location $\mathbf{s}_k$ is written

$$\hat{X}_0(\mathbf{s}_k) = \sum_{i=0}^{\mu} \mathbf{\Lambda}_i^{\mathrm{T}} \mathbf{X}_i \tag{6}$$

where

$$\mathbf{X}_i^{\mathrm{T}} = [X_i(\mathbf{s}_1), X_i(\mathbf{s}_2), \ldots, X_i(\mathbf{s}_{m_i})]$$

$$\mathbf{\Lambda}_i^{\mathrm{T}} = [\lambda_{i1}\lambda_{i2} \ldots \lambda_{im_i}], \qquad i = 0, 1, \ldots, \mu$$

and m_i is the number of locations where data are available for the natural process $X_i(\mathbf{s})$; the estimation system now includes generalized covariances and *cross-generalized covariances* between the elements of $\mathbf{X}(\mathbf{s})$ in Eq. (5).

5.7 Physical Significance of the Estimation System

Physical reasoning underlying the estimation system accounts for certain important factors, such as

(i) The *relative positions* between the data points and the points to be estimated [items $(\mathbf{s}_i, \mathbf{s}_k)$, $i = 1, 2, \ldots, m$].

(ii) The *geometry* of the data points configuration [items $(\mathbf{s}_i, \mathbf{s}_j)$, $i, j = 1, 2, \ldots, m$].

(iii) The *spatial variability* of the natural process [through the GSC-ν $k_x(\cdot)$ and the order of intrinsity ν].

(iv) The *support* V_k, in the case of volume estimation [items $(\mathbf{s}_i, V_k)$ and (V_k, V_k), $i = 1, 2, \ldots, m$].

(v) The *measurement errors*, when they exist (through the measurement model).

(vi) Spatial *correlations* between the natural process of interest and other natural processes (through the vector form of the estimation scheme).

5.8 The Notion of Neighborhood

Estimation systems call for the notion of *neighborhood.* If a unique neighborhood is used, so all the available data are taken into account when estimating each value, only the right-hand sides of Eqs. (13) of Section 4 above change. Thus the covariance matrix must be inverted only once. Moving neighborhoods, on the other hand, restrict calculations to a *local* model of the spatial RF; that is, only data at a certain distance from the estimated point are used each time.

There are several reasons for choosing moving neighborhoods instead of a unique one:

(i) Reduced computational effort and better precision of the results, due to the smaller number of estimation equations.

(ii) More reliable hypotheses of statistical inferences are made at small distances where the fitting of structural models is easier. Indeed, it is much more realistic to expect homogeneity for SI-ν taken over a limited area than over the entire region. Also, the GSC-ν are best known at small lags **h**.

(iii) Neighborhoods are selected according to a physically motivated criterion, such as data quality or smooth trends.

(iv) In practice, after a certain number the taking of further data in estimating unknown values may have a negligible effect on the accuracy gained, and a unique neighborhood would be useless.

Example 6: This section constitutes a follow-up of the analysis of Example 2, Section 5 of Chapter 7. On the basis of the spatial variability identification results obtained in that example, we can estimate the shear strength over the entire region of interest. First we use only the 49 field measurements; the estimated shear strength map is shown in Fig. 9.10. The histogram of the estimates (Fig. 9.11) maintains the bimodal shape of the histogram of the data (Fig. 7.23 of Chapter 7), a fact that shows consistency of the estimated structure with the structure identified earlier on the basis of the field vane (FV) tests. In practice, the estimation map is always accompanied by the map of the estimation error variances (Fig. 9.12). The latter is very accurate in regions covered by a significant number of vane holes; the accuracy decreases with the distance from the data points. If we assume that the estimation error is approximately normally distributed, a good measure of the estimation precision may be offered by the confidence interval

$$[\hat{X}(\mathbf{s}_k)-2\sqrt{\sigma_x^2(\mathbf{s}_k)},\ \hat{X}(\mathbf{s}_k)+2\sqrt{\sigma_x^2(\mathbf{s}_k)}]$$

of the actual strength values, where $\hat{X}(\mathbf{s}_k)$ and $\sigma_x^2(\mathbf{s}_k)$ are readily available from Fig. 9.10 and 9.12, respectively. Taking advantage of the "data value-independence" property of the estimation error variance, we added three "fictitious" data points at the eastern part of the site. The new error variances show an improvement in the accuracy of the estimated strength in the area by about 25%. Of course, this percentage is rapidly reduced with the distance from the "fictitious" observation points.

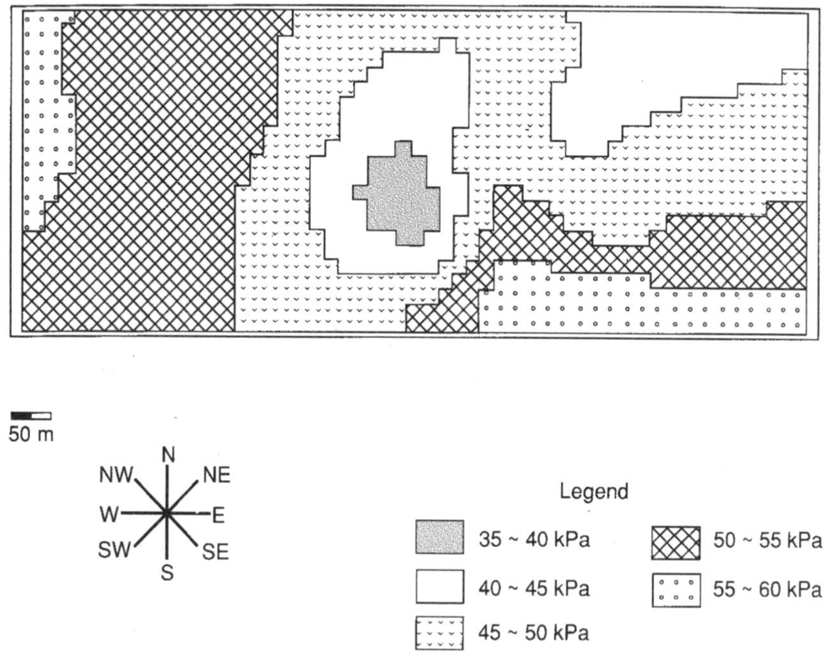

Figure 9.10 Map of shear strength estimates obtained from the field data

Next we incorporated into the analysis all 168 data (49 field measurements and 119 "subjective" data). (For the corresponding semivariogram see Fig. 7.25, Chapter 7.) The new estimation map is plotted in Fig. 9.13 and is very similar to the one obtained before on the basis of the 49 field measurements (Fig. 9.10). This should be expected, because the spatial variability identification has led in both cases to similar results (Example 2, Section 5 of Chapter 7). In addition to this consistency, which may be attributed partially to the adequate geometrical configuration of the installed vane holes, the new error map (Fig. 9.14) shows that the estimation accuracy improved significantly (mainly due to the reduction of the statistical error).

The insight gained by the soil strength data processing allowed a better understanding of the marine clay behavior, and it was used to guide engineering decisions and complement conventional design (Christakos, 1987b).

5.9 Estimation Error Variance Factorization

Depending on the analytical expression for the GSC-ν, further significant simplifications in the computation of estimation variances can be accomplished by factoring Eq. (13) of Section 4 above, as follows (Christakos, 1990b).

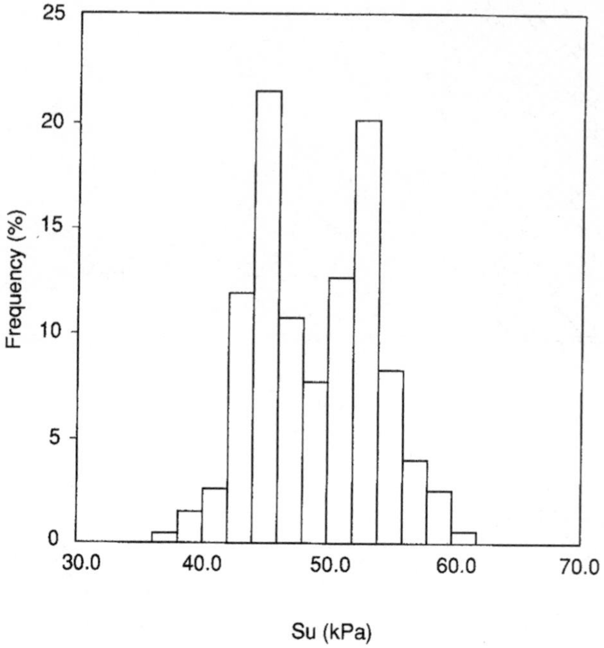

Figure 9.11 Histogram of the estimated shear strengths

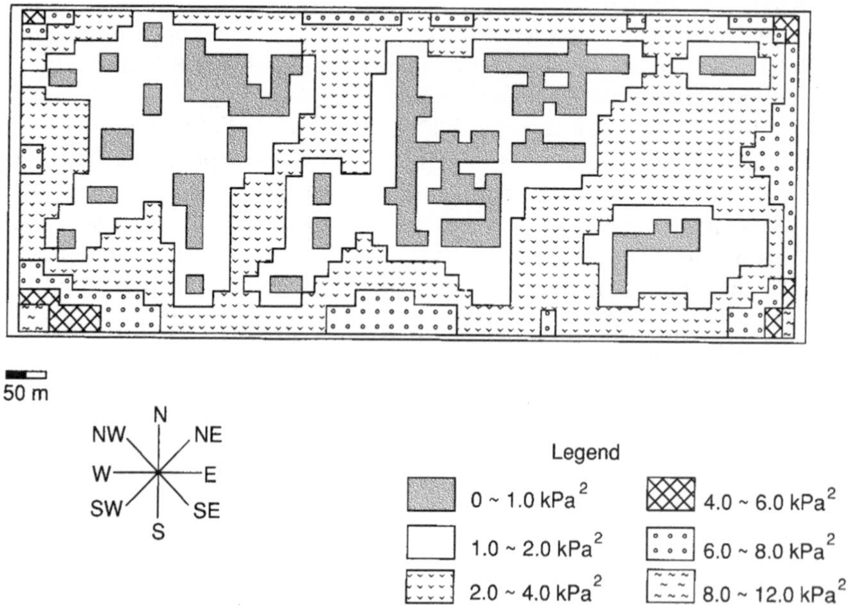

Figure 9.12 Spatial error variances in estimating the shear strength surface of Fig. 9.10

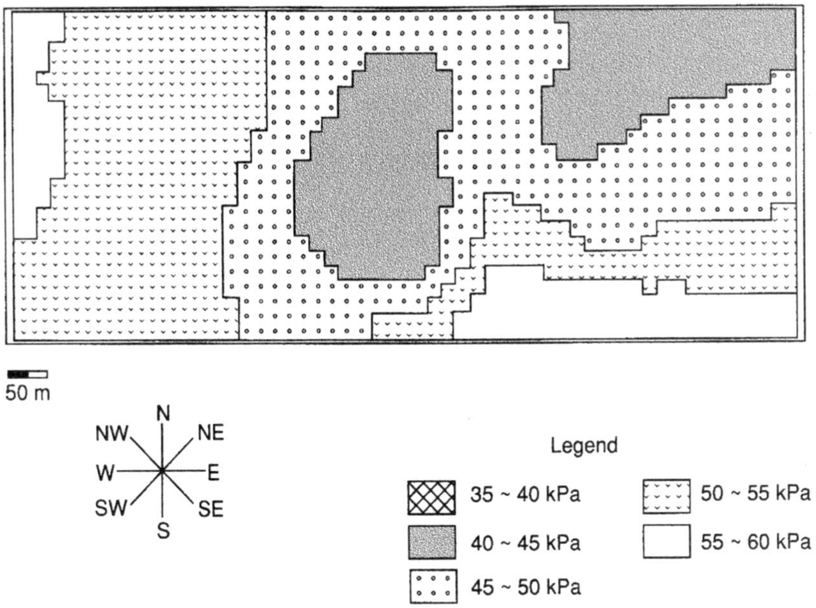

Figure 9.13 Map of shear strength estimates obtained on the basis of the 168 data

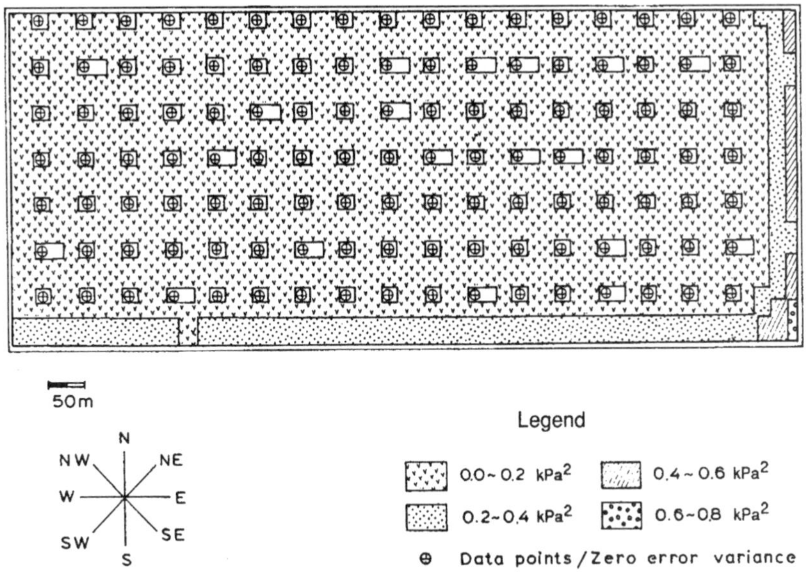

Figure 9.14 Spatial error variances in estimating the shear strength map of Fig. 9.13

Proposition 1: Let

$$k_x(r) = \mathbf{G}^{\mathrm{T}}\mathbf{R} \tag{7}$$

be a polynomial GSC-ν (Chapter 3), written in matrix form, where

$\mathbf{G}^{\mathrm{T}} = [c_0\, c_1\, c_3 \ldots c_{2\rho+1} \ldots c_{2\nu+1}]$ and $\mathbf{R}^{\mathrm{T}} = [\delta(r)\, r r^3 \ldots r^{2\rho+1} \ldots r^{2\nu+1}]$

where the coefficients $c_0 = a_0$ and $c_{2\rho+1} = (-1)^{\rho+1} c_\rho$ (watch for change in notation) satisfy the conditions depicted in Table 3.1 of Chapter 3. Then, for any fixed covariance coefficient c_t, $t = 0, 1, 3, \ldots, 2\nu+1$, the estimation variance of Eq. (16), Section 4 is written as

$$\sigma_x^2(\mathbf{s}_k) = c_t s_x^2 \tag{8}$$

where $s_x^2 = s_x^2(\tilde{c}_q, \tilde{c}_t)$ is the estimation variance obtained when $\tilde{c}_t = 1$, and the covariance coefficients c_q, $q = 0, 1, 3, \ldots, 2\rho+1$ are replaced by $\tilde{c}_q = c_q / c_t$. Herein we will refer to such estimation variance as *reference estimation variance.*

Example 7: Let $\nu = 1$ and $k_x(r) = c_0\, \delta(r) + c_1 r + c_3 r^3$. For $t = 3$, Eq. (8) gives $\sigma_x^2(\mathbf{s}_k) = c_3 s_x^2(\tilde{c}_0, \tilde{c}_1, \tilde{c}_3)$, where $\tilde{c}_0 = c_0 / c_3$, $\tilde{c}_1 = c_1 / c_3$, $\tilde{c}_3 = 1$. Similar expressions are obtained if, instead of $t = 3$, one takes $t = 0$.

6. Nonlinear Estimation

6.1 A Class of Nonlinear Estimators

In this section we will consider *nonlinear* minimum mean square error estimators of the quite general class defined by

$$\hat{X}(\mathbf{s}_k) = \sum_{i=1}^{m} G[Y(\mathbf{s}_i), \mathbf{s}_i] \tag{1}$$

Note that special cases of Eq. (1) are, among others, (i) the *linear* estimators considered in the previous section; and (ii) the so-called *disjunctive kriging* estimator (Matheron, 1976; Marechal, 1976; Kim *et al.*, 1977)

$$\hat{X}(\mathbf{s}_k) = \sum_{i=1}^{m} g_i[Y(\mathbf{s}_i)] \tag{2}$$

On the basis of the above setting, the nonlinear estimation problem is stated below.

Problem 1: Find estimates $\hat{X}(\mathbf{s}_k)$ of the actual values $X(\mathbf{s}_k)$ at points $\mathbf{s}_k$, on the basis of measurements of $Y(\mathbf{s})$ at the points $\mathbf{s}_i$ $(i = 1, 2, \ldots, m)$. The estimator is assumed to be of the nonlinear form (1) above, and the optimality criterion to be satisfied is the minimization of the estimation error variance given by Eq. (8) of Section 3.

By substituting Eq. (1) into Eq. (8) of Section 3, and then minimizing with respect to $G[\cdot]$, the proposition below can be proven.

Proposition 1: The solution of Problem 1 is given by these functions $G[\cdot]$, which are the solutions of the system of integral equations below.

$$\int_{-\infty}^{\infty} \chi_k f_{XY}(\chi_k, \psi_i)\, d\chi_k = \sum_{j=1}^{m} \int_{-\infty}^{\infty} G(\psi_j, \mathbf{s}_j) f_Y(\psi_i, \psi_j)\, d\psi_j \tag{3}$$

for $i = 1, 2, \ldots, m$. Let us call this estimator the *G-nonlinear estimator.*

By integrating both sides of Eq. (3) with respect to ψ_i we find that

$$E[X(\mathbf{s}_k)] = \sum_{j=1}^{m} E\{G[Y(\mathbf{s}_j), \mathbf{s}_j]\} = E[\hat{X}(\mathbf{s}_k)] \tag{4}$$

That is, the G-nonlinear estimator is *unbiased.*

Remark 1: If we set $G[\psi_j, \mathbf{s}_j] = \lambda_j \psi_j$ into Eq. (3), then multiply by ψ_i, and finally integrate with respect to ψ_i we derive the system of equations of the unconstrained Wiener–Kolmogorov estimator of Section 4.1.

6.2 Orthogonal Expansions of Nonlinearities

According to the theory discussed in Section 2 of Chapter 4, if the function $G[\psi, \mathbf{s}]$, as well as the bivariate probability density functions $f_{XY}(\chi_k, \psi_i)$ and $f_Y(\psi_i, \psi_j)$, satisfy the necessary convergence conditions, the following series expansions are valid.

$$f_Y(\psi_i, \psi_j) = f_Y(\psi_i) f_Y(\psi_j) \sum_{a=0}^{\infty} \sum_{b=0}^{\infty} \phi_{ab}(\mathbf{s}_i, \mathbf{s}_j) p_a(\psi_i) p_b(\psi_j) \tag{5}$$

$$f_{XY}(\chi_k, \psi_i) = f_X(\chi_k) f_Y(\psi_i) \sum_{a=0}^{\infty} \sum_{b=0}^{\infty} \zeta_{ab}(\mathbf{s}_k, \mathbf{s}_i) p_a(\chi_k) p_b(\psi_i) \tag{6}$$

and

$$G[\psi, \mathbf{s}] = \sum_{b=0}^{\infty} \xi_b(\mathbf{s}) p_b(\psi) \tag{7}$$

where $p_a(\chi)$ and $p_b(\psi)$ $(a, b = 0, 1, 2, \ldots)$ are sets of complete polynomials that are orthogonal with respect to $f_X(\chi)$ and $f_Y(\psi)$, respectively; i.e.

$$E[p_a(\chi) p_b(\chi)] = \int p_a(\chi) p_b(\chi) f_X(\chi)\, d\chi = \delta_{ab} \tag{8}$$

$$E[p_a(\psi) p_b(\psi)] = \int p_a(\psi) p_b(\psi) f_Y(\psi)\, d\psi = \delta_{ab} \tag{9}$$

The corresponding expansion coefficients are

$$\phi_{ab}(\mathbf{s}_i, \mathbf{s}_j) = \iint f_Y(\psi_i, \psi_j) p_a(\psi_i) p_b(\psi_j)\, d\psi_i\, d\psi_j \tag{10}$$

$$\zeta_{ab}(\mathbf{s}_k, \mathbf{s}_i) = \iint f_{XY}(\chi_k, \psi_i) p_a(\chi_k) p_b(\psi_i)\, d\chi_k\, d\psi_i \tag{11}$$

and

$$\xi_b(\mathbf{s}) = \int f_Y(\psi) G[\psi, \mathbf{s}] p_b(\psi)\, d\psi \tag{12}$$

Remark 2: We now present a few specific expansion coefficients and polynomials that will be useful in the following:

$$\phi_{00}(\mathbf{s}_i, \mathbf{s}_j) = \zeta_{00}(\mathbf{s}_k, \mathbf{s}_i) = 1$$

$$\phi_{01}(\mathbf{s}_i, \mathbf{s}_j) = \phi_{10}(\mathbf{s}_i, \mathbf{s}_j) = \zeta_{01}(\mathbf{s}_k, \mathbf{s}_i) = \zeta_{10}(\mathbf{s}_k, \mathbf{s}_i) = 0$$

$$\phi_{11}(\mathbf{s}_i, \mathbf{s}_j) = \frac{c_Y(\mathbf{s}_i, \mathbf{s}_j)}{\sigma_Y(\mathbf{s}_i)\sigma_Y(\mathbf{s}_j)}$$

$$\zeta_{11}(\mathbf{s}_k, \mathbf{s}_i) = \frac{c_{XY}(\mathbf{s}_k, \mathbf{s}_i)}{\sigma_X(\mathbf{s}_k)\sigma_Y(\mathbf{s}_i)}$$

$$p_0(\psi_i) = p_0(\chi_k) = 1$$

$$p_1(\psi_i) = \frac{\psi_i - E[Y(\mathbf{s}_i)]}{\sigma_Y(\mathbf{s}_i)}$$

and

$$p_1(\chi_k) = \frac{\chi_k - E[X(\mathbf{s}_k)]}{\sigma_X(\mathbf{s}_k)}$$

Proposition 2: Retain the orthogonality assumptions above. Then, the equations (3) of Proposition 1 reduce to the system of equations

$$\sum_{b=0}^{\infty} \sum_{j=1}^{m} \xi_b(\mathbf{s}_j) \phi_{ab}(\mathbf{s}_i, \mathbf{s}_j) = \zeta_{a0}(\mathbf{s}_k, \mathbf{s}_i) E[X(\mathbf{s}_k)] + \zeta_{a1}(\mathbf{s}_k, \mathbf{s}_i) \sigma_X(\mathbf{s}_k) \tag{13}$$

for all $i = 1, 2, \ldots, m$, and $a = 0, 1, \ldots$.

Proof: By taking into consideration Eqs. (5), (6), and (7), Eq. (3) can be written as

$$f_Y(\psi_i)\sum_{b=0}^{\infty}\sum_{a=0}^{\infty}\sum_{c=0}^{\infty}p_b(\psi_i)\int p_a(\psi_j)p_c(\psi_j)f_Y(\psi_j)\,d\psi_j\sum_{j=1}^{m}\xi_c(\mathbf{s}_j)\phi_{ab}(\mathbf{s}_i,\mathbf{s}_j)$$
$$=f_Y(\psi_i)\sum_{b=0}^{\infty}\sum_{a=0}^{\infty}p_b(\psi_i)\zeta_{ab}(\mathbf{s}_k,\mathbf{s}_i)\int\chi_k f_X(\chi_k)p_a(\chi_k)\,d\chi_k \tag{14}$$

for all $i=1,2,\ldots,m$. On the other hand, from Eqs. (8), (9), and by taking into account Remark 2 we find that

$$\int p_a(\psi_j)p_b(\psi_j)f_Y(\psi_j)\,d\psi_j=\delta_{ab} \tag{15}$$

also

$$\int\chi_k p_a(\chi_k)f_X(\chi_k)\,d\chi_k=0$$

if $a\neq 0,1$; $=E[X(\mathbf{s}_k)]$ if $a=0$; and $=\sigma_X(\mathbf{s}_k)$ if $a=1$. Finally, the last two integral equations in combination with Eq. (14) yield Eq. (13). □

Equation (13) consists of an infinite number of integral equations whose solution will provide the optimum nonlinear estimator of the form of Eq. (1). The implementation, however, of the above nonlinear estimation process will require certain additional assumptions, such as (i) only a limited number of the series expansion terms will be retained; in this case, of course the resulting solution will be an approximation; and (ii) the RF involved will be considered spatially homogeneous or isotropic.

6.3 Nonlinear Spatial Estimation in the Light of the Theory of Factorable Random Fields

In this section we will assume that the spatial RF involved in the estimation process are factorable RF (see Chapter 4).

Proposition 3: Assume that in Proposition 2 the spatial RF $X(\mathbf{s})$ and $Y(\mathbf{s})$ are factorable RF. Then, the nonlinear estimator reduces to a *linear* one, and the estimation system (13) becomes

$$\sum_{j=1}^{m}\xi_1(\mathbf{s}_j)c_Y(\mathbf{s}_i,\mathbf{s}_j)=c_{XY}(\mathbf{s}_k,\mathbf{s}_i)\sigma_Y(\mathbf{s}_j) \tag{16}$$

for all $i=1,2,\ldots,m$.

Proof: Since $X(\mathbf{s})$ and $Y(\mathbf{s})$ are factorable RF, the expansion coefficients in Eqs. (5) and (6) become $\phi_{ab}(\mathbf{s}_i,\mathbf{s}_j)=\zeta_{ab}(\mathbf{s}_k,\mathbf{s}_i)=0$ for $a\neq b$; and

$\phi_{aa}(\mathbf{s}_i, \mathbf{s}_j) = \phi_a(\mathbf{s}_i, \mathbf{s}_j)$, $\zeta_{aa}(\mathbf{s}_k, \mathbf{s}_i) = \zeta_a(\mathbf{s}_k, \mathbf{s}_i)$ for $a = b$. Consequently, the following series expansions of the corresponding bivariate probability densities are valid (see also Chapter 4)

$$f_Y(\psi_i, \psi_j) = f_Y(\psi_i) f_Y(\psi_j) \sum_{a=0}^{\infty} \phi_a(\mathbf{s}_i, \mathbf{s}_j) p_a(\psi_i) p_a(\psi_j) \tag{17}$$

$$f_{XY}(\chi_k, \psi_i) = f_X(\chi_k) f_Y(\psi_i) \sum_{a=0}^{\infty} \zeta_a(\mathbf{s}_k, \mathbf{s}_i) p_a(\chi_k) p_a(\psi_i) \tag{18}$$

By taking into account the last two equations, the system (13) reduces to only one equation, namely,

$$\sum_{j=1}^{m} \xi_1(\mathbf{s}_j) \phi_1(\mathbf{s}_i, \mathbf{s}_j) = \zeta_1(\mathbf{s}_k, \mathbf{s}_i) \sigma_X(\mathbf{s}_k)$$

for each $i = 1, 2, \ldots, m$; the latter is Eq. (16). □

Remark 3: Note that in the case of the disjunctive kriging form (2), the system of Eq. (16) becomes the disjunctive kriging system. Some interesting applications of disjunctive kriging in earth sciences are discussed in Rendu (1980) and Yates *et al.* (1986), among others.

6.4 Recursive On-Line Estimation Using Factorable Random Fields

The nonlinear on-line problem (e.g., time series) involves the estimation of a state process on the basis of an on-line observation process. Both the state and the observation models are nonlinear and are assumed given on some probability space. The estimates are to be determined *recursively* (see Fig. 9.1 above) in the minimum squared error sense.

The nonlinear problem has been tackled in a number of ways, leading to a growing literature in recent years. The first attempts to solve the problem were directed toward the generalization of well-known results of the linear estimation theory (see, e.g., Bucy, 1965; Stratonovich, 1968). Despite the several interesting theoretical results obtained (e.g., exact solutions were derived, but they required an infinite-dimensional system), the problem of deriving a practical as well as mathematically more substantial estimation procedure was not solved.

In the area of approximate estimators, practically useful algorithms have been constructed by using series expansions of either the state nonlinearity (Athans *et al.*, 1968; Jazwinski, 1966), or the probability density of the state conditioned on all available measurements (Sorenson and Stubberud, 1968; Willman, 1981). Within the framework of approximate nonlinear estimators, statistical linearization techniques have been of some success, particularly when the nonlinear functions are not differentiable (Mahalanabis and Farooq, 1971; Gelb, 1974).

More recent approaches have as a starting point well-established stochastic differential equations for the nonlinear estimator, such as the Fujisaki-Kallianpur-Kunita and the Duncan-Mortensen-Zakai ones (Fujisaki *et al.*, 1972; Duncan, 1967). Such approaches include the application of martingale theory to nonlinear estimation problems (Kallianpur, 1980), the Lie algebraic methods (Brockett, 1981; Hazewinkel and Marcus, 1982), and functional integration and group representations (Mitter, 1980). Davis (1981) emphasizes the relevance of the pathwise estimation theory to practical applications, and Pardoux (1981) suggests a solution of the nonlinear estimation problem, based on a pair of stochastic partial differential equations, one backward and one forward.

In this section the nonlinear problem is considered in light of the theory of factorable random processes (FRP) developed in Chapter 4. The estimation algebra is optimal for a broad class of time series, all of which possess the property of factorability. The setting of the FRP algorithm is such that it is readily applicable in practice, its use requiring the same computational effort as the standard (linear) Kalman filter. At the same time the theory is much simpler and the prerequisites are significantly lesser than in the aforementioned approaches.

Let $\Psi_{s-\tau}$ be the σ field of the sample path $\{Y(u), 0 \leq u \leq s-\tau\}$ and let s be a fixed real number. Estimation is concerned with making estimates $\hat{X}(s/s-\tau)$ about the process $X(s)$ at s on the basis of $\Psi_{s-\tau}$. If $\tau = 0$, the estimation problem is called *filtering*; if $\tau > 0$, it is called *prediction*. As was shown in Corollary 2, Section 5 of Chapter 4, the optimum nonlinear estimator of an FRP generated by the nonlinear state, nonlinear observation system (NSNOS) (1) and (2), Section 5 of Chapter 4 is identical to the optimum linear estimator of the linear state, linear observation system (LSLOS) (8) and (9). As a matter of convenience, it turns out that these coefficients need not be calculated for application of the FRP estimator. The ordinary Kalman filter algorithm can be used with processes generated by the linear system (8), (9), where the noise and initial densities satisfy the factorability assumption. These results are summarized below (Christakos, 1989).

Proposition 4: The optimum recursive estimator of the NSNOS (1) and (2), Section 5 of Chapter 4 is as follows, with filtered state and error variance given by

$$\hat{X}(s/s) = \hat{X}(s/s-1) + \beta_s \varepsilon(s) \tag{19}$$

$$\sigma^2_{s/s} = [1 - B(\beta_s)]^2 \sigma^2_{s/s-1} + \beta^2_s E[V^2(s)] \tag{20}$$

where

$$\varepsilon(s) = Y(s) - B[\hat{X}(s/s-1)] \tag{21}$$

is the innovation process, which is a zero-mean white noise with variance

$$r_s = B^2(\sigma_{s/s-1}) + E[V^2(s)] \tag{22}$$

and

$$\beta_s = \frac{B(\sigma^2_{s/s-1})}{r_s} \tag{23}$$

is the so-called gain. The predicted state and error variance are given by

$$\hat{X}(s/s-1) = G[\hat{X}(s-1/s-1)] \tag{24}$$

$$\sigma^2_{s/s-1} = G^2(\sigma_{s-1/s-1}) + E[W^2(s-1)] \tag{25}$$

Corollary 3: On the strength of Proposition 1, Section 3 of Chapter 4, the estimation algorithm of Proposition 2 above is valid for any random process that can be expressed as a strictly monotonic function of an FRP.

Remark 4: Note that, to apply the nonlinear estimation scheme above, we need to assure the validity of the factorability property for the processes $X(s)$ and $Y(s)$. In accordance with the theory of FRP, the relevant theta functions must belong to L_2 classes of functions. For example, in the bivariate Gaussian case, the integrability condition (9), Section 2 of Chapter 4 becomes $r_2 = 1/(1-\rho^2) < \infty$, $|\rho| < 1$, where ρ is the correlation coefficient.

Remark 5: The bivariate Gaussian regression model underlies most of the important results of the theory of random processes. In the estimation context, the conclusion of this section is that several of these results can be extended to the significantly richer bivariate factorability model.

Example 1: To gain some insight about the theory developed above, and to compare the FRP-based algorithm and the well-established extended Kalman algorithm (EKA; Gelb, 1974), the following NSNOS proposed in Jazwinski (1970) was simulated over time (i.e., $s = t$)

$$X_t = 99.95 \times 10^{-2} X_{t-1} + 4 \times 10^{-4} X^2_{t-1} + W_{t-1} \tag{26}$$

$$Y_t = X_t^2 + X_t^3 + V_t \tag{27}$$

where the following notation is used: $X_t = X(t)$, $Y_t = Y(t)$, $W_{t-1} = W(t-1)$, and $V_t = V(t)$; the initial state is Gaussian with mean 2 and variance 10^{-2}; and W_{t-1}, V_t are zero-mean Gaussian white noises with variances 5×10^{-5} and 9×10^{-2}, respectively. NSNOS of Eqs. (26) and (27) may be used, for example, to generate soil profiles. The simulated measurements are shown (Fig. 9.15). On the basis of Eqs. (26) and (27), the EKA Jacobians are $a_{1,t} = 99.95 \times 10^{-2} + 8 \times 10^{-4} \times \hat{X}_{t/t-1}$, $b_{1,t} = 2\hat{X}_{t/t-1} + 3\hat{X}^2_{t/t-1}$. For both the FRP and the EKA, the initial conditions assumed are $\hat{X}_{0/0} = 2$, $\sigma^2_{0/0} = 10^{-2}$. The FRP estimated state $\hat{X}_{t/t}$ is shown in Fig. 9.16. Because the true state

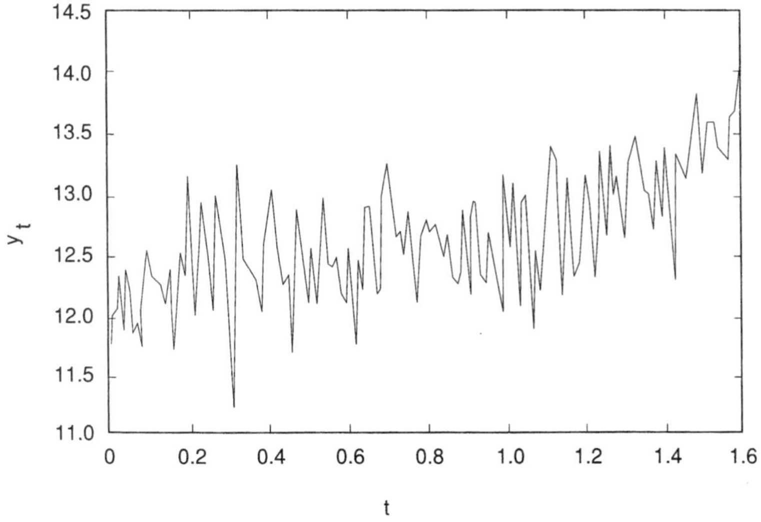

Figure 9.15 Simulated measurements

X_t is known, the estimation error $\tilde{X}_{t/t} = X_t - \hat{X}_{t/t}$ using $\pm 2\sqrt{\sigma^2_{t/t}}$ as confidence intervals can be plotted (Fig. 9.17). The error $\tilde{X}_{t/t}$ is small (its mean is about 2.4×10^{-4}) and only about 0.7% lies outside the bounds. The last FRP variance available is close to the sample variance (0.25×10^{-4} and

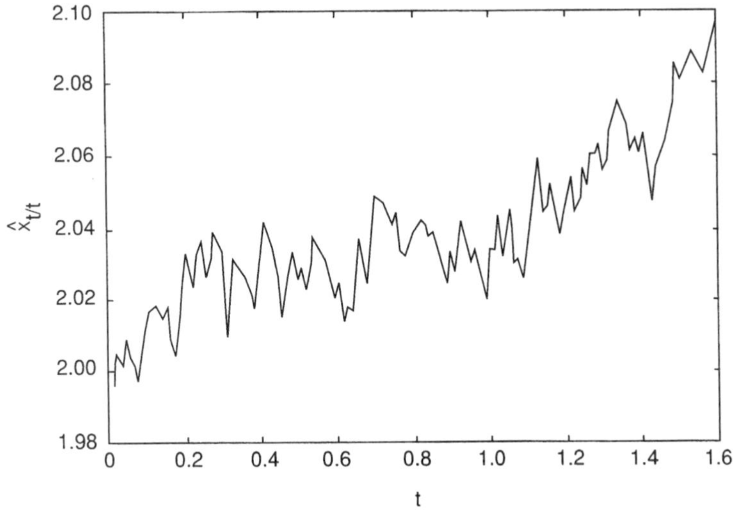

Figure 9.16 Estimated states (FRP)

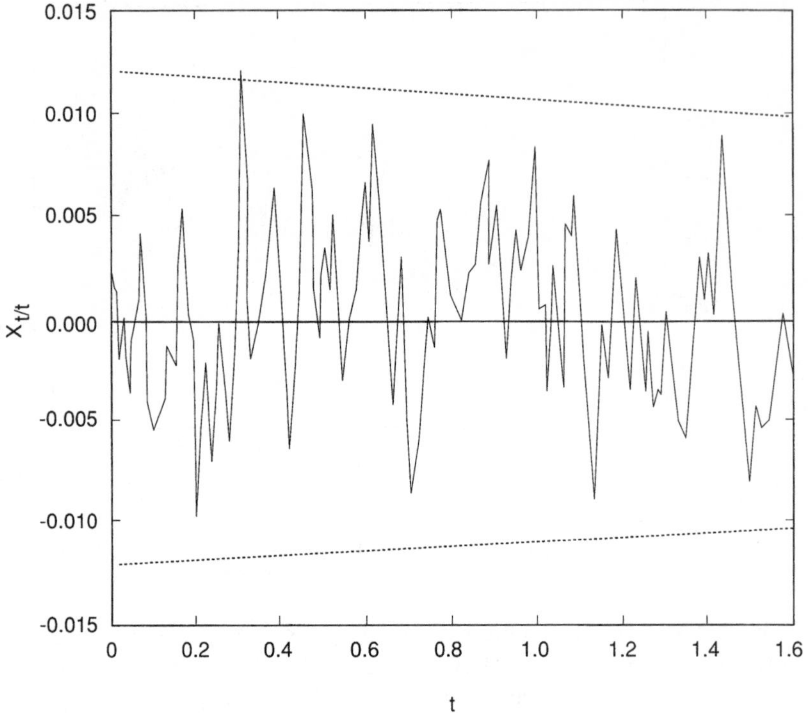

Figure 9.17 Error estimates and confidence intervals

0.20×10^{-4}, respectively). The innovation process ε_t, together with the confidence interval $\pm 2\sqrt{r_t}$ provided by the FRP (Fig. 9.18), shows none of the ε_t values exceeding the bounds, whereas the mean is close to zero ($\sim 3.4 \times 10^{-3} < 1.96\sqrt{\sigma_\varepsilon^2/n} = 5.84 \times 10^{-2}$, n is the number of simulated data). Only 2.67% of the sample correlation coefficient ρ_ε lies outside the bounds $\pm 1.96/\sqrt{n} = \pm 0.16$ (Fig. 9.19). Therefore, ε_t is a zero-mean white-noise process, which implies the proper functioning of FRP. The estimation error $\hat{X}_{t/t}$ (FRP) provided by the FRP is compared with that provided by the EKA, $\hat{X}_{t/t}$ (EKA) (Fig. 9.20); $\Delta\tilde{X}_{t/t} = |\tilde{X}_{t/t}(\text{FRP})| - |\tilde{X}_{t/t}(\text{EKA})|$ establishes the superior performance of FRP (note that negative $\Delta\tilde{X}_{t/t}$ values indicate more accurate FRP, whereas positive $\Delta\tilde{X}_{t/t}$ values indicate more accurate EKA). This superiority becomes more distinct as t increases.

To examine the sensitivity of the two algorithms to variations in the statistics of the state noise W_t, the NSNOS of Eqs. (26) and (27) again is considered where now the variance $E[W_t^2]$ takes the values of 10^{-5}, 5×10^{-5}, and 9×10^{-5}. The plots of Fig. 9.21 show the FRP estimation variance $\sigma_{t/t}^2$. Note the rapid reduction of the $\sigma_{t/t}^2$ with the number of observations processed, leading to increasingly accurate estimates. This reduction seems

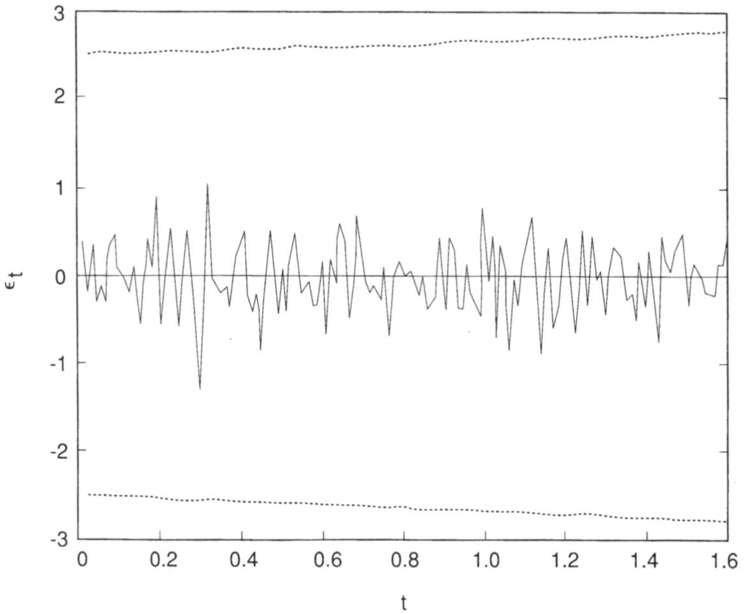

Figure 9.18 Innovation process and confidence intervals

to continue after the last observation available. The same setting for the EKA (Fig. 9.22) verifies that FRP also gives better results in this regard: smaller $\sigma^2_{t/t}$ values and less sensitivity to changes in $E[W_t^2]$.

7. Optimal Estimation of Spatiotemporal Random Fields

7.1 General Considerations

In Chapter 5 a theory of spatiotemporal random fields (S/TRF) was developed. Several interesting properties of such a class of random fields were examined and valuable insight into their mathematical structure was gained. In this section we will deal with the spatiotemporal estimation problem, which has various applications in almost any scientific discipline. In general the spatiotemporal estimation problem can be summarized as follows:

Problem 1: $X(q)$ be a GRS-ν/μ, and let $\mathscr{H}_{\nu/\mu}$ be the Hilbert space generated by the representations $X(\mathbf{s}, t)$ of $X(q)$ [the $X(\mathbf{s}, t)$ may represent, for instance, the precipitation, the atmospheric pollution, or a meteorologic element at position $\mathbf{s}$ at time t]. Let $X(\mathbf{s}_k, t_q) \in \mathscr{H}_{\nu/\mu}$. We want to find

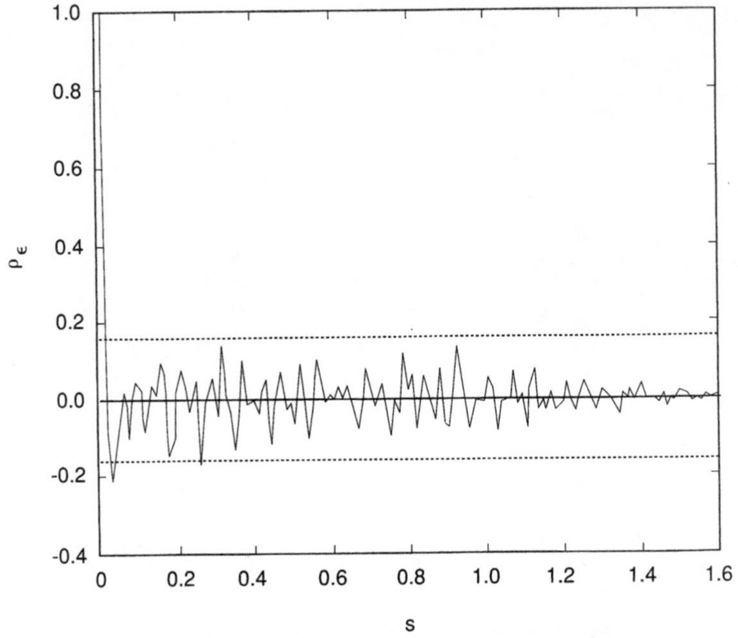

Figure 9.19 Sampled correlation and confidence intervals

estimates $\hat{X}(\mathbf{s}_k, t_q)$ of the actual values $X(\mathbf{s}_k, t_q)$ of the natural process of interest at unknown positions $\mathbf{s}_k$ and time instances t_q. The calculations are to be made on the basis of experimental data (observations) $X(\mathbf{s}_i, t_j)$, $i = 1, 2, \ldots, m$ and $j = 1_i, 2_i, \ldots, p_i$ (as before, the p_i denotes the number of time instances t_j used in estimation, given that we are at the spatial position $\mathbf{s}_i$). More precisely, an estimate $\hat{X}(\mathbf{s}_k, t_q)$ is defined as an element of $\mathcal{H}_{\nu/\mu}$, which fulfills the following requirements:

(i) *Linearity*, viz.,

$$\hat{X}(\mathbf{s}_k, t_q) = \mathbf{\Xi}^{\mathrm{T}}\mathbf{X} \tag{1}$$

where $\mathbf{\Xi}^{\mathrm{T}} = [\xi_{ij}]$ $(i = 1, 2, \ldots, m;\ j = 1_i, \ldots, p_i)$ is a vector of real coefficients ξ_{ij} to be calculated during the estimation process, and $\mathbf{X}^{\mathrm{T}} = [X(\mathbf{s}_i, t_j)]$ is a vector of known elements $X(\mathbf{s}_i, t_j) \in \mathcal{H}_{\nu/\mu}$, $(\mathbf{s}_i, t_j) \in \mathbf{A}$, where $\mathbf{A}$ is a compact set of data points/time instances. (Figure 9.23 illustrates the $R^1 \times T$ case of such a linear estimator.)

(ii) *Unbiasedness*, that is,

$$E[Z(\mathbf{s}_k, t_q)] = 0 \tag{2}$$

where $Z(\mathbf{s}_k, t_q) = \hat{X}(\mathbf{s}_k, t_q) - X(\mathbf{s}_k, t_q)$.

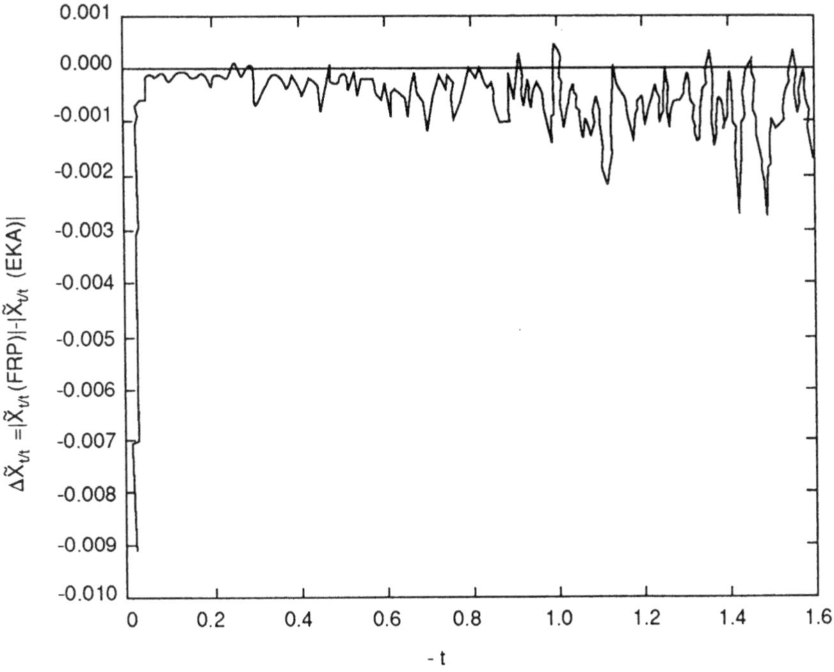

Figure 9.20 Differences in error estimates, FRP versus EKA

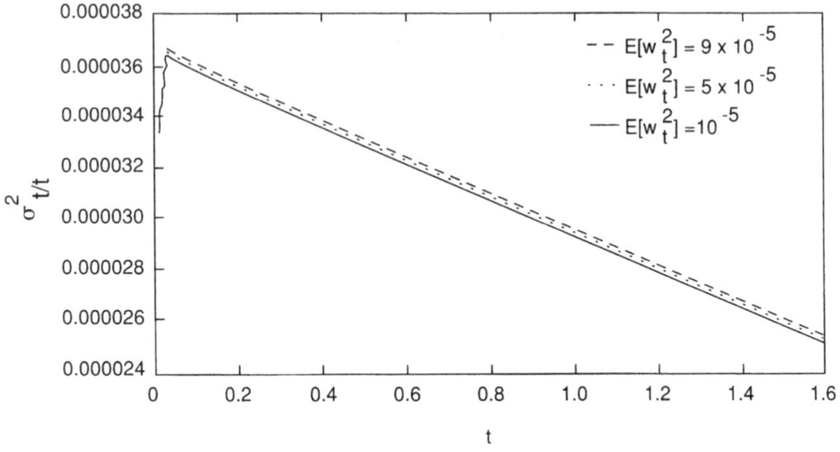

Figure 9.21 Error variances for FRP

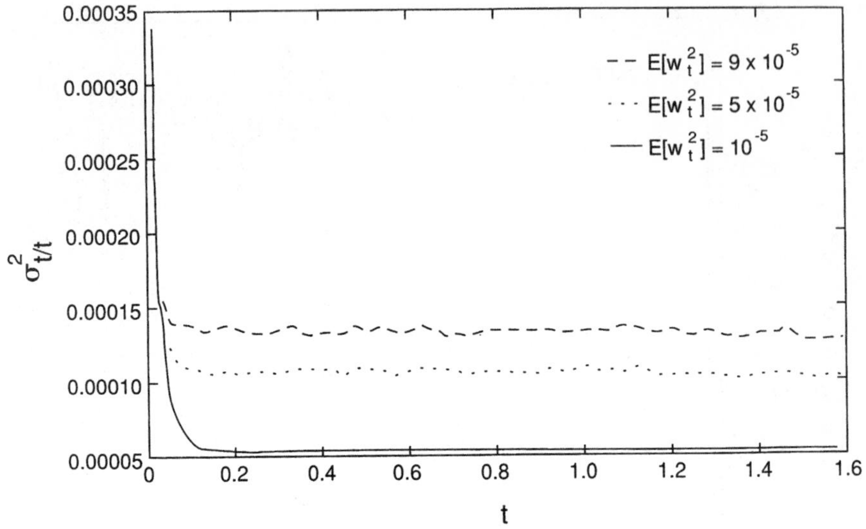

Figure 9.22 Error variances for EKA

(iii) *Optimality* (minimum mean square error); that is, it must minimize the estimation error

$$\sigma_x^2(\mathbf{s}_k, t_q) = E[Z(\mathbf{s}_k, t_q)]^2 \tag{3}$$

This is a constrained optimization problem in $R^n \times T$ whose solution depends on the regularity properties of the random field $X(\mathbf{s}, t)$ over space–time.

7.2 Optimal Estimation of Space-Homogeneous/Time-Stationary Processes

We assume here that the natural process of interest is represented by a space-homogeneous and time-stationary random field $X(\mathbf{s}, t)$, i.e. an S/TRF-$(-1/-1)$. Taking into account Eq. (1), Eq. (3) can be written as

$$\sigma_x^2(\mathbf{s}_k, t_q) = \sum_{i=1}^{m} \sum_{j=1_i}^{p_i} \sum_{i'=1}^{m} \sum_{j'=1_{i'}}^{p_{i'}} \xi_{ij}\xi_{i'j'}c_x(\mathbf{h}_{ii'}, \tau_{jj'})$$
$$-2\sum_{i=1}^{m} \sum_{j=1_i}^{p_i} \xi_{ij}c_x(\mathbf{h}_{ki}, \tau_{qj}) + c_x(\mathbf{0}, 0) \tag{4}$$

and the unbiased condition (2) is given by

$$\sum_{i=1}^{m} \sum_{j=1_i}^{p_i} \xi_{ij} = 1 \tag{5}$$

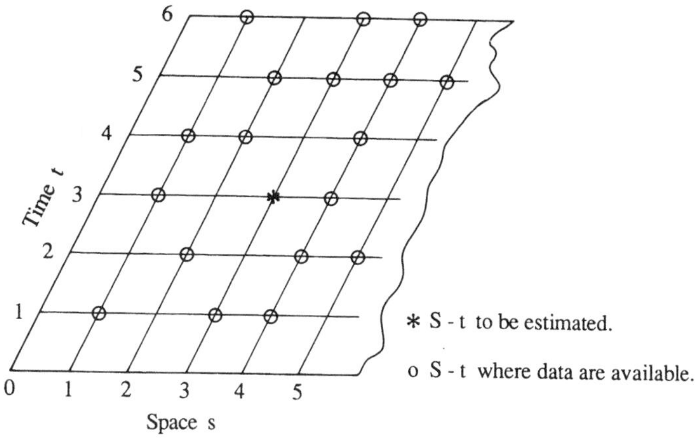

$$\hat{X}(s_3, t_3) = \sum_{i=1}^{5} \sum_{j=1_i}^{\ell_i} \xi_{ij} X(s_i, t_j). \quad \text{(LINEAR)}$$

SPACE s	TIME INSTANCES t	j	ℓ
1	1, 3, 4, 6	$1_1, 2_1, 3_1, 4_1$	4
2	2, 4, 5	$1_2, 2_2, 3_2$	3
3	1, 5, 6	$1_3, 2_3, 3_3$	3
4	1, 2, 3, 4, 5, 6	$1_4, 2_4, 3_4, 4_4, 5_4, 6_4$	6
5	2, 5	$1_5, 2_5$	2

Figure 9.23 The $R^1 \times T$ case of linear space–time estimation

By minimizing (4) with respect to ξ_{ij} subject to the condition imposed by Eq. (5), one finds the following system of equations written in matrix form

$$\mathbf{C}\mathbf{\Lambda} = \mathbf{H} \tag{6}$$

where $\mathbf{C} = [c_x(\mathbf{s}_i, t_j; \mathbf{s}_{i'}, t_{j'});\ i, i' = 1, 2, \ldots, m;\ j = 1_i, \ldots, p_i;\ j' = 1_{i'}, \ldots, p_{i'}]$ is a matrix of ordinary space-time covariances; $\mathbf{\Lambda}^{\mathrm{T}} = [\xi_{ij}, \mu;\ i = 1, 2, \ldots, m;$ $j = 1_i, \ldots, p_i]$ is a vector of estimation coefficients ξ_{ij} including the Lagrange multiplier μ; and $\mathbf{H}^{\mathrm{T}} = [c_x(\mathbf{s}_k, t_q; \mathbf{s}_i, t_j);\ i = 1, 2, \ldots m; j = 1_i, \ldots, p_i]$.

As soon as the estimation system (6) is solved with respect to ξ_{ij}, the latter can be inserted into Eqs. (1) and (4) to obtain the optimal space–time estimate and the associated error variance, respectively.

7.3 Optimal Estimation of Space-Nonhomogeneous/Time-Nonstationary Processes

Suppose now that the natural process can be represented by a S/TRF-ν/μ $X(\mathbf{s}, t)$ (Chapter 5). On the basis of Proposition 2, Section 7 of Chapter 5, the

$$Z(\mathbf{s}_k, t_q) = \hat{X}(\mathbf{s}_k, t_q) - X(\mathbf{s}_k, t_q)$$

is a S/TI-ν/μ, and its variance is given by

$$\sigma_x^2(\mathbf{s}_k, t_q) = \sum_{i=1}^{m} \sum_{j=1_i}^{p_i} \sum_{i'=1}^{m} \sum_{j'=1_{i'}}^{p_{i'}} \xi_{ij}\xi_{i'j'} k_x(\mathbf{h}_{ii'}, \tau_{jj'})$$
$$-2\sum_{i=1}^{m} \sum_{j=1_i}^{p_i} \xi_{ij} k_x(\mathbf{h}_{ki}, \tau_{qj}) + k_x(\mathbf{O}, \mathrm{O}) \tag{7}$$

($\xi_{kq} = -1$ and $\xi_{iq} = \xi_{kj} = 0$ $(i \neq k, j \neq q)$). The fact that $Z(\mathbf{s}_k, t_q)$ is a S/TI-ν/μ implies that

$$\sum_{i=1}^{m} \sum_{j=1_i}^{p_i} \xi_{ij} \mathbf{s}_i^{\boldsymbol{\rho}} t_j^{\zeta} = \mathbf{s}_k^{\boldsymbol{\rho}} t_q^{\zeta} \tag{8}$$

for all $0 \leq |\boldsymbol{\rho}| \leq \nu$ and $0 \leq \zeta \leq \mu$. [Note that Eq. (8) expresses the unbiasedness condition (2).] The minimization of Eq. (7) with respect to the ξ_{ij} subject to the constraint (8) yields the system of equations

$$\mathbf{K}\boldsymbol{\Xi}^* = \boldsymbol{\Theta} \tag{9}$$

where $\mathbf{K} = [k_x(\mathbf{s}_i, t_j; \mathbf{s}_{i'}, t_{j'}),\ s_i^{\boldsymbol{\rho}} t_j^{\zeta};\ i, i' = 1, 2, \ldots, m;\ j = 1_i, \ldots, p_i;\ j' = 1_{i'}, \ldots, p_{i'};\ |\boldsymbol{\rho}| \leq \nu, \zeta \leq \mu]$ is a matrix of GS/TC-ν/μ and space-time polynomials; $\boldsymbol{\Xi}^{*\mathrm{T}} = [\xi_{ij}, \psi_{\rho\zeta}, i = 1, 2, \ldots, m; j = 1_i, \ldots, p_i; \rho = |\boldsymbol{\rho}| \leq \nu; \zeta \leq \mu]$ is a vector of coefficients ξ_{ij} that includes the Lagrange multipliers $\psi_{\rho\zeta}$, and the vector $\boldsymbol{\Theta}^{\mathrm{T}} = [k_x(\mathbf{s}_k, t_q; \mathbf{s}_i, t_j),\ \mathbf{s}_k^{\boldsymbol{\rho}} t_q^{\zeta};\ i = 1, 2, \ldots, m;\ j = 1_i, \ldots, p_i,\ |\boldsymbol{\rho}| \leq \nu, \zeta \leq \mu]$.

7.4 Properties of Optimal Space–Time Estimation

The optimal space–time estimation scheme (7) through (9) above depends only on the GS/TC-ν/μ. Figure 9.24 presents the flowchart of the estimation scheme. *Physical reasoning* underlying the above estimation setting accounts for: (i) the spatiotemporal correlation structure of the data (through the matrix $\mathbf{K}$); (ii) the support and the space–time geometry of the data configurations (terms in i, j, i', and j'); and (iii) the relative positions/time instances between the data points/instances and the points/instances to be estimated (terms in i, j and k, q).

The *accuracy* of the estimation scheme will be significantly improved compared to that of the pure spatial estimation, since we now take into account important time-related information. The estimator is *consistent*, in

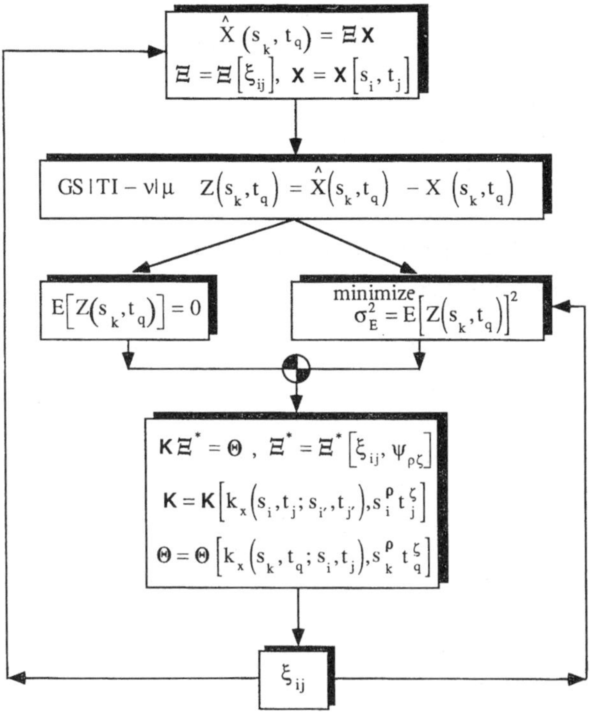

Figure 9.24 Flowchart of the spatiotemporal estimation approach

the sense that given the same data everyone using them should get the same result.

It is possible that the space-time data are interrupted by *measurement errors.* These errors may have various causes, such as inaccuracies in the instruments, errors in the primary processing and coding of the information, and errors in decoding and plotting the data. Naturally enough, the presence of such errors in the observations can have a considerable effect on the quality of a spatiotemporal analysis and estimation. In this situation suitable measurement models must be chosen by means of empirical research, and the necessary modifications in the estimation scheme should be made. In some applications one may seek the optimal estimator of some *functional* of the form

$$F[X(\mathbf{s}, t)] = \frac{1}{V_T} \int_U \int_T X(\mathbf{s}+\mathbf{s}', t+t')\, d\mathbf{s}'\, dt' \qquad (10)$$

where V_T is the support of the actual natural process $X(\mathbf{s}, t)$, U is a spatial volume, and T is a time interval. In this case certain modifications of the estimation scheme of Fig. 9.24 are necessary to obtain estimates $\hat{F}[X(\mathbf{s}, t)]$

of the functional $F[X(\mathbf{s}, t)]$. More specifically, the point/instant covariances $k_x(\mathbf{h}, \tau)$ should be replaced by the spatial volume/time averaged covariances $k_x(U, T)$, etc.

The following section discusses some interesting deviations from the fundamental assumptions of the minimum mean square error estimation methods developed above.

8. A Bayesian/Maximum-Entropy View of the Estimation Problem

8.1 General Considerations

An important feature of the minimum mean square error methods discussed in previous sections is that they rely primarily on the data values available. In principle, they do not incorporate into the analysis *prior information*, such as knowledge of the physics of the phenomena involved, geological interpretations, intuition, and experience with similar site conditions. This can be a disadvantage for an estimation method, especially in situations where the prior information is highly relevant to the spatial and/or temporal variability of the natural processes under estimation.

Within the framework of nonlinear geostatistics, important deviations from the above concept include disjunctive kriging and the multivariate Gaussian kriging (see, e.g., Journel and Huijbregts, 1978). Nonlinear estimation techniques require Gaussian-related hypotheses about the multivariate probability law of the underlying RF. In a more general context, an estimator of the considerably richer class of factorable random processes discussed in Chapter 4 was developed above. All these estimators, however, do not account for important sources of prior information like constraint intervals, inequality-type data, and "soft" qualitative data, which can specify different prior probability laws (Section 5.2.2 of Chapter 7). The most notable attempts to incorporate these issues into the estimation process have been made recently by Journel (see, e.g., Journel, 1986), who revived the importance of posterior cumulative distribution function determination as opposed to mere minimum mean square error methods. This approach, however, does not give a definite rule for setting up prior probabilities from the prior information available, and a certain amount of information may be lost due to the use of approximations in some stages of the estimation process.

In view of the foregoing, it will be interesting to search for an estimation method that will be capable of processing the highest possible prior information and of yielding insightful conclusions while, at the same time, it will not be lacking any of the important features of the traditional stochastic methods (i.e., the method should incorporate objective information, give consistent results, etc). To proceed in a stochastic context, we will first present four theses on which our analysis will be built. Herein we will refer to this estimation approach as the Bayesian–maximum-entropy approach (BME).

Thesis 1: Any prior information concerning the spatial structure of the natural process under estimation will be defined by reference to its prior probability. In fact, there is an inverse relation between the prior information and the prior probability: The more informative an assessment about the unknown value of a natural process, the less probable it is to occur.

In other words, the more narrowly the assessment restricts the possibilities it leaves open, the more informative it is. This is a widely accepted concept with deep philosophical roots, which have been extensively elaborated by Popper (1934; 1972), Carnap (1950), and others. It is not quite clear, however, how one may translate prior information into prior probability in an unambiguous, objective way. This leads to the next thesis.

Thesis 2: One stems high prior information (or, equivalently, low prior probability) about the spatial variability of the natural process of interest. In the stochastic context this can be achieved by considering the expected value of the appropriate information measure and then trying to maximize it.

Thesis 3: One's knowledge about the spatial distribution of the natural process after seeing the measurements (probability on evidence or posterior probability) is related to his knowledge before seeing the measurements (i.e., prior probability) by means of Bayes law.

Thesis 4: The high probability that one desires for his hypotheses is clearly posterior probability. In the estimation context this implies a posterior cumulative distribution function with minimum uncertainty attached to it.

From an intuitive point of view, it is reasonable to expect that the more improved are the emerging estimates, the more restrictive is the prior information. (Here, the word "restrictive" refers to the restraints imposed by the prior information on the solutions to the estimation problem.) In connection with this, the approach suggested by theses 1 to 4 tries to balance two requirements: high prior information about the natural process under estimation and high posterior probability for the resulting estimates.

8.2 The Estimation Approach

The estimation approach will be developed by applying the logical reasoning of the four foregoing theses. In doing so, let $X(\mathbf{s})$ be an SRF that represents a spatially distributed natural process, and assume that observed values χ_i of this process are available at points $\mathbf{s}_i$, $i=1, 2, \ldots, m$ in space. Let $\hat{X}(\mathbf{s}_k)$

be the estimator of $X(\mathbf{s}_k)$ at a point $\mathbf{s}_k$, $k \neq i$, where no observations are available. If $f_x(\chi_k, \chi_1, \ldots, \chi_m)$ is the joint probability density function of the associated random variables $x_k, x_1, \ldots, x_m$, prior to observing the data $x_i = \chi_i$, the fundamental normalization constraint is satisfied.

$$\underbrace{\int\int \cdots \int}_{m+1\,\text{times}} f_x(\chi_k, \chi_1, \ldots, \chi_m)\, d\chi_k\, d\chi_1 \ldots d\chi_m = 1 \tag{1}$$

where the integrations are carried out over the ranges of the random variables above. Herein, unless stated otherwise, these ranges will be assumed to vary from $-\infty$ to ∞.

The Rationale of the Estimation Procedure

In view of thesis 1, the probability density $f_x(\chi_k, \chi_1, \ldots, \chi_m)$ should be derived by means of an estimation process that takes into consideration *physical constraints* that represent either:

(a) the prior information and knowledge one may have about the natural process to be estimated, or
(b) the specific properties one wishes the estimator $\hat{X}(\mathbf{s}_k)$ to account for.

The choice, however, of an information measure is not, in general, an easy task. In the context of this study the traditional definition of the information measure (see, e.g., Shannon, 1948) has been chosen. More particularly, the information contained in the vector of random variables $\mathbf{X} = [x_k, x_1, \ldots, x_m]^T$ is assumed to be measured by (Chapter 2)

$$\text{Inf}[x_k, x_1, \ldots, x_m] = -\log[f_x(\chi_k, \chi_1, \ldots, \chi_m)] \tag{2}$$

The motivation underlying this choice is that Eq. (2) provides a mathematical expression of the inverse relation between the prior information and the prior probability of spatial variability as introduced by thesis 1 above: According to Eq. (2), the more informative the random vector $\mathbf{X}$, the less probable it will occur.

In relation to the information measure (2), the expected information considered in thesis 2 will be

$$\varepsilon(\mathbf{X}) = E\{\text{Inf}[x_k, x_1, \ldots, x_m]\} = E\{-\log[f_x(\chi_k, \chi_1, \ldots, \chi_m)]\}$$

$$= -\underbrace{\int\int \cdots \int}_{m+1\,\text{times}} \log[f_x(\chi_k, \chi_1, \ldots, \chi_m)]$$

$$\times f_x(\chi_k, \chi_1, \ldots, \chi_m)\, d\chi_k\, d\chi_1 \ldots d\chi_m \tag{3}$$

where $\varepsilon(\mathbf{X})$ is the entropy function (Chapter 2) of the natural process under estimation. Hence, the $\varepsilon(\mathbf{X})$ appears as a measure of the amount of uncertainty in the probability density $f_x(\chi_k, \chi_1, \ldots, \chi_m)$. By maximizing expected information (or, which is the same, by maximizing uncertainty) with respect to $f_x(\chi_k, \chi_1, \ldots, \chi_m)$, subject to constraints of the form (a) and (b) above, one essentially maximizes the corresponding entropy function. Remarkably, while the maximum entropy concept has been used extensively in a variety of fields such as pattern recognition, thermodynamics, information theory, statistics, and time series (see, e.g., Kullback, 1968; Burg, 1972; Shore and Johnson, 1980; Jaynes, 1982), its application in the context of spatial and spatiotemporal RF is extremely limited and the published literature is very fragmentary.

Since the estimates $\hat{X}(\mathbf{s}_k)$ are in principle expressed in terms of expectations of some function of the SRF $X(\mathbf{s})$, it is reasonable to consider prior physical constraints that can be expressed mathematically as

$$E[g_q] = \underbrace{\int\int\cdots\int}_{m+1\,\text{times}} g_q(\chi_k, \chi_1, \ldots, \chi_m) f_x(\chi_k, \chi_1, \ldots, \chi_m)\, d\chi_k\, d\chi_1 \ldots d\chi_m \tag{4}$$

where $g_q(\chi_k, \chi_1, \ldots, \chi_m)$, $q = 1, 2, \ldots, Q$, are suitable functions of $X(\mathbf{s})$. A few examples of such functions are given below (Christakos, 1990a; 1991a). Notice that we take $g_0(\chi_k, \chi_1, \ldots, \chi_m) = 1$, so that the $E[g_0] = 1$ defines the normalization constraint (1).

Example 1: Let $\hat{X}(\mathbf{s}_k)$ be the estimator of the homogeneous SRF $X(\mathbf{s}_k)$. In order that $\hat{X}(\mathbf{s}_k)$ takes into account the means $E[X(\mathbf{s}_i)]$ and covariances

$$c_x(\mathbf{s}_i, \mathbf{s}_j) = E\{[X(\mathbf{s}_i) - m_x(\mathbf{s}_i)][X(\mathbf{s}_j) - m_x(\mathbf{s}_j)]\}$$

of the data values $X(\mathbf{s}_i) = \chi_i$, the functions $g_q(\chi_k, \chi_1, \ldots, \chi_m)$ should be written

$$g_q(\chi_i) = \chi_i \tag{5}$$

where $i = 1, 2, \ldots, m$ and k; $q = 1, 2, \ldots, m+1$. Also,

$$g_q(\chi_i, \chi_j) = [\chi_i - m_x(\mathbf{s}_i)][\chi_j - m_x(\mathbf{s}_j)] \tag{6}$$

where $i, j = 1, 2, \ldots, m$ and k; $q = m+2, m+3, \ldots, (m+1)(m+4)/2$. Then, the resulting $E[g_q]$ yield the specified means $E[X(\mathbf{s}_i)]$ and covariances $c_x(\mathbf{s}_i, \mathbf{s}_j)$. Similarly, in the case that the information about spatial variability is available in the form of the semivariograms

$$\gamma_x(\mathbf{s}_i, \mathbf{s}_j) = \tfrac{1}{2}E[X(\mathbf{s}_i) - X(\mathbf{s}_j)]^2$$

the g_q functions should be written

$$g_q(\chi_i, \chi_j) = \tfrac{1}{2}[\chi_i - \chi_j]^2 \tag{7}$$

where $i, j = 1, 2, \ldots, m$ and k; $q = 1, 2, \ldots, m(m+1)/2$.

Example 2: To incorporate into the analysis information of the form of interval-type data, such as (Section 5.2.2 of Chapter 7)

$$\chi_i \in [0, \bar{\omega}_i] \subset R^1 \tag{8}$$

one may define the indicator function

$$g_q(\chi_i) = I_x(\chi_i, \bar{\omega}_i) = 1 \quad \text{if} \quad \chi_i \leq \bar{\omega}_i, \; = 0 \quad \text{otherwise} \tag{9}$$

where $q = 1, 2, \ldots, m+1$. Moreover one can write

$$g_q(\mathbf{s}_i, \mathbf{s}_j) = [I_x(\chi_i, \bar{\omega}_i) - m_I(\mathbf{s}_i, \bar{\omega}_i)][I_x(\chi_j, \bar{\omega}_j) - m_I(\mathbf{s}_j, \bar{\omega}_j)] \tag{10}$$

where $m_I(\mathbf{s}_i, \bar{\omega}_i) = E[I_x(X(\mathbf{s}_i), \bar{\omega}_i)]$, and $q = m+2, \; m+3, \ldots, (m+1) \times (m+4)/2$. Then, the $E[g_q(\mathbf{s}_i)] = 1$ would define constraints (8) while the

$$E[g_q(\chi_i, \chi_j)] = \text{cov}[I_x(X(\mathbf{s}_i), \bar{\omega}_i), I_x(X(\mathbf{s}_j), \bar{\omega}_j)]$$

is the indicator cross-covariance for the two threshold values $\bar{\omega}_i$ and $\bar{\omega}_j$. Also, let

$$g_q(\chi_i, \chi_j) = \tfrac{1}{2}[I_x(\chi_i, \bar{\omega}_i) - I_x(\chi_j, \bar{\omega}_i)][I_x(\chi_i, \bar{\omega}_j) - I_x(\chi_j, \bar{\omega}_j)] \tag{11}$$

where $i, j = 1, 2, \ldots, m$ and k; $q = 1, 2, \ldots, m(m+1)/2$. In this case the $E[g_q]$ defines the corresponding indicator cross-semivariograms.

Example 3: If $X(\mathbf{s})$ is an ISRF-ν (Chapter 3), the g_q-functions should be written, in general, as

$$g_q(\chi_i, \chi_j) = \left[\sum_k q_k S_{\mathbf{s}_i} \chi_k\right]\left[\sum_{k'} q_{k'} S_{\mathbf{s}_j} \chi_{k'}\right] \tag{12}$$

In particular, let $X(s)$ be an ISRF-ν in R^1 with $Y_q(s_i) = \Delta^{\nu+1} X(s_i)$, where

$$\Delta^{\nu+1} X(s_i) = \sum_{k=0}^{\nu+1} (-1)^k C_{\nu+1}^k X[s_i - (k - \nu - 1)\,\delta s]$$

is the finite difference operator of order $\nu+1$ and

$$C_{\nu+1}^k = \binom{\nu+1}{k}$$

(for simplicity and without loss of generality it will be assumed that $\delta s = 1$). Then

$$g_q(\chi_i, \chi_j) = [\Delta^{\nu+1} \chi_i][\Delta^{\nu+1} \chi_j] \tag{13}$$

the corresponding covariances turn out to be as follows

$$\begin{aligned} c_Y(s_i - s_j) &= \Delta^{\nu+1}\, \Delta^{\nu+1} c_x(s_i, s_j) \\ &= (-1)^{\nu+1} \sum_{k=0}^{2\nu+2} (-1)^k C_{2\nu+2}^k k_x(s_i - s_j + \nu + 1 - k) \end{aligned}$$

If the $X(\mathbf{s})$ is differentiable in the mean square sense, all

$$Y_k(\mathbf{s}_i) = \frac{\partial^{\nu+1}}{\partial s_{i_k}^{\nu+1}} X(\mathbf{s}_i) \qquad (k = 1, 2, \ldots, n)$$

are by definition homogeneous SRF. Also, the $Y(\mathbf{s}_i) = \nabla^{\nu+1} X(\mathbf{s}_i)$ is a zero-mean homogeneous SRF; in this case

$$g_q(\chi_i, \chi_j) = [\nabla^{\nu+1}\chi_i][\nabla^{\nu+1}\chi_j] \tag{14}$$

and the covariance of $Y(\mathbf{s}_i)$ is given by

$$\begin{aligned} c_Y(\mathbf{h}) &= \sum_{k=1}^{n} \sum_{k'=1}^{n} \frac{\partial^{2\nu+2}}{\partial s_{i_k}^{\nu+1}\, \partial s_{j_{k'}}^{\nu+1}} c_x(\mathbf{s}_i, \mathbf{s}_j) \\ &= (-1)^{\nu+1} \nabla^{2\nu+2} k_x(\mathbf{h}) \end{aligned}$$

Example 4: Let $X(\mathbf{s}, t)$ be an OS/TRF-ν/μ (Chapter 5). The g_q-functions are

$$\begin{aligned} g_q(\chi_{iv}, \chi_{j\eta}) = & \left[\sum_k \sum_p q_{kp} S_{\mathbf{s}_i, t_v} \chi_{kp}\right] \\ & \times \left[\sum_{k'} \sum_{p'} q_{k'p'} S_{\mathbf{s}_j, t_\eta} \chi_{k'p'}\right]. \end{aligned} \tag{15}$$

If $X(\mathbf{s}, t)$ is a differentiable OS/TRF-ν/μ, all

$$Y_k(\mathbf{s}_i, t_v) = \frac{\partial^{\nu+\mu+2}}{\partial s_{i_k}^{\nu+1}\, \partial t_v^{\mu+1}} X(\mathbf{s}_i, t_v) \qquad (k = 1, 2, \ldots, n)$$

are by definition zero-mean space-homogeneous/time-stationary random fields and so is

$$Y(\mathbf{s}_i, t_v) = \frac{\partial^{\mu+1}}{\partial t_v^{\mu+1}} \nabla^{\nu+1} X(\mathbf{s}_i, t_v)$$

In this case,

$$g_q(\chi_{iv}, \chi_{j\eta}) = \left[\frac{\partial^{\mu+1}}{\partial t_v^{\mu+1}} \nabla^{\nu+1} \chi_{iv}\right]\left[\frac{\partial^{\mu+1}}{\partial t_\eta^{\mu+1}} \nabla^{\nu+1} \chi_{j\eta}\right] \tag{16}$$

Experience and intuition will improve one's ability to translate qualitative knowledge into explicit mathematical constraints under the format of Eq. (4). In connection with this, it seems that the real power of subjective analysis is in fulfilling the need for normative rules according to which such translations will be carried out. Nevertheless, to assure the objectivity of the estimation approach, one should in general avoid the use of qualitative knowledge that is too vague to be expressed in a quantitative form.

The prior probability $f_x(\chi_k, \chi_1, \ldots, \chi_m)$ considered up to now includes a prior model for the relation between $X(\mathbf{s}_k)$ and the $X(\mathbf{s}_i)$ that refers to our knowledge regarding spatial variability before any specific measurements of the natural process have been taken into consideration. On the other hand, the posterior probability of thesis 3 (probability on evidence) $f_x^*(\chi_k|\chi_1, \ldots, \chi_m)$ refers to the (updated) knowledge, after these measurements have been incorporated into the estimation process. These two probabilities are related by the conditional probability law

$$f_x^*(\chi_k|\chi_1, \ldots, \chi_m) = \frac{f_x(\chi_k, \chi_1, \ldots, \chi_m)}{f_x(\chi_1, \ldots, \chi_m)} \tag{17}$$

which will be our basic tool in conducting sound stochastic inferences throughout the estimation process. Equation (17) constitutes a Bayes updating formulation in which any updating requires knowledge of the prior probability model $f_x(\chi_k, \chi_1, \ldots, \chi_m)$ of the unknown $X(\mathbf{s}_k)$ and the data $X(\mathbf{s}_i)$. In relation to this, it will be useful to set

$$\begin{aligned} B_x(\chi_k) &= \log[f_k^*(\chi_k|\chi_1, \ldots, \chi_m)] \\ &= \log[f_x(\chi_k, \chi_1, \ldots, \chi_m)] - \log[f_x(\chi_1, \ldots, \chi_m)] \end{aligned} \tag{18}$$

where $B_x(\chi_k)$ will be herein called the *BME function.*

The high posterior probability required by thesis 4 will be the final concern of the BME: The posterior probability $f_x^*(\chi_k|\chi_1, \ldots, \chi_m)$ or, equivalently, the $B_x(\chi_k)$ should be maximized with respect to χ_k. Note that in this case the χ_k is considered as a parameter of the posterior probability or the BME function of the underlying SRF.

The Estimation Scheme

In view of the above considerations, we can now formulate the BME version of the estimation problem as follows.

Problem 1: Let $X(\mathbf{s})$ be an SRF. Find estimates $\hat{X}(\mathbf{s}_k)$ of $X(\mathbf{s}_k)$ at points $\mathbf{s}_k$ given data $X(\mathbf{s}_i) = \chi_i$, $i = 1, 2, \ldots, m (i \neq k)$, such that

(I) the entropy $\varepsilon(\mathbf{X})$ of the prior model [Eq. (3)] is maximized with respect to $f_x(\chi_k, \chi_1, \ldots, \chi_m)$, subject to the normalization constraint (1) and the physical constraints (4); and

(II) the BME function $B_x(\chi_k)$ is maximized with respect to χ_k; the latter will be the desired value of the estimator $X(\mathbf{s}_k)$ at the point $\mathbf{s}_k$.

The BME solution to the Problem 1 is provided by the following proposition (Christakos, 1990a).

Proposition 1: The solution of Problem 1 above consists of the estimate $\hat{X}(\mathbf{s}_k) = \hat{\chi}_k$, which is the solution of the equation

$$\sum_{q=0}^{Q} \mu_q \frac{\partial g_q(\chi_k, \chi_1, \ldots, \chi_m)}{\partial \chi_k}\Big|_{\chi_k = \hat{\chi}_k} = 0 \tag{19}$$

where $g_0(\chi_k, \chi_1, \ldots, \chi_m) = 1$ and the μ_q $(q = 0, 1, \ldots, Q)$ are Lagrange multipliers to be determined from the set of equations below

$$\mu_0 = -\log\left\{\frac{1}{A}\underbrace{\int\int\cdots\int}_{m+1\,\text{times}} \times \exp\left[\sum_{q=1}^{Q} \mu_q g_q(\chi_k, \chi_1, \ldots, \chi_m)\right] d\chi_k\, d\chi_1 \ldots d\chi_m\right\} \tag{20}$$

and

$$\frac{\partial \mu_0}{\partial \mu_q} = \frac{1}{A}\underbrace{\int\int\cdots\int}_{m+1\,\text{times}} g_q(\chi_k, \chi_1, \ldots, \chi_m) \times \exp\left[\sum_{q=0}^{Q} \mu_q g_q(\chi_k, \chi_1, \ldots, \chi_m)\right] d\chi_k\, d\chi_1 \ldots d\chi_m = E[g_q] \tag{21}$$

where $q = 1, 2, \ldots, Q$ and

$$A = \underbrace{\int\int\cdots\int}_{m+1\,\text{times}} \exp\left[\sum_{q=0}^{Q} \mu_q g_q(\chi_k, \chi_1, \ldots, \chi_m)\right] d\chi_k\, d\chi_1 \ldots d\chi_m \tag{22}$$

The associated error variance can be determined by substituting the solution of Eq. (19) into

$$\sigma_x^2(\mathbf{s}_k) = E[\hat{X}(\mathbf{s}_k) - X(\mathbf{s}_k)]^2 \tag{23}$$

Remark 1: Note that requirement (I) of Problem 1 leads to the following general expression for the joint probability density:

$$f_x(\chi_k, \chi_1, \ldots, \chi_m) = \frac{1}{A}\exp\left[\sum_{q=0}^{Q} \mu_q g_q(\chi_k, \chi_1, \ldots, \chi_m)\right] \tag{24}$$

where A is given by Eq. (22). The probability density (24) is generally non-Gaussian. Moreover, according to requirement (II) of Problem 1 above we must have

$$\left.\frac{dB_x(\chi_k)}{d\chi_k}\right|_{\chi_k = \hat{\chi}_k} = 0$$

or

$$\left.\frac{d \log f_x(\chi_k, \chi_1, \ldots, \chi_m)}{d\chi_k}\right|_{\chi_k = \hat{\chi}_k} = 0 \tag{25}$$

In view of the above results, the proposed estimation approach will consist of the following steps:

(i) Solve the system of Eqs. (20) and (21) to determine the Lagrange multipliers μ_q $(q = 0, 1, \ldots, Q)$.

(ii) Substitute μ_q into Eq. (19) and solve with respect to $\hat{X}(\mathbf{s}_k)=\hat{\chi}_k$. This will lead to an estimate of the form $\hat{X}(\mathbf{s}_k)=\hat{\chi}_k=F[\chi_1,\chi_2,\ldots,\chi_m]$, where $F[.]$ is, in general a nonlinear function.

(iii) Substitute $\hat{X}(\mathbf{s}_k)$ into Eq. (23) to find the associated estimation error.

It is important to notice that the range of values of the natural processes involved in the estimation process determine the domains of integration in the entropy function as well as the constraints equation. Consequently, the solutions of Eqs. (20) and (21) depend on these ranges and so do the estimates obtained by means of Eq. (19). This should be intuitively expected, because the specific range of a natural process is valuable information that the BME incorporates into the estimation process.

In relation to this, the solutions of Eqs. (20) and (21) may have a closed analytical form, or a numerical method may have to be applied. The latter will probably be the most common situation in practical applications where the different information sources may lead to specific integral domains and a variety of g_q functions. Also, in certain cases a discretization process may be used.

8.3 Properties of the Bayesian/Maximum-Entropy Estimator

Given measurements of a natural process at a limited number of locations in space, the BME yields estimates of the process most likely to occur at unknown locations in space, subject to the *a priori* information about the spatial variability characteristics. In this regard, significantly different estimates can be derived for different prior information sources.

Depending on the form of the functions $g_q(\chi_k,\chi_1,\ldots,\chi_m)$, these estimates are, in general, nonlinear combinations of the data. Moreover, Eq. (19) does not involve the explicit form of the prior probability (24), which rather is a convenient vehicle for the development of the estimation scheme. This is a very useful feature of the estimation process, since the integral A of Eq. (22) may require tedious calculations.

It is natural to expect that as the number Q of the physical constraints (4) increases [that is, as the prior probability $f_x(\chi_k,\chi_1,\ldots,\chi_m)$ becomes more restrictive], the estimates $\hat{X}(\mathbf{s}_k)$ will be significantly improved. In the limiting case where there exists no physical constraint, the prior probability law obtained will be uniformly distributed over all possible $\hat{X}(\mathbf{s}_k)$ values, imposing no restrictions on the analysis. In this case the estimation problem has no unique solution. Therefore, for an estimation process to make sense, certain minimum prior information may be necessary, such as the spatial mean and covariance, the median, and any other quantile, or in general any spatial moment of any order.

In relation to this, an important part of the estimation problem is to identify that set of g_q constraints (4) which, together with the available measurements, assure the desired estimation accuracy (23). If the given set of constraints fully characterize the probabilistic structure of the underlying SRF, any further improvement of the accuracy (23) will depend mainly on two factors: (i) the experimentally calculated quantities $E[g_q]$, which are the operational parameters of the analysis, and (ii) the quality and number of the measurements available. On the other hand, if the given constraints do not provide a sufficient characterization of the underlying SRF, the estimation accuracy (23) will depend on the introduction of a more appropriate set of constraints (which will offer a better characterization of the SRF), as well as on the factors (i) and (ii) above.

The incorporation of the g_q-constraints, can improve significantly the characterization of the physical system under consideration. These constraints, which may be of a linear or a nonlinear form, in general, stress the model-dependent character of the BME concept. That is, its ability to incorporate as part of the analysis a body of information that should come from physical models. For illustration let us consider a few examples.

Example 5: When estimating piezometric heads it will be more realistic that some of the g_q constraints represent conditions imposed by the flow models and the head boundary conditions of an aquifer. For instance, in the case of a steady two-dimensional flow without sources or sinks, the stochastic groundwater flow model and the associated boundary conditions may be written as

$$\frac{\partial}{\partial s_1}\left[T(\mathbf{s})\frac{\partial h(\mathbf{s})}{\partial s_1}\right]+\frac{\partial}{\partial s_2}\left[T(\mathbf{s})\frac{\partial h(s)}{\partial s_2}\right]=0$$

and

$$f_1\frac{\partial h(\mathbf{s})}{\partial \eta}-f_2 h(\mathbf{s})=f_3$$

respectively, where $h(\mathbf{s})$ is the piezometric head, $T(\mathbf{s})$ is the transmissivity, $\partial h(\mathbf{s})/\partial\eta$ is a derivative normal to the boundary, and f_i, $i=1, 2$, and 3 are given functions on the boundary. These equations can lead to g_q constraints, in terms of the probability densities of $h(\mathbf{s})$ and $T(\mathbf{s})$, or their second-order moments.

Example 6: Constraints may be imposed by groundwater management models, where g_q should take into consideration inequalities of the form $\mathbf{Rf}\leq\mathbf{c}^*$, where $\mathbf{R}$ is a concentration response matrix which is developed on the basis of steady-state groundwater flow and solute transport equations, $\mathbf{f}$ is a vector whose elements correspond to solute disposal rates, and $\mathbf{c}^*$ is a vector of water quality standards (e.g., Gorelick, 1982).

Example 7: Other constraints arising in the context of water quality management may be of the form $P[c(\mathbf{s}, t) \leq c^*(\mathbf{s}, t)] \geq p$, where $c(\mathbf{s}, t)$ denotes spatiotemporal contaminant concentrations simulated on the basis of stochastic solute transport models, $c^*(\mathbf{s}, t)$ denotes the local water quality standards, and p are specified reliability levels (see "soft" data in Chapter 7; also, Wagner and Gorelick, 1987).

As we saw before, the Lagrange multipliers μ_q $(q = 0, 1, \ldots, Q)$ have certain interesting features linked to the spatial structure of the process under estimation. Particularly, Eq. (21), in addition to

$$-\frac{\partial \mu_0}{\partial \mu_q} = E[g_q]$$

entails that

$$-\frac{\partial^2 \mu_0}{\partial \mu_q^2} = E[g_q^2]$$

and

$$-\frac{\partial^2 \mu_0}{\partial \mu_q \, \partial \mu_{q'}} = E[g_q g_{q'}] \tag{26}$$

To fix ideas, consider the case of Example 1 above where for simplicity let $m = 2$. It holds true that

$$-\frac{\partial \mu_0}{\partial \mu_1} = m_x(\mathbf{s}_1), \qquad -\frac{\partial \mu_0}{\partial \mu_2} = m_x(\mathbf{s}_2), \qquad -\frac{\partial \mu_0}{\partial \mu_3} = m_x(\mathbf{s}_k)$$

$$-\frac{\partial \mu_0}{\partial \mu_4} = \sigma_x^2(\mathbf{s}_1) = -\frac{\partial^2 \mu_0}{\partial \mu_1^2}, \qquad -\frac{\partial \mu_0}{\partial \mu_5} = c_x(\mathbf{s}_1, \mathbf{s}_2)$$

$$-\frac{\partial \mu_0}{\partial \mu_6} = c_x(\mathbf{s}_1, \mathbf{s}_k), \qquad -\frac{\partial \mu_0}{\partial \mu_7} = \sigma_x^2(\mathbf{s}_2) = -\frac{\partial^2 \mu_0}{\partial \mu_2^2}$$

$$-\frac{\partial \mu_0}{\partial \mu_8} = c_x(\mathbf{s}_2, \mathbf{s}_k), \qquad -\frac{\partial \mu_0}{\partial \mu_9} = \sigma_x^2(\mathbf{s}_k) = -\frac{\partial^2 \mu_0}{\partial \mu_3^2}$$

Higher order moments may also be obtained; for example

$$-\frac{\partial^2 \mu_0}{\partial \mu_6 \, \partial \mu_8} = E\{[X(\mathbf{s}_1) - m_x(\mathbf{s}_1)][X(\mathbf{s}_2) - m_x(\mathbf{s}_2)][X(\mathbf{s}_k) - m_x(\mathbf{s}_k)]^2\}$$

These properties of the Lagrange multipliers μ_q may be useful in the context of multi-objective sampling strategies and economic analyses.

The BME reasoning does not require, in general, any Gaussian-type hypothesis. Moreover, the estimation scheme does not include any requirement of unbiasedness (as in ordinary kriging, for instance, where the sum of the weighting coefficients is required to be equal to unity). The lack of such a requirement allows the estimation scheme to consider additional information about the data.

Equation (17) is the expression approximated in the context of indicator kriging (Journel, 1986). There is, however, no maximization of entropy involved in that approximation. Some interesting connections between the Bayesian/maximum-entropy estimator and the kriging estimators will be discussed in the following section.

In the rather rare circumstance where the multivariate probability law $f_x(\chi_k, \chi_1, \ldots, \chi_m)$ is known *a priori*, it can be directly inserted into Eq. (25) to solve for the desired estimate $\hat{X}(\mathbf{s}_k) = \hat{\chi}_k$. Notice that the maximum loglikelihood equation, viz.,

$$\left.\frac{d \log f_x(\chi_1, \ldots, \chi_m | \chi_k)}{d\chi_k}\right|_{\chi_k = \hat{\chi}_k} = \frac{d \log f_x(\chi_k, \chi_1, \ldots, \chi_m)}{d\chi_k} - \left.\frac{d \log f_x(\chi_k)}{d\chi_k}\right|_{\chi_k = \hat{\chi}_k} = 0$$

yields the same estimates with Eq. (25) if the prior probability density $f_x(\chi_k)$ is uniform.

The Bayesian/maximum-entropy formalism can be extended to the case of S/TRF $X(\mathbf{s}, t)$, $(\mathbf{s}, t) \in R^n \times T$ (Chapter 5). For example, the information measure contained in a set of random variables in space-time, $\{x_{kp}, x_{11}, x_{12}, \ldots, x_{mr}\}$ with joint probability density $f_x(\chi_{kp}, \chi_{11}, \chi_{12}, \ldots, \chi_{mr})$ is given by

$$\operatorname{Inf}[x_{kp}, x_{11}, x_{12}, \ldots, x_{mr}] = -\log[f_x(\chi_{kp}, \chi_{11}, \chi_{12}, \ldots, \chi_{mr})] \tag{27}$$

The corresponding prior constraints involve expectations in space-time, spatiotemporal correlation functions, etc. Several of the space-time results just reflect analogous results of the SRF context discussed above, but there may be interesting differences as well (see, e.g., Chapter 5).

8.4 The Linear Case

To gain some insight about the BME estimator, Proposition 2 below considers a special but very important case (Christakos, 1990a).

Proposition 2: The solution to Problem 1 when the physical constraints are of the form (5) and (6)—that is, when the properties the estimator must account for are the spatial mean and covariance of the homogeneous random function $X(\mathbf{s})$—is given by

$$\hat{X}(\mathbf{s}_k) = m_x(\mathbf{s}_k) - \frac{\sigma_x^2(\mathbf{s}_k)}{a_{kk}} \sum_{i=1}^{m} a_{ki} \frac{\chi_i - m_x(\mathbf{s}_i)}{\sigma_x(\mathbf{s}_k)\sigma_x(\mathbf{s}_i)} \tag{28}$$

where a_{ki} are known functions of the covariances $c_x(\mathbf{s}_i, \mathbf{s}_j)$, $i, j = 1, 2, \ldots, m$ and k.

In relation to Proposition 2, the following points may be stressed:

(a) The input to the estimation approach is restricted to the spatial means and covariances.
(b) The probability law of the underlying SRF is found to be, in this case, multivariate Gaussian.
(c) The corresponding estimator is a linear combination of the data available.

On the basis of (a) and (b), if the underlying SRF $X(\mathbf{s})$ is Gaussian, constraints (5) and (6) are sufficient to fully characterize $X(\mathbf{s})$.

In view of these facts, it might be interesting to compare the estimator (28) with some kriging estimators: *Simple kriging* (see, e.g., Journel, 1989) in the presence of a multivariate Gaussian probability density amounts simply to consider for estimate

$$[\hat{X}(\mathbf{s}_k)]_{SK}$$

the corresponding conditional mean that happens to be linear in the data χ_i, $i = 1, 2, \ldots, m$. One the basis of Eq. (17), and since $f_x^*(\chi_k|\chi_1, \ldots, \chi_m)$ is Gaussian, its mean is also the value that maximizes the BME function $B_x(\chi_k)$ with respect to χ_k. Hence the following corollary holds true.

Corollary 1: Equation (28) coincides with the simple kriging estimator.

Example 8: Assume that the only information we have about the SRF $X(s)$ in R^1 is its mean $E[X(s)] = 0$ and its covariance function $c_x(s, s')$. From Eq. (28), the estimate of the SRF at point s, given data at points s' and s'', see Fig. 9.25, becomes

$$\hat{X}(s) = \frac{c_x(a)}{\sigma_x^2(s) + c_x(2a)} [X(s') + X(s'')] \tag{29}$$

The corresponding *ordinary kriging* estimator will be

$$\hat{X}_K(s) = \xi_1 X(s') + \xi_2 X(s'')$$

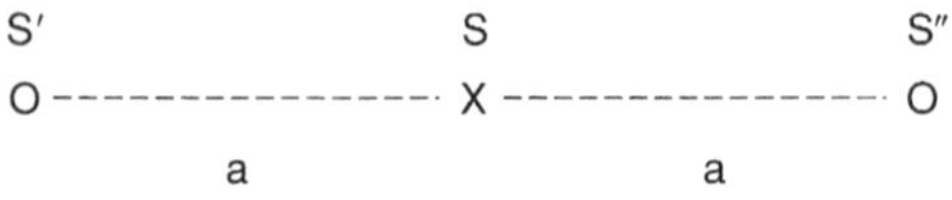

Figure 9.25 Estimation in R^1

where the weights ξ_1 and ξ_2 are determined by minimizing the estimation error variance $E[\hat{X}_K(s)-X(s)]^2$, subject to the unbiasedness condition $E[\hat{X}_K(s)-X(s)]=0$. The result is

$$\hat{X}_k(s)=\tfrac{1}{2}[X(s')+X(s'')] \tag{30}$$

It is obvious that while Eq. (29) accounts for spatial variability, Eq. (30) does not. The latter is, simply, a "naive" estimator, namely, the arithmetic mean of the two values available.

On the other hand, *disjunctive kriging* is a nonlinear estimator, which corresponds to reproduction of a particular type of Gaussian-related bivariate probability density (isofactorial densities). Constraints of the type (6) correspond to reproduction of the spatial covariance and, therefore, constraints (6) alone would fail to represent the disjunctive kriging approach.

Working along the lines of Proposition 2, the following result can be proven.

Proposition 3: The solution to Problem 1 when the physical constraints are of the form (7)—that is, when the property the estimator must account for is the spatial semivariogram of the homogeneous SRF $X(\mathbf{s})$—is given by

$$\hat{X}(\mathbf{s}_k)=\frac{\sum_{i=1}^{m} b_{ki}\chi_i}{\sum_{i=1}^{m} b_{ki}} \tag{31}$$

where b_{ki} are functions of the semivariograms $\gamma_x(\mathbf{s}_i, \mathbf{s}_j)$, $i, j=1, 2, \ldots, m$ and k that are determined from the constraints (7) and the normalized constraint (1).

Despite the fact that in the homogeneous case the covariance and the semivariogram functions are considered equivalent tools of statistical inference, the estimates (28) and (31) do not coincide. This is due to the additional sources of information—namely, the spatial means—incorporated in the derivation of the estimate (28).

8.5 Some Final Remarks

The Bayesian/maximum-entropy approach to the spatial estimation is very promising, for a variety of reasons. More specifically, the approach

(i) Takes into account not only the data, but also the prior information and knowledge that are highly relevant to the spatial variability of the natural process under estimation.

(ii) Leads to a posterior probability with minimum uncertainty attached to it.

(iii) Yields, in general, nonlinear estimators; it does not call for any Gaussian-type hypothesis or unbiasedness assumption.

(iv) Attributes great significance to procedures translating qualitative knowledge into appropriate quantitative constraints; in its present form, the Bayesian/maximum-entropy approach may not suffice to account for all sorts of qualitative prior information, but it does significantly restrict the range of arbitrariness.

(v) It can be applied in the case of spatiotemporal random fields as well.

(vi) Yields results similar to those derived by well-established estimation methods, when the same amount of information is used.

(vii) In addition, it has fruitful applications in sampling design and terminal decision analysis (see Chapter 10).

10

Sampling Design

"It is probably true quite generally that in the history of human thinking the most fruitful developments frequently take place at those points where two different lines of thought meet. These lines may have their roots in quite different parts of human culture, in different times or different cultural environments or different religious traditions; hence, if they actually meet, that is, if they are at least so much related to each other that a real interaction can take place, then one may hope that new and interesting developments may follow."

W. Heisenberg

1. Introduction

Many issues concerning water resources management, hazardous waste site exploration, industrial and municipal wastewater treatment systems are closely related to the quantitative evaluation of certain important characteristics of the underlying natural processes. For example, the classification of soil (contaminated versus noncontaminated) depends on the estimation of the concentration of contaminants in space. Groundwater quality monitoring in an aquifer can act as an early warning device in preventing groundwater contamination problems; the reliability of such monitoring is based on the quantitative assessment of the uncertainty in the model parameters. Subsurface pollution caused by industries and municipalities is a localized process and, hence, the design of any remedial measure requires information about the size, the direction, and the existing trends in the spatial variability of the contamination levels. In the context of site management and decision-making, some of the difficult problems are related to model uncertainties caused by the variability of model parameters and experimental procedures.

Issues such as the above need to be answered by the statistical analysis of site samples obtained on the basis of an exploration project. Particularly, exploration of natural processes is a procedure prepared and performed to get that amount of information that will allow an adequate physical understanding of the process. This is accomplished through partial sampling of extensive soil domains. In this chapter, the term *sampling design* refers to a mathematical procedure that provides the parameters associated with the arrangement of a number of observations over the particular site; such parameters are the sampling pattern (i.e., systematic, stratified, random, square, hexagonal, etc.), sampling density in space (i.e., number of samples per unit area), and sampling frequency in time (i.e., number of samples per unit time). Sampling leads to imperfect knowledge that is subject to sampling error. To reduce overall predictive uncertainty requires a more spatially disaggregate model than the one used by approaches based on classical statistics. Moreover, most natural processes exhibit a nonhomogeneous spatial variability with complex trends. This implies that a trade-off is being made between exploration uncertainty and spatial variability. The increase of the complexity and the cost of most sampling projects makes an optimum design of sampling necessary, in order to gain maximum information for a given cost.

Site exploration, on the other hand, may not be the actual or ultimate problem in need of a solution. Completing the exploration task produces information for solving other problems. The latter may be related, for example, to waste site management and planning; classification of water resources prospects and determination of whether the prospects warrant further study; and environmental policy decisions. In a realistic sampling design, the amount of information required, the kind of data to be collected, and the optimality criterion to be used are important factors that depend on the specific objectives of the problem at hand.

Certainly, there is not a universally best sampling design approach. In this chapter we do not intend to cover all possible aspects of sampling design in practice, such as management and planning objectives, social and economic factors, physical models, experimental conditions, and the multiple utilities of sampled data. Our main purpose will be to discuss several general methods that aim at the maximization of the efficiency of the sampling design in determining specific site exploration parameters, using concepts from the theory of random fields and stochastic estimation. It is then left to the scientists and engineers to judge the appropriateness of these methods for the specific conditions and objectives of the problem at hand.

The sampling design methods to be exploited in this chapter are prepared and performed

(a) to derive optimal values of the sampling parameters for the target site, such as sampling pattern, sampling density in space and frequency in time, and expected sampling accuracy;
(b) to construct predictive maps and identify boundaries between geologic formations, contaminated and uncontaminated soils, etc.; and
(c) to provide the model inputs necessary for site management, planning, and decision making.

Parts (a) through (c) above should take into consideration the physical structure of the natural processes involved and should be accomplished in a computationally efficient and cost-effective manner.

This chapter is mainly concerned with sampling in R^2, but certain results in R^n $(n>2)$ are also discussed. For practical applications, the sampling approaches are classified in a variety of ways. First, certain simple but useful practical solutions to the sampling problem, in terms of simple arithmetic mean estimators and dispersion variances, are discussed. Then, more sophisticated solutions are considered, based on the results of Chapter 9. The application of these results makes it possible to simplify the mathematics behind spatial sampling and, at the same time, to gain in generality in order to expand in applicability. On practical grounds, the establishment of systematic, step-by-step procedures is possible, where the required values of the accuracy parameters for the particular site are derived from readily available reference charts and entered into simple yet accurate formulas.

Within the more general framework of terminal decision analysis, the Bayesian/maximum-entropy approach of Chapter 9 allows us to incorporate into exploration strategies useful measures of information content, explanatory power, and equivocation of physical models.

2. About Sampling

2.1 General Considerations

In general, *sampling design* aims at an arrangement of observations of a spatially distributed natural process wherein certain objectives of site exploration are satisfied. In many cases the only objective is that the error in estimating the process is the smallest possible among a number of practically conceivable arrangements. In several other situations, however, additional factors, such as sampling cost and physical constraints, may need to be taken into account, as well. Finally, parameters related to site management, planning, and decision making may exert a major control on the effectiveness of the sampling design.

A sampling technique must account for the *spatial variability characteristics* of the natural process sampled. The analysis of spatial variability is the key to efficient sampling. Principles of sampling design can be understood by realizing how assumptions or knowledge about spatial structure have been taken into consideration in the design. After all a sampling case does not exist when a site is known to have no spatial variability: Increasing the sample size would not provide any new information. It is only when there exists some degree of spatial variability that an increase in the sample size tends to increase the accuracy of the estimation.

Classical sampling techniques are of limited help in analyzing this type of exploration survey. More specifically, the limitations of classical techniques for site characterization are as follows: classical techniques do not treat spatially correlated observations; they do not address geometric problems of search and pattern recognition; and they deal, essentially, with estimates of frequencies in large populations, not with decisions.

As we will see below, a much better approach is provided by the random field model and the stochastic estimation methods discussed in previous chapters. The random field approach can be used advantageously to specify important spatial variability characteristics, to design a sampling network, and to analyze and evaluate the results obtained. While in theory the sampling design methods to be discussed in this chapter are generally valid for spatially distributed natural processes that can be modeled as random fields, from a practical standpoint a number of important issues may arise. For example, practicing engineers and geologists know that it is not the same to measure depth to bedrock as it is to sample a particulate material; also, the experimental procedures to sample hydraulic conductivities at points in space are not the same with the ones used to measure transmissivities. Hence, although all the above natural processes can be modeled in terms of random fields, they drastically differ with respect to a variety of issues, such as experimental devices, instrumental biases, measurement errors, and scales of observation (e.g., Gy, 1982; Cushman, 1984). These important experimental issues will not be addressed in this chapter. Rather it will be assumed that observations can be obtained in space as a result of an experimental procedure that respects all the necessary rules of good sampling (controllable extraction biases, reproducible conditions, undisturbed samples, etc.). If this is not possible in practice, the mathematical sampling design methods cannot be used.

2.2 Optimal versus Suboptimal Solutions

In general, the derivation of *optimal sampling designs*, in a well-defined mathematical sense, is a difficult problem. In some cases, an optimal solution to this problem may not exist. Optimal sampling designs for fixed sample size N, when they exist, are in general difficult to determine, even in one dimension (e.g., Benhenni and Cambanis, 1990). Asymptotically optimal and approximate solutions are possible only under certain restrictive conditions (e.g., Ylvisaker, 1975).

From a statistician's point of view, a typical optimal sampling problem may be posed as follows. Let $X(\mathbf{s})$ be an SRF, and suppose that $\mathfrak{R} = \{S\colon S = \{\mathbf{s}_i;\ i = 1, \ldots, N\}\}$ (N is a fixed integer) is the space of all possible sets S of sampling points $\mathbf{s}_i$ within the domain $U(\mathbf{s})$ of $X(\mathbf{s})$. Choose a set

$$S^* = \{\mathbf{s}_i^*;\ i = 1, \ldots, N\} \in \mathfrak{R}$$

satisfying

$$\sigma^2(S^*) = \min_{\lambda_i} E\left[\int_{U(\mathbf{s}')} f(\mathbf{s})X(\mathbf{s})\, d\mathbf{s} - \sum_{i=1}^{N} \lambda_i X(\mathbf{s}_i^*)\right]^2$$

$$= \min_{S \in \mathfrak{R}} \min_{\lambda_i} E\left[\int_{U(\mathbf{s}')} f(\mathbf{s})X(\mathbf{s})\, d\mathbf{s} - \sum_{i=1}^{N} \lambda_i X(\mathbf{s}_i)\right]^2 \tag{1}$$

in which $f(\mathbf{s})$ is a known continuous function on $U(\mathbf{s}) \subset R^n$.

The above sampling problem can be put in a more general framework, by using the *loss function*

$$L[\hat{X}(\mathbf{s}), X(\mathbf{s})] = L\{F[X(\mathbf{s}_i), i = 1, \ldots, N];\ X(\mathbf{s})\}$$

(for notation see Section 3 of Chapter 9). In this case, the set S^* of sampling points must satisfy

$$\sigma^2(S^*) = \min_{F} E[L\{F[X(\mathbf{s}_i^*), \quad i = 1, \ldots, N];\ X(\mathbf{s})\}]$$

$$= \min_{S \in \mathfrak{R}} \min_{F} E[L\{F[X(\mathbf{s}_i), \quad i = 1, \ldots, N];\ X(\mathbf{s})\}] \tag{2}$$

Generally, depending on the choice of the loss function, to use criterion (2) one needs to know the underlying multivariate probability distributions. This usually makes the application of criterion (2) a more demanding proposition than that of (1).

In scientific applications, on the other hand, one faces a variety of sampling design problems. In a large group of sampling design problems, one is interested in the areal performance of the sampling network with respect to some efficiency indices (usually variances of estimates for linear functionals). This group is considered in detail in this chapter. In particular, the simplest *areal efficiency index* used is the variance

$$\sigma_A^2(S) = E[X_U - \hat{X}_U]^2 \tag{3}$$

in which

$$X_U = \frac{1}{U} \int_{U(\mathbf{s}')} X(\mathbf{s})\, d\mathbf{s} \tag{4}$$

is the mean and its estimator is given by

$$\hat{X}_U = \frac{1}{N} \sum_{i=1}^{N} X(\mathbf{s}_i) \tag{5}$$

where $S = \{\mathbf{s}_i;\ i = 1, \ldots, N\} \in \mathfrak{R}$. In this case, the objective of sampling design is to choose a set $S^* \in \mathfrak{R}$ so that the estimation variance (3) is minimized, viz.,

$$\sigma_A^2(S^*) = \min_{S \in \mathfrak{R}} \sigma_A^2(S) \tag{6}$$

Note that the expectation in Eq. (3) may be written as

$$E[\cdot] = E_x\{E_{sp}[\cdot]\} \tag{7}$$

where E_x means expectation with respect to the SRF $X(\mathbf{s})$ and E_{sp} denotes expectation with respect to the randomness of the sampling pattern, if any. Moreover, for $f(\mathbf{s}) = 1/U$ and $\lambda_i = 1/N$, expressions (1) and (6) coincide.

When the more sophisticated optimal estimation methods of Chapter 9 are used, the sampling efficiency may be evaluated through several areal efficiency indices, like the *average* estimation error over the domain U, viz.,

$$\sigma_A^2(S) = \frac{1}{K} \sum_{k=1}^{K} \sigma_x^2(\mathbf{s}_k) \tag{8}$$

where K is the number of the estimated points in U; $\mathbf{s}_k$ are the points (usually on a regular grid) where estimates are obtained by means of the methods of Chapter 9 and $\sigma_x^2(\mathbf{s}_k)$ are the point estimation errors. For example, one may write

$$\sigma_x^2(\mathbf{s}_k) = \min_{\lambda_{ki}} E\left[X(\mathbf{s}_k) - \sum_{i=1}^{m} \lambda_{ki} X(\mathbf{s}_i)\right]^2$$

where $m \leq N < K$. Another areal efficiency index is provided by the *maximum* estimation error, viz.,

$$\sigma_M^2(S) = \max_{\mathbf{s}_k} [\sigma_x^2(\mathbf{s}_k)] \tag{9}$$

Finally, the minimum estimation error is the trivial value zero at the sampling points. The objective of sampling design is to choose the set of sampling points that minimizes the areal efficiency index. When the areal efficiency index (8) is used one is looking for the set S^* that minimizes the average estimation error variance, viz.,

$$\sigma_A^2(S^*) = \min_{S \in \mathfrak{R}} \sigma_A^2(S) \tag{10}$$

In the case of (9) the set S^* is the one that minimizes the maximum estimation error; that is,

$$\sigma_M^2(S^*) = \min_{S \in \mathfrak{R}} \sigma_M^2(S) \tag{11}$$

Clearly, different areal efficiency indices may lead to different sampling designs. In earth science, criterion (10) is used in the majority of applications. In some special cases, however, different sampling criteria may be used, depending on the objectives of site exploration. For example, in Section 2.5 we consider criteria that account for the minimization of exploration cost and other constraints. In Section 5 we exploit some optimality criteria in the form of loss functions $L[\cdot]$; in many situations of practical importance, however, there are serious difficulties in assessing a realistic loss function. In Section 9 we study criteria based on the maximization of the sampling entropy. In general, one should first formulate the purposes of sampling explicitly, and then choose the appropriate criterion.

In most scientific applications, on the other hand, a mathematically optimal solution to the sampling problem is neither achievable nor necessary. Due to a variety of reasons, such as site uncertainties, economic factors, mathematical and computational difficulties, one is restricted to the *best suboptimal solutions*. There are, of course, many variations to the meaning of the term "best suboptimal solution." In practice it usually implies a sampling design that performs satisfactorily with respect to specific site exploration objectives and constraints.

2.3 Important Sampling Notions

The application of spatial estimation methods in sampling design calls for the notions of sampling pattern, sampling density, and estimation neighborhood.

Sampling pattern refers to the geometrical configuration of observations in space. The most important two-dimensional patterns fall into three major groups, as follows.

(i) *Systematic (regular) sampling patterns*; such as the hexagonal, the square, and the triangular patterns (Fig. 10.1). By convention, the names are assigned on the basis of partitioning the sampling area in a way that considers each sampling point as the centroid of a Voronoi polygon (see,

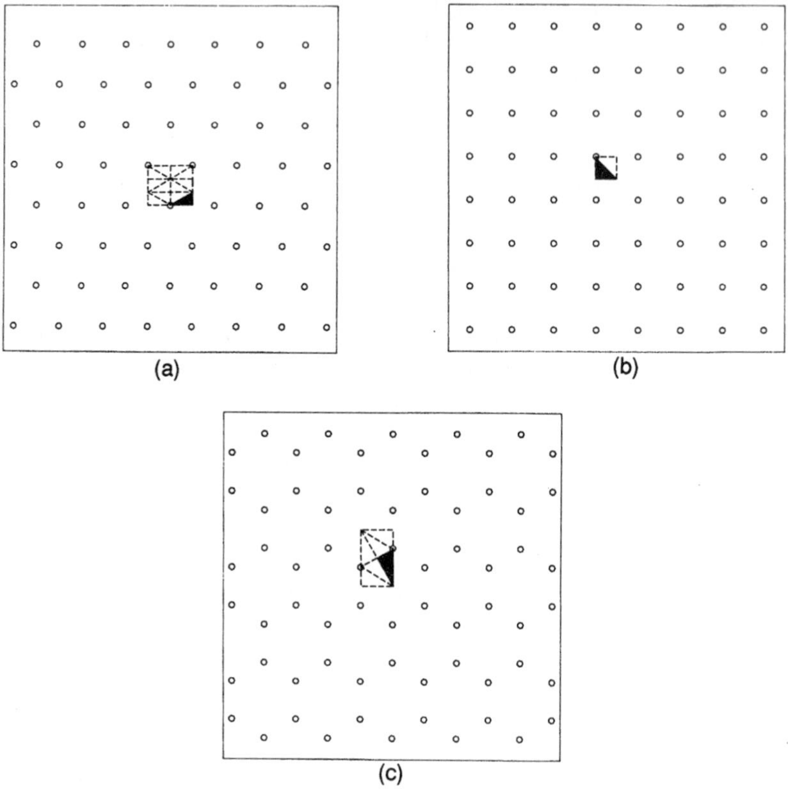

Figure 10.1 Systematic sampling patterns: (a) hexagonal, (b) square, and (c) triangular. The shaded areas are the basic repeating patterns

e.g., Fig. 10.2). As we will see below, a circular pattern (i.e., a pattern consisting of disjoint circles), although mathematically interesting, has no immediate practical consequence, since such a pattern is not easily implemented in practice.

(ii) *Stratified sampling patterns*; such as the hexagonal and the square patterns (Fig. 10.3). These patterns are obtained by dividing the sampling area into mutually exclusive partitions (e.g., polygons) and then selecting randomly a point from each partition.

(iii) *Random sampling patterns*; such as in Fig. 10.4. Other groups of sampling patterns are the clustered, the regular clusters, the bisymmetrical, and the orthogonal regular traverses (some examples are shown in Fig. 10.5; also, see Olea, 1984).

The *size of the sampling pattern* is an important factor in sampling. When this size is small it may produce a border effect. In theory, the sampling

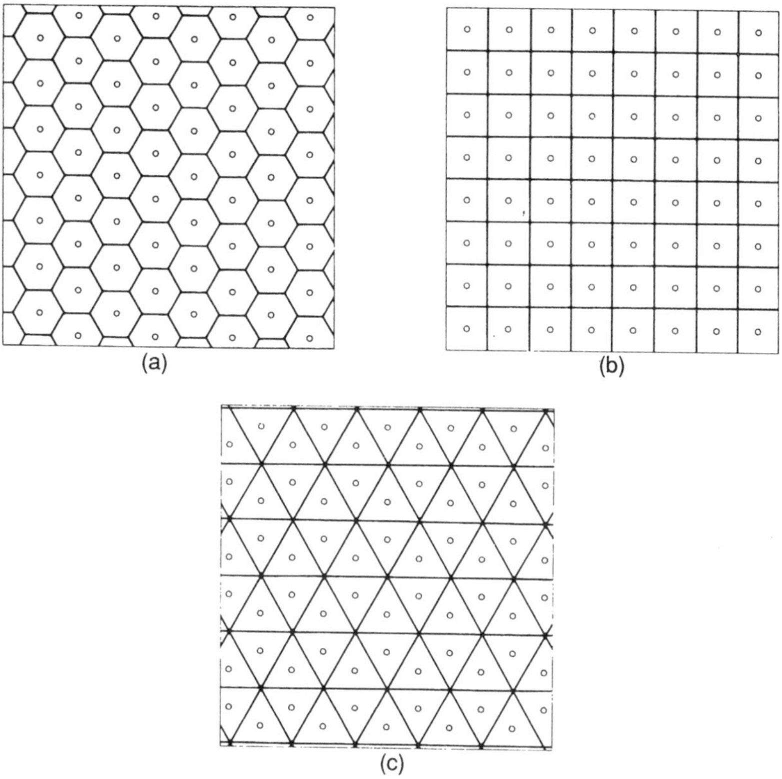

Figure 10.2 Voronoi polygons and systematic sampling patterns: (a) hexagonal, (b) square, and (c) triangular

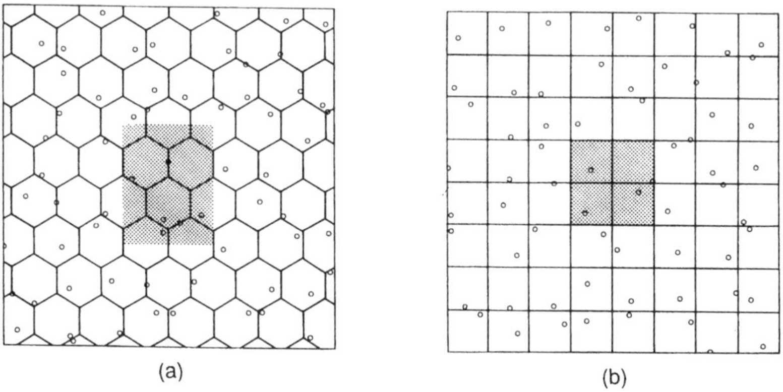

Figure 10.3 Sampling mechanisms for stratified sampling pattern: (a) hexagonal, (b) square. The shaded areas are the basic repeating patterns

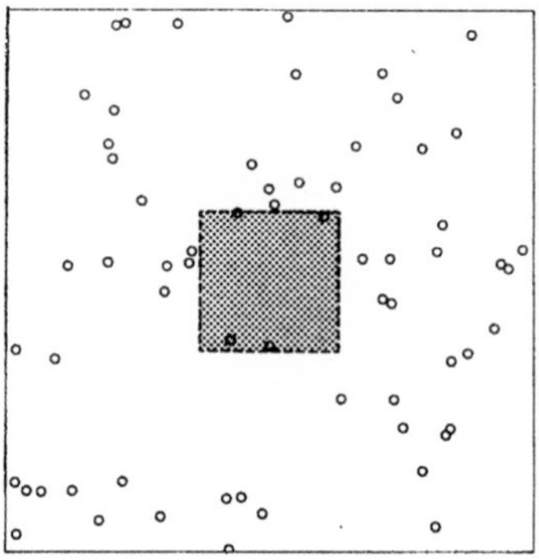

Figure 10.4 Random sampling pattern. The shaded area is the basic repeating pattern

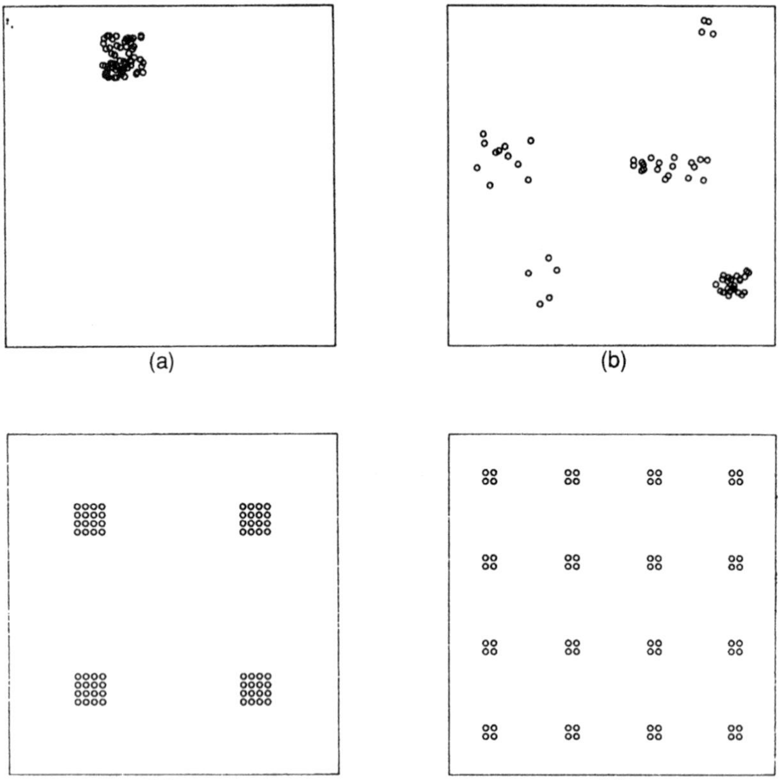

(a) (b) (c) (d)

Figure 10.5 Other examples of sampling patterns: (a) Clustered sampling pattern with one cluster; (b) clustered sampling pattern with five clusters; (c) regular clusters with sixteen points; and (d) regular clusters with four points per cluster

patterns are supposed to extend to infinity in order not to have to account for obvious bounding effects on the average estimation error. In practice, however, the average estimation error for the above, as well as for several of the following charts, has been calculated by considering samples of 64 points. More specifically,

(a) For systematic sampling patterns, the estimation variance shows a repeating pattern. Therefore it is required only to study the estimation variance within the *basic repeating pattern* to know the characteristics for an infinite pattern. The basic repeating pattern is always small enough, relative to the sample of 64 points, so that when up to 32 points are used to estimate there is no boundary effect in the estimations. In addition the solution is unique. In some cases, an integer multiple of the basic pattern may be necessary to focus attention within a rectangular or square area. In Figs. 10.1, 10.3, and 10.4 the shaded areas denote the basic repeating patterns; for example, in the case of a hexagonal sampling pattern the average estimation variance was computed over the rectangle comprising 12 basic repeating pattern units; in the case of a triangular sampling pattern the average estimation variance was computed over the rectangle comprising six basic repeating pattern units.

(b) On sampling patterns fully or partially random, there is no longer a small repeating pattern to represent the entire population. In this case judicious selection of the area to estimate is to be made in order to consider the largest area that will not have much of a boundary effect, taking advantage of the symmetry when possible. The solution is not unique. Each class, like a random pattern, involves an infinite number of possible point arrangements, each one with its own average estimation error.

Sampling density θ is the number of observations per unit area (e.g., the number of boreholes per square kilometer). In some cases sampling density and pattern completely define the pattern configuration (e.g., regular pattern). This is not the case, however, of a random or a stratified pattern where there is more than one configuration per pattern.

The *estimation neighborhood m* is a sampling factor that appears in the case of optimal linear estimation methods (Section 4 below), and denotes the minimum number of sampling points necessary for the estimation system to be solvable.

2.4 Search Algorithms

As we saw above, an important part of sampling design is the choice of a set of sampling points that minimizes an areal efficiency index or, more generally, that satisfies a given optimality criterion. This is a *search problem.*

In most cases, mathematically optimal solutions to the search problem are very hard or even impossible to determine, so one is restricted to the best suboptimal solution. Search algorithms play an important role in the context of local sampling methods; see Section 5 later in this chapter.

In R^2 the minimization of the areal efficiency indices $\sigma^2(S) = \sigma^2_{\mathrm{A}}(S)$ or $\sigma^2_{\mathrm{M}}(S)$, can be achieved by means of computational algorithms that yield good suboptimal solutions quickly. All search algorithms to be considered here assume that the number N of sampling points to be taken is fixed; moreover, the N sampling points must be selected from a domain U that is finite and well defined in space.

The most successful algorithms seem to be the *annealing* algorithms. These algorithms use random starting patterns to avoid a solution that is only locally minimum. An important ingredient of such algorithms is the probability π_k of going from a set of sampling points S_k to a set S_{k+1}. Examples of π_k include (Metropolis *et al.*, 1953; Geman and Geman, 1984; Sacks and Schiller, 1988): the probability

$$\pi_k = \min\{1, \exp[-\gamma_k(\sigma^2(S_{k+1}) - \sigma^2(S_k))]\} \tag{12}$$

where $\gamma_k = \log k/\gamma$ $(k \geq 2)$ and γ is a parameter to be specified by the analysis, and the probability

$$\pi_k = \begin{cases} 1 & \text{if} \quad \sigma^2(S_{k+1}) - \sigma^2(S_k) \leq 0 \\ 0.5 & \text{otherwise} \end{cases} \tag{13}$$

The search algorithm summarized in Table 10.1 is based on a method proposed by Sacks and Schiller (1988); it can be used for two-dimensional domains when the number of sampling points is fixed (say N). As is obvious from Table 10.1, the algorithm depends on certain parameters, namely η_0, δ', α, δ, and M, which need to be specified (experience plays an important role here). In practice, the algorithm of Table 10.1 seems to perform well for parameter values such as $\alpha = 0.01$, $\delta \in [0.3, 0.5]$, and $M \in [100, 250]$.

The search algorithm of Table 10.1 is still valid when the minimization of the areal efficiency indices is replaced by other optimality criteria; such as, for example, the maximum-entropy sampling criterion discussed in Section 9 below.

Less sophisticated, but nevertheless useful in practice, search algorithms are the greedy algorithm and the sequential exchange algorithm. These algorithms are used when one wants to pick N sampling points from among M candidate locations within the domain U, according to some optimality criterion. These locations are known *a priori* and their number M is much larger than N. Then, the number of possible combinations of N sampling points among M candidate locations is in the order of M^N.

Table 10.1 The Annealing Search Algorithm

Step 1: Start by picking a set S_0 of N sampling points at random from the domain U of the natural process, and let $\pi_0 = 0.7$.

Step 2: At stage $k+1$ pick a point $\mathbf{s}'$ at random from the set $U - S_k$, and find another point $\mathbf{s}'' \in S_k$ satisfying

$$\sigma^2(S_k \cup \mathbf{s}' - \mathbf{s}'') = \min_{\mathbf{s} \in S_k} \sigma^2(S_k \cup \mathbf{s}' - \mathbf{s})$$

If $\sigma^2(S_k \cup \mathbf{s}' - \mathbf{s}'') \leq \sigma^2(S_k)$, take $S_{k+1} = S_k \cup \mathbf{s}' - \mathbf{s}''$; if not, take $S_{k+1} = S_k \cup \mathbf{s}' - \mathbf{s}$ with probability π_k and $S_{k+1} = S_k$ with probability $1 - \pi_k$.

Step 3: In the case that $S_{k+1} = S_k$ pick a point $\mathbf{s}^*$ at random from $U - S_k - \mathbf{s}'$, and repeat step 2 using $\mathbf{s}^*$ instead of $\mathbf{s}'$.

Step 4: Repeat step 3, if necessary, η_0 times, $\eta_0 \leq K - N$ (K is the number of estimated points and N is the number of sampling points). If, after η_0 trials, no movement has occurred, take $S_{k+1} = S_k$ and $\pi_{k+1} = \min\{1, \pi_k/(1-\delta')\}$; change to stage $k+2$ and return to step 2.

Step 5: Let

$$\pi_{k+1} = \begin{cases} (1-\delta)\pi_k & \text{if } \sigma^2(S_{k+1}) \leq (1-\alpha) \min_{i \leq k} \sigma^2(S_i) \\ \pi_k & \text{otherwise} \end{cases}$$

Step 6: The sampling procedure is terminated when there have been M trials since a change in π_k.

The *greedy* algorithm (see Table 10.2) is very simple, and involves $N \times M$ possible combinations of N sampling points from the M candidate locations. It is by no means optimal, but it seems to perform well as a first guess in practice.

The *sequential exchange* algorithm (Table 10.3) does not get optimal solutions, in general. It involves $k \times N \times M$ combinations, where k is a small integer. According to Aspie and Barnes (1990), a reasonable range of values is $1 \leq k \leq 5$, while most often $k = 1$ or 2. Other interesting search algorithms include the genetic algorithm (e.g., Goldberg, 1989), the integer programming branch and bound algorithm (e.g., Garside, 1971), and the discretized partial gradient algorithm (e.g., Fedorov, 1972).

Table 10.2 The Greedy Search Algorithm

Step 1: Start by picking a single best sampling point (with respect to the given optimality criterion) among the M candidate locations.

Step 2: Choose as a second sampling point among the $M-1$ remaining locations the one that makes the best combination with the first point determined in step 1.

Step 3: Continue until all N sampling points have been selected.

Table 10.3 The Sequential Exchange Search Algorithm

Step 1: Start by picking an initial set S_0 of N sampling points among the M possible locations (e.g., the greedy algorithm can provide an initial choice S_0).

Step 2: Keep $N-1$ samples of the set S_0 fixed and find the best sampling location for the Nth sample; this will yield a new set S_1.

Step 3: Do the same for each one of the remaining $N-1$ points; let S_{N-1} be the last set of sampling locations, after all N points have been used.

Step 4: Repeat steps 1 through 3 starting with the set S_{N-1} instead of S_0. After k repeats the algorithm converges to a final set of sampling points, where no further improvement is possible.

2.5 Multiple Objectives in Sampling Design

As already mentioned, in many exploration projects the decision maker seeks to achieve more than one objective or goal in selecting the course of action. In addition, constraints imposed by site conditions, economic factors, and binding regulations might be incorporated into the analysis (see, e.g., Hwang and Masud, 1979; Zeleny, 1982; Christakos and Olea, 1988; 1992).

Within the framework of practical sampling strategies, the objectives above and at least the most important constraints may be effectively incorporated in terms of

(a) two major quantifiable objectives, minimization of estimation error (or, equivalently, maximization of estimation accuracy), and minimization of exploration cost;

(b) a set of constraints arising from the structural identification of a particular site as well as the budget of the exploration project;

(c) a preference (or loss, or cost) function which, given the objective levels attained, allows the decision maker to assess the relative value of the trade-off information between (a) and (b).

In mathematical terms, the issues (a), (b), and (c) may be summarized as follows:

$$\text{Minimize } P_F[f_1, f_2, \tau] \tag{14}$$

subject to

$$f_1, f_2 \in C \tag{15}$$

where P_F is the so-called preference, loss, or total cost function; f_1, f_2 are objective functions associated with the estimation accuracy and the cost of the sampling effort, respectively; τ is the trade-off between objectives; and

C is a set of constraints on f_1, f_2 imposed by sampling factors, subsurface conditions, or budget limitations.

2.6 Some Classifications of Sampling Design Methods

For application purposes, sampling approaches can be classified in a variety of ways.

The first, and perhaps most important, classification considers two major groups of sampling methods, namely, *global* (Section 3 and 4) and *local* methods (Section 5). The global approaches try to optimize the location of several observations at the same time by examining the efficiency of various sampling patterns by means of areal efficiency indices. They are fast and simple, and particularly useful when one needs to evaluate a group of possible sampling designs still on the drawing board or to perform extensive redesigns to maintain the efficiency of a sampling network. In the local approaches, the influence of additional sampling locations is analyzed separately. This is achieved either by means of variance reduction functions or in terms of the expected value of suitable loss functions. The local approaches may require the knowledge of multivariate probability distributions; they are used when one needs to expand an existing irregular sampling network or to partition a region into different zones.

The second kind of classification consists of methods based on simple averages, optimal linear estimators, and Bayesian/maximum-entropy analysis. The first two groups of methods (Section 3 and 4) apply minimum mean square error spatial estimation techniques and do not require the knowledge of multivariate probability distributions. On the contrary, to use the Bayesian/maximum-entropy sampling methods (Section 9) one needs to know the underlying multivariate probability distributions.

A third kind of classification distinguishes between methods for sampling homogeneous natural processes and methods for sampling nonhomogeneous processes.

Other groups of sampling design methods may be classified on the basis of the optimality criteria they apply. Each one of the above kinds of classification is independent of the others. For instance, a sampling approach based on optimal linear estimation may be used to sample homogeneous, as well as nonhomogeneous natural processes.

3. Simple Global Approaches to Sampling Design

3.1 Sampling Variances

The sampling problem to be examined here is the one described by Eqs. (3) through (6) in Section 2 above. The SRF are assumed to be homogeneous and, hence, $E[X_U - \hat{X}_U] = 0$. Let us first consider some interesting expressions of the estimation variances. By substituting Eqs. (4) and (5) of Section 2 into Eq. (3) of Section 2 it can be shown that

$$\sigma_A^2 = \bar{c}_x(U, U) + \bar{c}_x(N, N) - 2\bar{c}_x(U, N) \tag{1}$$

where, $\bar{c}_x(U, U)$ is the mean value of the covariance $c_x(\mathbf{h})$ when the two extremities of the vector $\mathbf{h} = \mathbf{s}_1 - \mathbf{s}_2$ independently describe the domain $U(\mathbf{s})$; that is,

$$\bar{c}_x(U, U) = \frac{1}{U^2} \int_{U(\mathbf{s})} \int_{U(\mathbf{s})} c_x(\mathbf{s}_1 - \mathbf{s}_2)\, d\mathbf{s}_1\, d\mathbf{s}_2 \tag{2}$$

$\bar{c}_x(N, N)$ is the mean value of the covariance $c_x(\mathbf{h})$ when the two extremities of the vector $\mathbf{h}$ independently describe the set of sample locations $\mathbf{s}_i$ $(i = 1, \ldots, N)$; that is,

$$\bar{c}_x(N, N) = \frac{1}{N^2} \sum_{i=1}^{N} \sum_{j=1}^{N} c_x(\mathbf{s}_i - \mathbf{s}_j) \tag{3}$$

$\bar{c}_x(U, N)$ is the mean value of the covariance $c_x(\mathbf{h})$ when one extremity of the vector $\mathbf{h}$ describes the set of sample locations $\{\mathbf{s}_i, i = 1, \ldots, N\}$ and the other extremity independently describes the domain $U(\mathbf{s})$; that is,

$$\bar{c}_x(U, N) = \frac{1}{NU} \sum_{i=1}^{m} \int_{U(\mathbf{s})} c_x(\mathbf{s}_i - \mathbf{s}_1)\, d\mathbf{s}_1 \tag{4}$$

A similar expression is valid in terms of semivariograms of SRF with homogeneous increments (Chapter 2), namely,

$$\sigma_A^2 = 2\bar{\gamma}_x(U, N) - \bar{\gamma}_x(U, U) - \bar{\gamma}_x(N, N) \tag{5}$$

with obvious notation.

Furthermore, if continuous observations are available within a domain υ, the variance in estimating X_U of Eq. (4) of Section 2 by

$$X_\upsilon = \frac{1}{\upsilon} \int_\upsilon X(\mathbf{u})\, d\mathbf{u} \tag{6}$$

is given by

$$\sigma_A^2 = \bar{c}_x(U, U) + \bar{c}_x(\upsilon, \upsilon) - 2\bar{c}_x(U, \upsilon) \tag{7}$$

where $\bar{c}_x(U, \upsilon)$, for example, is the mean value of the covariance $c_x(\mathbf{h})$ when one extremity of the vector $\mathbf{h}$ describes the domain $U(\mathbf{s})$ and the other extremity independently describes the domain $\upsilon(\mathbf{s}')$. Under the same circumstances, Eq. (5) yields

$$\sigma_A^2 = 2\bar{\gamma}_x(U, \upsilon) - \bar{\gamma}_x(U, U) - \bar{\gamma}_x(\upsilon, \upsilon) \tag{8}$$

with, again, obvious notation.

Remark 1: The sampling variance depends on the form of the sampling pattern. A well-known result states that for sampling patterns consisting of domains with different shapes but with equal areas the following ranking of sampling variances holds true:

$$\sigma_A^2(C) < \sigma_A^2(H) < \sigma_A^2(S) < \sigma_A^2(T) \tag{9}$$

where C, H, S, and T denote circle, regular hexagonal, square, and equilateral triangle sampling patterns, respectively.

Suppose now that the domain $U(\mathbf{s})$ is divided into N equal units υ_i, such that $U = \sum_{i=1}^{m} \upsilon_i = N\upsilon$. The *dispersion variance* $D_x^2(\upsilon/V)$ is defined as the mean value over $U(\mathbf{s})$ of the variance σ_x^2 in estimating

$$X_U(\mathbf{s}) = \frac{1}{U}\int_{U(\mathbf{s})} X(\mathbf{u})\, d\mathbf{u}$$

by

$$X_\upsilon(\mathbf{s}_i) = \frac{1}{\upsilon}\int_{\upsilon(\mathbf{s}_i)} X(\mathbf{u})\, d\mathbf{u}$$

of unit υ in $U(\mathbf{s})$; that is,

$$D_x^2(\upsilon/U) = \frac{1}{U}\int_{U(\mathbf{s})} \sigma_A^2(U(\mathbf{s}), \upsilon(\mathbf{s}_1))\, d\mathbf{s}_1 \tag{10}$$

Using Eqs. (7) and (8), Eq. (10) can be written as

$$D_x^2(\upsilon/U) = \bar{c}_x(\upsilon, \upsilon) - \bar{c}_x(U, U) = \bar{\gamma}_x(U, U) - \bar{\gamma}_x(\upsilon, \upsilon) \tag{11}$$

If υ consists of a *single point*, Eq. (11) gives

$$D_x^2(0/U) = c_x(\mathbf{O}) - \bar{c}_x(U, U) = \bar{\gamma}_x(U, U) \tag{12}$$

and the $D_x^2(0/U)$ may be viewed as the sample variance per sample point. As we shall see shortly, the dispersion variance $D_x^2(0/U)$ plays an important role in sampling.

Remark 2: It can easily be shown that

$$D_x^2(\upsilon/U) = D_x^2(0/U) - D_x^2(0/\upsilon) \tag{13}$$

The $D_x^2(\upsilon/U)$ can be calculated in terms of auxiliary functions available in Journel and Huijbregts (1978). More generally, the following well-known *Krige's formula* can be proven (Matheron, 1971).

$$D_x^2(\upsilon/V) = D_x^2(\upsilon/U) + D_x^2(U/V) \tag{14}$$

where $\upsilon \subset U \subset V$.

3.2 Some Important Sampling Patterns

Let us consider first the simplest case of *random* sampling in a domain U. The average sampling variance is given by

$$\sigma_A^2 = \frac{1}{N} D_x^2(0/U) \tag{15}$$

where N is the total number of samples taken, and $D_x^2(0/U)$ is the dispersion variance of Eq. (12) above.

In the case of a *stratified* sampling pattern, assume that the domain U is divided into k mutually exclusive partitions υ_i $(i = 1, \ldots, k)$ of the same size υ. Moreover, suppose that N/k samples are taken independently with uniform distribution within each υ_i. Then, the sampling variance is given by

$$\sigma_A^2 = \frac{1}{N} D_x^2(0/\upsilon) \tag{16}$$

where $D_x^2(0/\upsilon) = c_x(\mathbf{O}) - \bar{c}_x(\upsilon, \upsilon) = \bar{\gamma}_x(\upsilon, \upsilon)$. By taking into consideration Eq. (13) it is easily found that

$$\sigma_A^2(\text{SP}) < \sigma_A^2(\text{RP}) \tag{17}$$

where SP and RP denote stratified and random sampling patterns, respectively. For a given area υ, the best choice is the stratified sampling pattern with circular partitions υ_i, followed by the hexagonal and the square patterns (see also Remark 1 above).

Let us now discuss the case of *systematic* sampling, where the domain U is again divided into N equal units υ_i, with the same size υ. If one sampling point is taken within each unit, the sampling variance is given by

$$\sigma_A^2 = \frac{1}{N} \sigma_x^2(0/\upsilon_i) \tag{18}$$

where $\sigma_x^2(0/\upsilon_i)$ is the variance in estimating

$$X_\upsilon(\mathbf{s}_i) = \frac{1}{\upsilon} \int_{\upsilon(\mathbf{s}_i)} X(\mathbf{u}) \, d\mathbf{u}$$

by $X(\mathbf{s}_i)$; that is,

$$\sigma_x^2(0/\upsilon_i) = E\left[\frac{1}{\upsilon} \int_{\upsilon(\mathbf{s}_i)} X(\mathbf{u}) \, d\mathbf{u} - X(\mathbf{s}_i)\right]^2$$

It is assumed that the elementary error variances $\sigma_x^2(0/\upsilon_i)$ are independent. Equation (18) is a good approximation, even for small N. Other approximations are proposed in Matern (1960).

Table 10.4 Rates of Convergence of the Sampling Variance σ_A^2 in R^n for Certain Important Patterns

Random:	N^{-1}
Stratified:	$N^{-[1+(2/n)]}$
Systematic (midpoint):	$N^{-4/n}$

Finally, Table 10.4 (Tubilla, 1975) summarizes the asymptotic behavior of the sampling variance σ_A^2, as $N \to \infty$, assuming that $X(\mathbf{s})$ is a homogeneous SRF in R^n with a smooth covariance (i.e., a covariance whose partial derivatives of order k exist for all k).

Remark 3: Under certain conditions, several of the above results on purely spatial RF can be extended to *spatiotemporal* RF. Consider, for instance, the case of a zero-mean RF $X(\mathbf{s}, t)$ with space-time separable correlation structure of the form $c_x(\mathbf{h}, \tau) = \sigma_x^2 \rho_x(\mathbf{h}) \rho_x(\tau)$, where σ_x^2 is the point variance of $X(\mathbf{s}, t)$ and $\rho_x(\mathbf{h})$, $\rho_x(\tau)$ are the spatial and the temporal correlation functions, respectively. For fixed time t, the mean value of $X(\mathbf{s}, t)$ over the area U is given by

$$X_U(t) = \frac{1}{U} \int_{U(\mathbf{s})} X(\mathbf{s}, t)\, d\mathbf{s} \tag{19}$$

which has zero mean and covariance

$$c_{X_U}(t_1, t_2) = \sigma_x^2 \bar{\rho}_x(U, U) \rho_x(\tau) \tag{20}$$

where

$$\bar{\rho}_x(U, U) = \frac{1}{U^2} \int_{U(\mathbf{s}_1)} \int_{U(\mathbf{s}_2)} \rho_x(\mathbf{s}_1 - \mathbf{s}_2)\, d\mathbf{s}_1\, d\mathbf{s}_2$$

and

$$\tau = |t_1 - t_2|$$

4. Optimal Linear Estimation Approaches to Global Sampling Design

4.1 General Considerations

In Section 4 of Chapter 9 several optimal linear estimators were presented. All these estimators led to estimation variances that can be utilized as guidelines for sampling design. For example, one should give priority regarding future sampling to those parts of the site that have the highest estimation variances. Furthermore, these estimation variances are independent of the specific values of the natural process under consideration (data-independence property; Section 5.4 of Chapter 9). This property is of significant importance in sampling design, for it allows the assessment of the estimation error of an observation network before any observation is performed.

The estimation error values given in Chapter 9 are point efficiency indices throughout the sampling domain. As we saw above, however, in sampling one is rather interested in the areal performance of the sampling network, as measured through the areal efficiency indices. Just as for point estimation errors, the areal efficiency indices discussed in a previous section depend on

(i) The spatial variability characteristics of the natural process. For homogeneous natural processes these characteristics are the ordinary covariance $c_x(\mathbf{h})$ and the semivariogram $\gamma_x(\mathbf{h})$; for more complex, nonhomogeneous processes spatial variability is characterized in terms of the order of intrinsity ν and the generalized covariance $k_x(\mathbf{h})$. These are noncontrollable parameters for the designer; that is, they are inherent in the spatial distribution of the natural processes being sampled. The designer cannot alter such characteristics but instead must lead efforts to properly perform the modeling. Sampling designs may not be robust to changes in the spatial variability characteristics.

(ii) The estimation neighborhood m, which is a semicontrollable parameter. The latter means that depending on the spatial structure characteristics of the natural process, there is a minimum value that m must take for the estimation system to be solvable. Usually, it is not necessary to use all observations to evaluate a desired level of estimation accuracy. A subset might do as well. The break point in estimation is not a constant but a function of other factors.

(iii) The sampling pattern and the sampling density θ, which are controllable parameters; that is, they can be manipulated by the designer at will.

In global sampling design, the unknown locations $\mathbf{s}_i$ of the samples are assumed to belong to one of a series of possible sampling patterns (random, stratified, systematic square, hexagonal, etc.). Then, the set of sampling points $S=\{\mathbf{s}_i;\ i=1,\ldots,N\}$ is described by means of two parameters: the sampling pattern and the sampling density θ. In such a context, a number of possibilities exist, such as

(a) Both the sample size N and the sampling pattern are unknown.

(b) The sample size N is fixed *a priori* (e.g., due to limited budget requirements).

(c) Sampling is restricted to a few or even one specific type of pattern (e.g., due to physical constraints).

(d) The sample size N is not fixed and needs to be specified so that the resulting sampling network satisfies certain accuracy requirements; for example, the average estimation variance over the domain of interest U is below a certain value c, viz.,

$$\sigma_{\mathrm{A}}^2(S) < c \tag{1}$$

or the maximum estimation error is below a certain value c, viz.,

$$\sigma_{\mathrm{M}}^2(S) < c \tag{2}$$

In all these cases the required sampling parameters need to be determined as a result of the sampling design process. The sampling design approach, of course, will depend on the type of problem at hand. When, for example, the sample size N is fixed, one may use a search method (Section 2.4 above). Global sampling approaches based on linear spatial estimation techniques for homogeneous SRF models have been the most widely applied in various attempts to deal with practical sampling problems in hydrogeology, mining engineering, soil science, etc.

When the areal accuracy indices of Eqs. (8) or (9) of Section 2 are used, the point estimation errors $\sigma_x^2(\mathbf{s}_k)$ are obtained through one of the optimal estimation methods of Chapter 9. For instance, when the ordinary kriging estimator (Table 9.1 of Chapter 9) is used, the point sampling variance for homogeneous processes is given by

$$\sigma_x^2(\mathbf{s}_k) = c_x(\mathbf{O}) - \sum_{j=1}^{m} \zeta_j c_x(\mathbf{s}_k - \mathbf{s}_j) - \mu \tag{3}$$

This variance is independent of the specific values of the natural process; hence, assuming that the covariance $c_x(\mathbf{h})$ is known, $\sigma_x^2(\mathbf{s}_k)$ can be calculated before any observation is made. Then, depending on the particular problem (a) through (d) above, the best sampling design S^* is determined by applying the optimality criteria of Eqs. (10) or (11) of Section 2, Eqs. (1) and (2) above, etc.

Various interesting applications may be found in the literature (e.g., Newton, 1973; Jones *et al.*, 1979; McBratney *et al.*, 1981; McBratney and Webster, 1981; Webster and Burgess, 1984; Bogardi *et al.*, 1985; Bras and Rodriguez-Iturbe, 1985).

4.2 Sampling of Natural Processes Modeled as Intrinsic Spatial Random Fields of Order ν

A particularly attractive global sampling procedure is available for non-homogeneous natural processes, which are modeled in terms of ISRF-ν (Christakos and Olea, 1988; 1992). In this case, the application of the factorization property (Proposition 1, Section 5 of Chapter 9) can simplify significantly the mathematics behind sampling design. The effect of sampling pattern and sampling density on estimation is implicitly imposed by the generalized covariance matrices **K**.

Let $r_{ij} = |\mathbf{s}_i - \mathbf{s}_j|$ be the distance between locations $\mathbf{s}_i$ and $\mathbf{s}_j$ on a given two-dimensional pattern with sampling density θ points per unit area, and let $\tilde{r}_{ij}$ be the corresponding distance for the same type of pattern when the density is $\tilde{\theta}$. Then we can write

$$\frac{\theta}{\tilde{\theta}} = \left(\frac{\tilde{r}_{ij}}{r_{ij}}\right)^2 \tag{4}$$

(An example is given in Fig. 10.6.) The case $\tilde{\theta} = 1$ (unit density) will be the *reference pattern* of the given sampling pattern (with density θ). Equation (4) yields

$$\theta = \left(\frac{\tilde{r}_{ij}}{r_{ij}}\right)^2 \tag{5}$$

Now, the following proposition can be proven (Christakos, 1990b).

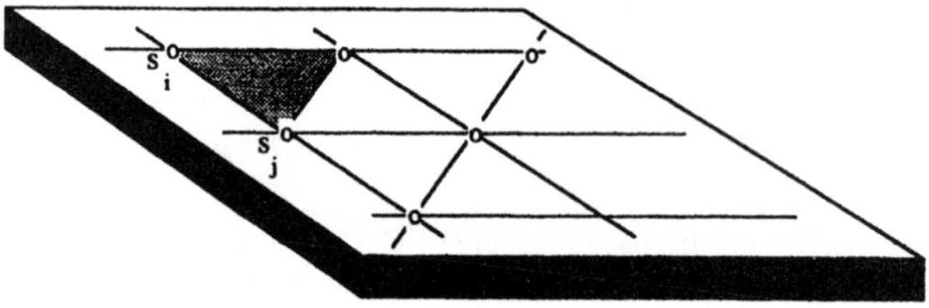

(a)

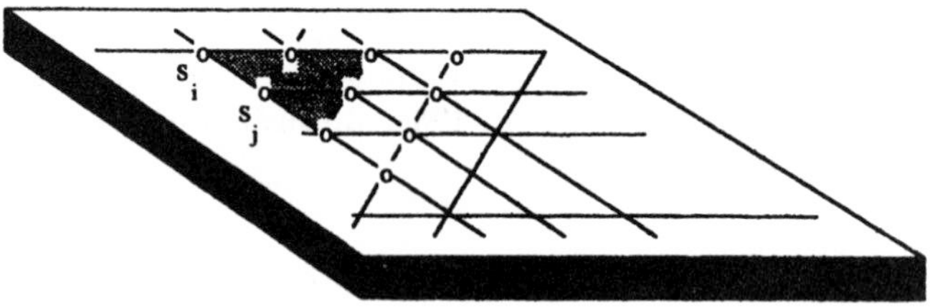

(b)

Figure 10.6 An illustration of Eq. (4). (a) $\tilde{r}_{ij} = 1$ (unit of distance) and $\tilde{\theta} = 0.5$ (points/unit area); (b) $r_{ij} = 0.5$ (units of distance) and $\theta = \tilde{\theta} \times (\tilde{r}_{ij}/r_{ij})^2 = 2$ (points/unit area)

Proposition 1: Assumptions of Proposition 1, Section 5 of Chapter 9. Let the pattern density be θ. For any fixed coefficient c_t $(t=0, 1, 3, \ldots, 2\nu+1)$, of the GSC-$\nu$ $k_x(\mathbf{h})$, the average sampling error variance is given by

$$\sigma_A^2 = \frac{c_t}{\theta^{t/2}} s_A^2 \tag{6}$$

where $s_A^2 = s_A^2(\tilde{c}_q, \tilde{c}_t, \tilde{\theta})$ is the *reference sampling error variance* obtained when the coefficients c_q of the polynomial generalized covariance model are replaced by $\tilde{c}_q = c_q \theta^{(t-q)/2}/c_t$ $(q \neq t)$, $\tilde{c}_t = 1$, and the pattern density is assumed to be $\tilde{\theta} = 1$.

Example 1: Consider Example 7 of Section 5, Chapter 9. In this case, Eq. (6) becomes

$$\sigma_A^2 = \frac{c_3}{\theta^{3/2}} s_A^2$$

where

$$\tilde{c}_0 = \frac{c_0 \theta^{3/2}}{c_3}, \qquad \tilde{c}_1 = c_1 \theta / c_3, \qquad \tilde{c}_3 = 1$$

Table 10.5 summarizes these results for several GSC-ν of practical interest. Table 10.5 is graphically displayed in Fig. 10.7.

A maximum of $m=32$ points were used in the estimation. The average estimation error σ_A^2 depends on the order of intrinsity ν and the GSC-ν $k_x(r)$ (noncontrollable parameters), the estimation neighborhood m (semi-controllable parameter), and the sampling pattern and the sampling density θ (controllable parameters). As shown in Fig. 10.8, other parameters being the same, the higher the degree of intrinsity ν, the higher the reference

Table 10.5 Estimation Variance Factorization

Case 1:
If $\nu = 0, 1$, or 2 and $k_x(r) = c_0\, \delta(r)$,
then $\sigma_A^2 = c_0 s_A^2$, where $s_A^2 = s_A^2(\tilde{c}_0 = \tilde{\theta} = 1)$.

Case 2:
If $\nu = 0, 1$, or 2 and $k_x(r) = c_t r^t$, $t = 1, 3$, or 5,
then $\sigma_A^2 = \dfrac{c_t}{\theta^{t/2}} s_A^2$, where $s_A^2 = s_A^2(\tilde{c}_t = \tilde{\theta} = 1)$.

Case 3:
If $\nu = 0, 1$, or 2 and $k_x(r) = c_0\, \delta(r) + \sum_{q=1,3}^{5} c_q r^q$,
then $\sigma_A^2 = \dfrac{c_t}{\theta^{t/2}} s_A^2$, where $s_A^2 = s_A^2\left(\tilde{c}_q = \dfrac{c_q \theta^{(t-q)/2}}{c_t}, \tilde{\theta} = 1\right)$.

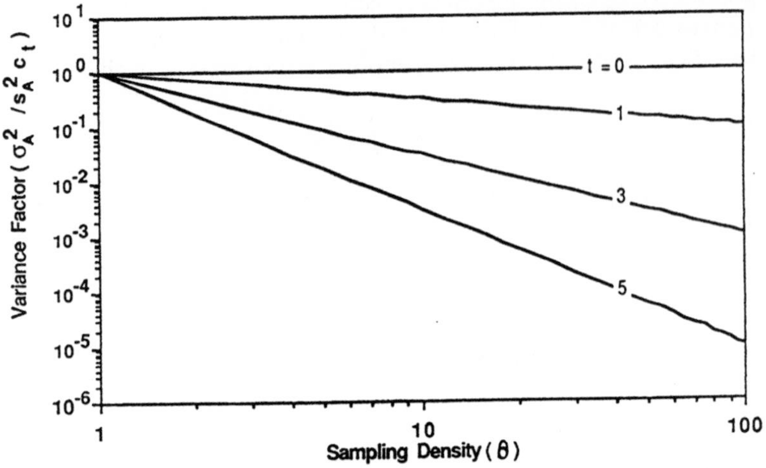

Figure 10.7 Graphical display of the results of Table 10.5

estimation variance. Figure 10.8 also shows that a specific number of observations is needed to achieve a desired level of estimation accuracy. Figure 10.9 presents the average reference estimation variances for a random scheme (R) and all possible two-dimensional systematic patterns: triangular (T), square (S), and hexagonal (H). On the other hand, the effect of density can be directly observed in Table 10.5. Notice that the estimation error is

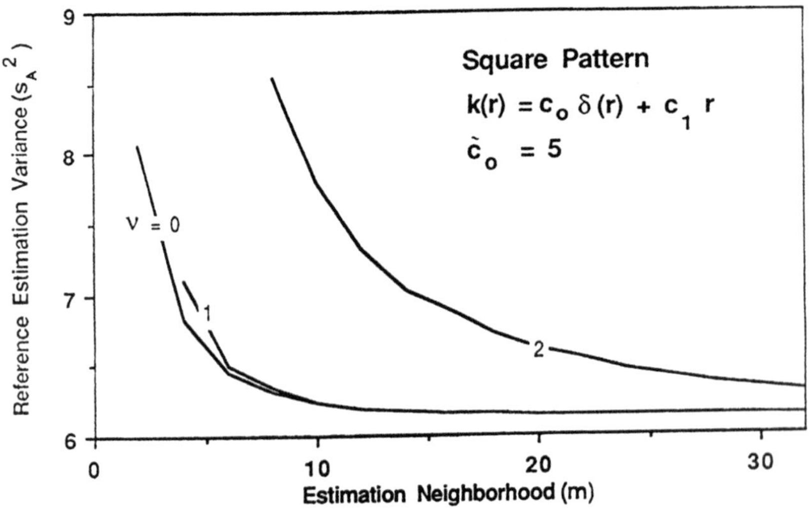

Figure 10.8 Change of average estimation variance with the order ν and the estimation neighborhood m

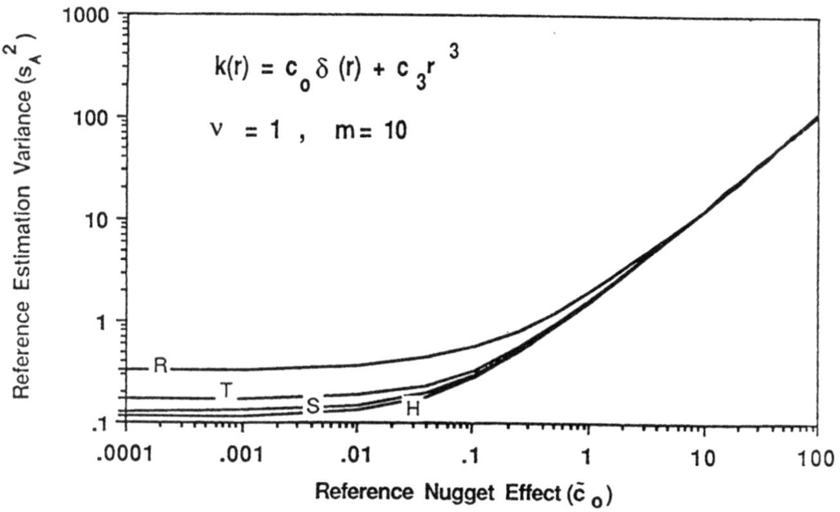

Figure 10.9 Change in average reference estimation variance with the sampling pattern and the reference nugget effect c_0

independent of density only in the case that the spatial variation is purely random (the white-noise SRF).

4.3 Reference Charts and a Sampling Design Procedure

Reference charts, like those of Figs. 10.8 and 10.9, apply in any case, independent of the particular site conditions. Additional reference charts are presented in Figs. 10.10 through 10.15. Figure 10.10 describes the variation of the estimation variance in a particular case: $k_x(r) = c_0\, \delta(r) + c_1 r$, the $\tilde{c}_0$ varies between 0.0001 and 100, $\nu = 0$, 1 or 2; square sampling pattern; sampling density $m = 10$.

In Fig. 10.11 the order of intrinsity is zero while the sampling density takes the values $m = 3$, 10, and 32. In Fig. 10.12 we plot the reference estimation variance as a function of the estimation neighborhood m and the sampling pattern, assuming a linear GSC-0. The hexagonal sampling pattern is the most accurate for the spatial structure data of Fig. 10.13. Figures 10.14 and 10.15 concern systematic square and random sampling patterns, respectively. In both charts we assume a GSC-1 of the cubic form $k_x(r) = c_0\, \delta(r) + c_1 r + c_3 r^3$ and $m = 10$.

Provided that the appropriate reference charts are at hand, sampling design can be performed according to the sampling design procedure of Table 10.6.

Example 2: Assume that the spatial structure of a natural process is described

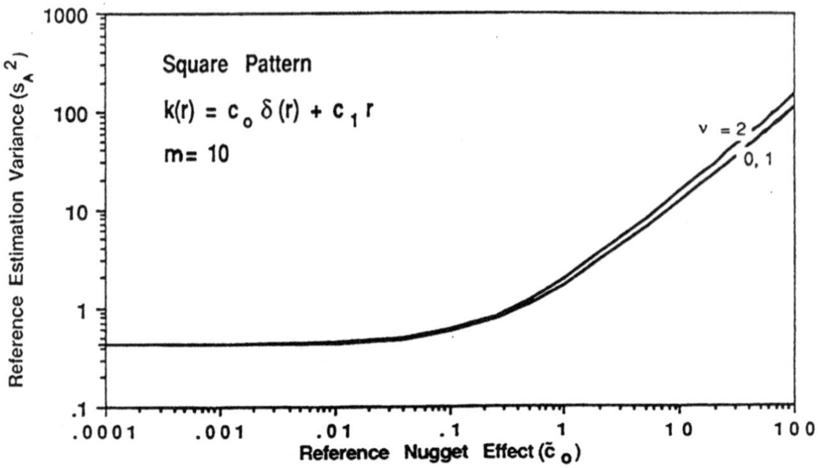

Figure 10.10 Change in average reference estimation variance with the order ν and the reference nugget effect $\tilde{c}_0$

by $\nu=0$ and $k_x(r)=-5.5r$. Suppose that in making decisions an accuracy at the level of $\sigma_A^2=6.35$ is required, while a random sampling pattern is by far the most comprehensive, given the morphology of the ground. From these data we need to find the optimal sampling density θ. By applying the procedure of Table 10.6, we first go to Fig. 10.12 to find that $s_A^2=0.57$. Then from case 2 of Table 10.5 we obtain

$$\theta = c_1^2\left(\frac{s_A^2}{\sigma_A^2}\right)^2 = 0.24 \quad \text{samples/unit area}$$

Let us now suppose that $\nu=1$, $k_x(r)=3.79\,\delta(r)+17.61r^3$, $m=10$ points. Suppose that we can afford a sampling density of $\theta=0.60$ points/unit area. We want to find the sampling pattern that provides the highest degree of accuracy. The answer to such a problem is immediately obtained from Fig. 10.9, which shows that the best pattern is the systematic hexagonal one.

From the same chart and for $\tilde{c}_0=(c_0/c_3)\theta^{3/2}=0.1$ we see that the corresponding reference estimation variance is $s_A^2=0.28$; and the estimation error of the particular sampling network will be $\sigma_A^2=10.60$.

Example 3: Let us revisit Example 4 of Section 4, Chapter 9. The 226 wells are arbitrarily distributed over an area of 800 square miles, with sampling density $\theta=0.28$ wells/square mile. Due to the uneven distribution of the observation wells (about 80% of the observation wells are located in the southern half of the Equus Beds area), which yields relative small number of data at some parts of the site, one should expect rather large estimation errors there. These locally large error variances have the most significant

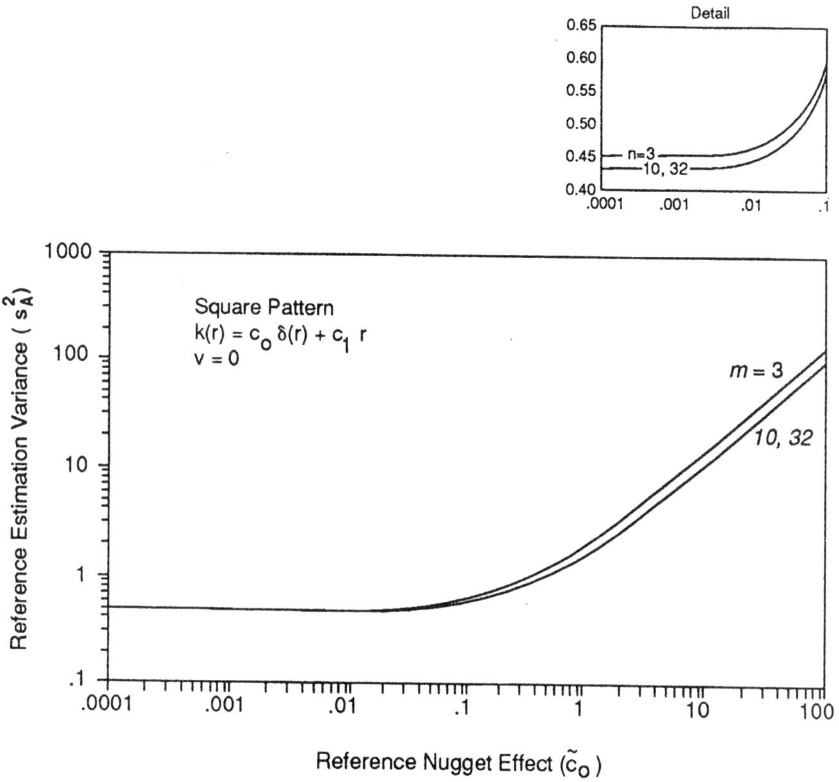

Figure 10.11 Change in average reference estimation variance with the estimation neighborhood m and the reference nugget effect $\tilde{c}_0$

contribution to the rather large value of the average estimation variance over the whole area, $\sigma_A^2 = 99.60\ \text{ft}^2$. Should the number of data points increase in these parts, the estimation accuracy would be significantly improved.

The evaluation of the effectiveness of a sampling network is well correlated with the spatial variabilities of the natural processes involved. If the parameters characterizing the spatial variability are known one may optimize the choice of the sampling pattern (systematic, random, stratified; square, triangular, hexagonal; etc.) that offers the desired accuracy with the lowest possible sampling density.

In the case of the Equus Beds the sampling design method developed above shows that the present observation well network is not optimal and that a regular or even a stratified observation well network should offer higher accuracies in the estimation of the form of the water table elevation. In particular, given the spatial variability characteristics of the water table

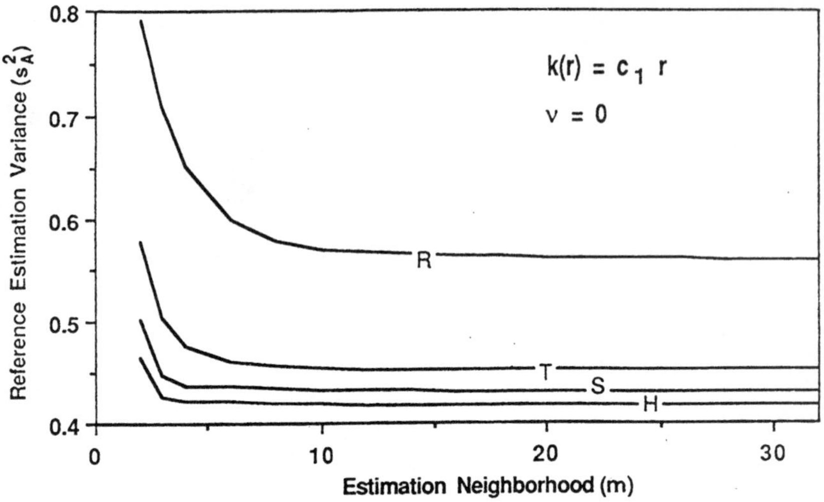

Figure 10.12 Change in average reference estimation variance with the sampling pattern and the estimation neighborhood m (H, S, T are systematic hexagonal, square, and triangular sampling patterns; R is a random pattern)

elevation process, $\nu = 1$ and $k(r) = -24.22r$, and the sampling neighborhood $m = 16$, we find that the reference estimation error is $s_A^2 = 0.40$ (systematic hexagonal sampling pattern), $=0.41$ (systematic square pattern) and $=0.44$ (systematic triangular pattern). Therefore the systematic hexagonal sampling pattern should be the most efficient pattern, and since the sampling density is $\theta = 0.28$ wells/square mile, the corresponding average estimation variance will be

$$\sigma_A^2 = \frac{24.22 \times 0.40}{\sqrt{0.28}} = 18.30 \text{ ft}^2$$

This average error variance is significantly smaller than the one obtained above on the basis of the existing well network (i.e., 99.60 ft^2); i.e., the average error variance of the existing network has decreased by 81.63%. It can be also shown that the level of accuracy of the existing network could have been achieved by using only 48 observation wells on a systematic hexagonal pattern over the Equus beds area.

Furthermore, if we assume that the estimation error is approximately normally distributed, then we can establish the 95% confidence interval

$$[\hat{X}(\mathbf{s}_k) - 2\sqrt{\sigma_x^2(\mathbf{s}_k)}, \hat{X}(\mathbf{s}_k) + 2\sqrt{\sigma_x^2(\mathbf{s}_k)}]$$

Here the estimate $\hat{X}(\mathbf{s}_k)$ at each point $\mathbf{s}_k$ and the estimation error variance $\sigma_x^2(\mathbf{s}_k)$ are obtained from Figs. 9.4 and 9.5 of Chapter 9.

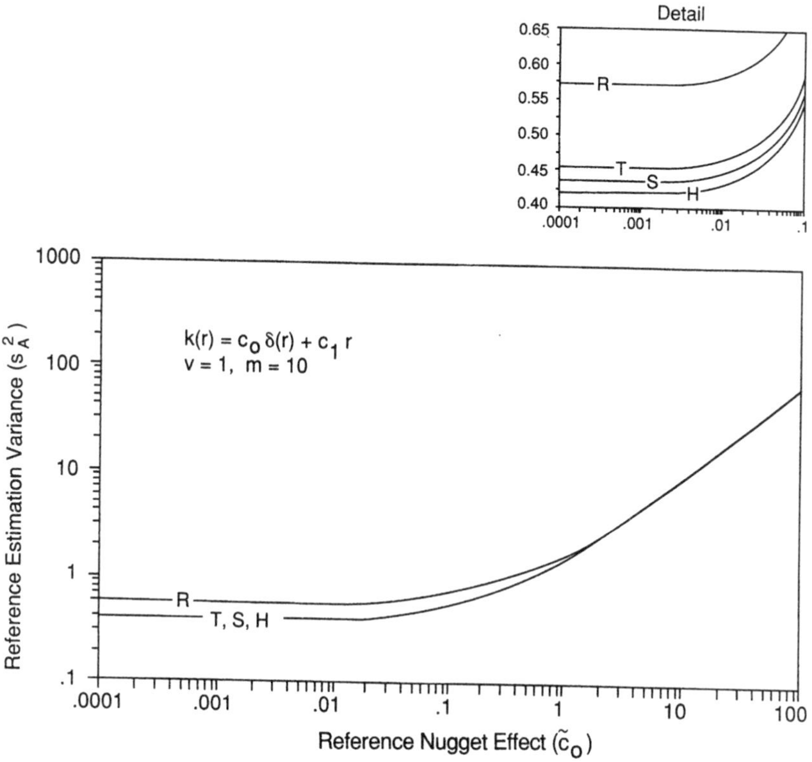

Figure 10.13 Change in average reference estimation variance with the sampling pattern and the reference nugget effect $\tilde{c}_0$

4.4 Multiobjectives in the Sampling Design of Intrinsic Spatial Random Fields of Order ν

Let us now discuss the case of more than one objective. For illustrative purposes, a simple but quite realistic multiobjective optimization problem is considered next:

$$\text{Minimize}[P_F = \theta + \tau\sigma_A^2] \tag{7}$$

where τ expresses the trade-off of increases in θ and decreases in σ_A^2. The optimal sampling procedure above provides all the elements to solve the problem for a variety of alternatives. Assume, for example, that the GSC-ν is of the linear polynomial form ($k_x(r) = c_t r^t$). Then we have a simple expression in θ, which, when differentiated and equated to zero, provides the solution for the optimal sampling density θ^*, viz.,

$$\theta^* = \left(\frac{t}{2} c_t \tau s_A^2\right)^{2/(t+2)} \tag{8}$$

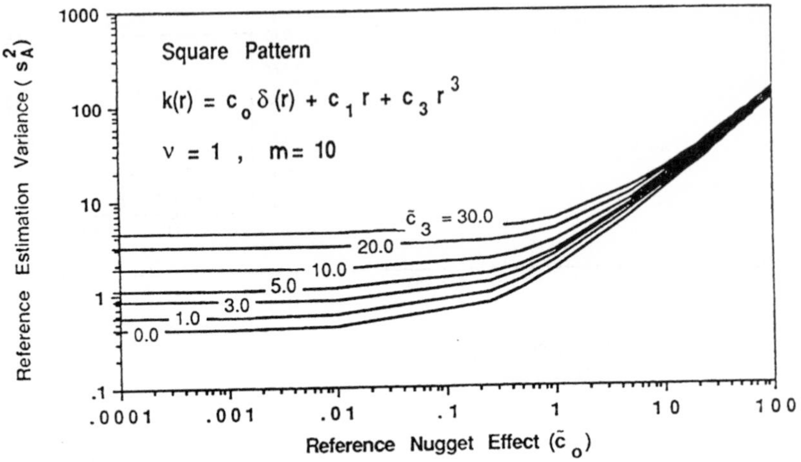

Figure 10.14 Change in average reference estimation variance for a systematic square sampling pattern with the reference coefficient $\tilde{c}_3$ and the reference nugget effect $\tilde{c}_0$

The corresponding average estimation variance σ_A^{*2} and minimum total cost P_F^* will be

$$\sigma_A^{*2} = (c_t s_A^2)^{2/(t+2)} \left(\frac{t\tau}{2}\right)^{t/(t+2)} \tag{9}$$

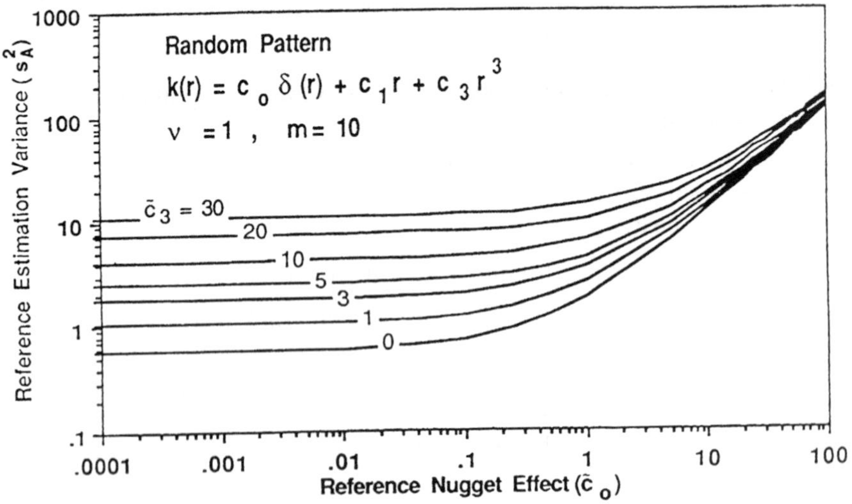

Figure 10.15 Change in average reference estimation variance for a random sampling pattern with the reference coefficient $\tilde{c}_3$ and the reference nugget effect $\tilde{c}_0$

Table 10.6 A Sampling Design Procedure

Step 1: Apply stochastic inference of random field representations to assess the spatial variability characteristics ν and $k_x(r)$ of the natural processes of interest (see also Chapter 7). At this point, data and experience from previous sites worked out will play a very important role.

Step 2: Depending on the results of step 1, choose the proper reference charts and find the reference values for the sampling parameters (like s_A^2).

Step 3: Apply formulas of Table 10.5 to obtain the required site values of the sampling parameters for the target site.

and

$$P_F^* = (c_t s_A^2 \tau)^{2/(t+2)} \left[\left(\frac{t}{2}\right)^{2/(t+2)} + \left(\frac{t}{2}\right)^{-2/(t+2)} \right] \tag{10}$$

respectively. Equations (8) through (10) are graphically displayed in Figs. 10.16 through 10.18.

Example 4: Assumptions of Example 2 above ($\nu = 0$ and $k_x(r) = -5.5r$). In addition suppose that the designer is willing to trade $\tau = 3$ units of increased θ for one unit of decreased error σ_A^2. In this case we first repeat the procedure applied in Example 2 to find $s_A^2 = 0.57$. Then by applying Eq. (8) or by using the chart of Fig. 10.16 we get $\theta^* = 2.8$ samples/unit area.

Remark 1: A flowchart summarizing the various stages of the sampling design procedure above is shown in Fig. 10.19. Of course, the implementation of the sampling approach in practice requires some initial assessment

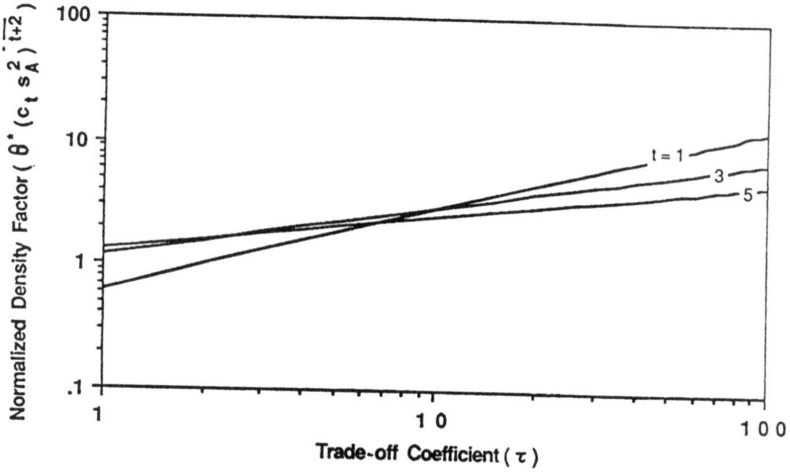

Figure 10.16 Change in normalized optimal sampling density θ^* with the degree of the GSC and the trade-off coefficient τ

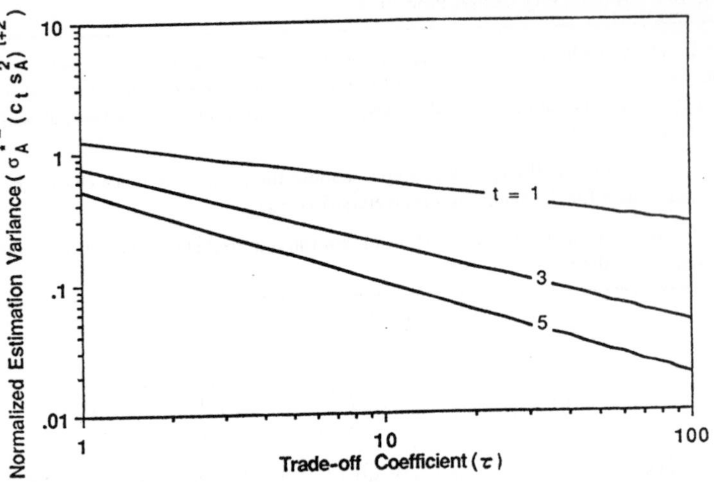

Figure 10.17 Change in normalized optimal estimation error σ^* with the degree of the GSC and the trade-off coefficient τ

of the spatial variability characteristics. This important issue concerns most sampling methods and is discussed in Section 6 below. As already mentioned, neither the actual implementation of *in-situ* sampling nor the solution of large systems of equations are expected to be necessary at the early stages of exploration. Hence, the sampling design procedure of Fig. 10.19 will be particularly useful when the analyses of sampling networks are still on the

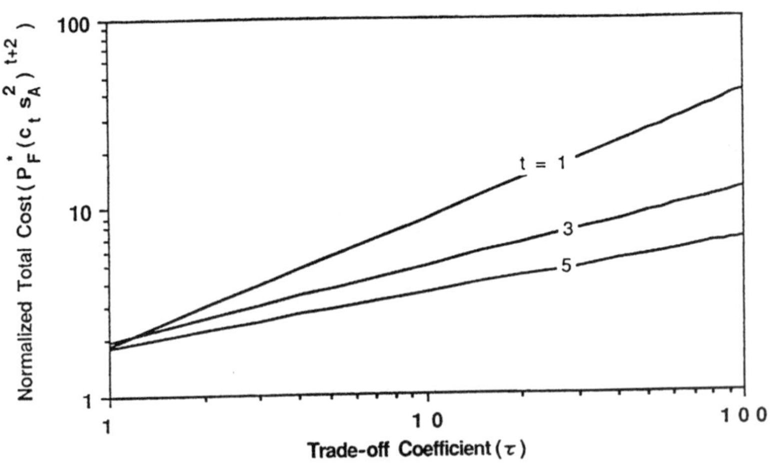

Figure 10.18 Change in normalized minimum total cost P_F^* with the degree of the GSC and the trade-off coefficient τ

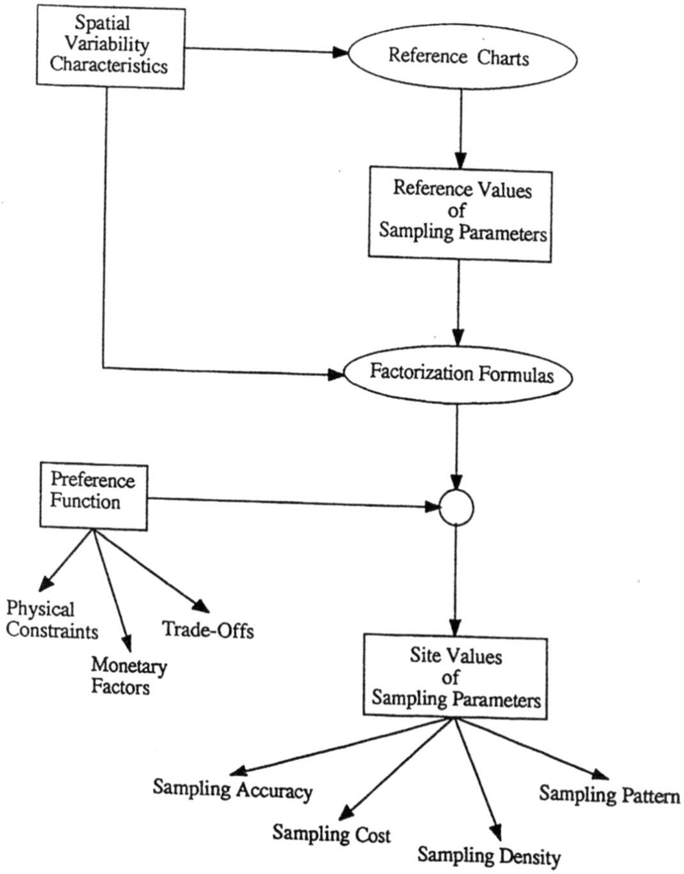

Figure 10.19 A sampling design flowchart

drawing board. However, in later stages of the sampling project the application of the above procedure will be a feedback process. That is, the current stage of sampling design will be improved as soon as new data become available during the sampling project.

5. Local Sampling Design Approaches

5.1 Sampling Design for a Set of Additional Observations

Local sampling design is best suited to expand an already existing network of M observations within the domain of interest U, by adding a set of new observations $S = \{\mathbf{s}_i; i = 1, \ldots, N\}$, where the sample size N is fixed *a priori.*

Given the appropriate optimality criterion, most local sampling approaches are essentially straightforward applications of the search algorithms discussed in Section 2.4.

Example 1: Assume that the optimality criterion chosen is the minimization of the areal estimation variance over U [see Eqs. (8) and (10) of Section 2]. Then, the best locations of the N new observations could be determined by applying the search algorithm of Table 10.1 in terms of the minimum areal estimation variance criterion. Just as for the global sampling approaches of Section 4, the point estimation variances are calculated here by means of the optimal spatial estimation methods of Chapter 9. In this case both the M available observations and the N additional sampling locations are used.

An interesting formulation of the local approach arises when the objective of site exploration is the partitioning of the region U into different zones. For example, in hazardous waste applications one may seek the partitioning of a site into zones of high contamination and zones of low contamination. In this case a suitable optimality criterion must be chosen. In some cases, the minimization of an areal efficiency index like that above may suffice. In many other situations, however, the incorporation of a loss function may be necessary. For illustrative purposes let us discuss the following example.

Example 2: Suppose that the SRF $X(\mathbf{s}_j)$ represents a contamination process in a region U, and let $\hat{X}(\mathbf{s}_j)$, $\mathbf{s}_j \in U$ be its estimator (obtained through one of the optimal spatial estimation methods of Chapter 9). Suppose that M samples are available over U. The problem is where to locate N additional samples in order to obtain a better identification and characterization of the contamination zones. We will follow the methodology proposed by Aspie and Barnes (1990). Let α denote the predetermined threshold limit value of contamination, and let $\beta = \beta(\mathbf{s}_j)$ denote the classification cutoff at point $\mathbf{s}_j$.

Based on these values the following indicator RF (see, also, Chapter 7) are defined

$$\begin{aligned} I_x^0(\mathbf{s}_j, \alpha, \beta) &= 1 \quad \text{if} \quad X(\mathbf{s}_j) < \alpha \quad \text{and} \quad \hat{X}(\mathbf{s}_j) > \beta, \\ &= 0 \quad \text{otherwise} \end{aligned} \tag{1}$$

and

$$\begin{aligned} I_x^u(\mathbf{s}_j, \alpha, \beta) &= 1 \quad \text{if} \quad X(\mathbf{s}_j) > \alpha \quad \text{and} \quad \hat{X}(\mathbf{s}_j) < \beta, \\ &= 0 \quad \text{otherwise} \end{aligned} \tag{2}$$

A reasonable, though rather simplistic, loss function is

$$L[X] = \sum_{\mathbf{s}_j \in U} \{c_0 I_x^0(\mathbf{s}_j, \alpha, \beta) + c_u I_x^u(\mathbf{s}_j, \alpha, \beta)\}\, \Delta(\mathbf{s}_j) \tag{3}$$

where $\Delta(\mathbf{s}_j)$ is an elementary area in U centered at $\mathbf{s}_j$, c_0 is the over-classification cost per unit area, and c_u is the underclassification cost per unit area. The expected value of the loss function (3) over the domain of interest U is given by

$$E[L] = \sum_{\mathbf{s}_j \in U} \{c_0 E[I_x^0(\mathbf{s}_j, \alpha, \beta)] + c_u E[I_x^u(\mathbf{s}_j, \alpha, \beta)]\} \, \Delta(\mathbf{s}_j) \tag{4}$$

where

$$E[I_x^0(\mathbf{s}_j, \alpha, \beta)] = P[X(\mathbf{s}_j) < \alpha \quad \text{and} \quad \hat{X}(\mathbf{s}_j) > \beta(\mathbf{s}_j)] \tag{5}$$

and

$$E[I_x^u(\mathbf{s}_j, \alpha, \beta)] = P[X(\mathbf{s}_j) > \alpha \quad \text{and} \quad \hat{X}(\mathbf{s}_j) < \beta(\mathbf{s}_j)] \tag{6}$$

The optimal classification cutoff at each point $\mathbf{s}_j$, $\beta(\mathbf{s}_j)$, is determined by minimizing $E[L]$; that is, we set $\partial E[L]/\partial\beta = 0$, which yields

$$P[X(\mathbf{s}_j) > \alpha \,|\, \hat{X}(\mathbf{s}_j) = \beta(\mathbf{s}_j)] = \frac{c_o}{c_o + c_u} \tag{7}$$

The practical application of the analysis above is achieved only under specific probability distribution assumptions on $X(\mathbf{s})$. In particular, the latter is assumed to be a multivariate Gaussian SRF, or an SRF that has a probability distribution that can easily be transformed to a multivariate Gaussian distribution (for example, a multivariate lognormal distribution). Under this assumption, the sampling design consists of the following stages:

(i) The first and second-order statistics of $X(\mathbf{s}_j)$ and $\hat{X}(\mathbf{s}_j)$, $\mathbf{s}_j \in U$ are estimated by means of an optimal spatial estimation method (e.g., simple kriging; see Chapter 9), where both the M available observations and the N additional sampling locations are used.

(ii) Then, the quantities (5) and (6) as well as the cutoff coefficients $\beta(\mathbf{s}_j)$ in Eq. (7) can be calculated in terms of these statistics.

(iii) Finally, the best sampling design S^* is the one that minimizes Eq. (4). Again, the implementation of the sampling design process (i) through (iii) is achieved with the aid of a search algorithm (Section 2.4 above).

5.2 The Variance Reduction Approach

Variance reduction is another local approach that is particularly useful when our goal is to expand an existing, usually irregular, network. Its main feature is that it examines the influence of a single additional sampling point on the efficiency of the existing sampling network. In particular, the improvement in the efficiency of the existing sampling network is measured by means of the variance reduction function. This is discussed in the following proposition due to Rouhani (1985).

Proposition 1: Let $X(\mathbf{s})$ be an ISRF-ν, representing a natural process with domain U. Assume that $S=\{\mathbf{s}_i, i=1,\ldots,N\}$ is the set of existing sampling points in U. Then, the *estimation variance reduction*, $\mathrm{VR}_{k/q}$, at point $\mathbf{s}_k \in S$ due to a new observation at point $\mathbf{s}_q \in S' \neq S$ is given by

$$\mathrm{VR}_{k/q} = \frac{1}{V_q(N)}\left[k_x(\mathbf{s}_k, \mathbf{s}_q) - \sum_{i=1}^{N} \lambda_i k_x(\mathbf{s}_i, \mathbf{s}_k) - \sum_{|\boldsymbol{\rho}|\leq \nu} \mu_\rho p_\rho(\mathbf{s}_k)\right]^2 \tag{8}$$

where $k_x(\cdot)$ is the generalized covariance; $V_q(N)$ is the estimation variance at $\mathbf{s}_q$ prior to the new observation; λ_i, μ_ρ, and $p_\rho(\mathbf{s}_k)$ are the intrinsic kriging parameters [estimation coefficients, Lagrange multipliers and monomials (for notation see Chapter 9)] in estimating $X(\mathbf{s}_k)$ prior to the new observation.

Similar expressions of $\mathrm{VR}_{k/q}$ are valid for homogeneous SRF, as well as for other methods of estimation. The total gain in accuracy or the information gain due to the new observation at $\mathbf{s}_q \in S'$ is measured by the *total reduction variance (TRV)* due to sampling, viz.,

$$\mathrm{TVR}_q = \sum_{\mathbf{s}_j \in U} \mathrm{VR}_{j/q} \tag{9}$$

The potential sampling points $\mathbf{s}_q$ can be ranked in terms of the values of the corresponding TVR_q; that is, one should pick the sampling point with the maximum TVR_q.

Under certain conditions, the local approach can account for observation costs by introducing a function that measures the expected economic benefit of further sampling. In particular, let $Z(\mathbf{s}_i) = \hat{X}(\mathbf{s}_i) - X(\mathbf{s}_i)$ be the sampling error, and consider the loss function

$$L[Z(\mathbf{s}_i)] = c_o Z(\mathbf{s}_i)_+ + c_u Z(\mathbf{s}_i)_- \tag{10}$$

where $Z(\mathbf{s}_i)_+ = Z(\mathbf{s}_i)$ if $Z(\mathbf{s}_i) \geq 0$, $=0$ otherwise; $Z(\mathbf{s}_i)_- = Z(\mathbf{s}_i)$ if $Z(\mathbf{s}_i) \leq 0$, $=0$ otherwise; c_u is the loss per unit of underestimating $X(\mathbf{s}_i)$; and c_o is the loss per unit of overestimating $X(\mathbf{s}_i)$. (For example, if $X(\mathbf{s}_i)$ is the actual piezometric head in feet, c_u is the loss per foot of underestimation and c_o is the loss per foot of overestimation.) The expected loss will be simply

$$E\{L[Z(\mathbf{s}_i)]\} = \int L(\zeta) f_z(\zeta)\, d\zeta \tag{11}$$

where $f_z(\zeta)$ is the probability density of $Z(\mathbf{s}_i)$.

Assuming that the $f_z(\zeta)$ is Gaussian, Eq. (11) gives

$$E\{L[Z(\mathbf{s}_i)]\} = \frac{c_o + c_u}{\sqrt{2\pi}} \sigma_x(\mathbf{s}_i) \tag{12}$$

Using Eq. (12), the *total loss reduction (TLR)* due to an additional measurement at $\mathbf{s}_q$ is defined by

$$\mathrm{TLR}_q = \frac{c_o + c_u}{\sqrt{2\pi}} \sum_{\mathbf{s}_j \in U} \left[\sigma_x(\mathbf{s}_j) - \sqrt{\sigma_x^2(\mathbf{s}_j) - \mathrm{VR}_{j/q}}\right] \tag{13}$$

and the *net expected benefit (NEB)* of a new sampling point is

$$\mathrm{NEB}_q = \mathrm{TLR}_q - \mathrm{MC}_q \tag{14}$$

where MC_q is the cost of the additional observation at $\mathbf{s}_q$. Clearly, all points with negative NEB_q should be eliminated as potential sampling locations (Rouhani, 1985).

A few important notes concerning the variance reduction approach above are worth mentioning here. First, the new sampling location $\mathbf{s}_q$ must be selected from a set M of candidate location within the domain U; this set M must be finite and its elements well defined in space. Second, the analysis applies when only one additional sampling point is considered; in other words, the variance reduction caused by a set of additional sampling points is not equal to the sum of the corresponding $\mathrm{VR}_{j/q}$. Third, the additional observations are not expected to significantly affect the assumed generalized covariance $k_x(\cdot)$; however, some sensitivity studies have shown that even a small level of fluctuations in the additional observations may cause large changes in the parameters of the estimated generalized covariance. Nevertheless, in a more recent work Rouhani and Fiering (1986) show that their impact on selected points may not be too significant. Fourth, the loss function (10) may be too simplistic for many applications; in addition, with only very few exceptions (such as the Gaussian assumption), the calculation of the expected loss function (11) is prohibitively tedious.

Other interesting works on local sampling design include Davis and Dvoranchik (1971), Veneziano and Kitanidis (1982), and Barnes (1988).

6. Statistical Inference Problems in Sampling Design

Clearly, before a sampling design approach can be used in a particular problem, the latter must be developed beyond the "initial phase" to the point where enough prior information is available to determine the needed input parameters, such as $c_x(\mathbf{h})$, $\gamma_x(\mathbf{h})$, ν, and $k_x(r)$.

However, in many practical applications very few observations are available, or no observations at all. Then, the following guidelines may assist in the determination of the input parameters above:

(a) Use knowledge about the physical laws governing the sampled natural processes, information provided by historic data, experience from previous sites worked out, as well as intuition and intelligent guessing. This important matter was discussed in Chapter 7.

(b) Choose input parameters [$c_x(\mathbf{h})$, $\gamma_x(\mathbf{h})$, ν, and $k_x(r)$], that, in addition to being compatible with (a) above, give a sampling design that performs satisfactorily for a wide range of actual (but unknown) parameters.

For example, one may consider several possible combinations of ν and $k_x(r)$, and use the reference charts to obtain the corresponding values of the sampling parameters (expected sampling error, cost, etc.). Then, on the basis of these values, useful conclusions regarding the most appropriate sampling design can be derived. This strategy is particularly fruitful when the linear form of the GSC-ν makes the implementation of such a strategy most efficient. Under certain circumstances, it may be shown that for most possible combinations of ν and $k_x(r)$ the systematic hexagonal sampling pattern offers the lowest sampling error. Or, in the case of a nugget effect, it may turn out that the choice of the sampling pattern has a rather small influence on the resulting values of the sampling error variance. Or, one may use the fact that under specific site conditions, certain natural processes exhibit similar ν and GSC-ν; in connection with this, tables of ν values and $k_x(r)$ models for important natural processes will be of great help in sampling design.

7. The Design of Spatiotemporal Sampling Networks

Many natural processes change in space and time, such as rainfall, top soil pH values, the concentration of a chemical species dissolved in groundwater, and the earth's surface temperature. These processes are modeled by means of spatiotemporal random fields (Chapter 5).

The design of sampling networks for observing natural processes varying in space–time can be achieved in terms of the spatiotemporal estimation theory developed in Chapter 9. The main problem here is to find the set S^* of sampling points in space–time $\{\mathbf{s}_i, t_j;\ i=1,2,\ldots,N$ and $j=1_i,2_i,\ldots,p_i\}$ that minimizes the average estimation error in the domain of interest, viz.,

$$\sigma_A^2(S^*) = \min_S \sigma_A^2(S) \tag{1}$$

where

$$\sigma_{\mathrm{A}}^2(S) = \frac{1}{K} \sum_{i=1}^{M} \sum_{j=1_i}^{\Lambda_i} \sigma_x^2(\mathbf{s}_i, t_j) \tag{2}$$

$K = \sum_{i=1}^{M} \Lambda_i$ is the number of estimated points in space-time, and $\sigma_x^2(\mathbf{s}_i, t_j)$ is the optimal estimation variance at location $\mathbf{s}_i$ and time t_j obtained in Section 7 of Chapter 9. Other objectives can be taken into consideration in a manner similar to that for spatial sampling designs.

When the hypothesis of space-time separable correlation structure is physically realistic, the sampling design process can be simplified significantly. As we saw in Example 8, Section 3 of Chapter 5, in hydrology the point-precipitation intensity at location **s** during the time t is considered as an S/TRF $X(\mathbf{s}, t)$ with separable covariance of the form $c_x(\mathbf{h}, \tau) = \sigma_x^2 \rho_x(\mathbf{h}) \rho_x(\tau)$ [σ_x^2 is the point variance of $X(\mathbf{s}, t)$, $\rho_x(\mathbf{h})$ is the spatial correlation, and $\rho_x(\tau)$ is the temporal correlation]. By taking advantage of the covariance separability, the sampling variance of the regional mean rainfall

$$X_{U,T} = \frac{1}{NT} \sum_{i=1}^{N} \sum_{t=1}^{T} X(\mathbf{s}, t) \tag{3}$$

(N is the number of stations over the region U and T is the number of months, years, or seasons that the network is in operation) can be expressed as (Bras and Rodriguez-Iturbe, 1985)

$$\mathrm{Var}[X_{U,T}] = \sigma_x^2 F_1(N) F_2(T) \tag{4}$$

where $F_1(N)$ and $F_2(T)$ are reduction factors due to sampling in space and in time, respectively. $F_1(N)$ depends on the spatial correlation structure $\rho_x(\mathbf{h})$, the sampling pattern and the sampling density. $F_2(T)$ depends on the temporal correlation structure $\rho_x(\tau)$ and the length of time that the sampling network has been in operation, and is independent of the number of stations. When Eq. (4) is applied, trading temporal sampling versus spatial sampling, while possible, is an expensive approach.

Interesting case studies of space-time sampling designs may be found in Rodriguez-Iturbe and Mejia (1974) and Loaiciga (1989).

8. A Taxonomy of Site Exploration Tasks

On the basis of the preceding analysis, a typical temporal division of site exploration is presented in Table 10.7. For many scientists the strength of such a taxonomy is that it produces an organizing reference for the problems of exploration, in which functionally similar tasks are grouped together (e.g., Baecher, 1980). Decision models are logical structurings of possible

Table 10.7 Taxonomy of Site Exploration Tasks

(1) *Reconnaissance*: Reviewing existing information to identify spatial or spatiotemporal variability. Selecting the appropriate values of sampling parameters to optimally allocate exploration effort.

(2) *Actual sampling of natural processes*: Using field and laboratory tests to infer *in situ* natural processes related to the site characterization.

(3) *Pattern recognition and reconstruction*: Recognizing the spatial or spatiotemporal distributions of the natural processes of interest, and estimation of their values in areas not actually observed.

(4) *Decision-making parameters*: Providing the inputs to decision processes, strategy modeling, risk analysis, etc.

site exploration outcomes and the decisions regarding management and planning that particular outcomes would lead to.

9. Terminal Decision Analysis and Sampling Design

9.1 Extensive Form Analysis

Suppose that $(\mathrm{E}, \mathrm{Z}, \hat{\mathrm{X}}, \mathrm{X})$ is a quadruple comprising spaces of experiments e, observations (samples) ζ of the SRF $Z(\mathbf{s})$, estimates $\hat{X}(\mathbf{s}) = \hat{\chi}$ and actual values χ of the SRF of interest $X(\mathbf{s})$ (see Fig. 10.20). In the decision analysis context one may seek answers to questions such as

- What amount of information is contained in experiment e of $Z(\mathbf{s})$ regarding the process $X(\mathbf{s})$? (For example, $Z(\mathbf{s})$ and $X(\mathbf{s})$ may represent specific capacity and aquifer transmissivity, respectively; E may represent a well field.)
- Which is the most informative experiment e or which is the "best" field of experiments E? (For example, which is a cost-effective well field E that assures water supply demands, maintains certain groundwater levels, etc.?)
- How good is a model whose undetermined parameter is $X(\mathbf{s})$? (For example, how good is a simulation model of a groundwater system for studying management-decision variables, such as hydraulic heads, pumping rates, etc.?)

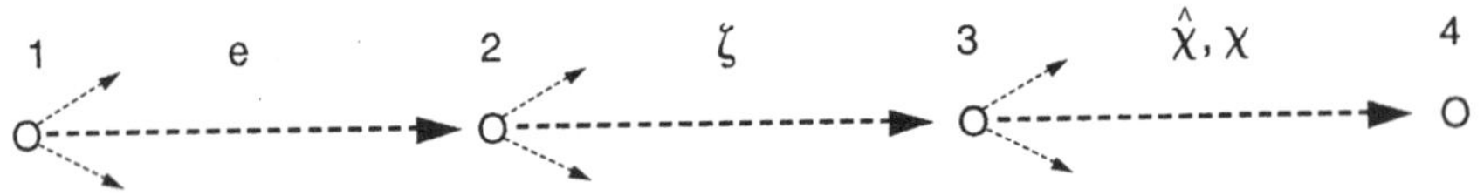

Figure 10.20 The terminal decision analysis setting

- How is a model $Z(\mathbf{s})$ affecting a strategy about $X(\mathbf{s})$? (For example, how can hydraulic conductivity affect the dewatering strategy of a site?)

The first part of the decision analysis that follows can be associated with the *extensive form analysis* (e.g., Raiffa and Schlaifer, 1961; Rosenkrantz, 1970), which involves working backward from node 4, successively expecting out the $X(\mathbf{s})$ and $Z(\mathbf{s})$, to node 2. More specifically (see also Christakos, 1991a):

Node 4: This is the *posterior* state, where a particular experiment e has been performed, ζ has been observed, and the actual value of $X(\mathbf{s})$ turned out to be χ.

Node 3: The utility function chosen is the information measure of performing e and observing ζ, viz.,

$$U(e, \zeta, x) = \mathrm{Inf}[x \mid \zeta] = -\log[f_{x|\zeta}(\chi, \zeta)] \tag{1}$$

where $f_{x|\zeta}(\chi, \zeta)$ is the posterior probability density of $X(\mathbf{s})$. The specific utility for each couple (e, ζ) is the expected posterior value

$$\bar{U}(e, \zeta) = E_{x|\zeta}\{\mathrm{Inf}[x \mid \zeta]\} = \varepsilon_{\zeta}(x) \tag{2}$$

The $\hat{\chi} = \hat{\chi}_{\zeta}$: $\mathrm{Max}_{\chi} f_{x|\zeta}$ is the Bayes estimate against $f_{x|\zeta}$. For instance, it may be the estimate $\hat{\chi}_{\zeta}$ of transmissivity at an unmeasured location, given observations ζ of specific capacity.

Node 2: The utility after experimentation is

$$\bar{\bar{U}}(e) = E_z[\varepsilon_{\zeta}(x)] = \overline{\varepsilon_z}(x) \tag{3}$$

Note that while $\varepsilon_{\zeta}(x)$ is the entropy of $X(\mathbf{s})$ conditional on the specific value ζ, $\overline{\varepsilon_z}(x)$ is the average entropy of $X(\mathbf{s})$ conditional on the random variable z. It is true that [Eq. (8), Section 13 of Chapter 2]

$$\overline{\varepsilon_z}(x) = \varepsilon(x, z) - \varepsilon(z) \tag{4}$$

Node 1: Let $U(e_0, \zeta_0, x) = \mathrm{Inf}[x] = -\log[f_x]$ [f_x = prior probability density of $X(\mathbf{s})$, e_0 = null experiment, and ζ_0 = dummy observation). The utility of immediate action without the benefit of experimentation is

$$\bar{\bar{U}}(e_0) = E_x\{\mathrm{Inf}[x]\} = \varepsilon(x) \tag{5}$$

Here, $\hat{\chi} = \hat{\chi}_0$: $\mathrm{Max}_{\chi} f_x$ is the prior Bayes estimate against f_x. For instance, it may be the estimate of transmissivity at an unmeasured location without using any specific capacity observations.

9.2 Information Measures

In the light of the analysis above, let us denote the experiment e by $e = [Z(\mathbf{s}), X(\mathbf{s}), f_x, f_{z|x}]$. As we saw, $\varepsilon(x)$ is a measure of the amount of uncertainty in f_x. In connection with this, the function

$$I(e) = \varepsilon(x) - \varepsilon_\zeta(x) \tag{6}$$

is called *conditional sample information (CSI)* and provides a measure of the amount by which an observation ζ on $Z(\mathbf{s})$ is expected to reduce uncertainty about $X(\mathbf{s})$. Then, the function [see Eq. (10), Section 13 of Chapter 2]

$$\bar{I}(e) = E_z[I(e)] = \varepsilon(x) - E_z[\varepsilon_\zeta(x)] = \varepsilon(x) - \overline{\varepsilon_z}(x) \tag{7}$$

is called *expected sample information (ESI)* and is a measure of the average amount by which the experiment e reduces the uncertainty about $X(\mathbf{s})$. In other words, $\bar{I}(e)$ is the amount of average information contained in e and, hence, it provides a means for choosing the most informative experiment. Clearly, an experiment e_1 is more informative than e_2 if $\bar{I}(e_1) \geq \bar{I}(e_2)$ for all f_x, and strict inequality holds for some f_x.

Just as $\bar{I}(e)$ is a measure of the evidence that e of $Z(\mathbf{s})$ provides, the function

$$\bar{I}^*(e) = \varepsilon(z) - E_x[\varepsilon_\chi(z)] = \varepsilon(z) - \overline{\varepsilon_x}(z) \tag{8}$$

is a measure of the *explanatory power* of the model used to study $X(\mathbf{s})$. Note that the $\bar{I}^*(e)$ is minimum (=0) when the SRF $X(\mathbf{s})$ and $Z(\mathbf{s})$ are independent $[\varepsilon(z) = \overline{\varepsilon_x}(z)]$; and it is maximum $[=\varepsilon(z)]$ in the case of a decisive experiment e $[\overline{\varepsilon_x}(z) = 0]$. The $\bar{I}^*(e)$ provides the means for deciding if the assumed model is satisfactory in studying $X(\mathbf{s})$, or whether another model will do a better job. For example, suppose that f_x^1 and f_x^2 are the two probability densities regarding a hydrologic magnitude $X(\mathbf{s})$, which are derived by means of two different hydrogeologic models of the aquifer. Then, the explanatory powers of these two models can be evaluated and compared by means of Eq. (8).

When we want to compare the explanatory powers of two models, which are associated with different experiments, a more meaningful measure of the explanatory power may be provided by

$$R(e) = \frac{\bar{I}^*(e)}{\varepsilon(z)} \tag{9}$$

where $\varepsilon(z)$ expresses the initial uncertainty. Again, note that $R(e) = 0$ when the $X(\mathbf{s})$ and $Z(\mathbf{s})$ are independent; $R(e) = 1$ when the experiment e is decisive. In relation to this, the quantity $Q(e) = 1 - R(e)$ expresses the *equivocation* of the model whose undetermined parameter is $X(\mathbf{s})$.

From a stochastic point of view, a physical model maximizes explanatory power in the sense of measure $\bar{I}^*(e)$, if the most probable hypotheses regarding its parameters $X(\mathbf{s})$ impose the strongest possible constraints on the possible outcomes of the relevant set of experiments E. Note that for practical considerations, Eq. (8) may be written as

$$\bar{I}^*(e) = [\max \varepsilon - \overline{\varepsilon_x}(z)] - [\max \varepsilon - \varepsilon(z)] \tag{10}$$

where max ε is the entropy of the uniform probability density. This formula shows that $\bar{I}^*(e)$ results from two separable constraints: (i) the constraint $c_1 = \max \varepsilon - \varepsilon(z)$, which reflects the *a priori* predictability of the outcomes of the experimental field E (say, $E =$ a well field), and (ii) the constraint $c_2 = \max \varepsilon - \overline{\varepsilon_x}(z)$ reflecting the predictability of the outcomes by the hydrogeologic model per se. Clearly, $\bar{I}^*(e)$ is decreasing in c_1; however, it must be noted that the latter may depend on the number of possible experimental outcomes selected, the sampling strategy, etc., which have nothing to do with the explanatory power of the hydrogeologic model per se. In fact, this is sometimes one of the reasons for using Eq. (9), which measures the explanatory power by the magnitude of the transmitted information relative to that of the initial uncertainty.

When the experiments of the field E are carried out in such a manner that before performing a new one, we know the result of the last one, some additional measures of information may be introduced. For example, the experiment e_2, which is carried out after the experiment e_1 was performed and the outcome ζ_1 was observed, may be denoted as

$$e_2(\zeta) = [Z(\mathbf{s}), X(\mathbf{s}), f_{x|\zeta_1}, f_{z|X}]$$

The average information contained in e_2 with respect to $X(\mathbf{s})$ after e_1 has been performed is

$$\bar{I}(e_2|e_1) = E_z[I(e_2(\zeta))] \tag{11}$$

Then, the information contained in the whole field E of m experiments, say $E = e^m = (e_1, e_2, \ldots, e_m)$, will be given by

$$\bar{I}(E) = \bar{I}(e_1) + \sum_{i=1}^{m-1} \bar{I}(e_{i+1}|e^i) \tag{12}$$

Remark 1: Similarly, when several natural processes or models are involved in a problem, the above measures will have to be extended appropriately. For example, in the case of two processes $Z_1(\mathbf{s})$ and $Z_2(\mathbf{s})$ associated with the experiments e_1 and e_2, respectively, Eq. (7) will be written as $\bar{I}(e_1, e_2) = \varepsilon(x) - \overline{\varepsilon_{z_1,z_2}}(x)$, etc.

9.3 The Sysketogram Function

On the basis of the foregoing analysis, as well as the results in Chapter 2, we can define a powerful measure of spatial correlation as follows.

Definition 5: The *sysketogram* function $\beta_x(\mathbf{s}, \mathbf{s}')$ of an SRF $X(\mathbf{s})$ is defined by

$$\beta_x(\mathbf{s}, \mathbf{s}') = \varepsilon_x(\mathbf{s}) - \bar{\varepsilon}_{x|x'}(\mathbf{s}, \mathbf{s}') \geq 0 \tag{13}$$

where $\varepsilon_x(\mathbf{s})$ is the entropy of $X(\mathbf{s})$ and $\bar{\varepsilon}_{x|x'}(\mathbf{s}, \mathbf{s}')$ is the average conditional entropy function at point $\mathbf{s}$ given the value of the SRF at point $\mathbf{s}'$.

Notice that in practical applications we usually consider the discrete case where the entropy is given by $\varepsilon_x(\mathbf{s}) = -\sum_\chi f_x(\chi) \log f_x(\chi)$ and the average conditional entropy is defined as

$$\bar{\varepsilon}_{x|x'}(\mathbf{s}, \mathbf{s}') = -\sum_{\chi'}\sum_{\chi} f_{x,x'}(\chi; \chi') \log f_{x|x'}(\chi|\chi')$$

The equality in Eq. (13) holds if and only if the $X(\mathbf{s})$ and $X(\mathbf{s}')$ are independent. For homogeneous SRF this is true when $\mathbf{h} = |\mathbf{s} - \mathbf{s}'| \to \infty$; moreover, when $|\mathbf{h}| = 0$, $\beta_x(\mathbf{s}, \mathbf{s}') = \varepsilon_x(\mathbf{s})$, which is a maximum. The sysketogram provides a measure of spatial correlation information, that is, a measure of the amount of information on $X(\mathbf{s})$ that is contained in $X(\mathbf{s}')$. In other words, $\varepsilon_x(\mathbf{s})$ in Eq. (13) is the information obtained by observing the SRF $X(\mathbf{s})$ at point $\mathbf{s}$ only, while $\bar{\varepsilon}_{x|x'}(\mathbf{s}, \mathbf{s}')$ is the average conditional information obtained by observing the SRF $X(\mathbf{s})$ when its value at point $\mathbf{s}'$ is already known.

In general, $\varepsilon_x(\mathbf{s}) \neq \bar{\varepsilon}_{x|x'}(\mathbf{s}, \mathbf{s}')$, since the knowledge about $X(\mathbf{s}')$ can provide a certain amount of information about the value of $X(\mathbf{s})$ if the latter is spatially correlated with the former. Hence the difference, as expressed by Eq. (13), is the part that is no longer considered as new information when the SRF is observed at point $\mathbf{s}$ after it has been observed at point $\mathbf{s}'$. As a consequence, $\beta_x(\mathbf{s}, \mathbf{s}')$ measures the strength of spatial correlation. Also notice that in contrast to the entropy $\varepsilon_x(\mathbf{s})$, the $\beta_x(\mathbf{s}, \mathbf{s}')$ allows a convergent counterpart in the continuous case.

A useful expression for $\beta_x(\mathbf{s}, \mathbf{s}')$ is given by

$$\beta_x(\mathbf{s}, \mathbf{s}') = \varepsilon_x(\mathbf{s}) + \varepsilon_x(\mathbf{s}') - \varepsilon_{x,x'}(\mathbf{s}, \mathbf{s}') \tag{14}$$

The sysketogram function has certain important properties that in many physical situations may favor its use instead of the traditional covariance function $c_x(\mathbf{s}, \mathbf{s}')$. More specifically: (a) The $\beta_x(\mathbf{s}, \mathbf{s}')$ is zero if and only if the $X(\mathbf{s})$ and $X(\mathbf{s}')$ are independent, while the $c_x(\mathbf{s}, \mathbf{s}')$ may be zero even when they are not independent; hence, the sysketogram contains more information about spatial correlation than the covariance. (b) $\beta_x(\mathbf{s}, \mathbf{s}')$ depends only on the probability laws, while the $c_x(\mathbf{s}, \mathbf{s}')$ depends on both the probability laws and the numerical values of $X(\mathbf{s})$ and $X(\mathbf{s}')$. (c) It can be shown that

$$\beta(\mathbf{s}, \mathbf{s}') = \beta(\phi, \phi') \tag{15}$$

where $\phi = \phi[X(\mathbf{s})]$ and $\phi' = \phi'[X(\mathbf{s}')]$ are one-one functions; that is, the sysketogram is not affected if the $X(\mathbf{s})$ is replaced by some function of it, provided that the latter is one-one. An important implication of (15) is that the sysketogram is completely independent of the scale of measurement of $X(\mathbf{s})$. This property is extremely useful in stochastic hydrogeology, for example, where the concepts of "scale of measurement" and "instruments window" play a crucial role (e.g., Cushman, 1984; Dagan, 1989). Finally, one may define the *relative sysketogram* function by

$$r\beta_x(\mathbf{h}) = \frac{\beta_x(\mathbf{0}) - \beta_x(\mathbf{h})}{\beta_x(\mathbf{0})} \in [0, 1] \tag{16}$$

where $\mathbf{h} = |\mathbf{s} - \mathbf{s}'|$. It is worth noticing that Eq. (16) experiences some similarities with the so-called *relative semivariogram* function

$$r\gamma_x(\mathbf{h}) = \frac{c_x(\mathbf{0}) - c_x(\mathbf{h})}{c_x(\mathbf{0})} \in [0, 1]$$

The sysketogram function defined above can be extended in the space-time context. In particular, the *sysketogram* function $\beta_x(\mathbf{s}, t; \mathbf{s}', t')$ of the S/TRF $X(\mathbf{s}, t)$ is given by (Christakos, 1991b)

$$\beta_x(\mathbf{s}, t; \mathbf{s}', t') = \varepsilon_x(\mathbf{s}, t) - \bar{\varepsilon}_{x|x'}(\mathbf{s}, t|\mathbf{s}', t') \tag{17}$$

where $\varepsilon_x(\mathbf{s}, t)$ is the entropy function at $(\mathbf{s}, t)$ and $\bar{\varepsilon}_{x|x'}(\mathbf{s}, t|\mathbf{s}', t')$ is the average conditional entropy function at $(\mathbf{s}, t)$ given the value of the S/TRF at $(\mathbf{s}', t')$.

The space-time sysketogram (17) has similar properties with its spatial equivalent. For example, the sysketogram is not affected if the S/TRF $X(\mathbf{s}, t)$ is replaced by some function of it, provided that the latter is one-one. This property implies that the sysketogram (17) is an absolute quantity rather than a relative one, in the sense that the space-time correlation defined by the sysketogram is independent of the scale of measurement used. For example, assume that we are interested about the probability law of the distribution of aerosol particles whose space-time coordinates are random variables. The correlation information as expressed by Eq. (17) is independent of the coordinate system chosen. *Mutantis mutandis*, the absoluteness property of the sysketogram brings in one's mind a basic result of modern physics according to which only absolute quantities (i.e., quantities independent of the coordinate system) can be used as the ingredients of a valid physical law (a law built according to such specifications is called covariant).

9.4 Sampling Design

The analysis above provides the means for developing a sampling design process, as follows (see also Shewry and Wynn, 1987):

Suppose that Π is the finite set of all possible observation points, $S \subset \Pi$ is the target set of sampling points, and S' is its complement with respect to Π. Then, we can write [Eq. (8), Section 13 of Chapter 2]

$$\varepsilon(\mathbf{X}_\Pi) = \varepsilon(\mathbf{X}_S) + \overline{\varepsilon_S}(\mathbf{X}_{S'}) \tag{18}$$

where $\mathbf{X}_\Pi$ is the vector of all possible random variables for $\mathbf{s}_i \in \Pi$, $\mathbf{X}_S$ is the vector of sampled variables (for $\mathbf{s}_i \in S$), and $\mathbf{X}_{S'}$ is the vector of unsampled variables ($\mathbf{s}_i \in S'$).

The amount of information about $\mathbf{X}_{S'}$ contained in $\mathbf{X}_S$ is given by

$$\bar{I}_S(\mathbf{X}_{S'}) = \varepsilon(\mathbf{X}_{S'}) - \overline{\varepsilon_S}(\mathbf{X}_{S'}) \tag{19}$$

Optimum sampling design, therefore, seeks the minimization of

$$\overline{\varepsilon_S}(\mathbf{X}_{S'}) = \varepsilon(\mathbf{X}_\Pi) - \varepsilon(\mathbf{X}_S)$$

But since $\varepsilon(\mathbf{X}_\Pi)$ is fixed and finite, this is equivalent to the maximization of $\varepsilon(\mathbf{X}_S)$ with respect to the sampling locations $\mathbf{s}_i \in S$, that is,

$$\max_{\mathbf{s}_i \in S} \varepsilon(\mathbf{X}_S) \tag{20}$$

Of course, to apply the criterion (20) a search algorithm such as those discussed in Section 2 must be used.

Example 1: Suppose that the random vector $\mathbf{X}_\Pi$ is multivariate Gaussian. In this case, Eq. (20) reduces to the simple expression

$$\max_{\mathbf{s}_i \in S} \log|\mathbf{C}_S| \tag{21}$$

where $\mathbf{C}_S$ is the covariance matrix of $\mathbf{X}_S$ and $|\cdot|$ denotes the determinant of the matrix.

References

"La critique est la vie de la science."

Cousin

Ababou, R., and Gelhar, L. W. 1990. "Self-similar randomness and spectral conditioning: Analysis of scale effects in subsurface hydrology." In *Dynamics of Fluids in Hierarchical Porous Media.* J. H. Cushman (ed.). Academic Press, New York. 393–428.

Adler, R. J. 1980. *The Geometry of Random Fields.* J. Wiley and Sons, New York.

Ahmed, S., and De Marsily, G. 1987. "Comparison of geostatistical methods for estimating transmissivity using data on transmissivity and specific capacity." *Water Resource Research* **23**, 9, 1717–1737.

Alabert, F. 1987. "The practice of fast conditional simulations through the LU decomposition of the covariance matrix." *Math. Geology* **19**, 5, 369–387.

Alexits, G. 1981. *Convergence Problems of Orthogonal Series.* Pergamon Press, New York.

Aoki, M. 1967. *Optimization of Stochastic Systems.* Academic Press, New York.

Armstrong, M. 1984. "Improving the estimation and modeling of the variogram." In *Geostatistics for Natural Resource Characterization.* G. Verly, M. David, A. G. Journel, and A. Marechal (eds.). Reidel, Dordrecht, 1–19.

Armstrong, M., and Matheron, G. 1986. "Disjunctive kriging revisited: Part 1 and 2." In *Math. Geology* **18**, 8, 711–742.

Arnold, L. 1974. *Stochastic Differential Equations: Theory and Applications.* J. Wiley, New York.

Aspie, D., and Barnes, R. J. 1990. "Infill-Sampling Design and the Cost of Classification Errors." In *Mathematical Geology* **22**, 8, 915–932.

Athans, M., Wishner, R. P., and Bertolini, A. 1968. "Suboptimal state estimation for continuous-time nonlinear systems for discrete noisy measurements." In *IEEE Trans. on Automatic Control* **AC-13**, 504–514.

Baecher, G. B. 1980. "Analyzing exploration strategies." Research Notes, Civil Engin. Dept., M.I.T., Cambridge, Massachusetts.

Bakr, A. A., Gelhar, L. W., Gutjahr, A. L., and MacMillan, J. R. 1978. "Stochastic analysis of spatial variability in subsurface flows: 1. Comparison of one- and three-dimensional flows." In *Water Resource Research* **14**, 2, 263–271.

Barnes, R. J. 1988. "Sample design for geologic site characterization." In *Proceedings of the 3rd Int. Geostatistical Congress*, Avignon, France.

Barrett, J. F., and Lampard, D. G. 1955. "An expansion for some second-order probability distributions and its applications to noise problems." In *IRE Trans. on Information Theory* **1**, 10–15.

Bartlett, M. S. 1955. *An Introduction to Stochastic Processes.* Cambridge, England.

Belyaev, Yu.K. 1972. "Point processes and first passage problems." In *Proceed. 6th Berkeley Symp. Math. Statist. Prob.* **2**, 1–17, Univ. of California Press, Berkeley.

Bell, T. L. 1987. "A space-time stochastic model of rainfall for satellite remote-sensing studies." In *Geophysical Research* **92**, D8, 9631-9643.

Benenson, F. C. 1984. *Probability, Objectivity and Evidence.* Routledge & Kegan Paul, London, England.

Benhenni, K., and Cambanis, S. 1990. "Sampling designs for estimating integrals of stochastic processes using quadratic mean derivatives." Techn. Report n.293, Center for Stochastic Processes, Dept. of Statistics, Univ. of North Carolina.

Bennett, R. J., and Chorley, R. J. 1978. *Environmental Systems.* Princeton: Princeton Univ. Press, New Jersey.

Beran, M. J. 1968. *Statistical Continuum Theories.* Monographs in Statistical Physics, n.9. Interscience Publ., New York.

Beveridge, G., and Schechter, R. S. 1970. *Optimization—Theory and Practice.* McGraw Hill, New York.

Bilonick, R. A. 1985. "The space-time distribution of sulfate deposition in the Northeastern United States." In *Atmospheric Environment* **19**, 11, 1829–1845.

Bishop, A. W., Green, G. T., and Skinner, A. E. 1973. "Strength and deformation measurements on soils." In *Proceed. of 8th Int. Conf. Soil Mech. and Found. Eng.* Moscow. 57–64.

Bjerrum, L. 1973. "Problems of soil mechanics and construction of soft clays and structurally unstable soils." In *Proceed. 8th Int. Conf. Soil Mech. Found. Eng.* **3**, 111–159, Moscow.

Bochner, S. 1959. *Lectures on Fourier Integrals.* Princeton Univ. Press, New Jersey.

Bogardi, I., Bardossy, A., and Duckstein, L. 1985. "Multicriterion network design using geostatistics." In *Water Resources Research* **21**, 2, 199–208.

Bohr, N. 1963. *Atomic Theory and Human Knowledge.* J. Wiley, New York.

Boll, M. 1941. *Les Certitudes du Hasard.* Paris.

Borel, E. (ed.) 1925. *Traite du Calcul des Probabilites et ses Applications.* 4 vols. Gauthier-Villars, Paris.

Borel, E. 1950. *Probabilite et Certitude.* Presses Univ. de France, Paris.

Borgman, L. E. 1969. "Ocean wave simulation for engineering design." In *J. Waterways and Harbors Div., Proc. ASCE* **95**, 4, 556–583.

Box, G. E. P. and Jenkins, G. M. 1970. *Time Series Analysis, Forecasting and Control.* Holden-Day, San Franscisco.

Bras, R. L., and Georgakakos, K. P. 1980. Real-time nonlinear filtering techniques in streamflow forecasting: A statistical linearization approach." In *Proceed. 3rd Intern. Symposium on Stochastic Hydraulics.* Tokyo, 95–105.

Bras, R. L., and Rodriguez-Iturbe, I. 1976a. "Network design for the estimation of areal mean of rainfall events." In *Water Resources Research* **12**, 6, 1185–1196.

Bras, R. L., and Rodriguez-Iturbe, I. 1976b. "Rainfall network design for runoff prediction." In *Water Resources Research* **12**, 6, 1197–1208.

*Bras, R. L., and Rodriguez-Iturbe, I. 1985. *Random Functions and Hydrology.* Addison-Wesley, New York.

Bratley, P., Fox, B. L., and Schrage, L. E. 1983. *A Guide to Simulation.* Springer-Verlag, New York.

Brockett, R. W. 1981. "Nonlinear systems and nonlinear estimation theory." In *Stochastic Systems.* M. Hazewinkel and J. C. Willems (eds.). Reidel, Holland. 441–478.

Bucy, R. S. 1965. "Nonlinear filtering." In *IEEE Trans. on Automatic Control* **AC-10**, 198.

Burg, J. P. 1972. "The relationship between maximum-entropy spectra and maximum likelihood spectra." In *Geophysics* **38**, 375–376.

Byrne, E. F. 1968. *Probability and Opinion.* Martinus Nijhoff, The Hague.

Cacko, J., Bily, M., and Bukoveczky, J. 1988. *Random Processes: Measurement, Analysis and Simulation.* Elsevier, New York.

Carnap, R. 1950. *Logical Foundations of Probability.* Univ. of Chicago Press, Chicago.

Carr, J. R. 1990. "Application of spatial filter theory to kriging." In *Mathematical Geology* **22**, 8, 1063–1079.

Chandrasekhar, S. 1943. "Stochastic problems in physics and astronomy." In *Rev. Mod. Phys.* **15**, 1–89.

Chiles, J. P. 1979. "Le variogramme generalise." Research Rep. N-612, Centre de Geostatistique, Ecole des Mines de Paris.

Christakos, G. 1982. *Mathematical Modeling of Advanced Estimation Approaches for Geotechnical Systems.* M.S. thesis, Dept. of Civil Engineering, M.I.T., Cambridge, Massachusetts.

Christakos, G. 1984a. "The space transformations and their applications in systems modeling and simulation." In *Proceed. 12th Int. Conf. on Modeling and Simulation (AMSE)* **1**, 3, 49–68, Athens.

Christakos, G. 1984b. "On the problem of permissible covariance and variogram models." In *Water Resources Research* **20**, 2, 251–265.

Christakos, G. 1985a. "Recursive parameter estimation with applications in earth sciences." In *J. Mathematical Geology* **17**, 5, 489–515.

Christakos, G. 1985b. "Modern statistical analysis and optimal estimation of geotechnical data." In *Engineering Geology* **22**, 2, 175–200.

Christakos, G. 1986a. "Space transformations in the study of multidimensional functions in the hydrologic sciences." In *Advances in Water Resources* **9**, 1, 42–48.

Christakos, G. 1986b. "Space Transformations: An Approach for studying Spatial Functions in Geosciences." In *Inst. of Geology and Mineral Exploration.* Athens.

Christakos, G. 1986c. *Recursive Estimation of Nonlinear State—Nonlinear Observation Systems (part 1: on-line data).* Res. Rep. OF.86-29, Kansas Geological Survey, Lawrence, Kansas, 85 p.

Christakos, G. 1987a. "Stochastic simulation of spatially correlated geoprocesses." In *J. Mathematical Geology* **19**, 8, 803–827.

Christakos, G. 1987b. "A stochastic approach in modeling and estimating geotechnical data." In *Int. J. of Numerical and Analytical Methods in Geomechanics* **11**, 1, 79–102.

Christakos, G. 1987c. "The space transformation in the simulation of multidimensional random fields." In *J. Mathematics and Computers in Simulation* **29**, 313–319.

Christakos, G. 1987d. "Stochastic approach and expert systems in the quantitative analysis of soils." In *Proceed. 5th Int. Conf. on Application of Statistics and Probability in Soil and Structural Engineering* **2**, 741–748, Vancouver.

Christakos, G. 1988a. "On-line estimation of nonlinear physical systems." In *J. Mathematical Geology* **20**, 2, 111–133.

Christakos, G. 1988b. "Modeling and Estimation of nonlinear systems." In *Proceed. 12th IMACS* **2**, 377–379.

Christakos, G. 1989. "Optimal estimation of nonlinear-state nonlinear-observation systems." In *J. Optimization Theory and Appl.* **62**, 1, 29–48.

Christakos, G. 1990a. "A Bayesian/Maximum-Entropy view to the spatial estimation problem." In *J. Mathematical Geology* **22**, 7, 763–776.

Christakos, G. 1990b. "Random Field Modeling and Its Applications in Stochastic Data Processing." Ph.D. thesis, Div. of Applied Sciences, Harvard Univ., Cambridge, Massachusetts.

Christakos, G. 1990c. "On a frequency domain approach to spatiotemporal stochastic hydrodynamics." *Research Notes RP10.90*, Dept. of Environmental Sciences and Engineering, Univ. of North Carolina, Chapel Hill, NC.

Christakos, G. 1991a. "Some applications of the Bayesian, maximum-entropy concept in Geostatistics." In *Fundamental Theories of Physics.* Kluwer, The Netherlands, 215–229.

Christakos, G. 1991b. "A theory of spatiotemporal random fields and its application to space–time data processing." In *IEEE Trans. Systems, Man, and Cybernetics* **21**, 4, 861–875.

Christakos, G. 1991c. "Certain results on spatiotemporal random fields, and their applications in environmental research." In *Proceed. NATO Advanced Studies Inst.* Lucca, Italy.

Christakos, G., and Panagopoulos, C. 1992. "Space transformation methods in the representation of random fields." In *IEEE Trans. Remote Sensing and Geosciences* **30** (1), 55–70.

Christakos, G., and Olea, R. A. 1992. "Sampling design for spatially distributed hydrogeologic and environmental processes." *Advances in Water Resources,* in press, to appear April 1992.

Christakos, G., and Paraskevopoulos, P. N. 1986. "On the functional optimization of a certain class of nonstationary spatial functions." In *J. Optimization Theory and Applications* **52**, 2, 191–208.

Christakos, G. and Pantelias, E. 1987. Optimal sampling strategy and its applications in geologic and mineral exploration." In *Proceed. 2nd Congress on the Greek Mineral Resour.* Athens, **2**, 149–161.

Christakos, G., and Olea, R. A. 1988. "A multiple-objective optimal exploration strategy." In *J. Math. and Comp. Modeling* **1**, 413–418.

Cramer, H. 1946. *Mathematical Methods of Statistics.* Princeton Univ. Press, Princeton, New Jersey.

Cressie, N. 1985. "Fitting variogram models by weighted least squares." In *J. Math. Geology* **17**, 563–586.

Cressie, N. 1991. *Statistics for Spatial Data.* J. Wiley, New York.

Cressie, N., and Hawkins, D. M. 1980. "Robust estimation of the variogram." *Mathematical Geology* **12**, 115–125.

Cushman, J. H. 1984. "On unifying the concepts of scale, instrumentation, and stochastics in the development of multiphase transport theory." *Water Resources Research* **20**, 11, 1668–1676.

Dagan, G. 1989. *Flow and Transport in Porous Formations.* Springer-Verlag, New York.

Daniel, C., Wood, F. S., and Gorman, J. W. 1971. *Fitting Equations to Data.* J. Wiley-Interscience, New York.

Da Prato, G., and Tubaro, L. 1987. *Stochastic Partial Differential Equations and Applications.* Lecture Notes in Mathematics, n. 1236. Springer, New York.

David, M. 1977. *Geostatistical Ore Reserve Estimation.* Elsevier, Amsterdam.

Davis, D. R., and Dvoranchik, W. M. 1971. "Evaluation of the worth of additional data." In *Water Resources Research* **7**, 700–707.

Davis, E. H., and Poulos, H. G. 1967. "Laboratory investigations of the effects of sampling." *Civil Engin. Trans., The Inst. of Engineers* **GE9**, 1.

Davis, J. C. 1973 *Statistics and Data Analysis in Geology.* J. Wiley, New York.

Davis, M. H. A. 1981. "Pathwise nonlinear filtering." In *Stochastic Systems.* M. Hazewinkel and J. C. Willems (eds.). Reidel, Dordrecht. 505–528.

Davis, P. J. 1975. *Interpolation and Approximation.* Dover, New York.

Davis, R. C. 1952. "On the theory of prediction of nonstationary stochastic processes." In *J. Appl. Physics* **23**, 1047–1053.

Deans, S. R. 1983. *The Radon Transform and Some of Its Applications.* J. Wiley and Sons, New York.

De Finetti, B. 1974. *Theory of Probability.* J. Wiley, New York.

Delhomme, J. P. 1977. *Modeles de Simulation et de Gestion des Resources en Eau des Basins de l'Orne, la Dives, et la Seulles.* CIC, Centre de Geostatistique, France.

Delfiner, P. 1976. "Linear estimation of nonstationary spatial phenomena." In *Adv. Geostatistics in the Mining Industry.* M. Guarascio *et al.*, eds. Reidel, Dordrecht. 49–68.

Delfiner, P. 1979. "Basic introduction to geostatistics." Research Rep. C-78, Centre de Geostatistique, Ecole des Mines de Paris, Fontainebleau, France.

Deutsch, C. V., and Journel, A. G. 1991. "The Application of Simulated Annealing to Stochastic Reservoir Modelling," Report 4. Stanford Center for Reservoir Forecasting, Stanford University, Stanford, California.

DeWitt, B., and Graham, R. N. 1973. *The Many-Worlds Interpretation of Quantum Mechanics.* Princeton Univ. Press, Princeton, New Jersey.

Dimitrakopoulos, R. 1990. "Conditional simulation of intrinsic random functions of order *k*." *Mathematical Geology* **22**, 3, 361–380.

Doob, J. 1960. *Stochastic Processes.* J. Wiley, New York.

Duncan, T. E. 1967. *Probability Densities for Diffusion Processes with Applications to Nonlinear Filtering Theory and Diffusion Theory.* Ph.D. thesis, Stanford Univ., Stanford, California.

Einstein, A. 1905. "Concerning the notion as required by the molecular-kinetic theory of heat of particles suspended in liquids at rest." In *Ann. Phys.* (Leipzig) **17**, 549. Also, *Investigations on the Theory of Brownian Motion.* Dover, New York (1956).

Elishakoff, L. 1983. *Probabilistic Methods in the Theory of Structures.* J. Wiley, New York.

* Ewing, G. M. 1969. Calculus of Variations with Applications. Norton, New York.

Fedorov, V. V. 1972. *Theory of Optimal Experiments.* Academic Press, New York.

Feller, W. 1966. *An Introduction to Probability Theory and Its Applications.* Vols. 1 and 2. J. Wiley, New York.

Fiering, M. B. 1964. "Multivariate techniques for synthetic hydrology." In *J. Hydraulic Div., ASCE* 43–60.

Fisher, R. A. 1959. *Statistical Methods and Scientific Inference.* Oliver, London.

Francois-Bongarcon, D. 1981. "Les coregionalisations le cokrigeage." Research Rep. C-86, Ecole des Mines de Paris, CGMM, Fontainebleau, France.

Friedman, A. 1975. *Stochastic Differential Equations and Applications: Vol. 1.* Academic Press, New York.

Friedman, A. 1976. *Stochastic Differential Equations and Applications: Vol. 2.* Academic Press, New York.

Fujisaki, M., Kallianpur, G., and Kunita, H. 1972. "Stochastic differential equations for the nonlinear filtering problem." In *Osaka J. Math.* **9**, 19–40.

Gajem, Y. M., Warrick, A. W., and Myers, D. E. 1981. "Spatial dependence of physical properties of a typic torrifluvent soil." *Soil Sci. Am. J.* **45**, 709–715.

Gandin, L. S. 1963. *Objective analysis of meteorological fields.* Gidrometerorologicheskoe Izdatel'stvo, Leningrad.

Garside, M. J. 1971. "Some computational procedures for the best subset problem." *Applied Statistics* **20**, 8–15.

Gelb, A. 1974. *Applied Optimal Estimation.* MIT Press, Cambridge, Massachusetts.

Gel'fand, I. M. 1955. "Generalized random processes." In *Dokl. Akad. Nauk SSSR* **100**, 853–856.

Gel'fand, I. M., and Shilov, G. E. 1964. *Generalized Functions,* **1**. Academic Press, New York.

Gel'fand, I. M., and Vilenkin, N. Y. 1964. *Generalized Functions,* **4**. Academic Press, New York.

Geman, D. and Geman S. 1984. "Stochastic relaxation, Gibbs distributions, and the Bayesian restoration of images." In *IEEE Trans. PAMI* **6**, 721–741.

Gendre, J.-L. 1947. *Introduction a l'Etude du Jugement Probable.* Presses Univ. de France, Paris.

Gihman, I. I., and Skorokhod, A. V. 1974a. *The Theory of Stochastic Processes: 1.* Springer-Verlag, New York.

Gihman, I. I., and Skorokhod, A. V. 1974b. *The Theory of Stochastic Processes: 2.* Springer-Verlag, New York.

Gihman, I. I., and Skorokhod, A. V. 1974c. *The Theory of Stochastic Processes: 3*. Springer-Verlag, New York.

Gihman, I. I., and Skorokhod, A. V. 1972. Stochastic Differential Equations. Springer-Verlag, Berlin.

Gilbert, R. O. 1987. *Statistical Methods for Environmental Pollution Monitoring*. Van Nostrand-Reinhold, New York.

Goda, Y. 1980. *Random Seas and Design of Maritime Structures*. Univ. of Tokyo Press, Tokyo.

Goldberg, D. E. 1989. *Genetic Algorithms in Search, Optimization, and Machine Learning*. Addison-Wesley, Reading, Massachusetts.

Gorelick, S. M. 1982. "A review of distributed parameter groundwater management modelling methods." In *Water Resource Research* **19**, 2, 305–319.

Gradshteyn, T. S., and Ryzhik, I. M. 1965. *Tables of Integrals, Series, and Products*. Academic Press, New York.

Granger, C. W. J. 1975. "Aspects of the analysis and interpretation of temporal and spatial data." In *The Statistician* **24**, 197–210.

Granger, C. W. J., and Hatanaka, M. 1964. *Spectral Analysis of Economic Time Series*. Princeton University Press, Princeton, New Jersey.

Granger, C. W. J., and Newbold, P. 1986. *Forecasting Economic Time Series*. Academic Press, New York.

Grenander, U., and Rosenblatt, M. 1957. *Statistical Analysis of Stationary Time Series*. J. Wiley, New York.

Gy, P. M. 1982. *Sampling of Particulate Materials*. Elsevier, New York.

Haldar, A., and Tang, W. H. 1979. "Probabilistic evaluation of liquefaction potential." *Jour. Geotech. Eng. Div., ASCE* **105**, GT2, 145–163.

Haslett, J., and Raftery, A. E. 1989. "Space–time modeling with long-memory dependence: Assessing Ireland's wind power resource." In *Appl. Statist.* **38**, 1, 1–50.

Hazewinkel, M., and Marcus, S. I. 1982. "On lie algebras and finite-dimensional filtering." *Stochastics* **7**, 29–62.

Heisenberg, W. 1930. *The Physical Principles of the Quantum Theory*. Dover, New York.

Helgason, S. 1980. *The Radon Transform*. Birkhauser, Boston.

Hohn, M. E. 1988. *Geostatistics and Petroleum Geology*. Van Nostrand-Reinhold Co., New York.

Horn, R. A., and Johnson, C. R. 1985. *Matrix Analysis*. Cambridge Univ. Press, Cambridge, England.

Hwang, C. L., and Masud, A. S. M. 1979. *Multiple-Objective Decision Making Methods and Applications*. Lecture notes in Economics and Mathematical Systems, Springer-Verlag, Berlin.

Infeld, E., and Rowlands, G. 1990. *Nonlinear Waves, Solitons, and Chaos*. Cambridge Univ. Press, Cambridge, UK.

Isaaks, E. H., and Srivastava, R. M. 1989. *An Introduction to Applied Geostatistics*. Oxford Univ. Press, New York.

Ito, K. 1954. "Stationary random distributions." In *Univ. Kyoto Memoirs*, Series A, **XXVIII**, 3, 209–223.

Ivanov, A. V., and Leonenko, N. N. 1989. *Statistical Analysis of Random Fields*. Kluver, The Netherlands.

*Jackson, D. 1941. *Fourier Series and Orthogonal Polynomials*. The Mathematical Assoc. of America.

Jaynes, E. T. 1957. "Information theory and statistical mechanics." In *Phys. Rev.* **106**, 620–630.

Jaynes, E. T. 1968. "Prior probabilities." In *IEEE Trans. Syst. Sci., Cybern.* **4**, 227–241.

Jaynes, E. T. 1982. "On the rationale of maximum-entropy methods." In *Proceed. IEEE*, **70**, 9, 939–952.

Jazwinski, A. H. 1966. "Filtering of Nonlinear Systems." In *IEEE Trans. on Automatic Control* **AC-11**, 765–766.

Jazwinski, A. H. 1970. *Stochastic Processes and Filtering Theory.* Academic Press, New York.

Jeffreys, H. 1939. *Theory of Probability.* Oxford Univ. Press, Oxford, England.

Johnson, M. E. 1987. *Multivariate Statistical Simulation.* J. Wiley, New York.

Johnson, N. L., and Kotz, S. 1972. *Distributions in Statistics—Continuous Multivariate Distributions.* Wiley, New York.

Jones, D. A., Gurney, R. J., and O'Connell, P. E. 1979. "Network design using optimal estimation procedures." In *Water Resources Research* **15**, 6, 1801–1812.

Journel, A. G. 1974. "Geostatistics for conditional simulation of ore bodies." In *Economic Geology* **69**, 673–687.

Journel, A. G., and Huijbregts, Ch. 1978. *Mining Geostatistics.* Academic Press, London.

Journel, A. G. 1984. *Indicator Approach to Toxic Chemical Sites.* Report of Project n. CR-811235-02-0, EPA, EMSL-Las Vegas.

Journel, A. G. 1986. "Constrained interpolation and qualitative information—the soft kriging approach." In *Mathematical Geology* **18**, 3, 269–286.

Journel, A. G., and Alabert, A. 1988. "Focusing on spatial connectivity of extreme-valued attributes: Stochastic indicator models of reservoir heterogeneities." *SPE* paper no. 18324.

Journel, A. G. 1989. *Fundamentals of Geostatistics in Five Lessons.* American Geophysical Union, Washington, D.C. 40 p.

Journel, A. G., and Zhu, H. 1990. "Integrating soft seismic data." Research Rep., *Stanford Center for Reservoir Forecasting*, Stanford, California.

Kailath, T. 1974. "A review of three decades of linear filtering theory." In *IEEE Trans. on Information Theory* **IT-20**, 2, 146–181.

Kafritsas, J., and Bras, R. L. 1980. *The Practice of Kriging.* Research Report, M.I.T., Massachusetts.

Kallianpur, G. 1980. *Stochastic Filtering Theory.* Springer-Verlag, New York.

Kanwal, R. P. 1983. *Generalized Functions—Theory and Technique.* Academic Press, New York.

Karlin, S., and McGregor, T. 1960. "Classical diffusion processes and total positivity." In *Jour. of Mathematical Analysis and Applications* **1**, 163–183.

*Keynes, J. M. 1921. *A Treatise on Probability.* Oxford Univ. Press, Oxford, England.

Khinchin, A. 1934. "Korrelations theorie des stationaren stochastischen prozesse." *Math. Anal.* **109**, 604–615.

Khinchin, A. 1949. *Mathematical Foundations of Statistical Mechanics.* Dover, New York.

Kim, Y. C., Myers, D. E., and Knudsen, H. P. 1977. "Advanced geostatistics in ore reserve estimation and mine planning (practitioner's guide)." Rep. Subcont. n. 76-003-E, Phase II, U.S. Energy Res. and Developm. Admin., Tucson, Arizona.

Kitanidis, P. K. 1983. "Statistical estimation of polynomial generalized covariance function and hydrologic applications." In *Water Resource Research* **19**(4), 909–921.

Kolmogorov, A. N. 1933. *Grundbegriffe der Wahrscheinlichkeitrechnung.* Ergebnisse der Mathematik. (English translation: *Foundations of the Theory of Probability*, Chelsea, New York. 1950.)

Kolmogorov, A. N. 1941. "The distribution of energy in locally isotropic turbulence." *Dokl. Akad. Nauk SSSR* **32**, 19–21.

Kraft, L., and Mukhopadhyay, J. 1977. "Probabilistic analysis of excavated earth slopes." In *Proceed 9th Int. Conf. Soil Mech. and Found. Eng.* Tokyo. **2**, 109–116.

Kullback, S. 1968. *Information Theory and Statistics.* Dover Publ., New York.

Ladd, C. C., Foot, R., Ishihara, K., Schlosser, F., and Poulos, H. G. 1977. *State of the Art Report. 9th Int. Conf. Soil Mech. and Found. Eng.* Tokyo. 421–494.

Lakatos, I. 1970. "Falsification and methodology of scientific research programmes." *In* Lakatos, I., and A. Musgrave (eds.). *Criticism and the Growth of Knowledge.* Cambridge University Press, Cambridge, England. 91–196. (Republished in Lakatos, I. *The Methodology of Scientific Research Programmes*, Cambridge University Press, Cambridge, England. 1978.)

Lake, L. W., and Carrol, H. B. 1986. *Reservoir Characterization.* Academic Press. Orlando, Florida.

Lambe, T. W., and Whitman, R. V. 1969. *Soil Mechanics.* J. Wiley, New York.

Lancaster, H. O. 1958. "The structure of bivariate distributions." In *Annals of Mathematical Statistics* **29**, 719–736.

Langevin, P. 1908. "Sur la theorie du mouvement brownien." *C.R. Acad. Sci., Paris.* **146**, 530–533.

Levy, P. 1948. *Processus Stochastiques et Mouvement Brownien.* Gauthier-Villars, Paris.

Lindzen, R. S. 1989. "Greenhouse warming: Science vs. Consensus." Notes, *Center for Meteorology and Physical Meteorology,* M.I.T., Massachusetts.

Locaiciga, H. A. 1989. "An optimization approach for groundwater quality monitoring network design." In *Water Resource Research* **25**, 8, 1771–1782.

Loeve, M. 1953. *Probability Theory.* Van Nostrand, Princeton.

Mahalanabis, A. K., and Farooq, M. 1971. "A second-order method for state estimation of nonlinear dynamic systems." In *Int. J. Control* **14**, 631–639.

Mantoglou, A., and Wilson, J. L. 1982. "The turning bands method for simulation of random fields using line generation by a spectral method." In *Water Resource Research* **18**, 5, 1379–1394.

Mantoglou, A. 1987. "Digital simulation of multivariate two- and three-dimensional stochastic processes with a spectral turning bands method." *Mathematical Geology* **19**, 2, 129–149.

Marchuk, G. I. 1986. *Mathematical Models in Environmental Problems.* North-Holland, Amsterdam.

Marechal, A. 1970. "Cokrigeage et regression en correlation intrinseque." Research Rep. N-205, Ecole des Mines de Paris, CGMM. Fontainebleau, France.

Marechal, A. 1976. "Selecting mineable blocks. Experimental results observed on a simulated orebody." In *Advanced Geostatistics in the Mining Industry.* M. Guarascio *et al.* (eds.). Reidel, Dordrecht, 253–276.

Marshall, R. J., and Mardia, K. V. 1985. "Minimum norm quadratic estimation of components of spatial covariance." *Mathematical Geology* **17**, 517–525.

Matern, B. 1960. Spatial Variation. Medd. Fran. Stat. Skogsf. **49**, 5. Stockholm.

Matheron, G. 1965. *Les variables regionalisees et leur estimation.* Masson, Paris.

Matheron, G. 1971. The theory of regionalized variables and its applications." Cahier 5, Centre de Geostatistique, Ecole des Mines de Paris. Fontainebleau, France.

Matheron, G. 1973. "The intrinsic random functions and their applications." In *Adv. Appl. Prob.* **5**, 439–468.

Matheron, G. 1975. "Forecasting block grade distributions. The transfer functions." *Geostat. 75*, 237–251.

Matheron, G. 1976. "A simple substitute for conditional expectation. The disjunctive kriging." In *Advanced Geostatistics in the Mining Industry.* M. Guarascio *et al.* (eds.). Reidel, Dordrecht. 221–236.

Matheron, G. 1978. *Estimater et Choisir.* Ecoles des Mines de Paris, Fontainebleau, France. (English translation: *Estimating and Choosing.* Springer-Verlag, New York, 1989.)

Matheron, G. 1984. "Isofactorial models and change of support." In *Adv. Geostatistics for the characterization of natural resources.* Verly *et al.* (eds.). **1**, 449–468. Reidel.

McBratney, A. B., Webster, R., and Burgess, T. M. 1981. "The design of optimal sampling schemes for local estimation and mapping of regionalized variables—I." In *Computers and Geosciences* **7**, 4, 331–334.

McBratney, A. B., and Webster, R. 1981. "The design of optimal sampling schemes for local estimation and mapping of regionalized variables—II." In *Computers and Geosciences* **7**, 4, 335–365.

Mein, R. G., Laurenson, E. M., and McMahon, T. A. 1974. "Simpler nonlinear model for flood estimation." In *ASCE J. Hydraulic* **100**, 1507–1518.

Mejia, J. A., and Rodriguez-Iturbe, I. 1974. "On the synthesis of random field sampling from the spectrum: An application to the generation of hydrologic spatial processes." In *Water Resource Research* **10**, 4, 705-711.

* Messiah, A. 1965. *Quantum Mechanics*. North-Holland, Amsterdam.

Metropolis, N., Rosenbluth, A. W., Rosenbluth, M. N., Teller, M. N., and Teller, E. 1953. "Equations of state calculations by fast computing machines." In *J. Chem. Phys.*, **21**, 1087-1091.

Milovic, D. M. 1970. "Effect of sampling on some soil characteristics." In *ASTM*, STP **483**, 164-179.

Mitter, S. K. 1980. "On the analogy between mathematical problems of nonlinear filtering and quantum physics." In *Richerce di Automatica* **10**, 163-216.

Mizell, S. A., Gutjahr, A. L., and Gelhar, L. W. 1982. "Stochastic analysis of spatial variability in two-dimensional steady groundwater flow assuming stationary and nonstationary heads." In *Water Resource Research* **18**, 4, 1053-1067.

Monin, A. S., and Yaglom, A. M. 1971. *Statistical Fluid Mechanics*. MIT Press, Cambridge, Massachusetts.

Myers, D. E. 1982. "Matrix formulation of co-kriging:" *Mathematical Geology* **14**, 3, 249-257.

Nagel, E. 1939. *Principles of the Theory of Probability*. In *Intern. Encyclopedia of Unified Sciences* **1**, 6, Univ. of Chicago Press, Chicago.

Newton, M. J. 1973. "The application of geostatistics to mine sampling patterns." *Int. Symposium in Computer Appl. in the Mineral Industry*. 11th session. Tucson, Arizona.

NRC. 1991. *Opportunities in the Hydrologic Sciences*. National Research Council, Nation. Acad. Press. Washington, D.C.

Olea, R. A. 1984. "Sampling design optimization for spatial functions." In *Mathematical Geology* **16**, 4, 369-392.

Omatu, S., and Seinfeld, J. H. 1981. "Filtering and smoothing for linear discrete-time distributed parameter systems based on Wiener-Hopf theory with application to estimation of air pollution." *IEEE Trans. on Systems, Man, and Cybernetics* **11**, 12, 785-801.

Orfeuil, J. P. 1972. "Simulation du Wiener-Levy et de ses integrales." Internal rep. N-290. Centre de Geostatistique, Fontainebleau, France.

Orlowski, A., and Sobczyk, K. 1989. "Solitons and shock waves under random external noise." *Reports on Mathematical Physics* **27**, 1, 59-71.

Panchev, S. 1971. *Random Functions and Turbulence*. Pergamon Press, New York.

Pardoux, E. 1981. "Nonlinear filtering, prediction and smoothing." In *Stochastic Systems*. M. Hazewinkel and J. C. Willems (eds.). Reidel, Dordrecht. 529-557.

Pearson, K. 1901. "Mathematical contributions to the theory of evolution, VII: On the correlation of characters not quantitatively measurable." In *Philosophical Trans. Royal Soc. of London*, Series A **195**, 1-47.

Pinsker, M. S. 1955. "Theory of curves in a Hilbert space with stationary increments of order *n*." *Izv. Akad. Nauk SSSR, Ser. Mat.* **19**, 319-344.

Poincare, H. 1912. *Calcul de Probabilites*. G. Carre, Paris.

Poincare, H. 1929. *La Science et l'Hypothese*. Flammarion, Paris.

Polanyi, M. 1958. *Personal Knowledge: Toward a Post-Critical Philosophy*. Univ. of Chicago Press, Chicago.

Polya, G. 1954. *Mathematics and Plausible Reasoning*. Vols. 1 and 2. Princeton Univ. Press, Princeton, New Jersey.

Popper, K. R. 1934. *Logik der Forschung*. Springer, Vienna. (English translation, *The Logic of Scientific Discovery*. Hutchinson, London, 1959. Also, *The Logic of Scientific Discovery*. Harper Torchbooks, New York, 1965.)

Popper, K. R. 1972. *Objective Knowledge—An Evolutionary Approach*. Oxford Univ. Press, Oxford, England, 395 p.

Priestley, M. B. 1981. *Spectral Analysis and Time Series.* Vols. 1 and 2. Academic Press, New York.

Priestley, M. B. 1988. *Nonlinear and Nonstationary Time Series Analysis.* Academic Press, New York.

Pugachev, V. S., and Sinitsyn, I. N. 1987. *Stochastic Differential Systems.* J. Wiley, New York.

Quenouille, M. H. 1957. *The Analysis of Multiple Time Series.* Griffin and Co., London.

Radon, J. 1917. "Uber die Bestimmung von funktionen durch ihre integralwerte langs gewisser mannigfaltigkeiten." *Berichte Sachs. Aked. der Wissenschaften. Leipzig, Math.-Phys. Kl.* **69**, 262–267.

Raiffa, H., and Schlaifer, R. 1961. *Applied Statistical Decision Theory.* Harvard Business School, Boston.

Rao, C. R. 1973. *Linear Statistical Inference and Its Applications.* J. Wiley, New York.

Reichenbach, H. 1935. *The Theory of Probability.* Leiden, Germany.

Rendu. J.-M. 1980. "Disjunctive kriging: Comparison of theory with actual results." In *Math. Geology* **12**, 4, 305–320.

Rice, S. O. 1954. "Mathematical analysis of random noise." In *Selected Papers on Noise and Stochastic Processes.* N. Wax (ed.). Dover, New York.

Riordan, J. 1983. *Combinatorial Identities.* J. Wiley, New York.

Ripley, B. D. 1981. *Spatial Statistics.* J. Wiley, New York.

Ripley, B. D. 1988. *Statistical Inference for Spatial Processes.* Cambridge Univ. Press, New York.

Rodriguez-Iturbe, I., and Mejia, J. M. 1974. "The design of rainfall networks in time and space." In *Water Resource Research* **10**, 4, 713–728.

Rosenblatt, M. 1956. "A central limit theorem and the strong mixing conditions." In *Proc. Nat. Acad. Sci., USA* **42**, 43–47.

Rosenkrantz, R. 1970. "Experimentation as communication with nature." In *Information and Inference.* J. Hintikka and P. Suppes (eds.). D. Reidel, Dordrecht. 58–93.

Rouhani, S. 1985. "Variance reduction analysis." In *Water Resource Research* **21**, 6, 837–846.

Rouhani, S., and Fiering, M. B. 1986. "Resilience of a statistical sampling scheme." *J. of Hydrology* **89**, 1–11.

Rouhani, S., and Hall, T. J. 1989. "Space-time kriging of groundwater data." In *Geostatistics* **2**, M. Armstrong (ed.). Kluwer Acad. Publ., 639–650.

Russel, B. 1962. "Probability." In *Human Knowledge: Its Scope and Limits.* Simon and Schuster, New York. 333–418.

Rytov, S. M., Kravtsov, Yu. A., and Tatarskii, V. I. 1989. *Principles of Statistical Radiophysics.* 4 Vols. Springer-Verlag, New York.

Sacks, J., and Schiller, S. 1988. "Spatial designs." In *Statistical Decision Theory and Related Topics IV* **2**. S. S. Gupta and J. O. Berger (eds.). Springer-Verlag, New York. 385–399.

Savage, L. J. 1954. *The Foundation of Statistics.* J. Wiley, New York.

Schoenberg, I. J. 1938. "Metric spaces and completely monotone functions." In *Ann. Math.* **39**, 811–841.

Schowengerdt, R. A. 1983. *Techniques for Image Processing and Classification in Remote Sensing.* Academic Press, New York.

Schwartz, L. 1950–51. *Theorie des Distributions 1, 2.* Paris.

Servien, P. 1949. *Hasard et Probabilite.* Presses Univ. de France, Paris.

Shannon, C. E. 1948. "A mathematical theory of communication." In *Bell System Tech. J.* **27**, 379–423 and 623–656.

Shewry, M. C., and Wynn, H. P. 1987. "Maximum entropy sampling." In *J. Appl. Statistics* **14**, 2, 165–170.

Shinn, J. H., and Lynn, S. 1979. "Do man-made sources affect the sulfur cycle of northeastern states?" In *Envir. Sci. Technol.* **13**, 1062–1067.

Shinozuka, M. 1971. "Simulation of multivariate and multidimensional random processes." In *J. Acoust. Soc. Am.* **49**, 357-367.

Shinozuka, M., and Jan, C.-M. 1972. "Digital simulation of random processes and its applications." In *J. Sound and Vibration* **25**, 1, 111-128.

Shore, J. E., and Johnson, R. W. 1980. "Axiomatic derivation of the principle of maximum entropy and the principle of minimum cross-entropy." In *IEEE Trans. Inform. Theory* **26**, 26-37.

Singh, A., and Lee, K. L. 1970. "Variability of soil parameters." *8th Annual Symposium on Eng. Geology and Soils Eng.* Pocatello, Idaho.

Sobczyk, K. 1991. *Stochastic Differential Equations.* Kluwer, Dordrecht.

Soong, T. T. 1973. *Random Differential Equations in Science and Engineering.* Academic Press, New York.

Sophocleous, M. 1983. "Groundwater observation network design for the Kansas groundwater management districts, USA." *J. of Hydrology* **61**, 371-389.

Sorenson, H. W., and Stubberud, A. R. 1968. "Nonlinear filtering by approximation of the posterior density." In *Int. J. Control* **8**, 33-51.

Spinazola, J. M., Gillespie, J. B., and Hart, R. J. 1985. "Groundwater flow and solute transport in the Equus Beds area, South Central Kansas." *U.S. Geological Survey, Water-Resources Investigations,* Report 85-4336, Lawrence, Kansas.

Stavroulakis, P. (ed.). 1983. *Distributed Parameter Systems 1,2.* Hutchinson Ross Publ. Co., USA.

Stein, M. L. 1987. "Minimum norm quadratic estimation of spatial variograms." *J. of the American Statistical Assoc.* **82**, 387-405.

Stein, A., Hoogerwerf, M., and Bouma, J. 1988. "Use of soil-map delineations to improve (co-) kriging of point data on moisture deficits." In *Geoderma* **43**, 163-177.

Stern, A. C., Fox, D. L., Boubel, R. W., and Turner, B. 1973. *Fundamentals of Air Pollution.* Academic Press, New York.

Stratonovich, R. L. 1968. *Conditional Markov Processes and their Application to the Theory of Optimal Control.* Elsevier, New York.

Syski, R. 1967. "Stochastic differential equations." In *Modern Nonlinear Equations.* T. L. Saaty (ed.). McGraw-Hill, New York, 346-456.

Tavenas, F. A., Ladd, R. S., and Larochelle, P. 1972. *The Accuracy of the Relative Density Measurements: Results of a Comparative Test Program.* Presented at Symposium on Evaluation of Relative Density and Its Role in Geotechnical Projects involving Cohesionless Soils, ASTM, CA.

Tubilla, A. 1975. "Error Convergence Rates for Estimates of Multidimensional Integrals of Random Functions. Techn. Report 72, Dept. of Statistics, Stanford Univ., California.

Veneziano, D. 1980. *Random Processes for Engineering Applications.* Course notes, Dept. of Civil Eng., M.I.T., Cambridge, Massachusetts.

Veneziano, D., and Kitanidis, P. K. 1982. "Sequential sampling to contour an uncertain function." In *Math. Geology* **15**, 5, 387-404.

Von Storch, H., Weese, U., and Xu, J. S. 1989. "Simultaneous analysis of space-time variability: Principal oscillation patterns and principal interaction patterns with applications to the southern oscillation." Report 34, Max Planck Institut fur Meteorologie, Hamburg.

*Von Mises, R. 1928. *Probability, Statistics and Truth.* Vienna, Austria.

Von Neumann. 1955. *Mathematical Foundations of Quantum Mechanics.* Princeton Univ. Press, Princeton, New Jersey.

Wagner, B. J., and Gorelick, S. M. 1987. "Optimal groundwater quality management under parameter uncertainty." In *Water Resource Research* **23**, 7, 1162-1174.

Wainstein, L. A., and Zubakov, V. D. 1962. *Extraction of Signals from Noise.* Dover, New York.

Webster, R. 1985. "Quantitative spatial analysis of soil in the field." In *Advances in Soil Sci,* 3, 1-70.

Webster, R. and Burgess, T. M. 1984. "Sampling and bulking strategies for estimating soil properties of small regions." In *J. Soil Sci.* **35**, 127–140.

Whittle, P. 1954. "On stationary processes in the plane." In *Biometrika* **41**, 434–449.

Whittle, P. 1963. *Prediction and Regulation.* English Univ. Press, England.

Whittle, P. 1965. "Recursive relations for predictors of nonstationary processes." In *J. R. Statist. Soc.*, Ser. B, 27, 523–532.

Wiener, N. 1930. "Generalized harmonic analysis." In *Acta Math.* **55**, 2–3, 117–258.

Wiener, N. 1949. *Time Series.* MIT Press, Cambridge, Massachusetts. 163 p.

Wiener, N., and Massani, P. 1958. "The prediction theory for multivariate stochastic processes." *Acta Math.*, **98**, pp. 111–150 and **99**, pp. 93–137.

Wilde, D. J., and Beightler, C. S. 1967. *Foundations of Optimization.* Prentice-Hall, New Jersey.

Willman, W. W. 1981. "Edgeworth expansions in state perturbation estimation." In *IEEE Trans. on Automatic Control* **AC-26**, 493–498.

Wittig, L. E., and Sinha, A. K. 1975. "Simulation of multicorrelated random processes using the fast Fourier transform algorithm." In *J. Acoust. Soc. of America* **58**, 3, 630–634.

Wosten, J. H. M., Bannink, M. H., and Bouma, J. 1987. "Land evaluation different scales: you pay for what you get." In *Soil Survey and Land Evaluation* **7**, 13–24.

Yaglom, A. M., and Pinsker, M. S. 1953. "Random processes with stationary increments of order *n*." In *Dokl. Acad. Nauk USSR* **90**, 731–734.

Yaglom, A. M. 1955. "Correlation theory of processes with stationary random increments of order *n*." In *Mat. USSR Sb.* 37–141. (English translation in *Am. Math. Soc. Trans.* Ser. 2, 8–87, 1958.)

Yaglom, A. M. 1957. "Some classes of random fields in *n*-dimensional space, related to stationary random processes." In *Theory Probab. Its Appl.* English transl. **3**, 273–320.

Yaglom, A. M. 1962. *Stationary Random Functions.* Prentice-Hall, Englewood Cliffs, New Jersey.

Yaglom, A. M. 1986. *Correlation Theory of Stationary and Related Random Functions: Basic Results.* Springer-Verlag, New York.

Yates, S. R., Warrick, A. W., and Myers, D. E. 1986. "Disjunctive kriging 2. Examples." In *Water Resource Research* **22**, 10, 623–630.

Yates, S. R., and Warrick, A. W. 1987. "Estimating soil water content using co-kriging." In *Soil Sci. Soc. of America J.* **51**, 23–30.

Ylvisaker, D. 1975. "Designs of random fields." In *A Survey of Statistical Design and Linear Models.* J. N. Srivastava (ed.). North-Holland Publ. Co., 593–607.

Yucement, M. S., Tang, W. H., and Ang, A. H.-S. 1973. *A Probabilistic Study of Safety and Design of Earth Slopes.* Civil Engin. Studies, Struct. Res. Ser., 402, Univ. of Illinois, Urbana, Illinois.

Zeleny, M. 1982. *Multiple Criteria Decision Making.* McGraw-Hill, New York.

Index

A CATALOG OF SELECTED

DOVER BOOKS

IN SCIENCE AND MATHEMATICS

Mathematics

FUNCTIONAL ANALYSIS (Second Corrected Edition), George Bachman and Lawrence Narici. Excellent treatment of subject geared toward students with background in linear algebra, advanced calculus, physics, and engineering. Text covers introduction to inner-product spaces, normed, metric spaces, and topological spaces; complete orthonormal sets, the Hahn-Banach Theorem and its consequences, and many other related subjects. 1966 ed. 544pp. 6⅛ x 9¼. 40251-7

ASYMPTOTIC EXPANSIONS OF INTEGRALS, Norman Bleistein & Richard A. Handelsman. Best introduction to important field with applications in a variety of scientific disciplines. New preface. Problems. Diagrams. Tables. Bibliography. Index. 448pp. 5⅜ x 8½. 65082-0

VECTOR AND TENSOR ANALYSIS WITH APPLICATIONS, A. I. Borisenko and I. E. Tarapov. Concise introduction. Worked-out problems, solutions, exercises. 257pp. 5⅝ x 8¼. 63833-2

THE ABSOLUTE DIFFERENTIAL CALCULUS (CALCULUS OF TENSORS), Tullio Levi-Civita. Great 20th-century mathematician's classic work on material necessary for mathematical grasp of theory of relativity. 452pp. 5⅝ x 8¼. 63401-9

AN INTRODUCTION TO ORDINARY DIFFERENTIAL EQUATIONS, Earl A. Coddington. A thorough and systematic first course in elementary differential equations for undergraduates in mathematics and science, with many exercises and problems (with answers). Index. 304pp. 5⅜ x 8½. 65942-9

FOURIER SERIES AND ORTHOGONAL FUNCTIONS, Harry F. Davis. An incisive text combining theory and practical example to introduce Fourier series, orthogonal functions and applications of the Fourier method to boundary-value problems. 570 exercises. Answers and notes. 416pp. 5⅜ x 8½. 65973-9

COMPUTABILITY AND UNSOLVABILITY, Martin Davis. Classic graduate-level introduction to theory of computability, usually referred to as theory of recurrent functions. New preface and appendix. 288pp. 5⅜ x 8½. 61471-9

ASYMPTOTIC METHODS IN ANALYSIS, N. G. de Bruijn. An inexpensive, comprehensive guide to asymptotic methods–the pioneering work that teaches by explaining worked examples in detail. Index. 224pp. 5⅜ x 8½ 64221-6

APPLIED COMPLEX VARIABLES, John W. Dettman. Step-by-step coverage of fundamentals of analytic function theory–plus lucid exposition of five important applications: Potential Theory; Ordinary Differential Equations; Fourier Transforms; Laplace Transforms; Asymptotic Expansions. 66 figures. Exercises at chapter ends. 512pp. 5⅜ x 8½. 64670-X

INTRODUCTION TO LINEAR ALGEBRA AND DIFFERENTIAL EQUATIONS, John W. Dettman. Excellent text covers complex numbers, determinants, orthonormal bases, Laplace transforms, much more. Exercises with solutions. Undergraduate level. 416pp. 5⅜ x 8½. 65191-6

TENSOR CALCULUS, J.L. Synge and A. Schild. Widely used introductory text covers spaces and tensors, basic operations in Riemannian space, non-Riemannian spaces, etc. 324pp. 5⅜ x 8¼. 63612-7

ORDINARY DIFFERENTIAL EQUATIONS, Morris Tenenbaum and Harry Pollard. Exhaustive survey of ordinary differential equations for undergraduates in mathematics, engineering, science. Thorough analysis of theorems. Diagrams. Bibliography. Index. 818pp. 5⅜ x 8½. 64940-7

INTEGRAL EQUATIONS, F. G. Tricomi. Authoritative, well-written treatment of extremely useful mathematical tool with wide applications. Volterra Equations, Fredholm Equations, much more. Advanced undergraduate to graduate level. Exercises. Bibliography. 238pp. 5⅜ x 8½. 64828-1

FOURIER SERIES, Georgi P. Tolstov. Translated by Richard A. Silverman. A valuable addition to the literature on the subject, moving clearly from subject to subject and theorem to theorem. 107 problems, answers. 336pp. 5⅜ x 8½. 63317-9

INTRODUCTION TO MATHEMATICAL THINKING, Friedrich Waismann. Examinations of arithmetic, geometry, and theory of integers; rational and natural numbers; complete induction; limit and point of accumulation; remarkable curves; complex and hypercomplex numbers, more. 1959 ed. 27 figures. xii+260pp. 5⅜ x 8½. 42804-4

POPULAR LECTURES ON MATHEMATICAL LOGIC, Hao Wang. Noted logician's lucid treatment of historical developments, set theory, model theory, recursion theory and constructivism, proof theory, more. 3 appendixes. Bibliography. 1981 ed. ix+283pp. 5⅜ x 8½. 67632-3

CALCULUS OF VARIATIONS, Robert Weinstock. Basic introduction covering isoperimetric problems, theory of elasticity, quantum mechanics, electrostatics, etc. Exercises throughout. 326pp. 5⅜ x 8½. 63069-2

THE CONTINUUM: A Critical Examination of the Foundation of Analysis, Hermann Weyl. Classic of 20th-century foundational research deals with the conceptual problem posed by the continuum. 156pp. 5⅜ x 8½. 67982-9

CHALLENGING MATHEMATICAL PROBLEMS WITH ELEMENTARY SOLUTIONS, A. M. Yaglom and I. M. Yaglom. Over 170 challenging problems on probability theory, combinatorial analysis, points and lines, topology, convex polygons, many other topics. Solutions. Total of 445pp. 5⅜ x 8½. Two-vol. set.
Vol. I: 65536-9 Vol. II: 65537-7

INTRODUCTION TO PARTIAL DIFFERENTIAL EQUATIONS WITH APPLICATIONS, E. C. Zachmanoglou and Dale W. Thoe. Essentials of partial differential equations applied to common problems in engineering and the physical sciences. Problems and answers. 416pp. 5⅜ x 8½. 65251-3

THE THEORY OF GROUPS, Hans J. Zassenhaus. Well-written graduate-level text acquaints reader with group-theoretic methods and demonstrates their usefulness in mathematics. Axioms, the calculus of complexes, homomorphic mapping, *p*-group theory, more. 276pp. 5⅜ x 8½. 40922-8

Math–Decision Theory, Statistics, Probability

ELEMENTARY DECISION THEORY, Herman Chernoff and Lincoln E. Moses. Clear introduction to statistics and statistical theory covers data processing, probability and random variables, testing hypotheses, much more. Exercises. 364pp. 5⅜ x 8½. 65218-1

STATISTICS MANUAL, Edwin L. Crow et al. Comprehensive, practical collection of classical and modern methods prepared by U.S. Naval Ordnance Test Station. Stress on use. Basics of statistics assumed. 288pp. 5⅜ x 8½. 60599-X

SOME THEORY OF SAMPLING, William Edwards Deming. Analysis of the problems, theory, and design of sampling techniques for social scientists, industrial managers, and others who find statistics important at work. 61 tables. 90 figures. xvii +602pp. 5⅜ x 8½. 64684-X

LINEAR PROGRAMMING AND ECONOMIC ANALYSIS, Robert Dorfman, Paul A. Samuelson and Robert M. Solow. First comprehensive treatment of linear programming in standard economic analysis. Game theory, modern welfare economics, Leontief input-output, more. 525pp. 5⅜ x 8½. 65491-5

PROBABILITY: An Introduction, Samuel Goldberg. Excellent basic text covers set theory, probability theory for finite sample spaces, binomial theorem, much more. 360 problems. Bibliographies. 322pp. 5⅜ x 8½. 65252-1

GAMES AND DECISIONS: Introduction and Critical Survey, R. Duncan Luce and Howard Raiffa. Superb nontechnical introduction to game theory, primarily applied to social sciences. Utility theory, zero-sum games, n-person games, decision-making, much more. Bibliography. 509pp. 5⅜ x 8½. 65943-7

INTRODUCTION TO THE THEORY OF GAMES, J. C. C. McKinsey. This comprehensive overview of the mathematical theory of games illustrates applications to situations involving conflicts of interest, including economic, social, political, and military contexts. Appropriate for advanced undergraduate and graduate courses; advanced calculus a prerequisite. 1952 ed. x+372pp. 5⅜ x 8½. 42811-7

FIFTY CHALLENGING PROBLEMS IN PROBABILITY WITH SOLUTIONS, Frederick Mosteller. Remarkable puzzlers, graded in difficulty, illustrate elementary and advanced aspects of probability. Detailed solutions. 88pp. 5⅜ x 8½. 65355-2

PROBABILITY THEORY: A Concise Course, Y. A. Rozanov. Highly readable, self-contained introduction covers combination of events, dependent events, Bernoulli trials, etc. 148pp. 5⅜ x 8¼. 63544-9

STATISTICAL METHOD FROM THE VIEWPOINT OF QUALITY CONTROL, Walter A. Shewhart. Important text explains regulation of variables, uses of statistical control to achieve quality control in industry, agriculture, other areas. 192pp. 5⅜ x 8½. 65232-7

Math–Geometry and Topology

ELEMENTARY CONCEPTS OF TOPOLOGY, Paul Alexandroff. Elegant, intuitive approach to topology from set-theoretic topology to Betti groups; how concepts of topology are useful in math and physics. 25 figures. 57pp. $5\frac{3}{8}$ x $8\frac{1}{2}$. 60747-X

COMBINATORIAL TOPOLOGY, P. S. Alexandrov. Clearly written, well-organized, three-part text begins by dealing with certain classic problems without using the formal techniques of homology theory and advances to the central concept, the Betti groups. Numerous detailed examples. 654pp. $5\frac{3}{8}$ x $8\frac{1}{2}$. 40179-0

EXPERIMENTS IN TOPOLOGY, Stephen Barr. Classic, lively explanation of one of the byways of mathematics. Klein bottles, Moebius strips, projective planes, map coloring, problem of the Koenigsberg bridges, much more, described with clarity and wit. 43 figures. 210pp. $5\frac{3}{8}$ x $8\frac{1}{2}$. 25933-1

CONFORMAL MAPPING ON RIEMANN SURFACES, Harvey Cohn. Lucid, insightful book presents ideal coverage of subject. 334 exercises make book perfect for self-study. 55 figures. 352pp. $5\frac{5}{8}$ x $8\frac{1}{4}$. 64025-6

THE GEOMETRY OF RENÉ DESCARTES, René Descartes. The great work founded analytical geometry. Original French text, Descartes's own diagrams, together with definitive Smith-Latham translation. 244pp. $5\frac{3}{8}$ x $8\frac{1}{2}$. 60068-8

PRACTICAL CONIC SECTIONS: The Geometric Properties of Ellipses, Parabolas and Hyperbolas, J. W. Downs. This text shows how to create ellipses, parabolas, and hyperbolas. It also presents historical background on their ancient origins and describes the reflective properties and roles of curves in design applications. 1993 ed. 98 figures. xii+100pp. $6\frac{1}{2}$ x $9\frac{1}{4}$. 42876-1

THE THIRTEEN BOOKS OF EUCLID'S ELEMENTS, translated with introduction and commentary by Thomas L. Heath. Definitive edition. Textual and linguistic notes, mathematical analysis. 2,500 years of critical commentary. Unabridged. 1,414pp. $5\frac{3}{8}$ x $8\frac{1}{2}$. Three-vol. set. Vol. I: 60088-2 Vol. II: 60089-0 Vol. III: 60090-4

GEOMETRY OF COMPLEX NUMBERS, Hans Schwerdtfeger. Illuminating, widely praised book on analytic geometry of circles, the Moebius transformation, and two-dimensional non-Euclidean geometries. 200pp. $5\frac{5}{8}$ x $8\frac{1}{4}$. 63830-8

DIFFERENTIAL GEOMETRY, Heinrich W. Guggenheimer. Local differential geometry as an application of advanced calculus and linear algebra. Curvature, transformation groups, surfaces, more. Exercises. 62 figures. 378pp. $5\frac{3}{8}$ x $8\frac{1}{2}$. 63433-7

CURVATURE AND HOMOLOGY: Enlarged Edition, Samuel I. Goldberg. Revised edition examines topology of differentiable manifolds; curvature, homology of Riemannian manifolds; compact Lie groups; complex manifolds; curvature, homology of Kaehler manifolds. New Preface. Four new appendixes. 416pp. $5\frac{3}{8}$ x $8\frac{1}{2}$. 40207-X

Physics

OPTICAL RESONANCE AND TWO-LEVEL ATOMS, L. Allen and J. H. Eberly. Clear, comprehensive introduction to basic principles behind all quantum optical resonance phenomena. 53 illustrations. Preface. Index. 256pp. 5⅜ x 8½. 65533-4

QUANTUM THEORY, David Bohm. This advanced undergraduate-level text presents the quantum theory in terms of qualitative and imaginative concepts, followed by specific applications worked out in mathematical detail. Preface. Index. 655pp. 5⅜ x 8½. 65969-0

ATOMIC PHYSICS: 8th edition, Max Born. Nobel laureate's lucid treatment of kinetic theory of gases, elementary particles, nuclear atom, wave-corpuscles, atomic structure and spectral lines, much more. Over 40 appendices, bibliography. 495pp. 5⅜ x 8½. 65984-4

A SOPHISTICATE'S PRIMER OF RELATIVITY, P. W. Bridgman. Geared toward readers already acquainted with special relativity, this book transcends the view of theory as a working tool to answer natural questions: What is a frame of reference? What is a "law of nature"? What is the role of the "observer"? Extensive treatment, written in terms accessible to those without a scientific background. 1983 ed. xlviii+172pp. 5⅜ x 8½. 42549-5

AN INTRODUCTION TO HAMILTONIAN OPTICS, H. A. Buchdahl. Detailed account of the Hamiltonian treatment of aberration theory in geometrical optics. Many classes of optical systems defined in terms of the symmetries they possess. Problems with detailed solutions. 1970 edition. xv+360pp. 5⅜ x 8½. 67597-1

PRIMER OF QUANTUM MECHANICS, Marvin Chester. Introductory text examines the classical quantum bead on a track: its state and representations; operator eigenvalues; harmonic oscillator and bound bead in a symmetric force field; and bead in a spherical shell. Other topics include spin, matrices, and the structure of quantum mechanics; the simplest atom; indistinguishable particles; and stationary-state perturbation theory. 1992 ed. xiv+314pp. 6⅛ x 9¼. 42878-8

LECTURES ON QUANTUM MECHANICS, Paul A. M. Dirac. Four concise, brilliant lectures on mathematical methods in quantum mechanics from Nobel Prize–winning quantum pioneer build on idea of visualizing quantum theory through the use of classical mechanics. 96pp. 5⅜ x 8½. 41713-1

THIRTY YEARS THAT SHOOK PHYSICS: The Story of Quantum Theory, George Gamow. Lucid, accessible introduction to influential theory of energy and matter. Careful explanations of Dirac's anti-particles, Bohr's model of the atom, much more. 12 plates. Numerous drawings. 240pp. 5⅜ x 8½. 24895-X

ELECTRONIC STRUCTURE AND THE PROPERTIES OF SOLIDS: The Physics of the Chemical Bond, Walter A. Harrison. Innovative text offers basic understanding of the electronic structure of covalent and ionic solids, simple metals, transition metals and their compounds. Problems. 1980 edition. 582pp. 6⅛ x 9¼. 66021-4

CATALOG OF DOVER BOOKS

QUANTUM MECHANICS: Principles and Formalism, Roy McWeeny. Graduate student-oriented volume develops subject as fundamental discipline, opening with review of origins of Schrödinger's equations and vector spaces. Focusing on main principles of quantum mechanics and their immediate consequences, it concludes with final generalizations covering alternative "languages" or representations. 1972 ed. 15 figures. xi+155pp. 5⅜ x 8½. 42829-X

INTRODUCTION TO QUANTUM MECHANICS WITH APPLICATIONS TO CHEMISTRY, Linus Pauling & E. Bright Wilson, Jr. Classic undergraduate text by Nobel Prize winner applies quantum mechanics to chemical and physical problems. Numerous tables and figures enhance the text. Chapter bibliographies. Appendices. Index. 468pp. 5⅜ x 8½. 64871-0

METHODS OF THERMODYNAMICS, Howard Reiss. Outstanding text focuses on physical technique of thermodynamics, typical problem areas of understanding, and significance and use of thermodynamic potential. 1965 edition. 238pp. 5⅜ x 8½. 69445-3

TENSOR ANALYSIS FOR PHYSICISTS, J. A. Schouten. Concise exposition of the mathematical basis of tensor analysis, integrated with well-chosen physical examples of the theory. Exercises. Index. Bibliography. 289pp. 5⅜ x 8½. 65582-2

THE ELECTROMAGNETIC FIELD, Albert Shadowitz. Comprehensive undergraduate text covers basics of electric and magnetic fields, builds up to electromagnetic theory. Also related topics, including relativity. Over 900 problems. 768pp. 5⅝ x 8¼. 65660-8

GREAT EXPERIMENTS IN PHYSICS: Firsthand Accounts from Galileo to Einstein, Morris H. Shamos (ed.). 25 crucial discoveries: Newton's laws of motion, Chadwick's study of the neutron, Hertz on electromagnetic waves, more. Original accounts clearly annotated. 370pp. 5⅜ x 8½. 25346-5

RELATIVITY, THERMODYNAMICS AND COSMOLOGY, Richard C. Tolman. Landmark study extends thermodynamics to special, general relativity; also applications of relativistic mechanics, thermodynamics to cosmological models. 501pp. 5⅜ x 8½. 65383-8

STATISTICAL PHYSICS, Gregory H. Wannier. Classic text combines thermodynamics, statistical mechanics, and kinetic theory in one unified presentation of thermal physics. Problems with solutions. Bibliography. 532pp. 5⅜ x 8½. 65401-X

Paperbound unless otherwise indicated. Available at your book dealer, online at **www.doverpublications.com**, or by writing to Dept. GI, Dover Publications, Inc., 31 East 2nd Street, Mineola, NY 11501. For current price information or for free catalogs (please indicate field of interest), write to Dover Publications or log on to **www.doverpublications.com** and see every Dover book in print. Dover publishes more than 500 books each year on science, elementary and advanced mathematics, biology, music, art, literary history, social sciences, and other areas.